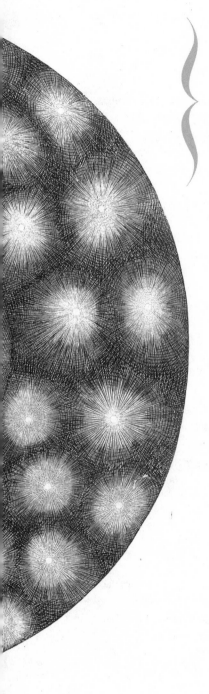

The Newton Wars
and the Beginning of the
French Enlightenment

J. B. SHANK

The University of Chicago Press

CHICAGO & LONDON

J. B. SHANK is associate
professor of history at the
University of Minnesota.

The University of Chicago Press, Chicago 60637
The University of Chicago Press, Ltd., London
© 2008 by The University of Chicago
All rights reserved. Published 2008
Printed in the United States of America

17 16 15 14 13 12 11 10 09 08   1 2 3 4 5

ISBN-13: 978-0-226-74945-7 (cloth)
ISBN-10: 0-226-74945-2 (cloth)

Library of Congress Cataloging-in-Publication Data

Shank, J. B.
    The Newton wars and the beginning of the French
Enlightenment / J. B. Shank.
        p. cm.
    Includes bibliographical references and index.
    ISBN-13: 978-0-226-74945-7 (cloth : alk. paper)
    ISBN-10: 0-226-74945-2 (cloth : alk. paper)
    1. Science—France—History—18th century.
2. Enlightenment—France. 3. Newton, Isaac, Sir, 1642–
1727—Influence. 4. Science—Philosophy—History—18th
century. I. Title.
    Q127.F8S53 2008
    509.44'09033—dc22

                                                2008006414

♾ The paper used in this publication meets the minimum
requirements of the American National Standard for
Information Sciences—Permanence of Paper for Printed
Library Materials, ANSI z39.48-1992.

Contents

Abbreviations

Ac. Sc.	Archives de l'Académie des sciences, Paris
A.N.	Archives nationales, Paris
B.N.	Bibliothèque nationale, Paris
FJPC	Fonds Jean-Pierre de Crousaz, 14 vols., Bibliothèque cantonale et universitaire, Lausanne, Switzerland
HARS	*Histoire de l'Académie royale des sciences. Avec les mémoires de mathématique et de physique . . . tirés des registres de cette académie (1699–1790)*, 92 vols. (Paris, 1702–1797). Each volume comprises two separately paginated parts, referred to as *Hist.* and *Mem.*, respectively.
PVARS	Procès verbaux de l'Académie royale des sciences. Archives de l'Académie des sciences, Paris
Voltaire, *Corr.*	*Correspondence and Related Documents,* vols. 83–135 in *Oeuvres complètes de Voltaire,* ed. Theodore Besterman, 135 vols. (Geneva, Banbury, and Oxford: Voltaire Foundation at the Taylor Institution, 1968–1977)

Illustrations

Acknowledgments

No book comes to life ex nihilo, and as a first book, mine arrives carrying a particularly heavy load of debt. Work on this project began as a doctoral dissertation completed at Stanford University in 2000, and the early steps were guided in countless constructive ways by the community of faculty and graduate students that I worked with there. Keith Michael Baker supervised the dissertation, offering many incisive interventions then, and many in the years since my graduation. Each has made my thinking sharper and this book better. He was also a consummate mentor, guiding me into the world of the academy with a care and wisdom that I am only now beginning to truly appreciate. Paula Findlen arrived at Stanford after the dissertation was well under way, yet she accepted me as an advisee nevertheless and contributed enormously to the improvement of my work. She especially guided me generously into the world of early modern science studies, a space that I now call home. I was grateful then, and still am now, to have her enthusiasm and sharp mind in my camp. Paul Robinson initiated me into the fold of modern European intellectual historians while also making me a better writer in the process, and Tim Lenoir always made sure I felt at home among Stanford's historians of science. Michael Marrinan did the same with the field of art history, and in countless ways each helped to make the author of this book a better and more thoughtful historian. John Bender, Philippe Buc, and Mary Louise Roberts also contributed in important ways toward making this book what it is.

Stanford also provided abundant material support that made the initial steps in this project possible. Especially fruitful was my year as a fellow at the Stanford Humanities Center in 1994–1995 at a time when the details of my project were still incubating. A Fulbright fellowship in 1995–1996 allowed for a crucial year of research in France, and a dissertation fellowship from the Mellon Foundation gave me the time necessary to start drafting the text. Later, the Stanford Introduction to Humanities Program, together with the Freshman/Sophomore College at Sterling Quad, provided stimulating work and a comfortable living situation that made my final years at Stanford amidst the insanity of Silicon Valley in the 1990s a productive time of growth. These programs also provided me with a wealth of new colleagues and friends. Special debts of gratitude are owed to Renée Courey, my officemate, colleague, and good friend, and Cheri Ross, who extended unusual support and concern my way. Nancy Anderson, Marina Bonanno, Erin Carlson, Richard Cushman, Mariatte Denman, Andy

Dimock, Rodney Koeneke, Shari Palmer, Steve Johnstone, and Ben Robinson further kept me stimulated intellectually when I was not dissertating. Also wonderful was my exceptional cohort of graduate student colleagues, especially Steve Schloesser, whose intellect, passion and commitment continue to inspire me, and Amir Alexander, Jordanna Bailkin, Dan Colman, Christine Holbo, Bruce Lidl, Tara Nummedal, Eric Oberle, Sara Pritchard, Mary Salzman, Robert Scafe, Janet Kobrin Watson, and Gillian Weiss. Although my contact with these friends and colleagues is more infrequent than I would like, their contributions to this book remain indelible.

In 2000, I left the Bay Area for the prairie and glacial lakes of the upper Midwest. Thankfully, leaving behind good friends and stimulating colleagues at Stanford did not mean leaving behind such things altogether. Instead, at the University of Minnesota I have found a community of scholars and students that have contributed enormously to the quality of this book. Tom Wolfe was one of the first people I met at Minnesota, and with an office dangerously close to my own we began a series of distracting conversations that have continually proven to be more intellectually fruitful than most of my "serious work." Without the presence of his insightfulness, mindfulness, and pythonesque spirit, this book may never have been completed. Along with Tom, Kirsten Fischer, Chris Isett, Pat McNamara, M.J. Maynes, Ajay Skaria, Eric Weitz, Barbara Welke, and the rest of my colleagues in the Minnesota history department made the Social Sciences Tower, against all architectural odds, a stimulating place to complete a book. For a pathological dilettante and disciplinary misfit such as myself, however, a university is measured by the quality of its transdepartmental communities, and my intellectual work at Minnesota has benefited enormously from the unusually vibrant and vigorous community of early modernists that exists there. Before I was even hired, Juliette Cherbuliez presented me with the proposal for what has since become the Theorizing Early Modern Studies Research Collaborative (http://www.tems.umn.edu). Working with her in this space and in others has provided some of my most meaningful and productive intellectual exchanges, along with some of my most congenial. Michael Gaudio, who joined our group in 2003, has become an equally crucial intellectual colleague and collaborator, and this book bears the legacies of our work together in countless visible and invisible ways. My work also owes a debt to other early modernists at Minnesota such as Dan Brewer, Jim Parente, Alan Smith, and John Watkins, each of whom contributed to my thinking in ways big and small.

Minnesota has also proven to be an equally generous financial supporter of my work, and I am grateful to the history department, the Center for Early Modern History, and the Grant-in-Aid for Research program of the University of Minnesota Graduate School for the research funds that made completing this book possible. Special thanks also need to go to the McKnight Foundation for its generous support of the University of Minnesota, support that made possible a summer research fellowship and a two-year McKnight Land-Grant professorship that allowed for the freedom to finalize and edit the text. These same funds also made two exceptional research assistants available to me, Mike Sizer and Chris Freeman, whose work was crucial in bringing this book to completion. Among the institutions that provided the images for this book was the James Ford Bell Library at the University of Minnesota, and I am glad to have this occasion to thank Carol Urness, the curator emeritus, Maggie Ragnow, the current curator, and Susan Stekel, the curatorial assistant, for their unfailing support of whatever research request I have made.

Beyond Stanford and Minnesota, I have also received generous support from a number of institutions and colleagues. In Paris, the Centre Koyré generously welcomed me when I was still a very green student, and I am grateful to its director at the time, Roger Chartier, to Ernst Coumet, who invited me to attend his seminar, and to Eric Brian and Michel Blay who generously gave their time to discuss my work. The staff at the Archives de l'Académie des Sciences, especially former head archivist Christiane Demeulenaere and staff members Claudine Pouret, Anne-Sylvie Guenoun, Josette Grobon, and Pierre Leroi, made this not only a crucial archival source for me, but my favorite place to work in Paris. Special thanks also to Françoise Blèchet, Irène Passeron, Rhoda Rappaport, and Alice Stroup for their willingness to talk with me in and around the pleasant confines of the Institut. I am also grateful to John Detloff and to Mike Lynn, who I came to know in Paris at this time. The many fruitful intellectual interactions and beer-drinking sessions (the two are not mutually exclusive) that we had then and have continued to have ever since were invaluable. Special thanks also to International Summer School in the History of Science, and especially to the faculty and students who attended the 1998 session in Uppsala, Sweden, for inviting me to exchange my ideas with this stimulating workshop, and to the helpful staff at the Bibliothèque Universitaire et Cantonale de Lausanne who made my work on Jean-Pierre Crousaz such a pleasant and fruitful experience. An NEH Summer Seminar in 2004 further gave me the opportunity to deepen my understanding of Leibniz, and I am thankful to Roger Ariew and Daniel Garber, the directors of the seminar, and to the other participants, for the insights that improved my discussion of Leibniz in this book. More recently in 2005–2006, a long-standing admiration for the scholarship of Niccolò Guicciardini turned into a deep, and I hope long-lasting, intellectual friendship. Niccolò also arranged for a very fruitful year as a visiting scholar at the University of Siena and the Institute and Museum of the

History of Science in Florence, and my book benefited enormously from the interactions I had in Italy, and especially from my conversations with Niccolò. For that I am deeply grateful.

Over the decade and more that it has taken to produce this book I have had more transformative conversations about it than I can count or remember. Special thanks nevertheless need to be extended to Marco Beretta, Paola Bertucci, Tom Broman, Greg Brown, Robert Darnton, Loraine Daston, Suzanne Desan, Nick Dew, Tore Frängsmyr, Stephen Gaukroger, Peter Galison, Jan Goldstein, Dena Goodman, Anthony Grafton, Anita Guerrini, Philippe Hamou, Roger Hahn, Tom Hankins, Mary Henninger-Voss, Peg Jacob, Colin Jones, Matthew Jones, Tom Kavanagh, Larry Klein, Massimo Mazzotti, Scott Mandelbrot, Ron Numbers, Margaret Osler, Guliano Pancaldi, Mike Shank, David Sturdy, Mary Terrall, Steven Shapin, Kathleen Wellman, Elizabeth Williams, Judith Zinsser, and Joe Zizek for generously sharing invaluable thoughts and criticisms of my work with me. Thanks also to the many conference and symposium audiences that offered criticisms and useful suggestions as well. Most recently the editorial staff at the University of Chicago Press, especially the late Susan Abrams, who originally arranged my contract, and more recently Karen Merikangas Darling, who shepherded the book into print, have also earned my gratitude. Without their patience with my authorial naiveté and their wisdom about the material details of book production this book would have been far worse.

Most important in sustaining this book over its very long and ever crooked path to completion was my family. They therefore deserve special recognition. My late father, John Shank, did not live to hold the book in his hands, but had he done so he would have held an object that materially embodies his unfailing commitment to learning, as well as his willingness to give generously of his financial means to make such goals tangible. My stepmother Diane Shank never questioned this commitment, and her support made this book possible as well. My mother, Sharon Ronan, likewise contributed enormously through her unfailing support and conviction that the book would indeed get done, and that it would be good once it was completed. She was certainly right in part, and her unfailing optimism and love helped me to make this book what it is. My in-laws John and Elaine McPhail also offered unending support for this enterprise despite the burdens it imposed on their daughter and grandchildren. No son could have asked for more from his parents. My own brothers and my two brothers-in-law were also never-ending supporters, especially my brother-in-law Brian McPhail, who always expressed exceptional enthusiasm for this undertaking while further converting his zeal, along with generous portions of his salary, into gifts of wine, cheese, fruit tarts, and other California delicacies, along with a lot of sincere friendship, during the "years of struggle" in the 1990s. This Epicurean support was not incidental in moving the book toward completion.

My first son Ian was born in 1993, within hours of the project itself. My second son Bryn was born in 1996, at the end of the year in Paris when the project really took flight. Each has likewise grown together with this book from infancy into early adolescence (hopefully the book has grown a bit more mature than that). Throughout this journey they brought the unquestioned love, joy, and (perhaps most important of all) distraction that only children can provide. Perhaps someday they will read "Daddy's story" and agree that the sacrifices they have endured were worth it (they already know from experience that the story will be long-winded and discursive). Whether they read it or not, I am glad that I had two such beautiful reminders of why real children are so much more meaningful and important than surrogate children like books. My wife Alison has sacrificed most of all toward the realization of this work and I hope she knows how thankful I am for all that she has given to it. Thankfully, she never read a single page of it, nor engaged in any constructive criticism of its argument. Yet by making sure that there was always plenty of laughter, diversion, and loving supportive company around, she contributed in vastly more important ways to its completion. In ways intended and accidental, she also made sure that I never forgot a fundamental truth: that a book, after all, is just a book. Out of gratitude for her singular love and devotion, I dedicate this mere book to her.

{ *Provincializing Newton,*
or Building in the Ruins
of a Grand Narrative of
Modernity [1]

Étienne-Louis Boullée personified the sophisticated, urban elite that recent historians have located at the center of French society in the final decades of the ancien régime.[2] A successful and affluent architect, he designed townhouses for some of Paris's wealthiest denizens, including the elegant d'Evreux family, whose Parisian *hôtel* Boullée designed in the 1770s. Today, this masterpiece of eighteenth-century domestic architecture is called the Elysée and serves as the official residence of the president of France. Boullée also designed churches, provincial *châteaux*, the façade of the Parisian stock exchange, and a new building (never built) to house the Royal Library. Yet while much of this work exists only on paper, Boullée's status as one of eighteenth-century France's most distinguished architects remains rock solid. An acclaimed scholar and teacher as well, Boullée further trained an entire generation of architectural successors from his chair, awarded when he was only nineteen years old, at the prestigious Parisian *École des Ponts et Chaussées*. He also served as a member of the Royal Academy of Architecture from 1762 until its dissolution during the French Revolution, and then, when the royal academies were replaced by the new Institut de France, he served as a founding member of this institution as well. Since, by all accounts, Boullée was also a modest, affable, and sociable man, in this way too he exemplified the enlightened values typical of the French elite at the end of the eighteenth century.[3]

Yet when the distinguished architect, at the height of his career in the 1780s, sat down to compose his magnum opus, a treatise of modern architectural

1. My title is, of course, an appropriation of Dipesh Chakrabarty's brilliant postcolonial work, *Provincializing Europe: Postcolonial Thought and Historical Difference* (Princeton, 2000).

2. See Daniel Roche, *France in the Enlightenment,* trans. by Arthur Goldhammer (Cambridge, 1998); and Colin Jones, *The Great Nation: France from Louis XV to Napoleon* (London, 2003).

3. Helen Rosneau, *Boullée and Visionary Architecture* (London and New York, 1976), 7–13, idem, *Boullée's Treatise on Architecture* (London, 1953), 1–13.

theory, he did not turn to the patrimony of his native France for inspiration. Instead, he drew upon the memory of a man born one hundred and fifty years earlier to yeoman farmers in the small Lincolnshire village of Woolsthorpe, England. "Sublime intellect! Vast and profound genius! A divine being!"[4] These were the phrases that Boullée used to describe Isaac Newton, whom he sought to immortalize in the ideal city that he envisioned as the materialization of his architectural theory.

Everything in Boullée's imagined city was conceived on a heroic scale. It was also addressed to a public that, he hoped, would view these monumental edifices as worthy demonstrations of Enlightenment. In a representative monument called the Métropole, Boullée imagined "the most profound veneration being excited" as the assembled public gathered in a massive structure modeled loosely on the Roman Pantheon to collectively worship the natural majesty of God.[5] Rays of light radiating toward an altar placed at the building's center would confirm the divine enlightenment being celebrated, while comparably grand edifices housing a museum and a library would further spread the light of art and learning to the public at large. To further "perpetuate the memory of those who deserve consecration," a series of commemorative cenotaphs were also proposed to honor the public's heroes.[6] At the center of this commemorative complex was to be a cenotaph for Isaac Newton, the man who Boullée believed best personified these Enlightenment ideals.

"By the extent of your enlightenment and the sublimnity of your genius, you determined the figure of the earth," Boullée wrote in praise of Newton.[7] Accordingly, he imagined placing his hero's tomb inside a massive orb that conveyed the image of a planetary body. Inside the sphere, a planetarium-like display would evoke Newton's legacy by immersing visitors in the celestial mysteries that his science had illuminated. His tomb would also be placed at this imagined celestial system's mathematical center of gravity so as to put visitors in an accurate position to contemplate Newton's scientific achievement. In his text, the architect admitted that his vision exceeded his technical abilities since he saw no way to create the imagined structure using contemporary artistic technologies. He defended his artistic aspirations nevertheless, saying that only in this way would the memory of Newton and the image of an all-powerful nature be fused in the appropriate way. He also allowed that if his cenotaph were ever built, a dark space illuminated by a single, solitary lamp

4. Resneau, *Boullée's Treatise*, 83.
5. Ibid., 39.
6. Ibid., 83.
7. Ibid.

FIGURE 1. *Étienne Louis Boullée, Métropole, interior view.*
Courtesy of the Bibliothèque nationale de France.

could serve as an appropriate substitute when representing the significance of
this great man.[8]

Boullée's ideal city was never built, of course, nor was his utopian vision
even published until the twentieth century. Yet from its folder buried within
the architect's papers, the manuscript echoed a common theme of the period in
its use of Newton to venerate nature, reason, and Enlightenment. In all corners
of late eighteenth-century France, in fact, similar manifestations of Newton's
iconic status were commonplace. In many, his name was simply offered as a
synonym for superhuman genius itself. The abbé Gerard, who offered a representative celebration of Newton in his 1778 poem *Le comte de Valmont,* described him as "an eternal honor to the human spirit," a man who "had raised
himself far above the common sphere" and whose "sparkling domination of
the empire of sciences" earned him "the respect of all modern philosophers."[9]

8. Ibid., 84–85.

9. Philippe-Louis Gerard, *Le comte de Valmont, où Les égarements de la raison* (Paris,
1787), 197.

FIGURE 2. *Étienne Louis Boullée, cenotaph for Newton, exterior. Courtesy of the Bibliothèque nationale de France.*

Along the same lines Louis-Sebastien Mercier used the physicist's example to illustrate the mystery of singular inventive brilliance. "Newton saw an apple drop and after meditating on it conceived the system of universal gravitation. Another, lacking the ability to see the ties that bind the planets to their orbits, would simply have grabbed the apple and eaten it."[10] The mathematician Joseph-Louis Lagrange phrased the same point slightly differently, writing that "Newton was the greatest genius who ever lived, and the most fortunate, for we cannot find more than once a system of the world to establish."[11]

Others deployed the same language of genius while either humanizing or moralizing Newton's sublime intellect. For the Parisian *salonnière* Julie de Lespinasse, Newton's achievements exemplified the rewards of diligent persistence in the pursuit of worthy goals. In a letter to M. de Guibert written in 1776, she marveled at Newton's patience in pursuing the same scientific question for over thirty years, inviting her correspondent to give similar devotion to the

10. Louis-Sebastien Mercier, *Tableau de Paris,* 8 vols. (Amsterdam, 1782–1783), 1: 148.

11. Jean-Baptiste-Joseph Delambre, "Notice sur la vie et les ouvrages de M. le Comte J. L. Lagrange," in *Oeuvres de Lagrange,* 14 vols. (Paris, 1867–1892), 1: xx.

FIGURE 3. *Étienne Louis Boullée, cenotaph for Newton, interior view.*
Courtesy of the Bibliothèque nationale de France.

even more worthy goal of love.[12] Even those who found Newton's legacy negative and destructive nevertheless found reason to invoke the elevated nature of
his genius. In his reactionary, anti-philosophe tract *Les helviennes* written in
1781, the abbé Barruel recommended burning Newton's *Principia* along with
all the other treatises on physics that convey absurd and impious conceptions
of nature.[13] Thanks to Johannes Kepler, Newton, Leonhard Euler, and Johann
Bernoulli, he warned, "schools of atheism had appeared."[14] Yet when he exposed the absurdity of the materialist philosophy sanctioned by these thinkers, Barruel used Newton's genius to make his point. Could one really imagine
"the soul of Newton arising in an insect," he asked in an attack on materialist
conceptions of the human mind?[15] According to materialist doctrine "an oys-
ter has as much intellect as Newton," Barruel mocked. But while this created a

12. Julie de Lespinasse, *Lettres à M. de Guibert* (Paris, 1876), 259.
13. Augustin Barruel, *Les helviennes, ou Lettres provinciales philosophiques* (Paris, 1830), 37.
14. Ibid., 44.
15. Ibid., 378.

FIGURE 4. *Étienne Louis Boullée, cenotaph for Newton, interior view with orrery. Courtesy of the Bibliothèque nationale de France.*

persuasive argument against materialist conceptions of the human spirit, it also echoed the wider veneration of Newton that reinforced these very ideas.[16]

Pairings between Newton and other thinkers were also common as a way of associating his aura of genius with other people and topics. In some cases his name was used to construct a pantheon of recognized philosophical heroes, but just as often other, more discordant pairings were deployed. Chateaubriand, in *La génie du Christianisme,* likened Newton's "divine inspiration" to that of Francis Bacon, Euler, and Gottfried Leibniz, but he also compared it favorably to that of the great classical age cleric Bishop Bossuet. Chateaubriand similarly linked Newton's spirit to that of Bossuet's contemporary, Father Massillon, famous for his eloquent preaching at the royal court of Versailles.[17] De Jouy offered a different juxtaposition in his 1812 *L'hermite de la chaussée-d'Antin.* He paired Newton with the seventeenth-century French playwright Corneille when describing those people who had most manifestly possessed that "rare

16. Ibid., 477.
17. René-François de Chateaubriand, *La génie du Christianisme* (Paris, 1803), 9–11.

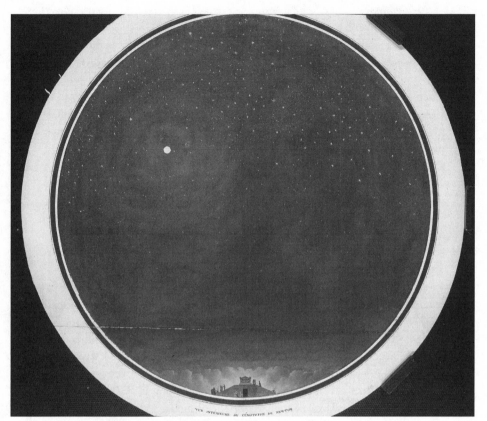

FIGURE 5. *Étienne Louis Boullée, cenotaph for Newton, interior view.*
Courtesy of the Bibliothèque nationale de France.

and precious faculty [of genius] denied by nature to the immense mob of humans."[18] Mercier went even further. He succeeded in linking the "august ghost" of Newton to Bacon and Galileo Galilei, but also to Charlemagne, Oliver Cromwell, Michelangelo, the Duc de Guise, Queen Elizabeth, John Calvin, William Shakespeare, Cardinal Richelieu, and the great French military leader the Vicomte de Turenne all in a single, enthusiastic sentence. "Oh, how I love to make myself feel small by surrounding myself with the idea of all these great men [*sic*] and to taste the pleasure that comes with admiring them," he gushed.[19]

18. Victor Joseph Étienne de Jouy, *La hermite de la chaussée-d'Antin, ou Observations sur les moeurs et l'usages parisiens au commencement du XIXe siècle* (Paris, 1815–1817), 121.

19. Mercier, *Tableau*, 1: 295.

Also common was using Newton's status as the founder of the true phys-ics of nature to celebrate achievements in other disciplines or sciences. Fran-çois Quesnay, the founder of the new science of economics, was often called the Newton of *économie*, and Bernardin de Saint-Pierre described the natu-ralist Joseph Tournefort in his 1814 *Harmonies de la nature* as "having as much knowledge in botany as Newton had in astronomy." [20] In his 1797 work *L'emigré*, Gabriel Senac de Meilhan offered a "two cultures" comparison, mak-ing Newton and the English novelist Richardson the two greatest geniuses of modern times. "The first divined the laws of the celestial bodies," he wrote, "and the second penetrated into the deepest abysses of the human heart." [21]

The connection between Newton's genius and the dawn of Enlightenment was also a common trope as French men and women found their own way to repeat Englishman Alexander Pope's famous epitaph that with "nature and nature's law lay hid at night: God said 'Let Newton be!' and all was light." In some cases the echoes were direct, such as in Charles Palissot de Montenoy's 1771 translation of Pope's *Le dunciade,* a poem that describes Newton as a Promethean voyager who "went into the heavens to disrobe its light." [22] For Cha-teaubriand, Newton was a thief who "stole so to speak the secret of nature from God." [23] Jacques Turgot echoed this same note of larceny in a Latin couplet he composed for the Jean-Antoine Houdon bust of the American Newtonian Ben-jamin Franklin. "He snatched lightning from the skies," Turgot wrote with reference to Franklin's Newtonian studies of electricity, and "the scepter from tyrants," a reference to Franklin's parallel participation in the recent United States Declaration of Independence. In J.-A. Roucher's 1779 poem *Les mois*, nature, having spent "a long night hiding its miracles in silence," finally discloses them to Newton, "that audacious eagle," who then uses them to "break the irons of antique ignorance." [24] "You, the pride of Albion, . . . placed so far away from human weakness. You alone were able to plumb every domain of the heavens. Through mad error, the mortals before you had disfigured the law of the uni-verse. You appeared, and suddenly the heavens belonged to you." [25]

20. Jacques-Henri Bernardin de Saint-Pierre, *Les harmonies de la nature,* in *Oeuvres posthumes de Bernardin de St.-Pierre,* 2 vols. (Paris, 1840), 2: 323.

21. Gabriel Senac de Meilhan, *L'emigré,* in *Romanciers du XVIIIe siècle,* ed. Étiemble, 2 vols. (Paris, 1965), 2: 1565.

22. Charles Palissot de Montenoy, *La dunciade* (London, 1771), 151.

23. René-François de Chateaubriand, *Essai sur la révolution,* in *Oeuvres complètes de Chateaubriand* (Paris, 1846), 1: 246.

24. Jean-Antoine Roucher, *Les mois* (Paris, 1779), 99.

25. Ibid., 154.

Similar language was often used to further attach Newton to broader accounts of human progress through science and reason. In his nine-stage *Esquisse,* or sketch, of the progress of the human mind, the Marquis de Condorcet traced how humans passed "by imperceptible gradations from the brute to the savage and from the savage to Euler and Newton." [26] Striking the same note about human progress in a 1777 text, while hitting it in a more explicitly colonial key, the playwright and author of the satirical play *Les Philosophes* Jean Delisle des Sales allegorically placed Newton in Senegal as a civilizing force of Enlightenment. [27] These formulations often functioned as simple narratives of historical significance, such as when Chateaubriand praised Newton for completing a hundred-year period where astronomy made more strides than in the previous three thousand years combined. [28] Framed this way, Newton was at once a titan of science who transformed medieval error into modern scientific truth and, at the same time, a world-historical genius who activated through physics the civilizing progress of mankind as a whole. Discourse such as this also performed an important cultural function in eighteenth-century France since it linked the precise history of Newton's scientific achievements, a history that was not without controversy, to the wider cause of Enlightenment and progress that was alleged to have followed from them. "Today," wrote Condorcet in his *Esquisse,* "a young man leaving our schools knows all the mathematics that it took Newton years of hard labor to learn or his genius to discover." [29] This was a marker of the advancement of French civilization, and it was also a pointer to the history whereby France had been put on the correct civilizing course by the genius of Newton and the zeal of his followers.

First in Jean Le Rond d'Alembert's 1751 *Discours préliminaire de l'Encyclopédie,* and then in dozens of other essays, poems, pamphlets, treatises, plays, letters, and other works, French men and women defined their status as modern, Enlightened subjects by associating themselves with the imagined legacy of Newton. They did so in particular by accepting and repeating the heroic story of France's release from error and superstition that occurred thanks to the intellectual efforts of Newton and his zealous followers. Not everyone saw this story in a positive light, of course. Barruel found in it the roots of France's descent

26. Cited in Frank E. Manuel and Fritzie P. Manuel, *Utopian Thought in the Western World* (Oxford, 1979), 492.

27. Jean Delisle des Sales, *De la philosophie de la nature,* 4 vols. (Paris, 1804), 4: 205. See Patricia Fara, *Newton: The Making of a Genius* (New York, 2002), figure 5.1.

28. Chateaubriand, *Génie du Christianisme,* 145.

29. Marquise de Condorcet, *Esquisse d'un tableau historique du progress de l'esprit humain,* ed. O. H. Prior (Paris, 1933), 231.

into atheism, impiety, and immorality. "Newton refutes Moses?" he fumed with respect to Voltaire's Newtonian critique of the Creation story offered in Genesis. "Our rabbis will find that a surprise." [30] Charles-Louis Richard turned this disgust into a warning in his 1785 *Exposition de la doctrine des philosophes modernes*, writing that "everywhere *philosophie* lights the torch of discord and of war, prepares poisons, sharpens swords, lays fires, orders murder, massacre, and carnage." For Richard, *philosophie* meant the broader complex of modern culture authorized by Newton and Enlightenment in eighteenth-century France, and for conservatives like him these developments were something to fight against. As Barruel sneered, "The school of Raynal, Voltaire, Jean-Jacques, Helvétius, and Diderot is one of rebellion, insubordination, and anarchy." [31]

Attacks such as these, which grew in force as the eighteenth century progressed, were part of a general reaction by clerics and conservatives against the positive equation made after 1750 of Newton, *philosophie*, and civilizational progress. However, in making these attacks, these "enemies of Enlightenment," as Darrin McMahon has called them, not only condemned the Newtonian Enlightenment defended by modern French subjects, they also mirrored it, and thus reinforced it, in reverse.

Also reinforcing were other, less blatantly polemical understandings of the Newton-Enlightenment dyad that connected each to modern, Enlightened subjectivity. One perspective viewed it all in terms of fashion, seeing in the connection between Newton and modern civilization nothing more than a vogue for things English. During the 1760s, the wealthy Marquis de Girardin redesigned the gardens of his estate at Erménonville, self-consciously injecting the spirit of Albion into his manorial grounds. Instead of the rigorous, mathematical arrangements of the traditional French garden, Girardin asked his garden architect to affect an English-inspired informality and rustic naturalness. He also ordered the construction of a properly ruined temple of philosophers at the center of his ersatz wilderness where commemorative statues of Newton and the great French Newtonian Voltaire were displayed as icons of worship. Nothing pleased Girardin more than using his garden to pass "a calm and tranquil morning *à l'anglaise*," reading Newton and Milton underneath the trees along with other Newtonian poets such as the Scotsman James Thomson. [32] The widely read conservative moralist Madame de Genlis satirized this sensibility in her *Adèle et Théodore* of 1782, describing an extreme "Anglomaniac,"

30. Barruel, *Les helviennes*, 243.

31. Cited in Darrin M. McMahon, *Enemies of the Enlightenment: The French Counter-Enlightenment and the Making of Modernity* (Oxford, 2001), 42–43.

32. See Fara, *Newton*, 141–143.

M. de Valcé, who "bragged incessantly about the genius and profundity of the English, and who hated the French with all his soul." "He rides only English horses," she mocked, "reads only English newspapers, makes his morning visits wearing boots and spurs, takes tea two times a day, and believes himself to be the measure of Locke or Newton." [33]

Whether sustained through fashion or some deeper cultural mechanism, Newton's image became a major iconic presence in eighteenth-century France after 1750. As such, his name and imagined legacy functioned as a kind of talisman that magically united a set of philosophical ideas and cultural values together with a set of related historical developments to produce a charm that conjured, either positively or negatively depending on the user, France's recent passage into modernity. When, for example, one of Paris's lesser known *salonniéres,* Mademoiselle de Ferrand, asked the leading portraitist of the day, Maurice Quentin de La Tour, to paint her seated in front of a volume by Newton, she was fashioning herself as a learned and modern woman attuned to the civilizing currents of progress. Since this portrait was also displayed publicly at the Paris Salon of 1753, less than four years after the death of the Marquise du Châtelet, Voltaire's famous partner and France's most famous *femme savante,* and only two years after the display of du Châtelet's portrait showing her at work with Newtonian science, Ferrand was using her self-representation to link herself publicly to a living history. [34] In 1756, du Châtelet's posthumous translation and commentary on Newton's *Principia* appeared (still the only complete French translation of the work ever published), [35] and it too reinforced the living culture of Enlightenment Newtonianism into which Ferrand was attempting to tap. The Jacobin firebrand Jean-Paul Marat made use of this understanding as well when he used a pamphlet to liken his own struggles with the Paris Academy of Sciences in 1792 to those that had raged between Newtonians and Cartesians in the 1730s. "Everything rests on fashion, opinion, and systems," he wrote. "Thirty years ago [*sic*] all the academy was Cartesian and it waged war against Newton. Today it is Newtonian and it wages war against Descartes." [36] Marat's chronology was off by several decades, but he was connecting himself accurately with the actual history of Newtonianism in France, a history that made the widespread veneration of Newton as an icon of Enlightenment a consequence of a set of hard-won, public battles fought earlier in the century.

33. Madame de Genlis, *Adèle et Théodore, ou Lettres sur l'éducation* (Paris, 1782), 328.

34. See Fara, *Newton*, 136–137.

35. The translation has recently been reissued (with a preface by Voltaire) as *Principia: Les principes philosophiques de la philosophie naturelle* (Paris, 2005).

36. Charles Vellay, ed., *Les pamphlets de Marat—1792* (Paris, 1911), 152.

FIGURE 6. *Maurice Quentin de La Tour, Portrait de Mlle. Ferrand.*
Courtesy of the Alte Pinakothek, Munich.

This initial and initiating encounter between Newton and eighteenth-century France is the subject of this book. In 1687, the year Newton's *Philosophiae naturalis principia mathematica* appeared, few Frenchmen even knew who the author was and even fewer actually read the book. Yet a century later he and his work were held by large numbers of them to be the very personification of modernity. Ever since the eighteenth century, moreover, nothing has been more

FIGURE 7. *Emilie de Breteuil, Marquise du Châtelet.*
Courtesy of the Chateau de Breteuil, France.

commonly emphasized in writing about the eighteenth-century Enlightenment than its genetic attachment to the scientific legacy of Isaac Newton. Certainly Peter Gay echoes a widely accepted view when he states that "in the deification of Newton, the Enlightenment of the philosophes and the age of Enlightenment were at one." [37]

Indeed, the seemingly natural equation between Newtonianism and Enlightenment remains so pervasive and unquestionably persuasive today that histories of the former are very often taken as histories of the latter without so much as a historiographical pause. When, for example, the American publisher Atlantic Highlands set upon the admirable project of publishing an English translation of Vincenzo Ferrone's fine book *Scienza natura religione: Mondo newtoniano e cultura italiana nel primo Settecento,* the editors did not translate the title literally as *Science, Nature, Religion: Italian Culture and the Newtonian World in the Early 1700s.* They instead titled the book *The Intellectual Roots of the Italian Enlightenment,* suggesting through their invented title that the spread of Newtonianism in Italy was synonymous with *Illuminismo.*[38] In a similar fashion, the larger narratives that link together Newton and Enlightenment with the origination of modernity, narratives that, as we have seen, begin in the eighteenth century, also commonly take for granted the naturalness of this linkage and the self-evidence of its progressive agency.

Nothing seems more obvious in these stories than Newton's status as the singular genius who "unraveled once and for all the riddle of the cosmos." Likewise nothing seems more natural than the French philosophes' adopting Newton of all people as their intellectual hero in their fight to defend Enlightenment. The present study, however, adopts a different perspective on these matters. It takes as its starting point the claim that these were not natural outcomes of a progressive and teleological Enlightenment modernity, but, rather, a set of particular historical outcomes produced by French men and women caught up in a complex web of temporal, spatial, and other local contingencies. In short, this is a book about a beginning that seeks to escape the spell of teleological origin stories. It is also a history of Newton and Enlightenment in France that self-consciously detaches itself from the living history that continually naturalizes their marriage as a self-evident feature of modernity. To state

37. Peter Gay, *The Enlightenment: An Interpretation; The Science of Freedom* (New York, 1969), 130.

38. Vincenzo Ferrone, *Scienza natura religione: Mondo newtoniano e cultura italiana nel primo Settecento* (Naples, 1982), idem, *The Intellectual Roots of the Italian Enlightenment: Newtonian Science, Religion, and Politics in the Early Eighteenth Century,* trans. by Sue Brotherton (Atlantic Highlands, 1995).

the same point another way, this is a postmodern and post-Enlightenment history of one crucial moment in the beginning of Enlightenment modernity, the moment when Newtonian science became linked to it as its genetic code and avatar.

In 1641 there was no such being called Isaac Newton in Woolsthorpe, England, and in 1686 the would-be hero was still a little-known forty-three-year-old Cambridge University mathematician struggling to make headway with some thorny problems in mathematics and mechanics. He was also engaged in other activities (including alchemy and the historical interpretation of Biblical prophecy, to name just two) that pointed toward destinies other than the one for which he became famous. Yet within a hundred years these other activities would either be forgotten or erased as Newton's name and image became synonymous with the solidification of modern physics and the Enlightenment that was said to follow from it. In France and throughout all of Europe, including its colonial outposts around the globe, Westerners also began to fashion themselves as "modern" and "Enlightened" by accepting and repeating a historical narrative that purported to link Newtonian science with the birth and spread of this progressive modernity. Westerners, in fact, still do this today, even if they are less self-conscious about the act than were their eighteenth-century counterparts. In this respect, the eighteenth-century historical knot that tied Newtonian science together with the rise and spread of Enlightenment modernity still remains tangled today.

Consider the legacy of d'Alembert's 1751 *Discours préliminaire de l'Encyclopédie,* one of the founding texts of this mythology of origins.[39] In this text, the academician, mathematician, man of letters, socialite, and would-be philosophe developed a systematic introduction to the compendium of scientific knowledge that the volumes of the *Encyclopédie* contained. In part 2 of the text, he also deployed a historical narrative that quickly became classic when linking Newton together with Enlightenment and modernity. D'Alembert began his history with the fifteenth-century "renaissance of letters" that he believed started the modern development of knowledge. "The invention of printing and the patronage of the Medici and of Francis I revitalized minds and enlightenment was reborn everywhere."[40] At first, however, knowledge was dominated by book learning and belles lettres while philosophy remained trapped in its "medieval darkness."[41] Yet out of this darkness emerged the "immortal Chan-

---

39. Jean Le Rond d'Alembert, *Preliminary Discourse to the Encyclopedia of Diderot,* trans. by Richard N. Scwab, 2nd ed. (Chicago, 1995).

40. Ibid., 62.

41. Ibid., 73–74.

cellor Bacon," and he became the first to "illuminate the world" by shining the true lamp of science upon it. "[Bacon] asserted that the scholastics had ener-vated science by their petty questions, and that the mind ought to sacrifice the study of general things for that of individual objects." Bacon also recommended the direct study of nature, and "he [made] known the necessity of experimental physics, of which no one was yet aware." [42] In this way, d'Alembert explained, mankind left behind its medieval ignorance while charting a new course along the civilizing path of empirical science.

Soon Bacon was joined by the pioneering work of another immortal, the illus-trious René Descartes. "One can view Descartes as a geometer or as a philoso-pher," d'Alembert wrote, and "mathematics, which he seems to have considered lightly, nevertheless today constitutes the most solid and least contested part of his glory." "Above all what immortalized the name of this great man [was] the application he was able to make of algebra to geometry, one of the grandest and most fortunate ideas that the human mind has ever had." [43] His physics, however, "appears today almost ridiculous." But "let us recognize that Descartes, who was forced to create a completely new physics, could not have created it better; that it was necessary, so to speak, to pass by the way of the vortices [a reference to the mechanisms of Descartes' cosmology] in order to arrive at the true system of the world; and that if he was mistaken concerning the laws of movement, he was the first, at least, to see that they must exist." [44] D'Alembert then framed the birth of modern science in terms of a negotiation between Baconian empiricism on the one hand and Cartesian mathematicization on the other. This paved the way for the climax of his story, the Newtonian synthesis of empiricism and mathematics that established modern science in its full maturity.

"Newton . . . appeared at last and gave philosophy a form that apparently it is to keep," d'Alembert asserted. "That great genius saw that it was time to banish conjectures and vague hypotheses from physics, or at least to present them only for what they are worth. . . . [His] science was uniquely suscep-tible to experiments and geometry." [45] Writing from the perspective of 1751, d'Alembert further claimed that Newton's "Theory of the World" had become generally accepted, but that this consensus had not always reigned. "It is true that Newton had the singular advantage of seeing his philosophy generally ac-cepted in England during his lifetime and of having all his compatriots for par-

42. Ibid., 75–76.
43. Ibid., 78.
44. Ibid., 79.
45. Ibid., 80–81.

tisans and admirers."[46] Elsewhere, however, things were different. "At that time it would have taken a good deal to make Europe likewise accept his works. Not only were they unknown in France, but scholasticism was still dominant there when Newton had already overthrown Cartesian physics." D'Alembert then reminded his French readers, with a note of ironic surprise, that only twenty years had passed since "we began to renounce Cartesianism in France."[47]

This allowed the author to turn his attention to his homeland and to narrate the story of France's passage into the light. For many of his contemporaries, Voltaire's *Lettres philosophiques* of 1734 marked the key step in this process. D'Alembert's protégé and fellow philosophe Condorcet adopted this view, explaining in his retelling of this mythology of origins: "This text initiated among us a period of revolution. It began to make us aware of English philosophy and literature, . . . [and] to cause us to replace our former indifference toward it with a childlike aptitude."[48] Condorcet claimed further that "neither Newton's philosophical opinions, nor his system of the world, nor even his experiments regarding light were known in France at this time." Voltaire, therefore, initiated the French into Enlightenment by introducing these Newtonian ideas to them.[49] D'Alembert actually told a different story, making Pierre-Louis Moreau de Maupertuis' 1732 *Discours sur les différentes figures des astres* the precise text that opened French eyes to the light of Newtonian science.[50] In every other respect, however, their narratives agreed. Each contended that France remained trapped in Cartesian ignorance until shaken from its spell by the outspoken opposition of the first French Newtonians. They also agreed that the French had only embraced the achievements of Newtonian science after being drawn forcibly to the light by the singular efforts of a group of young thinkers in the 1730s. As d'Alembert explained, "it is the young geometers in France . . . who have directed the fate of the two philosophies."[51] Condorcet pointed to "a small number of young geometers [who] alone had the courage to abandon Cartesianism."[52] They further agreed that the nation's conversion to Newtonianism had been the key step in the beginning of Enlightenment in France.

46. Ibid., 88.

47. Ibid.

48. Marquis de Condorcet, *Vie de Voltaire* (Geneva, 1787). My citations will be drawn from the modern edition of the work (Paris, 1994), 46.

49. Ibid., 45.

50. D'Alembert, *Preliminary Discourse*, 89.

51. Ibid., 90.

52. Condorcet, *Vie de Voltaire* 52.

Joined by these and other similarities, the two narratives merged into a single story that explained France's passage into Enlightened modernity as the offshoot of Newtonian science and the heroic efforts of certain Frenchmen to defend it. We have already seen how this narrative was further echoed and reinforced throughout the eighteenth century as Newton was publicly canonized the patron saint of Enlightenment. These reverberations continued throughout the next two centuries, and through them the link between Newtonian science and Enlightenment modernity was solidified and disseminated. Even today, this narrative complex remains pervasive (and stubbornly persuasive), and it also remains foundational in contemporary historical scholarship about these topics.

In a lecture delivered at the University of Chicago in 1948, Alexandre Koyré, one of the founding fathers of the modern discipline of the history of science, illustrated well how the classic narrative of Newtonian Enlightenment developed by the eighteenth-century French philosophes became foundational to academic literature in the history of science.[53] The title of his talk was simply "The Significance of the Newtonian Synthesis," yet each element of the presentation followed d'Alembert's formulation closely. "Significance," for example, stretched in two directions for Koyré. Looking backwards, he reinscribed d'Alembert's story of the Renaissance revolution in learning by placing Newton at the climax of the so-called Scientific Revolution. Newton's significance in the making of modern science similarly mirrored d'Alembert's account as Koyré explained how Newton's dual empirical/mathematical approach to science, articulated most famously in the *Principia*, effected a synthesis that thereby created the modern scientific method.

Koyré's account of Newton's eighteenth-century legacy likewise echoed a number of philosophe themes. "In spite of the rational plausibility and mathematical simplicity of the Newtonian [inverse-square law of universal gravitation] there was in it something that baffled the mind."[54] Given this mystery, Koyré explained, Newton's work was originally met with a host of serious objections. France in particular was a haven of resistance to Newtonianism, and here again Koyré echoed d'Alembert in positioning figures like Christiaan Huygens, Gottfried Leibniz, and Johann Bernoulli as "Cartesian" opponents to the emerging "Newtonian world-view."[55] Koyré also deployed philosophe historiography directly by allowing the famous and polemical contrast between Newtonians

53. Alexandre Koyré, "The Significance of the Newtonian Synthesis," in *Newtonian Studies* (Chicago, 1965), 3–24.

54. Ibid., 16.

55. Ibid.

and Cartesians in Voltaire's *Lettres philosophiques*—"readable even today," he enthused—to stand in for a detailed, historical analysis of Newton's actual French reception.

Anchoring this twentieth-century appropriation of Enlightenment mythology, moreover, was another piece of eighteenth-century technology: a footnote. This referred readers to a learned monograph supportive of Koyré's views.[56] Published in 1931, Pierre Brunet's *L'introduction des théories de Newton en France au dix-huitième siècle*, the book that Koyré cited, remains today the standard reference when explanations of Newton's introduction into the intellectual world of eighteenth-century France are offered.[57] It was a pioneering study for its time, and the account still remains authoritative today despite the sea changes that have occurred within the history of science over the last seventy-five years. Indeed, even Alan Sokal, the physicist who hoodwinked *Social Text* into publishing his parody of postmodern science studies, knows and admires Brunet. In a post-mortem to the *Social Text* controversy published in 1998, Sokal wrote that "the consensus of historians appears to be that the slow acceptance of Newtonian mechanics in France arose from a scholastic attachment to Cartesian theories as well as from certain theological considerations." He then referred readers to Brunet.[58] Sokal is in fact right about Brunet, whose central thesis, articulated in the first pages of the book and defended throughout, is that "the theories of Newton encountered over the course of the eighteenth century particularly violent resistance in France because they clashed there with Cartesian doctrines already solidly established."[59] Sokal is also right about most historians of science—they still largely accept Brunet's account, especially his conception of an intransigent war between Newton and Descartes as the dominant theme of Newton's French reception.

What few see, however, is how Brunet's book merely reproduces the eighteenth-century French Enlightenment's own glorious view of its own history through the idioms of twentieth-century academic scholarship. They also do not see how, in accepting and repeating these stories, modern scholars and readers are also continuing to participate in the now two-century-old repro-

56. On the history of the footnote, see Anthony Grafton, *The Footnote: A Curious History* (Cambridge, 1997).

57. Pierre Brunet, *L'introduction des théories de Newton en France au dix-huitième siècle* (Paris, 1931).

58. Alan Sokal, "What the *Social Text* Affair Does and Does Not Prove," in *A House Built on Sand: Flaws in the Cultural Studies Account of Science*, ed. Noretta Koertege (Oxford, 1998), 21 n. 37.

59. Brunet, *L'introduction*, v–vi.

duction of Enlightenment modernity itself. Consider A. R. Hall's *The Revolution in Science, 1500–1750*, first published in 1954, but most recently reissued in a third edition under a new title in 1983.[60] This book still serves as the standard introduction for many to the history of the Scientific Revolution, and in it the assumptions of the classic narrative are foundational. Hall writes, for example, that in Newton's work "the scientific revolution reached its climax, and a model for future natural philosophers [was] created. . . . The unity of nature was made manifest in a grand synthesis revealing the applicability of the same laws, the same principles of explanation, in the heavens and on the Earth. The planetary motions of Copernicus, Kepler's laws, the discoveries made by Galileo and Huygens relating to the phenomenon of gravity and motion, were all shown to follow from these laws and to be embraced with within the same synthesis."[61] This formulation reconfirms in even more emphatic terms d'Alembert's and Koyré's classic conception of Newton as the climactic, synthetic hero of the "Scientific Revolution."

Many other textbooks in the history of science, including some of the most recent, also repeat the same story, and in all of these accounts eighteenth-century France is similarly treated in formulaic terms. Many simply end the story with Newton, making all of eighteenth-century science, including that practiced in France, little more than the mop-up operation after the triumphant Newtonian synthesis. As Hall describes this approach in the first edition of *The Revolution in Science,* "the refinements of the [eighteenth-century] continental mathematicians [like d'Alembert] in no way modified the essential principles of the mathematical, mechanical method in physics, which were already fixed. . . . The many new and useful ideas that they put forward must, therefore, be ascribed to a second order of discovery, as being derivative rather than fundamental."[62] Koyré inflected the same basic point in a slightly different way by positing first a period of opposition (Cartesian-inspired in France) to Newton's triumphant physics, and then a resumption and completion of the latter's program by "the leading mathematicians of Europe—Maupertuis, Clairaut, d'Alembert, Euler, Lagrange, Laplace."[63] This version gives agency to the eighteenth-century epigones of Newton as builders who completed his edifice; however, it in no way decenters Newton from his triumphant status as the singular, synthetic architect of the modern scientific world. In the third edi-

60. A. R. Hall, *The Scientific Revolution, 1500–1800: The Formation of the Modern Scientific Attitude* (London, 1954), 2nd ed. (London, 1960), and 3rd ed. (London, 1983).

61. Hall, *Scientific Revolution* (3rd ed.), 306.

62. Hall, *Revolution in Science* (1st ed.), 342.

63. Koyré, "Significance," 17.

tion of *The Revolution in Science,* Hall adopted this approach as well, citing like Koyré Voltaire's *Letters concerning the English Nation* as the turning point that marked the passage from French ignorance and hostility toward Newton toward their gradual acceptance of his science.[64] Even recent works that have self-consciously set out to displace these traditional frameworks still reproduce them in unfortunate ways. Steven Shapin's otherwise admirable *The Scientific Revolution* says virtually nothing about European science after Newton, and while Peter Dear's equally nuanced *Revolutionizing the Sciences* devotes an entire chapter to eighteenth-century science, his title, " 'Newtonians' and 'Cartesians,' " suggests, along with Brunet and the classic narrative, that this dichotomy was the dominant one of the period.[65]

The historiography of the French Enlightenment, at least that which explicitly treats Newton's role in it, likewise remains in thrall to the classic narrative produced by the Enlightenment philosophes. Virtually every contemporary Western civilization textbook tells a similar story, and in the less specialized literatures and public discourses that permeate contemporary Western society, the tendency is even more pronounced. A 2004 exhibition at the New York Public Library, The Newtonian Moment: Science and the Making of Modern Culture, moved visitors through displays entitled "The Apprenticeship of Genius" and "The Gospel according to Newton." An opening presentation echoed Lagrange in offering the hero of Woolsthorpe as "the greatest and most fortunate of humans" while "the era of the Enlightenment and Revolution" were positioned as offshoots of the "Newtonian Moment." A display concerning the eighteenth-century reception of Newton's ideas also noted the "chauvinistic overtones" that permeated it. "Newtonian ideas were as likely to be accepted or rejected along nationalistic lines as on the merits of the case," the curators explained, "and this was true not only for the German proponents of Leibniz, but also among Frenchmen who balked at the spectacle of the dethroning of Descartes—the reigning scientific philosopher—by an Englishman." Neither Newton nor the Newtonians, of course, were accused of the same chauvinistic prejudices, and since Voltaire and d'Alembert would have represented things in exactly the same way, the library was ultimately offering contemporary cultural consumers an opportunity to reaffirm their modernity by reconnecting with the founding mythologies of Enlightenment.[66]

64. Hall, *Revolution in Science* (3rd ed.), 347–348.

65. Steven Shapin, *The Scientific Revolution* (Chicago, 1996); and Peter Dear, *Revolutionizing the Sciences: European Knowledge and Its Ambitions, 1500–1700* (Princeton, 2001).

66. See the companion volume to the exhibition: Mordechai Feingold, *The Newtonian Moment: Newtonian Science and the Making of Modern Culture* (Oxford, 2004).

To be fair, not everything said today about the Newtonian legacy and its connection to Enlightenment remains imprisoned in these eighteenth-century "Just So Stories." There are first of all vast and growing specialized literatures about Newton and Enlightenment that take as their starting point the deconstruction of these timeworn myths of origin. This is the literature upon which this book is built. These more recent and critically detached works attest to the changing climate of opinion today, one where the heroic story of Newton's Promethean role in the birth of scientific modernity is no longer accepted with such zealous credulity. They also reflect the way that self-conscious detachment from the ritual narratives of modernity has become more commonplace, both within Newton and Enlightenment scholarship and in Western society as a whole.

Each of these trends inspires the project of this book, for in its grandest conceptualization it attempts to offer a detached, postmodern counter-narrative of the Newton-Enlightenment linkage in eighteenth-century France, one that strives to replace the reigning mythology of origins with a critical genealogy of beginnings. For some, the ambition is misguided since, they argue, the entanglement itself cannot be treated without reproducing the traditional categories and periodization that created the knot in the first place. These skeptics further argue that the credibility of the older narratives has all but evaporated. Why not avoid the urge to renarrate this history altogether and simply let these dead narrative horses lie? Still others argue cogently that narrative itself is a problem since all narratives deploy homogenized unities and causal frameworks and these will inevitably replicate the mythologies of origin so problematic in the older accounts. These criticisms are potent, but the resiliency of the older narratives is formidable as well. Can one really expect the classic stories to simply wither away and die? Or should one consciously work to replace them through the articulation of credible substitutes?

The chapters that follow find their conceptual starting point in an affirmative answer to this last question. Moreover, since many theorists and critical historians have provided indispensable orientation in pointing the book in this direction, the debt owed to their work should not be ignored. Crucial throughout has been the work of critical, constructivist scholars of science—Pierre Bourdieu, Peter Galison, Bruno Latour, Simon Schaffer, and Steven Shapin, to name just five of the luminaries—scholars who, in a very real sense, have created the conceptual space and critical tools that have made this book possible.[67] Indeed,

67. Pierre Bourdieu, *Distinction: A Social Critique of the Judgment of Taste,* trans. by Richard Nice (Cambridge, 1987), idem, *Outline of a Theory of Practice,* trans. by Richard Nice (Cambridge, 1977); Peter Galison, *Image and Logic: A Material Culture of Microphysics* (Chicago, 1997); Bruno Latour, *Science in Action: How to Follow Scientists and Engineers*

those seeking a more precise outline of the methodology pursued in this book can do no worse than to read the "rules of method" offered by Bruno Latour in his book *Science in Action*.[68] Also important have been numerous empirical inquiries in science studies, Enlightenment studies, and eighteenth-century European history more generally, studies that have given concrete substance to the conceptual agendas pursued here. In fact, without the critical archival perspectives and detailed historical interpretations offered by these more precise works, this book could not have been attempted. Numerous other scholars and scholarly works have been equally influential, and their particular contributions will be noted throughout the text. However, to recall all the influences here would produce a list both tedious to read and inevitably incomplete. Instead, let me rather invoke all of this work collectively by discussing two thinkers whose writings crystallize the conceptual frameworks employed throughout.

"Where, when, or what is a beginning?" asks Edward W. Said in his book *Beginnings*, initiating what he calls "A Meditation on Beginnings." "Is a beginning the same as an origin?" and "of what value, for critical or methodological or even historical analysis, is [the concept of] 'the beginning'"?[69] Said's meditation yields a rich harvest of arguments justifying the use of "the beginning" as a guiding analytical category. Also stressed, however, are the pitfalls that emerge when the term is used too loosely. Purported synonyms such as "birth," "origin," or "genesis" must, in fact, be strictly avoided, Said argues, since a "beginning," or a "starting" to introduce a valid synonym, is different from an organic birth or teleological origin. Properly conceived and deployed, the term "beginning" demarcates not a becoming or a previously prepared development but rather a precise, empirical moment of change, one that carries with it no presuppositions about causal or genetic agency. In other words, the term "beginning," if used precisely, allows one to talk about temporal changes and historical arrivals while avoiding altogether presumptions about developmental, mechanical, or genetic causality.

This book seeks to deploy the concept of the beginning in this precise way, and if Said's analysis defines the terms, Michel Foucault's Nietzschean-inspired

---

through Society (Cambridge, 1987), idem, *We Have Never Been Modern*, trans. by Catherine Porter (Cambridge, 1991), idem, *Pandora's Hope: Essays on the Reality of Science Studies* (Cambridge, 1999); Simon Schaffer and Steven Shapin, *Leviathan and the Air Pump: Hobbes, Boyle and the Experimental Life* (Princeton, 1987); and Steven Shapin, *A Social History of Truth: Civility and Science in Seventeenth-Century England* (Chicago, 1994).

68. These rules are conveniently summarized in appendix 1 and 2 of Latour, *Science in Action*, 258–259.

69. Edward W. Said, *Beginnings: Intention and Method* (New York, 1975), 29.

conception of genealogy articulates their realization as a historical and narratological method.[70] For Foucault, critical genealogy emerges as an antidote to the normalizing and naturalizing tendencies of traditional, narrative-causal history. Whereas traditional narrative history sees origins and not beginnings in the past and narrates change through claims about the unfolding of inherent teleologies assumed to exist inside these originating causal structures, genealogy instead asserts "the chimera of origins" and "the dispersion and disparity of beginnings." It does not trace descent; "on the contrary, it disturbs what was previously considered immobile, it fragments what was thought unified; it shows the heterogeneity of what was imagined consistent with itself."[71]

In this way, genealogy does not "discover a forgotten identity, eager to be reborn, but a complex system of distinct and multiple events, unable to be mastered by the powers of synthesis." Temporal change, while central to the genealogical enterprise, is also treated differently by the genealogist than by the historian. The latter sees underlying causes, be they mechanical, organic, or some hybrid between them, and narrates change in terms of the causal agency of these structures. The genealogist, by contrast, recognizes that there are no historical acorns that evolve to become the oak trees of the present; she also knows that archives do not contain historical objects possessing secret essences or hidden teleologies whose causal agency the historian can reveal. Instead, the genealogist accepts that historical change, to use Foucault's terms, is "derisive and ironic," and that historical objects are "fabricated in a piecemeal fashion from alien forms." It is "the jolts" and "the surprises" that bring about historical change, and to understand these shifts one must be "scrupulously attentive to their petty malice" while cultivating an appreciation for "the details and accidents that accompany every beginning."[72]

At the center of this methodological ideal are a number of fundamental principles. The genealogist for Foucault is first of all meticulously empirical. "In short, genealogy demands relentless erudition," he writes. "It requires patience and a knowledge of details, and it depends on a vast accumulation of source material." In collecting and organizing this material, however, genealogy also self-consciously opposes itself to historical positivism by refusing to use the archive to engage in a search for origins. Instead, genealogy "records the singularity of events outside of any monotonous finality." It "rejects the

70. See especially Michel Foucault, "Nietzsche, Genealogy, History," in *The Foucault Reader*, ed. Paul Rabinow (New York, 1984), 76–100.

71. Ibid., 82.

72. Ibid., 78–80.

meta-historical deployment of ideal significations and indefinite teleologies."[73] Genealogy, therefore, is "gray, meticulous, and patiently documentary," but it is also diachronic in approach and narrativistic in structure while emphatically noncausal and nonteleological in conception. Foucault ultimately suggests three plausible modes for genealogical narration—the parodic, the dissociative, and the sacrificial—and it is the second that has been adopted for this book. Dissociative genealogies are directed against identities and unities. They operate by multiplying dispersion at the moment of historical change. Challenging the "respect for ancient continuities" characteristic of traditional, narrative-causal history, a dissociative genealogy instead systematically disrupts continuity. It does so, moreover, while deploying the parodic mode to emphasize the ironies, accidents, and alien hybrids that often factor into actual historical change.[74]

The success with which this book actualizes these methodological ideals can be judged by the chapters that follow. For those familiar with Foucault's own brilliant genealogical work, reassurances can be offered that his intensely self-reflexive and opaque manner of writing has not been adopted here along with his philosophy of history. Regrets, however, also need to be extended to those, like me, who actually admire Foucault's immanently subversive prose. I am no Michel Foucault, however, only an admirer of his philosophy of history. I have not, therefore, attempted to write a Foucaultian book so much as to write a traditional work of history that is nevertheless inspired at every step by his philosophy and historical sensibility. This book, therefore, makes no claim to being a Foucaultian genealogy; it is rather a work of Foucault-inspired revisionist historical scholarship, one that marshals extensive and carefully analyzed documentary evidence in support of a new and different genealogical understanding of the Newtonian legacy in eighteenth-century France. To further take stock of its real content, therefore, let us leave behind the critical-theoretical notion of genealogical beginnings that informs the text and instead turn to an introductory summary of the book's substantive claims.

Reframing the Newton-Enlightenment Linkage

What is meant by the term "Enlightenment," and what, once defined, does it mean to talk about the beginning of the French Enlightenment? Much ink has been spilled in recent years trying to define and delimit this phenomenon

73. Ibid., 76–77.
74. Ibid., 94–95.

in eighteenth-century Europe. One key touchstone has been the work of Ernst Cassirer.[75] He posited an idealist, intellectualist conception of Enlightenment defined in terms of an eighteenth-century "philosophy" that carried this name. Against the grain of this idealist approach, however, numerous scholars have argued instead for a social conception of the same thing rooted in new forms of living, writing, and being in the eighteenth century.[76] Peter Gay's conception of a European-wide "party of humanity" joined through self-conscious defense of *philosophie* effectively married Cassirer's philosophical approach to Enlightenment with the new call for a social history of Enlightenment thinkers and texts. As a result, it has proven to be a very influential formulation.

Against it, however, others have emerged, including those that argue not for an Enlightenment rooted in philosophically unified communities but instead for one rooted in shared manners and social values.[77] In another formulation, one that follows Jürgen Habermas, Enlightenment arises out of the new publicity of the eighteenth century. Here the term finds its definition in the critical spirit that this publicity spawns and in the new and transformative social mixing that it allowed.[78] In still another conception, radicalism, be it philosophical, political, or both, is seen to emerge out of what Paul Hazard called *la crise de la conscience européenne* after 1680 bringing about a European-wide move toward Enlightenment.[79] Connected to all of these accounts, moreover, are the institu-

75. Ernst Cassirer, *The Philosophy of the Enlightenment,* trans. by Fritz C. A. Koelln and James P. Pettegrove (Princeton, 1951).

76. See especially Robert Darnton, *The Literary Underground of the Old Regime* (Cambridge, 1982); and Daniel Roche, *France in the Enlightenment,* trans. by Arthur Goldhammer (Cambridge, 1998).

77. Daniel Gordon, *Citizens without Sovereignty: Equality and Sociability in French Thought, 1670–1789* (Princeton, 1994); and Dena Goodman, *The Republic of Letters: A Cultural History of the French Enlightenment* (Ithaca, 1994).

78. Jürgen Habermas, *The Structural Transformation of the Public Sphere: An Inquiry into a Category of Bourgeois Society,* trans. by Thomas Burger (Cambridge, 1989). Other works that either explicitly or implicitly advance this Habermasian perspective include Thomas Broman, *The Transformation of German Academic Medicine* (Cambridge, 1992); Thomas E. Crow, *Painters and Public Life in Eighteenth-Century Paris* (New Haven, 1985); and Margaret C. Jacob, *Living the Enlightenment: Freemasonry and Politics in Eighteenth-Century Europe* (New York, 1991). The Habermasian historiography is ably synthesized in James Van Horn Melton, *The Rise of the Public in Enlightenment Europe* (Cambridge, 2001).

79. Paul Hazard, *La crise de la conscience européenne (1680–1715)* (Paris, 1935); Margaret C. Jacob, *The Radical Enlightenment: Pantheists, Freemasons, and Republicans* (London, 1981); Jonathan Israel, *The Radical Enlightenment: Philosophy and the Making*

tions of the Republic of Letters, institutions that in one formulation are synonymous with Enlightenment, and in another are its incubator and eventual foil.[80] Complicating matters still further is the question of geography. Is Enlightenment a singular, European-wide or even global phenomenon with a single point of origin, or were there many Enlightenments each with a particular character and trajectory? If the latter, then what unifies these various Enlightenments, and how were they connected to each other across time and space?[81]

A survey of this historiography and its debates would require a book unto itself, and for this book such a review is not necessary. A simple definition of what the term "Enlightenment" will mean in this book will serve to introduce how the subsequent chapters propose to trace its genealogy. First, by "Enlightenment" I will mean only a French phenomenon even if the wider implications of Enlightenment in this precise national context were multiple and substantial. In other words, I will not perform the familiar and legitimately criticized act of reading the history of the Enlightenment as a whole off of the particular history of France, or worse yet Paris, in the eighteenth century. Rather I offer a precise, genealogical account of the beginning of the French Enlightenment alone while offering no hypotheses about its connection to the wider phenomena of Enlightenment elsewhere in the world. Second, my understanding of Enlightenment is similarly narrow and French-specific since it finds its definition neither in a general set of social changes nor in a general philosophical outlook but rather in a historically specific marriage of each in eighteenth-century France. Finally, and perhaps most important of all, my understanding of Enlightenment is deconstructive in nature since it begins by subjecting the history of the Enlightenment produced by the eighteenth-century French actors themselves to critical, genealogical scrutiny.

Ultimately, the term "Enlightenment" as it is used in this book refers to the philosophe movement in France that arose and took consciousness of itself

---

of Modernity, 1650–1750 (Oxford, 2000), and idem, *Enlightenment Contested: Philosophy, Modernity, and the Emancipation of Man, 1670–1752* (Oxford, 2006).

80. The first position is defended in Goodman, *Republic of Letters,* and in a very different way by Laurence Brockliss, *Calvet's Web: Enlightenment and the Republic of Letters in Eighteenth-Century France* (Oxford, 2002). The second position is argued in Anne Goldgar, *Impolite Learning: Conduct and Community in the Republic of Letters, 1680–1750* (New Haven, 1995).

81. Roy Porter, *The Creation of the Modern World: The Untold Story of the British Enlightenment* (New York, 2000); J. G. A. Pocock, *Barbarism and Religion,* vol. 1, *The Enlightenment of Edward Gibbon, 1737–1764* (Cambridge, 1999); Roger Chartier, *The Cultural Origins of the French Revolution,* trans. by Lydia G. Cochrane (Durham, 1991).

after 1750. In this respect, then, the book does deploy an eighteenth-century understanding of the term Enlightenment since it associates the phenomenon with the very people and social movement that claimed that term for themselves. As we have already seen, however, the philosophes also created their own self-congratulatory history, and on this point the book parts company with them as it actively seeks to deconstruct the philosophe-authored story that posits Newtonian science as the natural springboard for Enlightenment. It also works to problematize the philosophe claim that Voltaire's 1734 *Lettres philosophiques* constituted the sui generis origination of the Newtonian Enlightenment in France. The scandal provoked by the *Lettres philosophiques* did indeed mark a beginning in the history of the French Enlightenment. The scandal also marked a key moment in the creation of the crucial Newton-Enlightenment linkage in France. Yet importantly neither began in 1734 with Voltaire, nor even did they begin with Newton and Maupertuis in 1732.

The scandal triggered by the *Lettres philosophiques,* this book argues, was produced by several decades of prior preparation. Voltaire's self-conception as a philosophe, which certainly did begin to take shape in the wake of the scandal produced by the *Lettres philosophiques,* also succeeds rather than precedes the crises of the 1730s. If, therefore, the French Enlightenment is to be located in the birth and consolidation of the philosophe movement, then the crucial genealogical project involves examining how the changing intellectual environment in France made possible Voltaire's initiatives with the *Lettres philosophiques,* and how these transformations in turn gave birth to the philosophe identity that became the defining feature of the French Enlightenment.

Importantly, one factor that did not play a role in this transformation was some new, Voltaire-initiated, and philosophe-led introduction of Newtonian science into France. Indeed, as chapter 1 discusses, France absorbed Newton's science immediately and in substantial, if idiosyncratic, ways from as early as 1690. This chapter, and the rest that follow, therefore, deconstruct the philosophe mythology that gives their movement credit for initiating Newtonian science into France. Chapter 1 summarizes the outcomes that Newton's early French reception actually produced before 1715. It also creates a context for understanding the changes that ensued after this date, changes that would eventually prepare the way for Enlightenment a half century later. It focuses in particular on a set of institutional changes within French science that proved foundational for the struggles that followed. Chapter 2 continues this genealogical work by accounting for the sources that made possible the changes that then ensued. Two themes dominate this analysis. One was a change in the climate of public science in France and the wider Republic of Letters after 1715. In the older climate, the one in which Newton's work was initially received, open

intellectual contestation and dispute were anathema and partisan philosophical warfare almost nonexistent. By 1730, this climate had radically changed,
and chapter 2 traces the complex shifts that marked this transformation. It also
traces the rise of a new and recognizable discourse of Newtonianism within
the Republic of Letters after 1715. This discourse emerged as defenders of
Newton's physics, motivated by a variety of different agendas, began to become more assertive and publicly polemical on its behalf than they had in the
first quarter-century after the appearance of the *Principia*. The result was the
newly assertive Newtonianism that Voltaire used to great effect in defining his
philosophe identity.

If chapters 1 and 2 account for the context that made Voltaire's initiative
possible in 1734, chapter 3 completes this picture by looking at the prior eruption of the very disputes that he thereafter skillfully exploited. At the center of
this account is the famous priority dispute that pitted defenders of Newton's
independently invented calculus against Leibniz's claim to have been the first
to invent this new mathematics. Begun in 1713, this controversy activated the
critical energies of the Republic of Letters in a new and largely unprecedented
way. The outcome was a new set of battle lines that placed newly self-conscious
Newtonians into open, public confrontation with those perceived to be hostile
to them. Also activated were a host of new media structures—learned journals,
books, pamphlets, institutions of intellectual sociability—that prepared the
way for Voltaire's moves in the 1730s. Each of these factors—the new critical
climate, the new discourses of Newtonianism, the new eighteenth-century mediascape, and the new battlefield focused on Newton—made possible Voltaire's
innovations in 1734.

Among the arguments implicit in part 1 of this book, therefore, is that Voltaire did not single-handedly launch the French Enlightenment through the
singular effort of his own will. Rather, his successes after 1734 were prepared
by a host of antecedent developments. So what did produce the scandal triggered by the *Lettres philosophiques,* and how did this book and its reception in
turn provoke the philosophe movement? Part 2 takes up this question, offering
a detailed account of the passage from the initial Newtonian interventions of
Maupertuis and Voltaire in the 1730s to the watershed of the *Encyclopédie* and
the further changes it catalyzed after 1751. It begins by examining Maupertuis'
defense of Newtonianism in the *Discours sur les différentes figures des astres* of
1732 showing how he, like Voltaire, found his Newtonianism, if not his Newtonian science, in the new public culture of the Republic of Letters and the new
cultural possibilities that it offered. In fact, as chapters 4 and 5 stress, the parallel uses of Newtonianism by Maupertuis and Voltaire between 1732 and 1734
marked a pivotal turning point in the history of Newton's French reception.

More precisely, since Maupertuis, unlike Voltaire, was both a royal academician and an accomplished practitioner of the highly mathematical "analytical mechanics" pioneered during the early decades of Newton's French
reception, he possessed a dual attachment to what had become, by 1730, two
distinct outcomes of Newton's eighteenth-century reception. Analytical mechanics, initiated after 1690 (and discussed in chap. 1), channeled the science
of Newton's *Principia* into what became after 1715 an important strand of official French science. By 1730, however, analytical mechanics had long since
lost any direct association with Newton's name or legacy. At the same time,
the new discourse of physical and metaphysical Newtonianism was increasingly being proclaimed in the Republic of Letters as the "true philosophy of
Newton." This discourse provoked a different image of Newton than the one
that prevailed in France prior to 1715. That Maupertuis joined these discrete
scientific and cultural strands together, even though they were not intrinsically
connected to one another, marks one of his primary historical achievements.
Yet the reasons for this marriage were highly contingent and context-specific.
The discourses of Newtonianism centered on questions of physics and metaphysics, and they had little bearing on the ongoing work in analytical mechanics that Maupertuis pursued as well. However, since they had important
cultural meaning in the wider public sphere, and since this sphere engaged
Maupertuis' interest as well, he ultimately sought a union between them. The
resulting alignment was deeply influential, for it allowed analytical mechanics,
and the official French scientific establishment that supported it, to join with
the new and often provocative discourses of Newtonianism in ways that were
transformative for each.

The scandal of Voltaire's *Lettres philosophiques* derived in part from a deployment of these same charged Newtonian discourses two years later, and in
this respect Maupertuis and Voltaire were companions in these public efforts.
Chapters 4 and 5 together examine this joint beginning, stressing in particular the important ties that bound the two men together in the cultural space
of French public science. However, it also problematizes their relationship
as well by resisting the conventional understanding that locates these two savants within an allegedly coherent and singular philosophy of Newtonianism.
In fact, these chapters argue, Maupertuis had his own Newtonianism and his
own agenda in publicly defending it, one oriented toward his power and prestige within the Paris Academy of Sciences. This agenda often distanced him
from Voltaire, who likewise had his own Newtonianism and his own social reasons for pursuing it. The outcome in Maupertuis' case was newfound notoriety
and prestige within official French science, a prestige that made Newtonian

discourse more welcome in these circles and more authoritative in France as a whole. His efforts also prepared the way for the full acceptance of Newtonian physics within French science, an outcome that is chronicled fully in later chapters.

What these endeavors did not encourage, however, was the beginning of the French Enlightenment, even if Maupertuis' ongoing public scientific activities, also examined, did shape this development in important ways. Instead, it was Voltaire who led the way into the new critical, public philosophical campaigns that defined the Enlightenment as a social movement. Importantly, though, it was not through his defense of Newtonianism per se that these changes were accomplished, but through the new critical discourses that he deployed through this precise philosophy. In other words, Voltaire's *Lettres philosophiques* did not, pace Condorcet and the classic narrative, liberate the French from their Cartesian ignorance by delivering them the gospel of Newton. Nor even did the *Lettres* join with Maupertuis' *Discours* in effecting the same philosophical liberation collectively. New philosophical ideas themselves, in fact, be they Newtonian or otherwise, were not even as crucial as one might think to the beginning of the French Enlightenment since many of them were already commonplace when Voltaire put them to use. Rather, what opened the French Enlightenment in 1734 was the particular way that Voltaire deployed these philosophical ideas, and the particular self-fashioning he accomplished with them, a self-fashioning that led to the definition of a new kind of critical, libertarian intellectual in France. Ultimately, this new, libertarian intellectual style, and the social assumptions that made it possible, cohered in the construction of a new persona, the philosophe. This persona in turn made possible the French Enlightenment that followed in its wake.

Stating the same point another way, the shock of the *Lettres philosophiques* is not to be found in the philosophy it defended or the ideas it imparted per se but instead in the text's critical style; it was Voltaire's tone, his intellectual stance, and his critical voice much more than the precise ideas he defended that were truly provocative. One of Voltaire's contemporaries, the abbé Le Blanc, captured well the real scandal of the work when he offered this account of the text to his friend Jean de Bouhier, a judge at Dijon. Ignoring Voltaire's defense of Lockean sensationalism and materialism in the text, his defense of Newtonianism, and even his association of Newton and Locke with the heresy of Arianism, Le Blanc instead criticized Voltaire's overall intellectual manner. "I was shocked by the tone of disdain [*mépris*] that reigned everywhere," he declared. "This disdain [*mépris*] was directed equally against his country, our government, our ministers, and especially against that which is most respectable—religion. He asserts himself [*Il décide*] as cavalierly about these matters as he would about four lines of English verse. It is horribly indecent."[82]

Le Blanc's analysis was on the mark. The commencement triggered by the *Lettres philosophiques* was not provoked by the intellectual claims of the book even if its contents were not completely irrelevant to it. Rather, it was the intellectual persona that Voltaire crafted with his text that proved provocative. Later philosophes would have us believe that the *Lettres philosophiques* initiated the Enlightenment because it introduced new and controversial ideas into France. In fact, the work created an upheaval because it defined a new and provocative intellectual identity—that of the philosophe. This identity provoked a movement centered on defending and spreading this critical style in eighteenth-century France, and the real rupture that opened the French Enlightenment is to be found here. A new genealogical project also emerges when one takes this understanding of the French Enlightenment and its beginnings seriously. Rather than seeking a genealogy of a set of characteristic Enlightenment ideas or even a set of characteristic Enlightenment social formations, we must look instead for a genealogy of the philosophe as a cultural (i.e., intellectual *and* social) identity in eighteenth-century France.

What made the philosophe historically innovative was his claim to possess intellectual integrity and authority—in short, to possess the right and the duty to criticize—even though he was not connected to a recognized and authoritative institution of Old Regime French society. This was both an intellectual claim and a sociopolitical one, but how did the conditions for its possibility arise? The key, I will argue, resides in seeing the appearance of the philosophe in eighteenth-century France as the result of a two-step process. First, France witnessed the birth of a new public culture between 1699 and 1730 that prepared the ground for the birth of the philosophe identity after this date. The foundations for this argument are laid in part 1. Here the new public intellectual culture that emerged in France and throughout Europe after 1700, and especially the new mediascape of publicly oriented books, journals, academies, societies, and other institutions critical to Enlightenment, are analyzed in terms of the changes they permitted. Without the new sociological spaces for publicity that emerged during these years, and without the new discursive conception of publicness and publicly oriented criticism that arose simultaneously with them, neither the philosophes nor the movement they initiated could have emerged.

82. Jean-Bernard Le Blanc to Jean de Bouhier, in Voltaire, *Correspondence and Related Documents*, in *Œuvres complètes de Voltaire*, vols. 85–135, ed. Theodore Besterman (Oxford, 1968–1976), 86: D718; hereafter referred to as *Correspondence*.

Indeed, the Siamese twins "philosophe" and "public" constitute arguably the core element of the French Enlightenment itself. The public was a necessary correlate of the philosophe persona because it served to locate him socially, supporting the crucial autonomy that defined his intellectual integrity. The philosophe, therefore, was first and foremost a public citizen and spokesman. But the public also authorized the critical stance of the philosophe, for lacking the institutional authority of the academician, the professor, the priest, or the state official, the philosophe compensated by claiming to speak on behalf of the public. In this way, the philosophe was also a new kind of public advocate. However, because the term "public" in Old Regime France conventionally referred to the collective body of the commonwealth in such a way that traditional intellectuals were said to serve the public whenever they served the Crown, the public also offered a further justification for this stance. It allowed the philosophe to eschew traditional definitions and directly serve the public, defined as the assembly of all human beings. In this way, the philosophe represented a new kind of public servant as well.

In all these ways, then, the philosophe and the public were inseparably linked. However, since publicity did not naturally and automatically produce the critical self-conception of the philosophe that took hold after 1750, accounting for it requires more genealogical work. The cultural shifts that occurred within the newly public space of the Republic of Letters and France and elsewhere between 1715 and 1730 in fact prepared the ground for these new developments, and accordingly they are analyzed in terms of these outcomes in the early chapters of this book. Part 2 continues this account by closely examining Voltaire's exploitation of these new cultural opportunities in his *Lettres philosophiques* of 1734. Here Newtonianism is discussed in terms of its real contribution to the beginning the French Enlightenment by showing how Voltaire used it to publicly define a new critical and argumentative intellectual style. Public argument and criticism like this became the sine qua non of the French Enlightenment after 1750, driving the institutional struggles that engaged the philosophes in France and fueling the crusades that gave their movement meaning. Newton himself was not a direct source of this critical style, but the Newtonianism that his advocates developed after 1715 was, and by deploying these discourses the way that he did after 1734 Voltaire made Newtonianism a key component in the emergence of Enlightenment criticism.

In chapters 6 and 7, this genealogy is completed as the Newton wars that Voltaire and Maupertuis helped to trigger after 1734 are examined in terms of the French Enlightenment that followed from them. How are these genealogical strands ultimately tied together? First, chapters 5 and 6 examine Voltaire's struggle to stabilize his own philosophe identity by exploring his response to the scandal triggered by the *Lettres philosophiques*. Here the conflicting pulls of liberty, crucial to the integrity of the public, critical voice of the philosophe, and *honnêteté,* the French term that best describes the values of intellectual integrity and honor that were paramount in Old Regime French society, are examined in the context of Voltaire's efforts to define himself publicly as a reputable authority on Newtonian philosophy. Maupertuis' parallel, but socially different, activities in this respect are also explored in chapters 6 and 7, as are those pursued by Voltaire's partner, the Marquise du Châtelet, and others who become involved in the Newton wars of the late 1730s and 1740s. Overall, the goal is to show how these public maneuvers and contests converged in a crucible whereby a new set of discourses and a new set of identities and institutional associations were fused in ways that made possible the cultural reconfiguration constitutive of the French Enlightenment. In a coda to this book, the final crystallization of this movement is examined through a brief account of Diderot and d'Alembert's *Encyclopédie,* the project that transformed Voltaire's fledgling philosophe identity into a party that united men of letters around the now solidified cause of Enlightenment.

Overall, then, the seven chapters and coda of this book offer a detailed genealogy of the many and varied changes between 1715 and 1751 that ultimately married Newton with Enlightenment in eighteenth-century France. It furthermore accomplishes this marriage while self-consciously undermining the standard historical understanding that contemporary scholarship has inherited from the philosophes themselves. In his 1992 introduction to the modern critical edition of Voltaire's *Éléments de la philosophie de Newton,* William H. Barber revealed the continuing prevalence of the philosophe story in his description of the significance of the book. Voltaire's work was important, he wrote, because it led to the rejection of Cartesianism and to the acceptance and adoption of Newtonianism in France. Before Voltaire, he contended, France was under the "spell" of Descartes' system, which was "hypothetical and scientifically worthless." Newton's system, by contrast, "provided a trustworthy instrument for predicting the motions of the planets," and the "controversy between [the two systems], first in England, then in Holland and finally in France, led to the general decline of the Cartesian system and its total replacement by that of Newton."[83] Voltaire would not have written the presentation of his book any differently, for here again one sees how contemporary historiography and the narratives produced by the Enlightenment philosophes themselves are locked in an echo chamber that merely reinstantiates the same, timeworn stories.

83. Voltaire, *Eléments de la philosophie de Newton,* ed. Robert L. Walters and W. H. Barber (Oxford, 1992), 6–7.

But the Enlightenment is now over, and the time has come to extricate its history from the story it wrote for itself. The history of Newtonian science will be the first beneficiary. The French philosophes defined Enlightenment by constructing a highly politicized history of modern science as an act of self-justification. This narrative served their movement well. But because this eighteenth-century understanding of the Newtonian legacy has since become a foundational story in the modern historiography of science, it is now creating serious obstacles to our current understanding. Our historical understanding of Enlightenment will also be improved immensely by this same reconceptualization. Many continue to echo the discourses of the philosophes by making the reception of Newtonianism in France a triumph for the forces of truth in their battle against the armies of error and prejudice. They also make Enlightenment and modernity the natural consequence of this scientific arrival. The actual history of Newton's French reception, however, was far more complicated, and its connection to the French Enlightenment was similarly rife with ironies and contingencies. Only by severing the heroic tie that binds Newton to Enlightenment in the traditional Enlightenment story can this more genealogical understanding of their connection be put in its place. This is what this book proposes to offer.

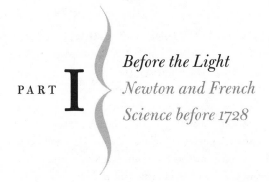

PART **I** *Before the Light*
Newton and French
Science before 1728

On May 28, 1728, a little-known member of the Paris Academy of Sciences rose before the assembly to deliver a paper on celestial mechanics. The academician was Joseph Privat de Molières, and in the spring of 1728, although fifty-two years old, Privat de Molières was still struggling to establish his reputation as a savant. He began his career as an Oratorian priest and teacher, studying mathematics with Father Reyneau at Angers in the 1690s and then serving as a priest and professor of mathematics at the Oratorian colleges of Saumur, Juilly, and Soissons from 1699 to 1704. He had come to both the Oratory and mathematics through a devotion to the writings of Nicolas de Malebranche, and in this way his intellectual trajectory mirrored that of many others in France in the same period.[1] Indeed, as will be discussed shortly, Privat de Molières' orientation toward Malebranche, Oratorianism, and mathematical mechanics pointed him toward the center of the scientific establishment that reigned in France in the decades around 1700.[2] At the center of this establishment was Malebranche himself and those who congregated in his "circle."[3] These included the Marquis de

---

1. A useful summary of Privat de Molières' career is found in E. Bonnardet, "Joseph Privat de Molières, Sr. L'Oratoire. Né à les Baux le 21 Mai 1676. Mort à Paris, collège royal, le 13 mai 1742. Notes bio-bibliographiques (Mai, 1936)," unpublished MS, Ac. Sci., Dossier Molières. See also Jean-Jacques Dortous de Mairan, "Éloge de M. de Molières," *HARS-Hist.* (1742): 195–205; and A. Saverien, *Histoire des philosophes modernes*, 8 vols. (Paris, 1761–1767), 6: 217–248.

2. See J. B. Shank, "'There Was No Such Thing as the 'Newtonian Revolution,' and the French Initiated It': Eighteenth-Century Mechanics in France before Maupertuis," *Early Science and Medicine* 9, no. 3 (September 2004): 257–292, idem, "Before Voltaire: Newtonianism and the Origins of the Enlightenment in France, 1687–1734" (Ph.D. diss., Stanford University, 2000).

3. The best account of the formation of the "Malebranche circle" is Pierre Costabel, introduction to André Robinet, ed., *Oeuvres complètes de Malebranche, Volume XVII–2: Mathematica* (Paris, 1968), 309–316. See also André Robinet, "Le groupe malebranchiste introducteur du calcul infinitésimal en France," *Revue d'histoire des sciences* (1960): 287–308, idem, *Malebranche, de l'Académie des sciences. L'oeuvre scientifique, 1674–1715*

l'Hôpital and Pierre Varignon, who helped to introduce Leibniz's new infinitesimal calculus into France in the 1690s.[4] It also included Bernard le Bovier de Fontenelle, who became the perpetual secretary of the Royal Academy of Sciences in 1697.[5] Fontenelle also helped to lead the academy through a series of reforms in 1699 that made the academy far more publicly oriented than it had ever been.[6] These changes too solidified the centrality of the Malebranche circle at the center of French science. In particular, they launched Fontenelle into a new public role as official spokesperson for French science, a job that he would keep for over four decades. And since he used this public platform to advocate for the new mathematical sciences that l'Hôpital, Varignon, and the other Malebranchians began to develop after 1690, his public ascent helped to ensure the notoriety of this work within the new public scientific discourse that he oversaw.

Privat de Molières entered the Oratorian order at this precise moment of change, when the alliance between Malebranche, the new analytical math-

---

(Paris, 1970), 47–62, and idem, "Les académiciens des sciences Malebranchistes," in *Oeuvres complètes de Malebranche*, 20: 162–174.

4. On Varignon and the Malebranche circle, see Pierre Costabel, *Pierre Varignon (1654–1722) et la diffusion en France du calcul différentiel et intégral* (Paris, 1965). Fontenelle discusses his relationship with Varignon in his "Éloge de M. Varignon," *Oeuvres de Fontenelle*, 7: 19–33. Varignon's achievement in mechanics is explored most fully in Michel Blay, *La naissance de la mécanique analytique: La science du mouvement au tournant des xviie et xviiie siècles* (Paris, 1992). Also helpful to my understanding of Varignon has been Michael S. Mahoney, "Pierre Varignon and the Calculus of Motion" (unpublished paper). I am grateful to Professor Mahoney for sharing this work with me and to John Detloff for facilitating this exchange.

5. On Fontenelle, see Alain Niderst, *Fontenelle* (Paris, 1991), idem, *Fontenelle à la recherche de lui-même* (Paris, 1972); and Alain Niderst, ed., *Fontenelle: Actes du colloque tenu à Rouen de 6 au 10 octobre 1987* (Paris, 1989); Leonard M. Marsak, *Bernard de Fontenelle: The Idea of Science in the French Enlightenment* (Philadelphia, 1959); Suzanne Delorme, ed., *Fontenelle, sa vie et son oeuvre, 1657–1757* (Paris, 1961); and Michel Blay, introduction to Bernard le Bovier de Fontenelle, *Eléments de la géometrie de l'infini*, ed. Michel Blay and Alain Niderst (Klinckseik, 1995); and *Les raisons de l'infini* (Paris, 1996), translated into English as *Reasoning with the Infinite*, trans. by M. B. DeBevoise (Chicago, 1998).

6. On the 1699 reform of the Royal Academy of Sciences, see Elmo Stewart Saunders, "The Decline and Reform of the *Académie des sciences à Paris*, 1676–1699" (Ph.D. diss., Ohio State University, 1980); and Shank, "Before Voltaire," chaps. 2–3. On Fontenelle's new roles as a result of these shifts, see also J. B. Shank, "Fontenelle's Calculus: The Cultural Politics of Mathematics in Louis XIV's France," in *Arts of Calculation: Quantifying Thought in Early Modern Europe,* ed. David Glimp and Michelle R. Warren (New York, 2004), 93–114.

ematical sciences, and the newly public Academy of Sciences converged to launch each into a position of public prominence. As an Oratorian mathematician and professor, Privat de Molières was closely connected to all of these developments, even if he was never more than a peripheral participant in any of them. He further manifested his allegiances openly in 1704 by resigning from the Oratorian order (a less drastic move than it might seem given the laxity of the congregation's disciplinary protocols) to pursue his scientific interests and perhaps win admission to the Royal Academy. His actual entrance into the academy, however, was anything but swift. Not possessed of the kind of intellect or connections that allowed figures like René Antoine de Réaumur, Jean-Jacques Dortous de Mairan, and Maupertuis to ascend rapidly to the top of the academic hierarchy in only a few years, Privat de Molières was forced to struggle for every opportunity he received. Only after five years of appeals to the abbé Jean-Paul Bignon, the royal official in charge of the academies and the "Maecenas" of royal cultural patronage, was the former priest awarded a modest position as the mathematical tutor of Bignon's nephew.[7] This was in 1709, and Privat de Molières' admission to the Royal Academy would require another decade of perseverance.

An opportunity arose in 1717, but, according to Privat de Molières, his admission was derailed by a plagiarism charge leveled against him by Joseph Saurin, another key player in the analytical mathematical developments of the previous decades.[8] Saurin claimed no knowledge of this attack later, but either for this reason or some other Privat de Molières was not admitted to the academy. By

7. On Bignon, see Jean-Jacques Dortous de Mairan, "Éloge de M. Bignon," in *HARS-Hist.* (1743): 189–197. The modern scholarship on Bignon remains astonishingly thin despite the valuable work by Françoise Bléchet. See especially Bléchet, "Un précurseur de l'Encyclopédie au service de l'état: L'abbé Bignon," in *Encyclopédisme: Actes du Colloque de Caen, 12–16 janvier 1987*, ed. Annie Beck (Caen, 1991), idem, "Le rôle de l'abbé Bignon dans l'activité des sociétés savantes au XVIIIe siècle," in *Actes du 100e congrès national des sociétés savantes, section d'histoire moderne et contemporaine et commission d'histoire des sciences et des techniques* (Paris, 1979), 31–41, and idem, "Fontenelle et l'abbé Bignon, du président de l'Académie royale des sciences au secretaire perpétuel: Quelques lettres de l'abbé Bignon à Fontenelle," *Corpus, revue de philosophie* 13 (1990): 51–61. See also Jack A. Clarke, "Abbé Jean-Paul Bignon: 'Moderator of the Academies' and Royal Librarian," *French Historical Studies* 8 (Fall 1973): 213–235. David Sturdy is at work on a short biography of Bignon, and I am grateful to him for sharing with me the results of his research thus far and for his interest and support of my work overall. I am also grateful to Françoise Bléchet for personally sharing with me her expertise about Bignon.

8. Privat de Molières made this accusation in an open letter to the members of the Royal Academy of Sciences dated August 9, 1721. Ac. Sci., Dossier Molières.

1717, the intellectual climate in the academy was beginning to change in ways that made it less disposed toward Malebranchians mathematicians than it once was. In particular, a new and more utilitarian approach to the mathematical sciences was beginning to eclipse the highly mathematical approach that had ascended to prominence in the academy after 1690. Whether this was another factor hindering Privat de Molières' entrance is not clear, but whatever the precise obstacles, they were overcome in 1721. In this year he was admitted as an *adjoint mécanicien,* the title created to replace the class of *éleves* when the original membership structure of the academy was reformed in 1716.[9]

Privat de Molières' frustrations, however, did not cease once he became an academician. Bignon remained a source of support, and when Varignon's distinguished career came to an end in December 1722 Privat de Molières was given his teaching post at the Collège Royale.[10] Inside the academy, however, similar rewards were not as forthcoming. In the first months of his membership, he appears to have confronted Saurin directly and initiated a controversy with him. This struggle resulted in an open letter of self-defense written by Privat de Molières and addressed to the members of the academy.[11] Typically, no other record of this dispute exists either in the academy registers or elsewhere, and thus it is hard to know what if any impact the letter had. It is clear from other evidence, however, that he struggled in this period to win the respect of his peers.

Privat de Molières was a fairly active member of the institution, delivering papers on a range of topics at least once or twice a year.[12] In this respect, he personified the professional ideals announced in the new academy regulations of 1699. In addition to its new public orientation, the post-1699 academy cultivated an esprit de corps that valued diligent labor, self-effacing service, professional honor, and devotion toward the public good.[13] Privat de Molières applied himself assiduously toward achieving these goals, yet before 1726 he appears to have exerted only a marginal influence inside the institution. Publication in the annual volumes of the academy (another innovation of 1699) was one marker of academic esteem since only the best papers were published each

9. PVARS, August 6, 1721.

10. Bonnardet, "Joseph Privat de Molières," in Dossier Molières.

11. See note 8 above.

12. See, for example, PVARS, November 22, 1721; April 29, 1722; February 17, 1723; and June 23, 1723.

13. The new regulations for the Royal Academy instituted in 1699 are found in Léon Aucoc, *Lois, statuts, et règlements concernant les anciennes académies et l'Institut de 1635 à 1889* (Paris, 1889), LXXVII–XCII. See also Shank, "Before Voltaire," esp. chaps. 2–3.

year. All but one of Privat de Molières' many papers, however, were rejected by this publication.[14] His one publication, a dissertation on the mechanism of muscles, also received no commentary in the annual *histoire*, or "year in science," that Fontenelle further began publishing after the 1699 reform. This annual history served as the preface to the academy's published *mémoires*, and given the eloquence and public notoriety of the history's author, the narrative quickly established itself as an important organ of public scientific commentary in France and throughout the Republic of Letters. To be ignored in its pages, therefore, as Privat de Molières largely was between 1721 and 1728, was tantamount to being a nonpresence in the important scientific discussions of the period. The former Oratorian was also rarely appointed to the academic committees that judged books, instruments, machines, and the annual prize contests in these years, even though this committee work became an increasing aspect of academic practice after 1715.[15] In sum, before 1728, Privat de Molières appears to have been at best a marginal figure in the Paris Academy and an even more obscure savant in the wider Republic of Letters to which the new public academy increasingly addressed itself.

Privat de Molières apparently sensed this marginalization, for in the spring of 1724 when an even less noteworthy academician, a savant named Beaufort, was promoted to the rank of *associé* ahead of him, Privat de Molières made his feeling of alienation manifest. According to the academy records, he was not even a nominee for this seat, yet in his own mind this was an injustice.[16] On May 9, three days after the vote was taken, he addressed a passionate letter to Bignon expressing his frustration. Privat de Molières began by noting his astonishment that the academy had voted for Beaufort, a man who in his estimation had "shown no sign of life since arriving in the academy." "He has not opened his mouth since moving to the rank of *mécanicien*," he wrote, "and he doesn't work at all, preferring instead to amuse himself with other things." By contrast, Privat de Molières continued, "you know, Sir, that from the very moment when the academy accorded me the honor of admission, I have not ceased to conform to its goals. I have certainly worked hard, and I have made this work known by the many *mémoires* that I have read during our assemblies. As a

14. Privat de Molières' published paper was "Mémoire sur l'action des muscles, dans lequel on tâche de satisfaire, par des voies simple et purement mechanique, aux difficultés proposé par M. Winslow dans son mémoire de 1720," *HARS-Mem.* (1726): 18–38.

15. On the increased burden of this kind of academic work in the eighteenth century, see Roger Hahn, *Anatomy of a Scientific Institution: The Paris Academy of Sciences, 1666–1803* (Berkeley, 1971), 117–126.

16. PVARS, May 6, 1724.

mécanicien, I have chosen general problems of mechanics as my focus because I believe it is shameful to the French nation and to the Academy in particular to cede to foreigners without any resistance the legacy of our ancestors who were the inventors of this science. In this endeavor I believe I have acquitted myself well."[17]

The members of the academy, however, did not see things this way, and elsewhere in his *apologia* Privat de Molières isolated what he perceived to be a central bias against his work. "In a word, I do not have the talent to express myself neatly and precisely," he complained. "But, Sir, do those who pull crude diamonds out of the mines and give them to the connoisseurs exactly as they have found them deserve to be punished because they do not possess the art of polishing them and making them fully brilliant?" The mandate of the Royal Academy of Sciences was to produce truth not eloquence, he continued, and since one spoke only to experts in this body, eloquence should not be as highly prized by the academicians as he believed it was. As he summed up: "In order to discover truths one must be so preoccupied with them that it is impossible to worry at the same time about the arrangements of words or even the arrangements of ideas. As soon as one begins playing with such things, one loses the truth from view." Privat de Molières believed that the academy had lost sight of this fundamental aspect of its mission, and he complained that his work was too rarely criticized for "the falsity or uselessness of the ideas" and too often for "the hardness of the language." He further isolated a reason for these misguided priorities. In his concluding paragraph, he criticized the way that the academy's orientation toward the public negatively influenced its scientific labors. "The public, which does not recognize the difference [between truth and rhetoric], . . . often reacts negatively to the academy's work," he averred. Yet the leading academicians, he believed, had achieved institutional success because they appealed directly to the public's bias rather than challenging it as was called for.[18]

Privat de Molières saw misplaced priorities rather than "personal injustices" at the root of his problems, and he thus ended his letter by issuing an appeal. "I only write in the hope of finding the forces necessary to turn the tides," he stated.[19] Yet to judge by his career over the next few years, his initiative did not effect much change. His work remained largely unpublished, and he was only promoted to *associé* in 1729. This promotion, however, marked a breakthrough

17. "Lettre à M. Bignon, sur les qualités requises à qui souhaite d'être membre de l'Académie des Sciences," May 9, 1724, Ac. Sci., Dossier Molières.

18. Ibid.

19. Ibid.

in his career, and it was the work on celestial mechanics that he began reading in May 1728 that triggered his ascendance. His paper was published in the
mémoires for that year, only his third paper to earn such an honor. Even more
important, Fontenelle devoted a significant portion of his *histoire* to a public
articulation of the work and its significance.[20] Later in 1728, Privat de Molières
was also appointed with Dortous de Mairan, the director of the academy for
that year, to judge a new work on the nature of the tides, a rare appearance for
him on an academic committee.[21] Privat de Molières continued his work in celestial mechanics in 1729, delivering a paper in March whose title translates as
"Physico-mathematical problem, the solution of which serves to respond to one
of the objections of M. Newton against the possibility of celestial vortices."[22]
This paper was also published in the academy's *mémoires,* accompanied by
another lengthy and praiseworthy synopsis in Fontenelle's *histoire.*[23] Finally,
when Beaufort's *associé* seat became vacant in August 1729, Privat de Molières
was elected to succeed him.[24] From this date until his death in 1742 he was a
major figure in the royal institution and in the wider public sphere to which it
was linked.

Privat de Molières may have felt a sense of justice in obtaining from Beaufort
in 1729 the very seat that had triggered his criticisms of the academy in 1724,
but his was a victory steeped in irony. For the public dimension of the academy
that Privat de Molières had placed at the center of his own professional stagna-
tion in 1724 was largely responsible for his ascent to prominence after 1728.
Introducing Privat de Molières' 1728 paper in his *histoire,* Fontenelle remarked
"it is not at all surprising that philosophers continually return to this matter.
Nothing is more interesting to them than to know whether Descartes' ingenious system of vortices, which presents itself so agreeably to the mind, will
fall victim to the many difficulties that are often posed against it." "Will [this
system] be replaced by another that has its own difficulties, including many
even more striking?" he asked with reference to Newton's nonvortical and nonmechanical alternative. "M. Molières, who declares himself to be a Cartesian,

20. Joseph Privat de Molières, "Loix générale du mouvement dans le Tourbillon
sphérique," *HARS-Mem.* (1728): 245–267. Fontenelle's discussion is found in *HARS-Hist.*
(1728): 97–103.

21. PVARS, July 17, 1728.

22. PVARS, March 16, 1729.

23. *HARS-Hist.* (1729): 87–91; Joseph Privat de Molières, "Problème physico-mathématique, dont la solution tend à servir de réponse à une des objections de M. Newton contre la
possibilité des tourbillons celestes," *HARS-Mem.* (1729): 235–243.

24. PVARS, August 6, 1729.

takes up the defense of the vortices, and begins here to develop his method for accounting for these difficulties."[25] The secretary's 1729 *histoire* framed the issue in even more pointed terms: "M. Molières here continues his program, first explained in the *histoire* of 1728, to conserve the vortical physics of Descartes, vigorously attacked by the formidable objections of M. Newton and his numerous sect."[26]

Historians of the eighteenth century will recognize in these declarations the famous pro-Cartesian/anti-Newtonian discourse that has since come to define the historiography of eighteenth-century French science and especially Fontenelle's place within it. This is no illusion. The year 1728 in fact marks the beginning of the open, public contestation between self-proclaimed Newtonians and Cartesians that defined much scientific discourse over the next three decades. It was this contestation that propelled Privat de Molières to the forefront of the French scientific world. His work after 1728 became a rallying point for the newly self-conscious Cartesian faction in France and accordingly a target for the self-proclaimed Newtonians who began to openly challenge them. Fontenelle played a significant role in foisting this role on Privat de Molières in 1730 (the year that his first *histoire* discussing Privat de Molières' Cartesian mechanics was published). Once offered the identity, however, the former Oratorian flew the Cartesian banner willingly. In his *Leçons de physiques,* which began appearing in 1734, Privat de Molières presented his theories as a deductive, geometric system, fueling the debate about the appropriate epistemology for science central to the Cartesian/Newtonian struggle. He also invoked his Cartesian identity aggressively, using Fontenelle's discourse to further his own ambitions.[27]

His ideas soon began to receive wide attention as a result of this precise positioning. Journalists in particular printed lengthy articles proclaiming either for or against Privat de Molières' Cartesian arguments. Other writers published pamphlets and treatises that did the same thing.[28] Responding to the surge

25. *HARS-Hist.* (1728): 97.

26. *HARS-Hist.* (1729): 87.

27. Joseph Privat de Molières, *Leçons de physique, contenant les éléments de la physique déterminées par les seuls lois de mécanique, expliqués au Collège Royale,* 4 vols. (Paris, 1734–1738).

28. A sympathetic set of reviews is found in the abbé Desfontaines' *OEM,* 13: 305–312; 14: 89–95, 138–145, 212–119, 233–241. A critical set of pamphlets was published by the university professor Pierre Sigorgne. The exchange began with Sigorgne's *Examen et réfutation des leçons de physique expliquées par M. de Moliéres au collège royal de France* (Paris, 1741). Privat de Molières responded to Sigorgne directly, and Sigorgne responded with

of public interest in Privat de Molières and the debates that he was entangled within, an independent man of letters named André-Pierre Le Guay de Prémontval also began offering a set of public lectures in Paris around 1737 based on Privat de Molières' work. Advertisements indicating the intended public for these courses appeared in literary periodicals such as *Nouveaux amusements du coeur et de l'esprit*.[29] Through these channels Privat de Molières became an ever more noteworthy player in the new public debates about celestial mechanics that he helped to initiate after 1728.

Writing to the astronomer Delisle in 1739, Privat de Molières assessed his triumph over the previous decade. "You will perhaps not be angry to learn of the success that [*Leçons de physique*] has had with the public," he enthused. "The entire University of Paris has adopted it and already ten philosophy professors have supported my sentiments publicly in their theses. In the academy as well, where M. le Cardinal de Polignac currently presides, my opinions are also supported. The Reverend Fathers of the Society of Jesus also teach the text at Paris, Caen, Reims, and Rouen, and they have told me personally that their entire society should be converted to the work very shortly. M. Cassini [II] also reported to the academy that in all the cities that he visited during his many voyages to measure the figure of the earth, people spoke only of my *Leçons de physique* and supported it publicly. I truly never imagined such a success. M. Mauvan having returned from England even reported to the academy that the majority of savants there have joined my party, and others say that the work has penetrated as far as Turin, Rome, and Naples."[30]

Not all of this was exaggerated. Privat de Molières was indeed a much talked about figure by 1740, even if the foundations of his victory were exceedingly shallow. As he lamented in the same letter to Delisle: "Despite it all, I have very little chance of winning a *pensionnaire* seat [in the academy] when one becomes

---

Réplique à M de Molières, ou démonstration physico-mathématique de l'insuffisance et de l'impossibilité des tourbillons (Paris, 1741). Privat de Molières was also defended by the abbé Launay in *Principes du système des petits tourbillons, ou abrégé de la physique de l'abbé de Molières* (Paris, 1743).

29. *Nouveaux amusements du coeur et de l'esprit*, 9: 471–475. Prémontval published two works based on these courses, *Discours sur l'utilité des Mathématiques. Pronocée par Monsieur de Prémontval, à l'ouverture de ses Conferences* (Paris, 1742) and *Discours sur la nature des quantités que les mathémathiques ont pour objet* (Paris, 1742). But the advertisement in the literary journal indicated that the course was based on Privat de Molières' *Leçons de physique*. On Prémontval, see *Le necrologes des hommes célèbres de France, par une société de gens de lettres*, 18 vols. (Paris, 1766-1784), 5: 95-118.

30. Lettre de Molières à Delisle, September 9, 1739, Archives national, Paris (hereafter referred to as AN), Archives de la Marine, 2JJ65.

vacant. They are always given to my competitors, men who have neither the age nor the experience that I have, but only a greater skill for intrigue than I will ever possess. This outcome has already occurred three times in such a way that despite my age I know that I have no reasonable chance of ever obtaining a pension."[31] His despair was justified. Privat de Molières died in 1742, still at the forefront of the Cartesian/Newtonian controversy that had brought him to prominence, but still lacking a pensioned seat in the Royal Academy.

The precise nature of Privat de Molières' work and its public reception is important to understanding the history of Newton's French reception. However, this story must be postponed since before it can be appreciated, one must first understand the beginnings of the debate that propelled him to fame in the first place. This is the project of part 1 of this book. Viewed in retrospect, 1728 marks an important moment of transformation in the history of eighteenth-century French science. Early in the century, Newton had been a noteworthy but easily dismissed figure in the public discourse of science in France. As a result, very little discussion of Newtonian science either for or against occurred in France in the decades around 1700. By 1728 this situation had changed dramatically. Fontenelle indicated the transformation in his precise public positioning of Privat de Molières' work, and several features of the newly polemical discourse he deployed are important to emphasize.

First, the secretary makes it clear that defense of Descartes' theory of celestial mechanics, a model that explains the motion of planetary bodies through recourse to the action of fluid vortices within which these bodies are alleged to sit, has now become a polemical cause, one defined by the reductive label "Cartesianism." During the century after the publication of Descartes' *Discours de la methode* in 1637, the term Cartesianism acquired many meanings and only a few of them were connected to this precise scientific usage.[32] For many, Cartesianism suggested a theory of matter and mind and especially a philosophical argument about their interaction. For others, the term suggested a complex and theologically provocative metaphysics grounded in a particular understanding of mind-body dualism. For still others, it implied a new method

31. Ibid.

32. On the history of Cartesianism in France, see Roger Ariew, *Descartes and the Last Scholastics* (Ithaca, 1999); François Azouvi, *Descartes et la France: Histoire d'une passion nationale* (Paris, 2002); Francisque Bouillier, *Histoire de la philosophie cartesienne,* 2 vols. (Paris, 1854); John Cottingham, ed., *The Cambridge Companion to Descartes* (Cambridge, 1992); Daniel Garber, *Descartes' Metaphysical Physics* (Chicago, 1992); Paul Mouy, *Le développement de la physique cartesienne, 1646–1712* (Paris, 1934); and Tad Schmaltz, *Radical Cartesianism: The French Reception of Descartes* (Cambridge, 2002).

of philosophical reasoning, one that led inexorably to these philosophical positions for either good or ill. Cartesianism could also refer to Descartes' laws of motion and to the rationalist, mechanical epistemology of science that he deployed in arriving at them. His cosmological theories were built upon these same physical and mechanical foundations, yet while these narrowly scientific understandings of the term Cartesianism were part of the intellectual lexicon from as early as 1660, they were not the most common meaning of this term in the decades around 1700, nor was any polemical urgency attached to them when they were deployed in this narrow and scientific way. By 1728, however, things had changed, and Fontenelle's precise, polemical use of the term Cartesianism in his description of Privat de Molières' work marks this new beginning.

When invoking Cartesianism in this disputatious manner, Fontenelle also gave its opponents an identity by indicating that Newton and his "numerous sect" now constituted a rival camp that opposed itself to Descartes and his scientific theories. No such polemical animosity had characterized the perpetual secretary's public discourse about Newton in the first decades of the eighteenth century since Newton's early reception had not generated this kind of controversy or acrimony. Nor did Fontenelle deploy the label Newtonianism to describe a precise philosophical position or sect of philosophers opposed to Descartes since no such position or sect existed in France or anywhere else before 1715. By 1728, however, a new climate of opinion was beginning to make itself felt, one that imagined the necessity of making an irreconcilable choice between Cartesian or Newtonian physics.

Two further developments were also embodied in this new polarization. First, Fontenelle explicitly identified the Newtonian enemy with the principle of gravitational attraction across empty space, indicating in this way that Newton's theory of universal gravitation had become an authoritative scientific position capable of generating sustained debate and direct refutation. This theory did not have this character in the decades around 1700. In fact, it was largely ignored until after 1715. Second, the appearance of a strong and irreconcilable opposition between Newton's work and the work of French savants has also emerged in Fontenelle's discourse, one that was not previously present. In his full public presentation of Privat de Molières' work, Fontenelle traced an intellectual pedigree that connected Varignon's work with Privat de Molières', implying that the Cartesian opposition to "the system of Newton" stretched back to 1700.[33] No such polemical opposition had existed in 1700, however, and Fontenelle's invocation of it here marks a new development. A third dimension of these changes was also implicitly revealed in Privat de Molières' letter

33. *HARS-Hist* (1728): 102–103.

to Bignon of 1724. Patriotic sentiments, Privat de Molières claimed, motivated his work in mechanics because French scientific glory, he believed, had been challenged in recent years by the work pursued in other nations. He therefore saw it as his mission to reclaim France's international prominence within the science of mechanics. In the cosmopolitan discourse of the Republic of Letters that prevailed circa 1700, by contrast, neither the "Englishness" of Newton's science nor the alleged "Frenchness" of Cartesian physics was point a consideration. After 1728, however, the scientific battles between self-proclaimed Cartesians and Newtonians increasingly adopted nationalist categories that made these struggles discursive vehicles for working out the perceived cultural differences between England and France. In this respect too, Privat de Molières' public debut marks a new beginning.

Neither these polarities, nor the labels used to describe them, were invented by Privat de Molières, nor were they invented by the academy secretary. Each emerged instead out of a complex set of historical developments that occurred in France and elsewhere between 1715 and 1728. Nevertheless, Fontenelle's use of the bipolar Cartesian/Newtonian labels in 1728 was decisive. By invoking this terminology, the secretary's public academic discourse implicitly unified the academy and positioned it squarely on one side of an emerging bipolar debate. It also made the institutional authority of the academy contingent upon the outcome of this highly public and emotionally charged contest. Nothing in the early history of Newton's French reception suggested that it would necessarily culminate in this result—a titanic culture war that pitted self-proclaimed Cartesians and Newtonians against one another in open, public opposition. But this is in fact what happened after 1728. And while Fontenelle helped to crystallize this opposition, no single factor or chain of factors explains its appearance. Rather, a number of different changes—some intellectual and cultural, others political and institutional—conspired to produce this outcome. Part 1 will offer a genealogy of this historical transformation.

1

Newton without Newtonianism

French Mathematical Science in

the Early Eighteenth Century

Crucial to the later emergence of the Newton wars in France was the reconfiguration of the institutional and cultural field that had originally mediated French discussion of Newton's science. From the moment of its appearance in 1687, Newton's *Philosophiae naturalis principia mathematica* shaped French scientific discussions in important ways. Yet viewed from the perspective of the Newton wars that erupted after 1728, Newton's early reception in France was quite different. Partisan warfare for and against Newton's work in the *Principia* was especially rare in the first two decades of the book's reception, and without it the text was instead absorbed comfortably, if not always calmly, into the preexisting structures of French science. A combination of institutional and intellectual alignments converged to produce this initial outcome.

The Initial Reception of Newton's *Principia*, 1687–1715

Newton's work in the *Principia* was radically innovative, and intellectually it was the way that preexisting French science both inspired Newton's innovations and served as their foil that shaped the book's initial reception. In the treatise, Newton used two books of exceedingly complicated mathematics to lay the foundation for a set of physical arguments about terrestrial and celestial physics in book 3.[1] As the initial French commentator on the *Principia*

---

1. My understanding of Newton's work in the *Principia* has been most influenced by the following works: I. B. Cohen, *Introduction to Newton's "Principia"* (Cambridge, 1971), idem, *The Newtonian Revolution* (Cambridge, 1983); Dana Densmore, *Newton's* Principia: *The Central Argument* (Santa Fe, 1995); François de Gandt, *Force and Geometry in Newton's Principia*, trans. by Curtis Wilson (Princeton, 1995); Niccolò Guicciardini, *Reading the Principia: The Debate on Newton's Mathematical Methods for Natural Philosophy from 1687–1736* (Cambridge, 1999); Richard Westfall, *Force in Newton's Physics: The Science of Dynamics in the Seventeenth Century* (New York, 1971), idem, *Never at Rest: A Biography*

noted in his review of text published just months after it appeared, it was the relationship between the mathematical and physical parts of the *Principia* that posed the greatest challenge.[2] In the mathematical sections of the book, Newton drew upon and expanded the prevailing science of mathematical mechanics to show a number of new and innovative mathematical truths. Especially noteworthy was his demonstration that bodies moving in elliptical orbits according to a uniform centripetal force obeyed a law whereby the force of the body's motion varied in a precise inverse relationship with the square of the distance between the body and its center of force. Since Kepler had already demonstrated a half century earlier that planetary bodies move in elliptical orbits centered upon the Sun at one focus, Newton's mathematical work was widely and immediately seen as giving Kepler's planetary system a new mathematical clarity and rigor. The first French commentator on Newton's work, writing in the *Journal des savants,* shared in this praise, calling the *mécanique,* or quantitative mathematical analysis, offered in the *Principia* "the most perfect imaginable."[3]

By celebrating the *Principia* in this precise way, this anonymous yet deeply learned French mathematician (for who else could even have read the *Principia*?) was also offering implicit praise for his fellow French mathematical colleagues. This is because the mathematical mechanics that Newton deployed in the *Principia* was at the leading edge of French science in this period. The mathematicians who practiced this mathematical science of *mécanique* were also located at the very center of the French scientific establishment. Many were leading members of the twenty-year-old French Royal Academy of Sciences, and the work they did from this position prepared the way for the advances that Newton offered in the first two books of the *Principia.* Christiaan Huygens had catalyzed this mathematical tradition in France when he was brought from the Netherlands to lead the new academy after its founding in 1666.[4] His work on

---

of Isaac Newton (Cambridge, 1980); D. T. Whitesides, *The Mathematical Principles Underlying Newton's Principia Mathematica* (Glasgow, 1970); and Curtis Wilson, "From Kepler's Laws So-called, to Universal Gravitation: Empirical Factors," *Archive for the History of Exact Sciences* 6 (1969–1970): 65–103.

2. Review of Isaac Newton, *Principia mathematica,* in *Journal des savants* (1688): 153–154.

3. Ibid.

4. On Huygens and the early Royal Academy, see Hahn, *Anatomy of a Scientific Institution,* 4–34; and C. D. Andriesse, *Huygens: The Man behind the Principle,* trans. by Sally Miedema (Cambridge, 2005), esp. chaps. 9–10. On Huygens's science more generally, see

the *mécanique* of pendulum motion was a direct influence on Newton's work in the *Principia*, and even if Huygens had since left Paris, his scientific legacy exerted a tremendous influence on how Newton's book was read in France. Oriented to understand and appreciate the stunning mathematical innovations that the *Principia* offered, these French mathematicians, therefore, absorbed and incorporated the book's mathematical arguments into their own work, including Newton's inverse square law of planetary motion.

In this way, books 1 and 2 of the *Principia* were immediately and uncontroversially absorbed into the French science of *mécanique* after 1687, along with an image of Newton that fit with this understanding of his achievement. In this picture, Newton was a transcendent mathematical genius who had made great strides in advancing this particular mathematical science. Book 3 of the *Principia*, however, concerned *physique* rather than *mécanique,* and the arguments of this book were not so easily assimilated.[5] In book 3, Newton marshaled a staggering amount of precise quantitative data, including much that he gathered himself through his own experimental research, to argue that the inverse square law of planetary motion was not just a quantitative mathematical description of observed phenomena, but a physical law of nature that described the action of a universal force of gravity operative in all bodies. From this perspective, the centripetal force, or force of attraction, developed mathematically in the first two books of the *Principia* was reimagined as a physical force existing and operating universally in all matter. Newton's treatise when viewed from this perspective was likewise transformed from an innovative work of mathematical mechanics into a work of physics and natural philosophy.

Everything rested, therefore, on the linkage between the first two books and book 3 of the *Principia*. Yet the problem with this construction was the very odd (by seventeenth-century standards) epistemological glue that secured the connection. For the linkage to hold, one needed to accept quantitative mathematical analyses as a foundation for making physical and natural philosophical claims about nature. Yet from the perspective of European science circa 1690 this was akin to asking readers to accept an epistemological category error. Physics and natural philosophy found their scientific authority through their

---

Joella G. Yoder, *Unrolling Time: Christian Huygens and the Mathematization of Nature* (Cambridge, 1988).

5. For a further discussion of this tension, see Alan Gabbey, "Newton's 'Mathematical principles of natural philosophy': A Treatise on Mechanics?" in *The Investigation of Difficult Things: Essays on Newton and the History of the Exact Sciences in Honour of D. T. Whiteside,* ed. Per M. Harman and Alan E. Shapiro (Cambridge, 1992), 305–322.

claim to reveal the causes inherent in nature. Mathematics, by contrast, dealt with the relations between quantitative and spatial magnitudes only, and since many questioned whether numbers and geometric shapes were even real at all, mathematics was not imagined by most seventeenth-century Europeans as a medium for making physical and philosophical claims.[6]

To be sure, this hierarchy had increasingly been challenged in the decades before 1687 by the new mathematical *mécanique* to which Newton was contributing in the *Principia*. The mechanics of Galileo, which had been picked up and advanced by Huygens and others after 1650, demonstrated in particular how advanced mathematical description could reveal the quantitative relations operative in nature. Such work especially showed the power of mathematical mechanics to quantitatively predict observed phenomena. Newton's mathematico-empirical proofs in book 3 of the *Principia* were a further, and to many stunning, demonstration of this particular power. Yet to mathematically predict an outcome was not at all the same thing in the seventeenth century as showing the physical and philosophical causes that explained the phenomenon scientifically. Newton offered none of the standard physical and philosophical explanations appropriate to seventeenth-century *physique,* and unsurprisingly, therefore, his first French commentator joined with virtually everyone else in calling the physical arguments of the *Principia* untenable. "Substitute true motions for those supposed," the reviewer wrote, and then "you will give us a *physique* as exact as the *mécanique*."[7]

By true motions, the reviewer meant motions that were understandable according to evident causal explanations, not Newton's suggestions about a force of gravity attracting bodies across empty space. When providing these evident explanations, the scholastic conception of causality derived from Aristotle's physics was still an available paradigm; however, by 1690 Descartes' antischolastic alternative had also become very influential. Descartes had eliminated Aristotle's four-part understanding of causality by reducing all physical change to a single, universal mechanism involving the physical impact of extended matter. In 1669, the members of the newly founded Paris Academy debated the physical question of how to understand terrestrial heaviness in material bodies (or their *pesanteur* in French), and their discussion illustrates well the spec-

6. On the tension between mathematical description and causal, physical / philosophical explanation in the seventeenth century, see Morris Klein, *Mathematics and the Physical World* (New York, 1959); and Peter Dear, *Discipline and Experience: The Mathematical Way in the Scientific Revolution* (Chicago, 1995).

7. *Journal des savants* (1688): 154.

trum between Aristotle and Descartes that defined legitimate physical theorizing in the seventeenth century.[8]

Several academicians explained weight in terms of a natural, if inexplicable, tendency for matter to seek its natural union with the earth. This tendency was used to explain why material bodies resisted being raised and why they naturally fell when dropped. Against these traditional, Aristotelian explanations rooted in "occult," or hidden, causes, Huygens offered a different explanation rooted in the new mechanistic theories of Descartes. He mounted a bucket of water on a turntable and placed a piece of buoyant wax in the center. He then rotated the bucket several times before bringing it to a halt, demonstrating to the academicians how the water continued to rotate while it pushed the wax particle toward the center of the fluid vortex that the moving water had created. This, he offered, was an experimental illustration of how Descartes had rightly conceived of the physics of *pesanteur*.

For Huygens and Descartes, Aristotle's "occult causes" were untenable because they did not stand up to the skeptical doubt that Descartes made the centerpiece of proper scientific reasoning. Only that which could be known with clear and distinct certainty was worthy of scientific credibility, and since the definition of matter as mere extension alone, and the conception of physical cause as a consequence of direct material impact met this epistemological standard, Descartes made these the founding principles of his physics. From this foundation, he then hypothesized the existence a system of fluid vortices in nature that were responsible, he believed, for both the mechanics of planetary motion and the phenomenon of *pesanteur* on earth. In this system, planets were understood to swim in a fluid vortex that pushed them along the elliptical orbits mathematically described by Kepler. Bodies on earth were likewise said to be heavy as a result of the centrifugal resistance that these celestial fluids exerted on bodies as they rose within them. In this way, as Huygens attempted to show with his bucket experiment, bodies did not fall to earth because they were naturally drawn to it, but fell because their rise was resisted by a fluid-mechanical force that physically pushed them back toward earth.[9]

---

8. The debate itself is found in PVARS, 4 August—20 November, 1669. Huygens's contribution to this debate is also analyzed in Ernst Mach, *The Science of Mechanics,* trans. by Thomas J. McCormack (La Salle, 1989), 199–200.

9. The Cartesian vortex theory of planetary motion is studied in detail in Eric J. Aiton, *The Vortex Theory of Planetary Motions* (London and New York, 1972). On Cartesian mechanistic physics more generally, see Garber, *Descartes' Metaphysical Physics;* and Mouy, *Le développement de la physique cartésienne.*

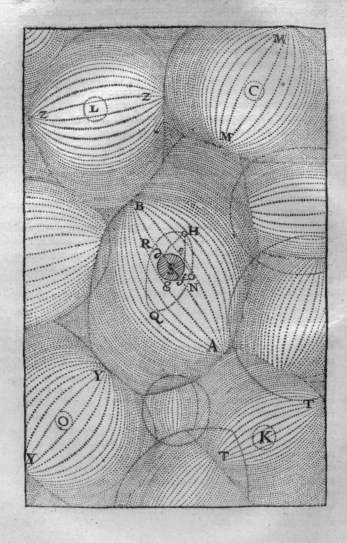

FIGURE 8. *Image of the celestial vortices, from René Descartes,*
Principia philosohiae *(Paris, 1644). Courtesy of the James Ford Bell Library,*
University of Minnesota Libraries.

Descartes' vortical system of celestial and terrestrial physics had won many adherents by 1690, even if several, including Huygens, saw difficulties with it. One involved comets, whose regular, recurring motions were not yet fully understood. What was the mechanics of comet motion if cometary bodies appeared to move with no regularity at all within the same vortices that regulated the motion of the planets with such consistency? Another problem was how to reconcile Cartesian mechanistic physics with the quantitative certainties of mathematical *mécanique*. Descartes reinforced the traditional hierarchy when he called Galileo's work a mere mathematical description of nature and not the true causal physics that he sought with his own science. From this perspective, the absence of a correlation between Descartes' vortical-mechanical explanations and the quantitative predictions of *mécanique* was simply an effect of comparing scientific apples and oranges. For someone like Huygens, however, this disconnect was deeply troubling, and he noted two fundamental problems with the Cartesian vortical system that would forever plague it. One involved the fact that vortical impacts pushed bodies (like Huygens's piece of wax) toward the center of the vortex, while terrestrial bodies fell toward the center of the earth. How could the vortical theory account for this anomaly? Another problem involved reconciling the imagined motions of the planetary vortices with Kepler's empirically proven quantitative laws of planetary motion. Making the empirical mechanics of fluid impact conform with Kepler's simple geometric laws required ever more complex vortical systems with variable speeds and other idiosyncrasies, and as the system grew more complex, its "evident" explanatory power diminished.[10]

Newton's inverse square law was every bit as mathematically and empirically proven in the *Principia* as Kepler's laws had been a half century earlier, and his law added, therefore, another quantitative measure that strained the explanatory power of the vortical theory. However, since Newton's critique of the vortical system was rooted primarily in quantitative mathematical demonstrations of the theory's inadequacy, and since he offered no rival, causal physical system that explained the force of universal gravitation in evident mechanical, or even Aristotelian, terms, his explanation was viewed by many as no explanation at all. Zealous lovers of the Cartesian system certainly found this rejection easiest to sustain, but the general and widespread rejection of Newtonian physics at first stemmed far more from the radically idiosyncratic scientific epistemology used in the *Principia* than from any blind adherence to Descartes or his vortical system. This was no less true in France than anywhere else where figures like Huygens could at once be critics of the Cartesian vortical system and great

10. See Aiton, *Vortex Theory of Planetary Motions*, 76–85; Brunet, *L'introduction de Newton en France*, chap. 1.1.

admirers of Newton's mathematical critique of it—book 2 of the *Principia* is largely about the mathematical inadequacies of vortical physics—yet still be unable to call the *Principia* a persuasive work of physics or natural philosophy.

Also supportive of this stance was how Newton, when read carefully, seemed to agree with these very same views. He was as aware as anyone that mathematical descriptions were not the same thing as physical and philosophical explanations, and accordingly he made a point throughout his treatise of carefully demarcating the limits of his argument. In the early mathematical books, for example, he repeated more than once that by the "force of attraction" that moved bodies, he meant only a mathematical entity and not a physical principle. As he wrote in one representative passage from book 1: "I use interchangeably and indiscriminately words signifying attraction, impulse, or any sort of propensity toward a center, considering these forces not from a physical but only from a mathematical point of view. Therefore, let the reader beware of thinking that by words of this kind I am anywhere defining a species or mode of action, or a physical cause or reason, or that I am attributing forces in a true and physical sense to centers."[11]

Moreover, since the title of his treatise (*The Mathematical Principles of Natural Philosophy* in English) announced emphatically the epistemological category error at the center of his book, Newton also took pains to precisely explain the goals and intentions of his treatise. In these statements, he shied away from aggressively calling his book a new kind of physics, arguing instead that his was merely a treatment of the mathematical principles of natural philosophy alone. He also avoided strong declarative statements (in prose at least; his use of mathematical language was another matter) indicting Cartesian vortical physics directly. He also resisted any strident defense of his own, implicit physics of universal gravitational attraction. Mathematical demonstrations and quantitative empirical proofs were his central argumentative weapons, and while the *Principia* cohered as a single book precisely because of the linkages that Newton built between mathematics and physics, the epistemological idiosyncrasy of such linkages allowed many to read the book in a more partial and disjointed way.

In France, where mathematical mechanics was both well established and highly esteemed, Newton not surprisingly entered the wider scientific consciousness through the influence of his *Principia* on the ongoing projects in *mécanique*. This assimilation occurred, moreover, while Newton's physical arguments in book 3 and the complex mathematical-philosophical epistemology that he used to anchor them were either ignored or quickly forgotten. Two

11. *Principia*, 408.

parallel developments further coincided to shape this early reception. One was the vogue for Malebranchian philosophy that gave a new impetus to the French science of mathematical mechanics after 1690.[12] Malebranche offered a philosophical justification for advanced mathematical mechanics that was either coincident with Newton's position or radically contradictory to it depending on how one interpreted Newton's precise intentions in the *Principia*. If Newton was indeed theorizing a new and essentially phenomenalist science of mathematico-physical description, one that would take strictly quantitative mathematical relations as those "upon which the study of philosophy should be based" (these are Newton's words from book 3 of the *Principia*),[13] then Malebranche argued much the same thing in his massive and influential *Recherche de la verité* that first began to appear in 1671. If, on the other hand, Newton was arguing that mathematical demonstrations, such as those offered in the *Principia*, proved that a universal attractive force existed in all matter, and that the action of this force explained the quantitative laws operating in planetary motion and terrestrial gravity, then Malebranche was arguing something quite different. Indeed, as a theorist of *physique* rather than as a mathematical natural philosopher, Malebranche was a convinced vorticist who contributed importantly to Cartesian celestial and terrestrial mechanics after 1690.[14]

Malebranchian mathematical philosophy, which was very influential in France, nevertheless reinforced reading Newton as a theorist of phenomenalist mathematical physics, and as a result this helped to push the initial French reception of the *Principia* in the direction that it eventually went. The other, simultaneous development that also pushed the reception this way was the arrival of Leibniz's new and conceptually powerful infinitesimal calculus. The French Malebranchian mathematicians who read Newton's *Principia* soon after it was published were the same ones who also started to incorporate Leibniz's new calculus into their mathematical mechanics. These initial students of Leibniz and Newton included Malebranche himself and a circle of his close followers, namely Pierre Varignon and the Marquis de l'Hôpital in Paris, and Johann Bernoulli in Basel. Newton too was part of the correspondence networks that linked this group, as was Leibniz.[15] Yet importantly,

12. See Shank, " 'There Was No Such Thing.' "

13. Newton, *Principia*, 793.

14. See Robinet, *Malebranche de l'Académie des sciences;* Aiton, *Vortex Theory of Planetary Motions;* and Mouy, *Le développement de la physique Cartésienne.*

15. See Robinet, "Le groupe Malebranchiste introducteur du calcul"; Blay, *La naissance de la mécanique analytique;* Costabel, *Varignon et la diffusion du calcul en France;* and Shank, "Before Voltaire," chap. 4.

while Newton was also in possession of his own differential calculus by this time, and while his new mathematics did shape his arguments in the *Principia*, his treatise systematically avoided its direct use. Newton also avoided publishing his calculus in another venue or advocating for its use among these mathematicians even though Leibniz was vigorously doing both. Thus, when the French Malebranchian mathematicians began reading the *Principia* and developing its ideas in terms of the new Leibnizian calculus, they were not so much translating Newton's work literally into a different mathematics as transforming it into a new science that better agreed with their guiding scientific assumptions. This is especially true in their complete avoidance of the physical arguments and empirical proofs of book 3 of the *Principia* during this period of transformation.

The first manifestation of this particular approach to Newton's *Principia* was a new and purely mathematical science of motion that Pierre Varignon, supported by his allies in the "Malebranche circle," began to develop after 1692. Newton's *Principia* exerted an important influence upon this science, which today we call "analytical mechanics" in recognition of the way that mathematical analysis (i.e., the differential and integral calculus) was essential to it. Varignon simply called it the "new science of motion." The new analytical mechanics quickly became a centerpiece of French scientific practice over the next decade, bringing with it an image of Newton as one of the founding fathers of this new science. He was certainly not the central figure in its development, for the French analytical turn, triggered as much by the philosophy of Malebranche and the calculus of Leibniz as the precise arguments of the *Principia*, effectively created a conceptual gap between Newton's work in the *Principia* and the related but distinct achievement of the French in the decades that followed its publication. Accordingly, no one ever called analytical mechanics Newtonian mechanics, nor was any discourse for or against Newton initiated by its inception.[16]

At this juncture, institutional changes further conspired to cement this particular French understanding of Newtonian science. As Varignon, l'Hôpital, Malebranche, Bernoulli, and their allies began to develop their new science of motion, a reformist wind hit the Royal Academy of Sciences carrying each to the center of a newly reoriented institution. Varignon and l'Hôpital were already academicians in 1699, the year that the reform was instituted, but Malebranche was not. He was added to the newly expanded company, along with a number of other, less famous Malebranchians. Bernoulli, Leibniz, and Newton also became members of the French Royal Academy, joining the first

16. Blay, *La naissance de la mécanique analytique;* and Shank, "'There Was No Such Thing.'"

FIGURE 9. *Pierre Varignon.*
Courtesy of the Smithsonian Institution Libraries.

class of foreign associates. Fontenelle, who had already shown his alliance with the mathematical Malebranchians in 1693 through the preface that he wrote to l'Hôpital's first-ever textbook on the differential calculus, also became a royal academician at this time.[17] He further became the institution's first formally appointed perpetual secretary of the Royal Academy, receiving as part of his appointment a new mandate to make the company more publicly oriented.

17. L'Hôpital, *Analyse des infiniment petits.*

FIGURE 10. *Louis Galloche, portrait de Bernard le Bovier de Fontenelle.*
Louvre, Paris, France. Courtesy of Art Resource, NY.

Fontenelle's charge included a new program of publication that would require him to write a history of the academy's scientific achievements each year
as part of a new annual volume in which the academy's best scientific papers
were published. The result was the new *Histoire de l'Académie royale des sciences. Avec les mémoires de mathématique et de physique . . .* that began to appear annually in 1700. Broader public outreach was also instituted through a
series of twice-annual public assemblies and a newly vigorous correspondence
program. At an early public assembly, Fontenelle also innovated within these
frameworks, creating a new practice of reading public funeral orations, or
éloges, for each recently deceased academician. Together with his annual *histoire,* these *éloges* allowed the perpetual secretary to use his new public office
to exert a decisive role in shaping the public discourse of science in France and
throughout the Republic of Letters.[18]

Fontenelle's work in this regard largely centered on presenting the academy
as a meritorious republic of science where *honnête* savants served the public
through their pursuit of natural knowledge. To the extent that this image dove-
tailed with the wider values of the Republic of Letters as a whole, it served to
implicate the company within this wider learned community. However, since
this discourse also channeled the wide esteem for both into an image of French
science satisfying the academy's royal patrons, his work also supported the
political agendas of royal administrators. Especially interested in these initiatives was the abbé Jean-Paul Bignon, the de facto cultural minister of France
during the first decades of the eighteenth century, and the man who initiated
the public reform at the Royal Academy of Sciences. All of the royal academies
were under Bignon's supervision, and each received a similar public reform
in 1699.[19] Bignon likewise reformed the official *Journal des savants* in 1702 in
order to make it too a more effective and authoritative public organ.[20]

Fontenelle was an eager ally in all of these political reforms; however, since
he was also a public intellectual with a particular set of scientific views, his
partisanship sometimes worked to complicate his alliance with Bignon.[21] This

---

18. Saunders, "Decline and Reform of the Académie des Sciences à Paris"; Shank, "Before Voltaire," chaps. 2–3.

19. For references to the scholarship on Bignon, see the introduction to part 1, n. 7
above.

20. See Bléchet, "Un précurseur de l'Encyclopédie," 407–408; and Raymond Birn, "*Le
Journal des savants* sous l'Ancien Régime," *Journal des savants* (January–March 1965):
15–35.

21. On the details of the Fontenelle-Bignon relationship, see Shank, "Before Voltaire,"
esp. chaps. 3, 6, 7.

was especially true with respect to his Malebranchian mathematical sympathies, and the difficulties began simultaneously with the new public reform of the academy. After 1699, as the new science of motion became an established feature of French academic science, Fontenelle used his public position to promote this science. Varignon presented his work on analytical mechanics at two of the first four public assemblies, and his papers were covered by the largely literary *Mercure galant*.[22] This fact is less surprising once one knows that the *Mercure,* through its publication of other, similar works in advanced mathematics, helped to fuel the vogue for Malebranchian mathematical philosophy that provided one context for Varignon's work.[23] Fontenelle, who had begun his literary career at this very journal, also opened the academy's first public assembly with an oration, addressed to the same audience that read the *Mercure,* arguing for the utility of advanced mathematics to public science as whole.[24] This oration, which was widely declared a masterpiece, circulated widely in print. Fontenelle's first *histoires* also celebrated this new science, yet problems arose when critics of infinitesimal calculus, the mathematics that anchored the new science of motion, rose in opposition to these Malebranchian perspectives.

Inside the academy, the "Affair of the Infinitesimals," as it eventually came to be called, raged with great force throughout the first decades of the eighteenth century.[25] The most intense period of disputation, however, occurred in the years immediately after 1699, and this meant that the new public academy came to life amidst a bitter public struggle. Divisive intellectual battles of this sort were anathema to the *honnête* image of science that elites in the Paris Academy and throughout the Republic of Letters shared. They were also polluting to the political agendas driving academic publicity. As a result, the royal authorities responsible for the institution, especially Bignon, used their supervisory power to suppress these disputes however possible. The newly public organs of sci-

22. PVARS, April 29 1699 and November 13, 1700. Varignon's public assembly papers were discussed in *Mercure galant* (May 1699): 9–11; (November 1700): 138–139.

23. See, for example, "L'ouvrage conçernant l'algèbre," in *Mercure galant* (April 1697): 42–83; and "Discours sur l'algèbre," *Mercure galant* (May 1698): 67–122. See also Shank, " 'There Was No Such Thing.' "

24. Bernard le Bovier de Fontenelle, "Préface sur l'utitlité des mathématiques et de la physique et sur les travaux de l'Académie des Sciences," in *Oeuvres de Fontenelle,* 6: 37–50.

25. On the calculus wars in France, see Paolo Mancosu, "The Metaphysics of the Calculus: A Foundational Debate in the Paris Academy of Sciences," *Historia Mathematica* 16 (1989): 224–248, idem, *The Philosophy of Mathematics and Mathematical Practice in the Seventeenth Century* (Oxford, 1996); and Shank, "Before Voltaire."

ence, however, made containment difficult, and consequently, despite vigorous efforts to resist such exposure, the French public came to witness an intense, academy-centered intellectual debate about the validity of infinitesimal calculus at the very same time that it was introduced to this new science and the new public academy that practiced it.

Fontenelle remained a vigorous defender of infinitesimal calculus throughout his life, yet confronted after 1699 with public criticisms about its validity, he used his official office less to fight on its behalf than to subdue the controversy in terms favorable to his "modern" allies. The "ancients" who resisted the new mathematics were also academicians, yet their criticisms, aired openly inside the academy, were not at first given public voice. Others, however, broke this silence. Especially vigorous on the side of the "ancients" was the new Jesuit-edited monthly, the *Journal de Trévoux,* which began to appear regularly in 1701.[26] In its first volumes, the Jesuit editors of this journal openly challenged the position of Varignon and the other mathematical "moderns" inside the academy, paving the way for Michel Rolle, a royal academician, to take his academic critique of infinitesimal analysis public. In launching his critique, Rolle did not use the new publications of the Royal Academy but the pages of the *Journal des savants.* His first public attack of infinitesimal calculus appeared in 1702, and this triggered others to join the fray. These included Joseph Saurin, who was enlisted in these public struggles by Varignon and Malebranche precisely because he was a "modern" mathematician who was not yet beholden to the disciplinary protocols of the Royal Academy. Saurin countered Rolle's work in the *Journal des savants* in 1703, and after further public skirmishes over the next three years, Bignon was compelled to officially silence the dispute. He invoked the new rules of the academy in a disciplinary way to prohibit Varignon and Rolle from publishing any further attacks while calling on Saurin's honor as a gentleman to do the same.[27]

In this way, the new science of motion, obliquely triggered by the initial French reception of Newton's *Principia,* did cause an open public controversy that was prescient of things to come. Yet in this case, it was the differences between these initial public battles and the Newton wars of the 1730s that were more important. First, and most important, the "Affair of the Infinitesimals"

26. On this journal, see Alfred R. Desautels, *Les Mémoires de Trévoux et le mouvement des idées au XVIIIe siècle* (Rome, 1956); Gustave Dumas, *Histoire du Journal de Trévoux depuis 1701 jusqu'en 1762* (Paris, 1936); and George R. Healy, "Mechanistic Science and the French Jesuits: A Study of the Responses of the *Journal de Trévoux* (1701–1762) to Descartes and Newton" (Ph.D. diss., University of Minnesota, 1956).

27. Shank, "Before Voltaire," chap. 6.

was not about Newton at all. His name was invoked by Varignon on a few occasions to add luster to his list of those who accepted the validity of infinitesimal analysis. The *Principia* was also cited as a text that showed the validity and value of this new mathematics, even if Newton's calculus was not obviously used in the text.[28] Yet such arguments in no way placed Newton or his work at the center of this dispute. Second, while the calculus wars did reveal the new critical possibilities implicit in the academy's new public stance inaugurated in 1699 and as such pointed to the critical public fights to come, they did not result in any open public divisions that challenged the academy as an institution. Indeed, while the authoritative resolution of the controversy did lead to the public victory of one faction (the moderns) and the defeat of the other (the ancients), the outcome ultimately worked to secure a new consensus and to reconfirm the perception of the institution as a neutral and *honnête* arbiter of scientific knowledge. Third and finally, the resolution of this controversy did not trigger a new period of critical, public contestation in French science; it rather led to a new and consensual peace—a *pax analytica,* it might be called— where the disciplinary structures of the new public academy worked to maintain a consensus forged out the calculus wars of the first years of the eighteenth century. Fontenelle played a singular role in effecting this final outcome, for by channeling his partisan advocacy for the new mathematics into the idioms and disciplinary structures of *honnête* public science, he helped to sustain a vision of the academy and its mathematical sciences that secured both the institutional power of his allies, the analysts, and the wider *pax analytica* at the same time.[29]

An image of Newton supported by this precise intellectual-institutional constellation also ascended to prominence in tandem with it. This image did not make Newton the opponent of Descartes; it rather made him the forerunner with him of the new analytical and mathematical approach to mechanics that Varignon, l'Hôpital, Leibniz, Bernoulli, and the other members of the "Malebranche circle" pioneered. With this understanding, Fontenelle could simultaneously celebrate the joint Cartesian-Newtonian heritage of analytical

28. The Malebranchian mathematician Charles-René Reyneau kept a diary of the Varignon-Rolle dispute over the infinitesimal calculus, and in it he noted Varignon's use of Newton's *Principia* as a weapon in his battles with the opponents of the calculus in the Royal Academy. Reyneau's diary is found in Bibliotèque nationale de France, Paris (hereafter cited as BN), fonds Fr. 25302, ff. 144–155. It is also reprinted in *Der Briefwechsel von Johann I Bernoulli. Band 2: Der Briefwechsel mit Pierre Varignon. Erster Teil, 1692–1702,* ed. Pierre Costable and Jeanne Pfeiffer (Basel, 1988), 352–376.

29. See Shank, "Before Voltaire," chaps. 6–7.

mechanics in the public sphere while upholding the Cartesian vortical physics criticized by Newton in the *Principia* and not sustain a contradiction. Others followed the academy secretary's lead in this respect, and in this way a consensus was publicly reinforced that embraced Newton's *Principia* without publicly contesting over it.

The *Pax Analytica* in France

The particular relationship to Newtonian science that this consensus produced can be illustrated by looking at the French discussions of the oceanic tides before 1720. The daily cycles of the tides were another topic explained by Newton's mathematico-empirical arguments in the *Principia*. In book 3, he used the theory of universal gravitation to connect the tides to the gravitational pulls of the moon and the earth. However, since Newton's theory depended on the physics of universal gravitation, a physics that few accepted around 1700, the French, like many, disregarded this theory along with the physics that anchored it. New French work on the tides, therefore, was not triggered by Newton's treatise, but by a new intitiative of the French Crown. Announced in 1701, this initiative proposed making the Royal Academy a clearinghouse for the collection and dissemination of new and more accurate tide data. Such a project had obvious utilitarian benefits for maritime navigation and commerce, and Fontenelle accordingly celebrated it in his *histoire,* calling it a worthwhile collaboration between the Crown, the academy, and the public.[30] In the years that followed, savants, provoked by the announcement, began to submit tide data to the Royal Academy, and the secretary, prodded by this success, made a point each year of publicly celebrating the program, its ministerial sponsors, the academy, and the savants who all contributed positively to it.

In 1710, the annual reporting of tide data was given a new twist by the intervention of the royal astronomer Jean-Dominique Cassini. This Cassini is often referred to as Cassini II because he was the son of Giovanni Domenico Cassini who came to France in 1671 to establish the first French Royal Observatory. Cassini II generally followed in his father's professional footsteps, as would his own son, Cassini III, after him. However, in this case the distinguished astronomer did a very uncharacteristic thing. He used the extant tide data to develop a physical theory of the tides that he presented at the academy's fall public assembly of 1710. Cassini II connected the "*flux et reflux*" of the oceans to the phases of the moon, cautiously offering a vortical explanation that made

30. *HARS-Hist.* (1701): 11–13.

the impact of the lunar vortex upon the earth's celestial vortex the cause for the rise and fall of the tides in twelve-hour cycles.[31] Varignon would later challenge Cassini II's theory in the academy, asking why barometers did not record a similar cycle if atmospheric pressure was at work in producing this tidal effect.[32] Yet in his public presentation of Cassini II's work, Fontenelle suppressed these challenges, adopting instead his familiar role as cheerleader. Presenting Cassini II's vortical theory of the tides as an important advance, the secretary declared that "nothing supports better the hypothesis of the lunar vortex than this fact."[33]

This apparent alliance between Cassini II and Fontenelle, however, was steeped in irony. As a royal astronomer, Cassini II was not at all accustomed to physical theorizing of this sort. Astronomy, in fact, was a mathematical science in the same class as mechanics, and it offered a no more acceptable ground for physical explanations than *mécanique*. Astronomers in the seventeenth century were empiricists concerned with producing ever more accurate data and ever more precise predictions of celestial phenomena. Mathematico-physical models of the cosmos were important to them since these models made possible the measurements and predictions upon which their science depended. However, astronomers *qua* astronomers almost never claimed these models as physical explanations of nature, nor did they engage in the kind of causal cosmology that defined celestial mechanics as physics or natural philosophy. To do so would have been to assume the identity of the physicist or natural philosopher, an identity that empirical astronomy in no way warranted.[34]

The French Royal Observatory that Cassini II's father founded in 1671 was a paragon of seventeenth-century astronomical science in this respect. Its large and well-funded staff, which included many royal academicians even though the two institutions were technically distinct, devoted itself to meticulous empirical research. What theory French royal astronomers did pursue involved new methods of calculation and instrumentation and new structures for transforming observations into reliable celestial knowledge. Such work rarely placed French royal astronomers at the center of the great cosmological debates of the period. The prestige of the French Royal Observatory, which was great, rested instead on its reputation as the authoritative center of empirical astro-

31. PVARS, November 12, 1710. For the public reception, see *Mercure galant* (November 1710): 13–15.

32. PVARS, April 27, 1712.

33. *HARS-Hist* (1710): 7.

34. See Robert S. Westman, "The Astronomer's Role in the Sixteenth Century: A Preliminary Study," *History of Science* 18 (1980): 105–147.

nomical science. Paris in the seventeenth century was zero-degree longitude, not Greenwich, and mariners everywhere relied on the annual celestial almanac produced by the French Royal Observatory, *Les connaissances des Temps*, to navigate the oceans of the world.[35] All the Cassinis embodied the scientific and institutional spirit that made French astronomy so esteemed. Thus, when Cassini II assumed the persona of the *physicien* before the French public with his vortical theory of the tides, he was breaking with the norms that defined the work of his astronomical colleagues.

Fontenelle's public celebration of Cassini II's theory was less exceptional. The secretary in fact made the Cartesian vortical mechanics a centerpiece of his public academic discourse from 1700 onward, and even Varignon's thoroughly nonvortical and purely mathematical mechanics was often explained to the public through pictures that deployed the Cartesian vortical system. Fontenelle continually used the terms of Cartesian cosmology to assimilate Cassini II's work with that of Varignon and others in his public explanation of each, and while Cassini II at least offered his vortical theory of the tides himself as a real, if tentatively advanced, physical hypothesis, the secretary made sure to accentuate this dimension in his public presentation of it.

With the new Cartesian framework in place, therefore, the continued reporting of the new tide data in Fontenelle's *histoires* took on a different significance. No longer simply a report of the recent empirical findings and the worthwhile work of the academy in collecting them, the tide data became evidence in support of Cassini II's vortical theory. Between 1712 and 1715, the secretary reported that the new data coming from Brest continued to confirm Cassini II's explanation, and in each case he opened the appropriate *histoire* with a glowing account of these findings, usually on the very first page.[36] "A new year of observations at Brest has confirmed everything established in the previous reports," the secretary wrote characteristically in 1714. "They offer proof of the excellence of all the other results obtained in the different ports and of the judiciousness of the conclusions that have been drawn from this data." Asserting an aggressiveness into Cassini II's claims that was not really contained in his theory itself, Fontenelle also concluded by boldly asserting that "the tides surely

35. On seventeenth-century French astronomy and the Royal Observatory, see Alice Stroup, *A Company of Scientists: Botany, Patronage, and Community at the Seventeenth-Century Parisian Royal Academy of Sciences* (Berkeley, 1990), chaps. 1–5; Monique Pelletier, *Les cartes des Cassini: La science au service de l'èétat et des régions* (Paris, 2002); and C. Wolf, *Histoire de l'Observatoire de Paris de sa fondation à 1793* (Paris, 1902)

36. *HARS-Hist.* (1712): 1–3; *HARS-Hist.* (1713): 1–6; *HARS-Hist.* (1714): 4–7; *HARS-Hist.* (1715): 1–5.

depend on [Cassini II's three] principles, which each depend on the moon, its phases, its distance from the earth, and its declension. Perhaps with an even greater number of observations we will soon know the *rapports* between the forces regulating these principles."[37]

Cassini II's vortical theory of the tides would serve as the opponent for the attractionist theory defended by the Newtonians after 1728. Thus, its emergence at the center of French scientific discourse after 1715 is important. However, just as important is the manner in which this theory moved to the center of the French theorization of this question. Begun as an explicitly state-sponsored project with concrete utilitarian goals, French work on the tides only became theoretical after the fact. Furthermore, while Cassini II certainly initiated the move to theory, he did so while remaining much more comfortable with the mere empirical demonstration of the correlation between the phases of the moon and the movements of the tides. More than anyone else, it was Fontenelle who insisted that Cassini II's work was proof of Cartesian vortical astronomy. He also went beyond Cassini II's own claims by arguing that the tides offer proof that the vortices in fact exist.

Writing of Cassini II's other geodesic work, Fontenelle dismissed the astronomer's zeal for empirical precision, calling it a "preoccupation satisfying to the mind but not otherwise important."[38] For Cassini II, empirical precision was the very essence of sound science since inductions from accurate measurements were, for him, the only legitimate way to arrive at general, theoretical conclusions. Fontenelle, however, believed that empirical induction was secondary to deductive certainty in the epistemological hierarchy of science. As he once said of Kepler's laws, "they are unreliable because they are founded only on an induction from multiple observations and not demonstrated a priori by the laws of motion."[39] In his public pronouncements about Cassini II's theory of the tides, the academy secretary applied these epistemological assumptions to the astronomer's very differently oriented work, in effect turning a very cautiously advanced empirical induction into an a posteriori proof of a rationally deduced theory.

Cassini II's relationship to vortical mechanics was very different. In the public discourse of the academy, however, these differences were effaced in such a way as to mask the diverse methodological positions that were actually present in French scientific practice. In this way, Fontenelle often conveyed the impression that vortical mechanics was more central to French academic

37. *HARS-Hist.* (1714): 4.
38. *HARS-Hist.* (1713): 66.
39. *HARS-Hist.* (1707): 97.

science than in fact it was. Indeed, as the Newtonians in France and elsewhere would make clear after 1730, the same empirical correlation between the lunar phases and the tides could be explained just as easily other ways. Fontenelle, however, never ceased to explain it in terms of the theory of the vortices. He did the same with Varignon's analytical mechanics, suggesting that the vortices were the cause of the celestial forces operative between bodies even though Varignon treated such forces as mathematical objects only, and never theorized the physical factors that may have produced them.[40]

Here the real significance of Fontenelle's continual employment of vortical cosmology in his public academic discourse emerges. By unifying a variety of different scientific approaches and epistemologies under the singular banner of Cartesian mechanistic science, and by further making each of these different scientific outlooks appear conformable to this one, master cosmology, Fontenelle in effect fostered the perception that a Cartesian unity existed in France when diversity was really the norm. A tripartite convergence forged in the immediate wake of the 1699 academy reform between the secretary's deep, epistemological commitment to Cartesian science, his commitment to projecting a consensual view of French science, and his need to develop clear and accessible images appropriate for presenting academic science to the wider public drove the secretary in this effort. The result was to make the Cartesian vortices appear as the principle of unity in French public science despite the presence of a variety of different scientific epistemologies within French scientific practice itself.[41]

Consensual harmony was indeed the leitmotif of French public science in the period of the *pax analytica*. Yet it was precisely this consensus, and the cultural forces that sustained it, that began to disintegrate after 1715. The change did not occur all at once. During its formative years in the 1690s, the new science of motion had provoked intense controversy. Yet in the face of this controversy Fontenelle worked hard after 1699 to make this science appear compatible with the values and decorum that new public academy hoped to project. His efforts were largely successful, and they allowed the practitioners of the analytical sciences to claim after 1710 a newfound authority as the spoils

40. Early in his career, Varignon did develop a Huygens-style vortical theory of *pesanteur*. His work was poorly received, however, and Varignon abandoned the theory and vortical mechanics altogether for the remainder of his career, focusing entirely on pure mathematical mechanics instead. See Pierre Varignon, *Nouvelles conjectures sur la pesanteur* (Paris, 1690). For a discussion of this shift in Varignon's work, see Yves Gingras, "What Did Mathematics Do to Physics?" *History of Science* 39 (2001): 383–416.

41. See Shank, "Before Voltaire," chap. 6.

of this victory. In 1710, Varignon was appointed for the first time to serve as one of the two directors of the Royal Academy. During the next five years either he or Fontenelle served in this capacity almost every year. In 1719, they served together, indicating the institutional prominence that the practitioners of the new analytical sciences had achieved by this date.[42] Other analysts reinforced this dominance, most notably Saurin, who became an important and respected member of the academy after his admission in 1707. The group was also rewarded with a decade and a half of triumphant peace secured according to the terms that they had fought to achieve. Varignon, Saurin, and their younger protégés continued to produce new work throughout the 1710s, and the academy secretary routinely sang its praises publicly from his elevated position within French society. In this way, the French program in analytical mechanics found itself basking in the glow of institutional and cultural prominence when the Sun King Louis XIV breathed his last breath in September 1715.

In the short term, the continuation of the *pax analytica* translated into a continuation of the attitudes about Newton and his *Principia* forged during the rise of analytical mechanics over the previous decades. In 1713, a newly revised second edition of Newton's *Principia* appeared replete with a new "General Scholium" by Newton and a provocative new "Preface" penned by his editor Roger Côtes.[43] The motivations behind these new provocations will be explored in the next chapter, but for here it is sufficient to know that each was directed, among other places, at the prevailing assumptions of French science. In the scholium, Newton asserted his opposition to Cartesian vortical astronomy in newly polemical terms, opening the piece with the famous statement that "the hypothesis of the vortices is pressed with many difficulties."[44] He then went on to list in devastating detail all of the inadequacies of the vortical theory of celestial mechanics. The scholium largely summarized arguments that were made in the first edition as well, but Newton added a new polemical force to them by grouping his criticisms into one sustained discursive (as opposed to mathematical) argument. He likewise asserted more categorically than before his belief that the gravitational attraction of bodies through empty space was a physical fact proven by abundant empirical evidence.

42. A useful chart listing the officers of the academy between 1699 and 1750 is found in David J. Sturdy, *Science and Social Status: The Members of the Académie des Sciences, 1666–1750* (Woodbridge, 1995), 421–423.

43. "General Scholium," in *Principia*, 939–944; and "Editor's Preface to the Second Edition," in *Principia*, 385–399. On Côtes, see Ronald Gowing, *Roger Côtes: Natural Philosopher* (Cambridge, 1983).

44. *Principia*, 939.

Côtes added further punch to these provocations in his preface. His targets were the critics of Newton's scientific method. "Those who have undertaken the study of natural science can be divided roughly into three classes," he began. First are the philosophers of the schools who derive their doctrine from Aristotle and the Peripatetics. "Although they affirm that individual effects arise from the specific natures of bodies, they do not tell us the causes of those natures, and therefore they tell us nothing."[45] The members of the second class, says Côtes, "have hoped to gain praise for greater carefulness by rejecting this useless hodgepodge of words." Here Côtes had in mind the followers of Descartes' mechanistic physics, even though he never explicitly used the label "Cartesian" to describe them. These philosophers hold "that all matter is homogenous and that the variety of forms that is discerned in bodies all arises from certain very simple and easily comprehensible attributes of the component particles." Such a view is not all wrong, he argued. "But when [these philosophers] take the liberty of imagining that the unknown shapes and sizes of the particles are whatever they please, and of assuming their uncertain positions and motions, and even further of feigning certain occult fluids that permeate the pores of bodies very freely, . . . they are drifting off into dreams and neglecting the true constitution of things."[46]

"Experimental philosophy," Côtes argued, offered the antidote to these misunderstandings. This philosophy "holds that the causes of all things are to be derived from the simplest possible principles," and experimental philosophers "assume nothing as a principle that has not yet been thoroughly proved from phenomena. They do not contrive hypotheses, nor do they admit them into natural science otherwise than as questions whose truth may be discussed." More precisely, correct philosophy followed a twofold method for Côtes, synthetical and analytical. "From some selected phenomena," Côtes explained, proper philosophers "deduce by analysis the forces of nature and the simpler laws of those forces." From these, they then give "the constitution of the rest of the phenomena by synthesis," or rational deduction. Singling out Newton's work in the *Principia* as the model application of this method, Côtes described it as "that incomparably best way of philosophizing which our most celebrated author thought should be justly embraced in preference to all others." Clarifying the consequences of this choice, he further added that Newton has given us "a most illustrious example" of the superiority of his method by explaining the system of the world through a synthetic deduction from the theory of universal gravitation. "That the force of gravity is in all bodies universally, others have

45. Ibid., 385.
46. Ibid., 386.

suspected or imagined," Côtes explained; however, "Newton was the first and only one who was able to demonstrate it from phenomena, and to make it a solid foundation for his brilliant theories."[47]

Each of these additions to the original *Principia* were new and polemical provocations, yet to judge by the reception offered Newton's second edition in the French press, the bombs were absorbed quietly within the existing assumptions about Newton's significance. Reviewing the first two books of the treatise in March 1715, the *Journal des savants* explained that "M. Newton treats the movement of bodies using the precision of geometry, but in such a way that the principles once established can be applied to a variety of physical problems. . . . We regret that we cannot offer a detailed summary of everything that Newton does here in order to expose the elevation of his genius."[48] Similar praise of Newton's mathematical work had become the norm in France after 1690, and in this review the writer was merely echoing the discourse about Newton's achievement that had become commonplace over the previous two decades.

The journal likewise joined this now familiar praise for Newton's mathematical achievements with a renewed apathy about his physical claims for universal gravitation. In 1688, the *Journal des savants* had chastised Newton for blurring the divide between mathematics and physics.[49] In 1715 no such charge was made even though the physical arguments of the treatise had been made more pronounced and provocative in the revisions of the second edition. Instead, the journal merely reported that Newton's third book "explains the system of the world by the same principles" and left the account there. Similarly, the reviewer expressed only silence with respect to Newton's famous attack on the vortices in book 2, his summary of this attack in the new "General Scholium," and Côtes' methodological critique of Cartesian physics in the new preface. Furthermore, what the reviewer did summarize pertained only to the mathematical treatment of central forces that had been a centerpiece of French science for over two decades.[50] The reviewer further avoided altogether the physics of Newtonian attraction, and in no way rose to either defend Newtonian physics or to refute it. In this way, the text echoed the consensus on Newton's significance that had formed in France between 1690 and 1715.

47. Ibid.
48. Review of Isaac Newton, *Principia mathematica* (2nd ed. of 1713), in *Journal des savants* (March 1715): 157.
49. Review of Isaac Newton, *Principia mathematica,* in *Journal des savants* (August 1688): 152–153.
50. Review of Isaac Newton, *Principia mathematica* (2nd ed.), in *Journal des savants* (March 1715): 153.

The Jesuit-edited *Journal de Trévoux* not surprisingly offered a different assessment. During the public debates over analytical mechanics, the Paris-based editors of this periodical had shown themselves to be vigorous and often provocative partisans in the controversies. However, as the *pax analytica* set in, the reporting in this journal also grew calmer. In their review of the 1713 edition of Newton's *Principia,* the Jesuits further confirmed the new consensus. The journal only ran its review of Newton's work in February 1718, almost five years after the new edition had appeared in print. The timing may not have been accidental, for the review appeared as new controversies about Newton's work were just beginning to emerge. Nevertheless, the editor's willingness to delay publication of the review until 1718 is the more telling fact. It confirms the lack of urgency that still surrounded the treatise in 1713.

The review that the Jesuit editors printed, moreover, although late in coming, also echoed many of the familiar opinions about Newton's work formed during the period 1690–1715.[51] The reviewer began by drawing the by then classic divide between Newton the *géomètre* and Newton the *physicien.* "The reputation of this work is constant among mathematicians who admire the force and depth of the author's genius," the reviewer wrote. "It is contested, however, among the physicists since the majority cannot accommodate themselves to the natural affections that the author alleges to exist between all bodies."[52] The text thus broke with convention in one respect by emphasizing the conundrums of gravitational attraction from the outset. Once raised, however, Newton's physics and the justifications for it were not dwelt upon, nor were they turned into a polemical vehicle for either criticizing or praising the treatise as a whole. Instead, offering a brief yet systematic account of the first two books of the *Principia* while largely ignoring the physical arguments of the third, the reviewer called Newton's work "an example of the finest geometry." He also called its writer "a most subtle author." Each of these features mark the review as reflective of the early French discourse about Newton.[53]

In this way, the years around 1715 were marked by continuity in the French discourse about Newton. Nevertheless, the foundations of the original consensus were also beginning to erode. One dimension of the shift was institutional. Gradually after 1710 the mathematical generation that had risen to prominence at the end of the seventeenth century began to die off without leaving successors in their place. L'Hôpital was the first member of the "Malebranche circle"

51. Review of Isaac Newton, *Principia mathematica* (2nd ed. of 1713), in *Journal de Trévoux* (February 1718): 466–475
52. Ibid., 466–467.
53. Ibid., 468, 469.

to depart, passing away in 1704. He was followed to the grave by Louis Carré in 1711, Malebranche in 1715, Leibniz and Joseph Sauveur in 1716, Jacques Ozanam in 1717, Guisnée in 1718, Pierre Rémond de Montmort in 1719, Varignon in 1722, and Bomie in 1727.[54] The opponents of the new analytical sciences also began to depart in these same years. The abbé Gallois, a leading critic, died in 1707, and his allies Philippe de La Hire, Michel Rolle, and the Jesuit Father Thomas Gouye passed away in 1718, 1719, and 1725, respectively. This further completed the departure of the original mathematical community in the Paris Academy. Fontenelle and Saurin, by contrast, lived very long lives and were thus figures of continuity well into the eighteenth century. In this respect, however, they were exceptional. For when Saurin retired from the academy in 1731 at the age of seventy-six, only two remaining academicians besides Fontenelle could claim a direct, personal link to the original struggles over analytical mechanics.

One of these remaining academicians was an obscure and thoroughly undistinguished figure named Christophe-Bernard de Bragelogne who did nothing to advance the analytical program. The other was François Nicole, a talented protégé of Malebranche and a well-respected analytical mathematician who especially earned the praise of Johann Bernoulli. By 1725, however, Nicole's mathematical colleagues in the academy were but a dismal reflection of the group that had entered in the 1690s. In 1724, Bernoulli described the infinitesimal calculus being practiced in the Paris Academy as "clumsy, awkward, cumbersome, and in some cases even misapplied."[55] He exempted Nicole from these criticisms, but his comments reflect how the intellectual climate in the institution had changed after 1710. A new marginality for the Swiss savant, who had once been a central figure in the work of the Paris Academy, was one result of this shift. Another was the marginalization of Nicole, who never assumed anything like the stature of l'Hôpital, Varignon, or Saurin even if he almost single-handedly preserved the legacy of their work into the 1730s.

When Fontenelle retired from the academy in 1740, at the age of eighty-three, the situation had drastically changed. The famous "French Newtonians"—Maupertuis, Alexis-Claude Clairaut, and Pierre Bouguer—were already in the academy by this date, and d'Alembert joined them the following year. The work of these mathematicians certainly reflected the continuing legacy of the

54. See Sturdy, *Science and Social Status,* 424–432.

55. Bernoulli to Jean-Jacques Dortous de Mairan, letter from 1724, cited in John L. Greenberg, *The Problem of the Earth's Shape from Newton to Clairaut: The Rise of Mathematical Science in Eighteenth-century Paris and the Fall of "Normal" Science* (Cambridge, 1995), 246.

FIGURE 11. *René Antoine de Réamur.*
Courtesy of the Smithsonian Institution Libraries.

analytical mechanics pioneered by the "Malebranche circle" in and around the Paris Academy after 1690; however, Nicole, who died in 1758, was the only one among them who was a direct protégé of the original cohort of analysts. Furthermore, he was a generation apart from this new class of mathematicians and only closely allied with Maupertuis, who further permanently left Paris for Berlin in 1745. Consequently—and crucially so—the second generation of analysts reached maturity without acquiring any attachment to the previous generation, or any consciousness of their status as intellectual successors to them. In this way, the identities of these two groups were never linked even though intellectually speaking their work could not have been more alike.

This outcome exerted a profound effect on Newton's French reception. Soon the new analysts who reached maturity after 1730 came to see their science as "Newtonian" while they distanced themselves from the assumptions and commitments that had shaped the work of the early pioneers of their science. The precise nature of this reorientation will be explored in a later chapter, but a set of institutional shifts between 1715 and 1730 were crucial to it and must be explored first. One decisive step was made unwittingly in 1707. During that year, Saurin made a meteoric three-month ascent from *éleve* to *pensionnaire* of the academy, a climb that points glaringly to the academic power the analysts held during this period. His promotion left Varignon, his initial patron, without an *éleve*. In Saurin's place, Varignon selected a young Parisian savant named René Antoine de Réaumur.

The Changing Character of French Academic Science after 1715

Réaumur was born in the Vendée, but together with his cousin, the future président Hénault, he moved to Paris in 1703 to make his fortune.[56] Witnessing upon their arrival the spectacle of the calculus wars, the two young provincials embraced the current vogue by taking mathematics classes with Varignon's protégé Guisnée. At some point in this period, either through the prestigious connections that Hénault possessed or through some other means, Réaumur also made the acquaintance of Bignon. The two provincials also frequented Parisian society in this period, a habit Hénault would retain even as Réaumur became more reclusive. They almost certainly encountered Fontenelle in these circles, and thus, either through his graces or those of Bignon or Guisnée, Vari-

---

56. The only modern biography of Réaumur is Jean Torlais, *Un esprit encyclopédique en dehors "l'Encyclopédie": Réaumur* (Paris, 1961). The story of Réaumur's entry into the academy is found here (27–32).

gnon was made aware of the young savant soon after his arrival. In March 1708, he announced to the academy that he would make Réaumur his *éleve*.[57]

Whether the senior academician harbored any hopes of making Réaumur a successor to his legacy is not clear. In his *éloge,* Fontenelle noted that Varignon had a reputation as a mentor who launched mathematicians into distinguished careers. Réaumur also showed signs of following in these footsteps during his first years in the academy. His initial academic paper, delivered in May 1708, focused on the analytical derivation of curves, and it employed the differential calculus.[58] The piece was published in the academy's *mémoires* for 1708 and merited a brief notice in Fontenelle's *histoire.*[59] It was followed by another mathematical work in May 1709 that was likewise published accompanied by a praiseworthy synopsis by the secretary.[60] In his report, Fontenelle positioned Réaumur at the center of the analytical mechanics program dear to him by describing the young man's achievement as "indicative of the taste for general [mathematical] theories that is now so pervasive."[61] "General solutions" had become Varignon's trademark by 1709, and by describing Réaumur this way he effectively subsumed him within the larger, Varignon-centered discourse about analytical mechanics that was then in ascendance. Some months later, the academician Jacques Ozanam, famous for his *mondain* books of mathematical recreations, posed a mathematical problem to the academy. Réaumur was among the group of mathematicians who rose to the challenge.[62] The others included Varignon, Rolle, Parent, Bomie, and Saurin, and to see Réaumur's name on this list is to recognize the extent to which he was an analytical mathematician at this time.[63]

57. PVARS, March 3, 1708.

58. PVARS, May 19, 1708.

59. René Antoine de Réaumur, "Manière générale de trouver une infinité de lignes nouvelles, en faisant parcourir une ligne quelconque donné, par une des extremités d'une ligne droit donné aussi, toujours placé sur une même point fixé," *HARS-Mem.* (1708): 197–212. Fontenelle's commentary is found in *HARS-Hist.* (1708): 83.

60. René Antoine de Réaumur, "Méthode générale pour déterminer le point d'intersection de deux lignes droites infiniments proches, qui rencontrent une courbe quelconque vers le même côté sous des angles égaux moindres, ou plus grands qu'un droit. Et pour connâitre la nature de la courbe décrite par une infinité de tels points d'intersection," PVARS, May 5, 1709 and *HARS-Mem.* (1709): 149–162. Fontenelle's commentary is found in *HARS-Hist.* (1709): 64–68.

61. *HARS-Hist.* (1709): 64.

62. On Ozanam, see Barbara Stafford, *Artful Science: Enlightenment Entertainment and the Eclipse of Visual Education* (Cambridge, 1994), 45–47.

63. PVARS, August 10, 1710.

During the same period, however, he also began to lay the foundations for what would become his more important life's work. In December 1708, he presented a paper to the academy on seashells, seabirds, and mussels. In November 1709 he delivered a similar paper as his inaugural public assembly address.[64] The biannual public assemblies, which were a key feature of the new academic publicity instituted in 1699, allowed the Parisian public to witness a carefully staged presentation of the academy's work, one that ideally projected the institution's dual identity as a learned body and a civilized company of *honnête* savants. Fontenelle's *éloges* had been invented with precisely this mission in mind, and they became a staple of the new public assemblies, along with the reading of several carefully selected academic papers.[65] In his public assembly debut in 1708, Réaumur demonstrated his instinctual understanding of these imperatives. He earned the praise of arguably the leading barometer of success for these sessions, the *Mercure Galant*. In its review, the journal fired both barrels, calling his paper "*savant et curieux.*" But the journalist reserved his greatest praise for Réaumur's overall scientific style. "His talk was very attentively followed," he declared. "It is not possible to speak about this topic with as much breadth. . . . Never has a study been so full of knowledge. And since this topic is not accessible to everyone and consists largely in experiments that few people have experienced, it demanded a wisdom that one rarely finds in similar works. . . . The curious will no doubt want to read this work when it is printed."[66]

The *Journal de Trévoux* further documented Réaumur's public success, stating that his dissertation "was found to be as *curieux* as any other ever written on the subject."[67] Réaumur followed this presentation with another public assembly address the next year about the silk production of spiders, and when

64. René Antoine de Réaumur, "De la formation et de l'accroissement des Cognilles des animaux, tant terrestres qu'aquatiques, soit de mer soit de riviere," PVARS, December 15, 1708; PVARS, November 13, 1709; and *Hars-Mem.* (1709): 364–399. To set the context for Réaumur's arrival, Bomie gave his first and only public assembly paper on the analytical mechanics of the planetary bodies in April 1708. Réaumur's arrival therefore corresponded almost exactly with the climax of the analytical mechanics discussions of the previous decade.

65. On Fontenelle's *éloges*, see Charles B. Paul, *Science and Immortality: The Éloges of the Paris Academy of Sciences (1699–1791)* (Berkeley, 1980).

66. *Mercure galant* (November 1709): 217–221.

67. Review of the public session of the Royal Academy of Sciences for November 1709, *Journal de Trévoux* (January 1710): 162–168.

it was received equally well, a tradition was established.[68] He soon gave up mathematical work altogether and began devoting himself exclusively to the practice of natural history. Just as important, he also became a popular fixture at the academy's public assemblies, and a widely praised figure in the Parisian public sphere, as a result of these choices. In 1734, the professional and public dimensions of Réaumur's career converged with the publication of his multi-volume *Mémoires pour servir a l'histoire des insectes*. Only the abbé Pluche's even more accessible *Le spectacle de la nature* published in 1732 sold more copies than Réaumur's treatise before 1750, and the reasons for this success were clear. As the *Journal historique de Verdun* explained, it was "not necessary to have a deep background in physics to find pleasure in reading this book." [69] Readers and audiences thus flocked to Réaumur, and this success confirmed his dual reputation as a serious academic savant and an eloquent and publicly accessible writer.[70]

His professional trajectory after 1710 followed this path, taking him very far away from his mathematical beginnings. The abbé Jean-Paul Bignon, who continued to supervise the French monarchy's cultural institutions until well into the 1730s, also appears to have supported Réaumur's precise choices at every step. Sometime around 1708, Bignon formed a relationship with Réaumur that was particularly intimate. These connections were no doubt a factor in the young savant's rapid ascent up the academic hierarchy. In 1711, Réaumur was appointed to the Malebranchian analyst Carré's *pensionnaire mécanicien* seat, completing his ascent to the top of the academic hierarchy in less than three years. In 1713, at only thirty years of age, he was further made a director of the academy.[71] From this date forward he became perhaps the single most powerful figure in the institution. Overall, he used this power to serve Bignon's agendas. Most important, in 1713 the minister placed the young *pensionnaire* in charge of the academy's still fledgling projects in the mechanical arts. Fontenelle had

68. René Antoine de Réaumur, "Examen de la soye des araignées," PVARS, November 12, 1710 and *HARS-Mem.* (1710): 386–407; review of the public session of the Royal Academy of Sciences for November 1710, *Mercure galant* (November 1710): 60–87.

69. Cited in Torlais, *Un esprit encyclopédique,* 124.

70. On Réaumur's literary reputation in relation to his science, see Daniel Mornet, *Les sciences de la nature en France au XVIIIe siècle* (Paris, 1911), and idem, "Les enseignements des bibliothèques privées (1750–1780)," *Revue d'histoire littéraire de la France* 17 (1910): 449–496.

71. An outline of Réaumur's academic appointments is found in Ac. Sci., dossier Réaumur.

promised the public a set of volumes describing the practices of all the mechanical arts in the kingdom in his "Preface" to the academy's first volume of *mémoires* published in 1700. The academy reform of the previous year had also created a new class of *mécaniciens,* hand chosen by Bignon, and the minister charged them with realizing this goal.[72] By 1713, however, none of the promised volumes had appeared, and Réaumur was assigned the task of reviving this work. From this date forward, the increased importance of the mechanical arts became manifest in many of his academic activities.[73]

During 1713 and 1714, for example, while Rolle, Saurin, and de La Hire engaged in yet another mathematical dispute, Réaumur began delivering papers on topics such as papermaking, leather tanning, and the manner of extracting gold and silver from the earth.[74] He paralleled this work with public assembly presentations on such natural history topics as the habits of sea animals, the color of false pearls, the reproduction of crabs, the sting of the torpedo fish, and the prodigious absorptive powers of certain natural substances. These public presentations reinforced Réaumur's wider image as an entertaining student of natural curiosities. Inside the academy, however, his work also became increasingly focused on more explicitly utilitarian topics related to arts and industry.[75] In January 1717, he gave his first paper on iron making, a topic that after this date occupied more and more of his attention.[76] Furthermore, as the decade progressed, this utilitarian work of Réaumur and his many protégés increasingly came to dominate the academy's public assemblies. In November 1715, Réaumur publicly presented a paper on "turquoise mines and the nature and color of the matter that one finds there," a topic that merged perfectly his administrative/utilitarian work with his public literary persona.[77] Gradually, however, he began delivering more purely utilitarian and industrial papers at

72. These first-ever academic *mécaniciens* were Gilles Filleau des Billettes, Jacques Jaugeon, and Father Sebastien Truchet. See Jacques Proust, *Diderot et l'Encyclopédie* (Paris, 1967), 177–188; Claire Solomon-Bayet, "Un préambule théorique à une académie des arts," *Revue d'histoire des sciences* 13 (1970): 229–250; and Shank, "Before Voltaire," chap. 2.

73. On Réaumur and the mechanical arts, see Proust, *Diderot et l'Encyclopédie,* 185–188; and Torlais, *Un esprit encyclopédique,* 57–105.

74. See, for example, PVARS, January–April 1714.

75. PVARS, April 15, 1711; November 12, 1712; November 15, 1713; November 14, 1715; and November 14, 1717.

76. PVARS, January 16, 1717. See also Torlais, *Un esprit encyclopédique,* 57–63, 69–74.

77. René Antoine de Réaumur, "Observations sure les mines de turquoises du royaume, sur la nature de la matière qui en y trouve, et sur la la manière dont on lui donne la couleur," PVARS, November 15, 1716 and *HARS-Mem.* (1715): 174–202.

FIGURE 12. *Abbé Jean-Paul Bignon.*
Courtesy of the Bibliothèque nationale de France.

the public assemblies as well. In 1721, for example, he delivered two different papers at the same public assembly dealing with aspects of the transformation of iron into steel. In 1726 he similarly spoke publicly on methods of cooling molten iron.[78] He further employed his abundant rhetorical skills in these presentations in order to make this more utilitarian work appealing to the *mondain* public. For example, on one representative occasion, he connected a set of new discoveries in metallurgy to the artistic potential they held for the construction of decorative banisters and *hôtel* architecture.[79]

In sum, between 1710 and 1725, Réaumur established himself at the center of French public science by virtue of his unwavering ability to appeal to the elite audience that was a prime target of academic publicity. The *Mercure Galant* captured his public aura elegantly when it declared in 1711 that "M. Réaumur reads the book of nature as easily as his works are read."[80] His service to Bignon and the state through his administration of the academy's ongoing efforts to improve industry, technology, and the mechanical arts was also crucial to his success. From as early as 1713, Bignon and Réaumur also began managing the public assemblies, a shift that worked to complicate Fontenelle's control over the public discourse of the academy. The correspondence between Réaumur and Bignon from this period indicates that they were in charge of managing the content of the sessions.[81] The academician Saulmon's 1714 public assembly paper, for example, went through several rounds of revisions with both Bignon and Réaumur before it was finally delivered in November.[82] Other academicians were subject to similar discipline. A letter to Réaumur from 1717 reveals Bignon defining the content of the next assembly. Réaumur's work is to be included, the letter states, but Bignon worries that it will be too long and asks that he shorten it. The chemist Geoffroy's work, which is "very useful and easily comprehensible to the public," is also to be included. Nevertheless, the minister directs Réaumur to arrange for revisions because, he declares, "if he

78. PVARS, November 12, 1721; May 4, 1726.

79. This tendency is still evident in Réaumur's official academic volume describing the art of iron and steel making published in 1762, *L'art de convertir le fer forgé en acier et l'art d'adoucir le fer fondu ou de faire des ouvrages de fer fondu aussi finis que de fer forgé* (Paris, 1762). I am grateful to John Detloff for informing me of this fact.

80. *Mercure galant* (February 1711): 51.

81. Much of this correspondence is found in Ac. Sci., dossier Réaumur, but some is found in Ac. Sci., dossier Bignon, and in the Bignon papers in BN, MSS, Nouv. Acq. Fonds Fr. 22226–22236.

82. Ac. Sci., dossier Saulmon.

does not find a means to reduce [the discourse] by half, I fear yawns."[83] Even Fontenelle was subject to this editorial control. In a letter to Réaumur from 1719, Bignon expresses his frustration at Fontenelle's inclination to dominate the public sessions with his *éloges*. He then explores options for keeping the discourses of the academy secretary in check.[84]

The Bignon / Réaumur alliance had a profound effect on the character and content of the academy's work. While mathematical papers, especially those by Varignon and other analysts, were a fixture at the early public assemblies before 1709, they all but disappeared after 1713. Furthermore, from this date forward, the work being done in analytical mechanics ceased to be featured in the public discourse of the academy save in the fragments that Fontenelle could squeeze into his *histoires*. New institutional arrangements within the company played a key role in this transformation, but at the same time the declining fortune of analytical mechanics was also a product of simple fatigue. In the first years after 1699, the public controversy triggered by the new mathematics attracted interest to it, as did the wider vogue for Malebranchianism that gave the new science wider cultural meaning. This confluence provided all the inspiration that Fontenelle needed to develop a powerful discourse on its behalf. Yet while his discourse accomplished its goal, it did not offer a justification for ongoing interest in the analytical sciences once their cultural moment had passed. Consequently, once analytical mechanics became established and fashions turned elsewhere, Fontenelle's discourse in support of it became redundant. Similarly, since nothing that Varignon and his colleagues produced after 1705 rivaled the astonishing innovations of the first decade, Fontenelle was forced into a repetitive mode.

Audiences both inside and outside of the academy also grew weary of advanced mathematics as time went on. In particular, the constant intellectual contestation, which in the first decade of the century undoubtedly triggered interest in the new science of motion, became tiresome to most by 1715. None of the epistemological conundrums that the new mathematical mechanics posed were ever resolved in the eighteenth century, and accordingly controversy was always ready to erupt whenever infinitesimal analysis was used.[85] Furthermore,

---

83. L'abbé Jean-Paul Bignon to René Antoine de Réaumur, 1717, Ac. Sci., dossier Bignon.

84. L'abbé Jean-Paul Bignon to René Antoine de Réaumur, 1719, Ac. Sci., dossier Réaumur.

85. On the long history of epistemological uncertainty with respect to infinitesimal analysis, see Morris Kline, *Mathematics: The Loss of Certainty* (Oxford, 1980).

the French practitioners of mathematical analysis, who had been formed during the bitter calculus dispute of 1699–1706, acquired a tendency toward polemic and partisan disputation that marked them out unfavorably in relation to other savants in this period. Saurin's copious polemical energies were often absorbed elsewhere, especially in 1711 when the Café Gradot's famous "quarrel of the couplets" sent the academician to prison (temporarily) on charges of libel.[86] Yet he nevertheless remained poised to respond to any perceived mathematical provocation. Varignon, who was always a leading pugilist in any mathematical battle, also went to his grave in 1722 immersed in a final struggle over the validity of his mathematical work. Bellicose tendencies were anything but positive values within the *honnête* climate of science that still prevailed in France before 1730. As result, the intellectual battles associated with advanced mathematics ultimately marginalized this science in France as the peace of the *pax analytica* set in.

These and other trends thus worked to push analytical mechanics and its practitioners to the margins of the French public science after 1715. At the same time, an inclination toward less mathematical and more directly empirical and utilitarian mechanics began to move to the center in its place. This rise was accompanied by a parallel surge in interest in more clearly practical and entertaining topics in natural history, botany, chemistry, anatomy, and astronomy. Similarly, academicians oriented in this new direction began to rise in the academy, while those with a more analytical or purely mathematical bent began to stagnate or decline in importance. Réaumur's early career exemplified this shift, and a set of new royal initiatives launched after the death of Louis XIV in 1715 further reveals the trajectory of this reorientation.

Soon after assuming control of the French monarchy, the regent Duc d'Orléans initiated a complete overhaul of the academy. The old membership structure was abolished, and in its place a new promotion ladder consisting of *adjoint, associé,* and *pensionnaire* was established.[87] This reorganization in and of itself did not change the power dynamics of the institution; however, imbedded in it was a new marginalization for certain academic mathematicians. The title of the analyst Bomie was changed from astronomer to mathematician to "better reflect the actual character of his work." Together with his analyti-

86. On Saurin, see Fontenelle, "Éloge de M. Saurin," in *Oeuvres de Fontenelle,* 7: 271–284; and Bernard de Casaban, "Joseph Saurin, member de l'Académie royale des sciences de Paris (1655-1737)," *Mémoires de l'Académie de Vaucluse,* 6th series, 1, no. 2 (1968): 187–310. The *"pieces justicatves"* pertaining to the "affair of the couplets" are found in A.N., X2B925–927.

87. Aucoc, *Lois statuts, et règlements,* XCIII–XCV.

cal colleagues Bragelogne and Nicole, he was also moved to a new and hybrid class within the academy, the *adjoint surnumeraires*. This new title brought the right of membership in the company, but it carried no pension and would not continue once the original holders resigned from the academy or died. It was thus a temporary rank that permitted a gradual downsizing of the membership. The presence of three analysts in this group, therefore, indicates their declining fortunes in the institution. Indeed, from this date forward Bomie and Bragelogne all but disappear from the academy records, and neither received an *éloge* from Fontenelle. The two new *adjoint géomètre* seats created in 1716, by contrast, went to Parent, a long-time participant in the Crown's mechanical arts initiatives, and Couplet, the treasurer of the academy who had ties to the Royal Observatory and was famous for his work on the hydraulics at Versailles.[88]

There are good reasons to believe that this was precisely as Bignon wanted it. Varignon believed that the academy's chief minister was not sympathetic to the new work in analytical mechanics, but the minister actually had no problem with it so long as it was pursued with *honnêteté* and did not produce squabbles that disrupted the work of the academy as a whole. As he wrote in 1713: "The partisans of *évidence* cease to be touched by it as soon as one disagrees with the views they are advancing. . . . This doesn't surprise me, for pride is of all the passions the last to be extinguished, especially among *gens des lettres.* However, what honor for the sciences if those who cultivated them also cultivated that spirit of gentleness [*douceur*] and modesty that is so rare today."[89] In addition to an honorable community of gentlemen, Bignon wanted to make the academy a center of socially useful and productive knowledge while using the institution's public organs to win wide approval for this administrative orientation. Varignon and Fontenelle had not necessarily positioned the academy in opposition to this program before 1709, but it was clear that Réaumur's conception of academic science was far more appealing to Bignon than that of the analysts. Réaumur's work, for example, was directly and obviously utilitarian while Fontenelle needed to perform rhetorical gymnastics to convince audiences that analytical mechanics was something other than an abstract *jeux d'esprit.* Furthermore, while analytical mechanics could be made palatable and appealing to the wider public when Fontenelle massaged it into his mellifluous public discourse, Réaumur's work did not need this kind of translation. Indeed, given the combination of serious science and popular appeal that he naturally affected, his work offered its own, direct appeals.

88. PVARS, January 8, 1716. On Couplet, see Sturdy, *Science and Social Status,* 133–137.

89. L'abbé Jean-Paul Bignon to Jean-Pierre de Crousaz, 12 August 1713, Fonds Jean-Pierre de Crousaz , 6: 131.

Given his larger priorities, therefore, Bignon likely encouraged the transfer of power away from the analysts and toward Réaumur in this period even if he did not accomplish this task in anything like an authoritarian or self-consciously conspiratorial way. An orientation toward direct, utilitarian knowledge had in fact characterized Bignon's ambitions for the Royal Academy from the outset, and Réaumur simply emerged after 1710 as the ideal fit for a position created unwittingly for him by the minister's work over the previous two decades. Furthermore, the ascent of Réaumur resulted in anything but a decline in Fontenelle's status or influence within the company. For one, he and Réaumur largely saw eye to eye on most intellectual matters, especially those related to science. The secretary also found his own intellectual proclivities mirrored in the young academician's combination of serious science and rhetorical *bel esprit*. For these and other reasons, then, the two academic colleagues quickly became intimate friends rather than rivals.

Fontenelle certainly had a deeper appreciation for the value and importance of advanced mathematics than Réaumur, but this was never a source of tension between them. Quite the contrary, the secretary expressed nothing but admiration for his younger colleague's work. Consequently, as the overall intellectual climate in the academy gradually began to shift toward the priorities of Réaumur and Bignon, the secretary simply found himself pulled into a new but equally comfortable position. He had always been a strong and sincere advocate for Bignon's agendas, and after 1715 he was simply called upon to play this role with increasing frequency while his parallel role as an advocate for advanced mathematics waned. Furthermore, since this transformation occurred very gradually and imperceptibly without any partisan rancor whatsoever, it transpired in a way that permitted Fontenelle to remain at the center of French public science despite the changes within the scientific culture in which he worked.

A set of larger pressures also pushed public science in these new directions, and these further worked to change the relationship that Bignon and Fontenelle had forged in the early 1690s. Unlike the academy secretary, the abbé answered primarily to the ministries of the French monarchy, and the relationship between these agencies and the Royal Academy underwent a reinvigoration in the wake of the Sun King's death. Scholars often overstate the significance of the Regency period in France (1715–1723), making it the rupture from which all future trends were born. In fact, there were as many if not more continuities throughout this period even if the departure of Louis XIV did catalyze certain trends while thwarting others.

Ministerially, the fate of the Phélypeaux clan illustrates well the complex relationship between continuity and change characteristic of the Regency period.

The career of Jerôme Phélypeaux de Pontchartrain, the royal minister above Bignon who supervised the academy's affairs after 1691 and the minister who had initiated the key reforms of 1699, came to an abrupt end in 1715 when the new regent the Duc d'Orléans removed him from his ministerial posts in disgrace. Pontchartrain's teenage son, however, the future Comte de Maurepas, quickly received many of his portfolios, including the crucial scientific ones at the Navy and the Royal Household. He would hold these posts until his own disgrace in 1749. In this way, the Regency period in no way disrupted the overall program of neo-Colbertian administrative rationalization that the Phélypeaux had begun in 1691 and applied at the Royal Academy in its public reforms after 1699. Similarly, after an early experiment with government by polysynodic councils, the regent returned in 1718 to the familiar Bourbon pattern of centralized government through ministerial bureaucracy. Thus, when young King Louis XV reached his maturity in 1723 and appointed Fleury to be his chief minister, few noticeable differences existed between his monarchy and the one which his grandfather had developed a half-century earlier.[90]

From the perspective of the Royal Academy of Sciences, this continuity meant an ever more systematic professionalization of its operations rooted in an ever growing and intensifying relationship between it and the administrative monarchy. Indeed, rather than the death of the Sun King in 1715, it was the termination of the Crown's war making in 1713 that was the real catalyst for change. Freed of the political, fiscal, and cultural constraints imposed by the almost constant wars of Louis XIV, the French Crown was able to use the "Long Peace" that stretched from the Treaty of Utrecht in 1713 to the outbreak of the War of Austrian Succession in 1741 to solidify more strongly than before the administrative initiatives begun in the previous century. For the academy, this meant an ever-growing dominance of its administrative identity after 1713.[91] An earlier courtly orientation that conceived of academic service in terms of

90. On French political history in the early eighteenth century, see Michel Antoine, *Le coeur de l'état: Surintendance, contrôle général et intendances des finances, 1552–1791* (Paris, 2003); Peter R. Campbell, *Power and Politics in Old Regime France, 1720–1745* (London and New York, 1996); James Collins, *The State in Early Modern France* (Cambridge, 1995); Emmanuel Le Roy Ladurie, *The Ancien Régime: A History of France 1610–1774*, trans. by Mark Greengrass (Oxford, 1996); and David Kammerling Smith, "'*Au Bien du Commerce*': Economic Discourse and Visions of Society in France" (Ph.D. diss., University of Pennsylvania, 1995).

91. The increasing focus on utilitarian, state-sponsored work within the Paris Academy is a central theme of Hahn's *Anatomy of a Scientific Institution.* See esp. chap. 5, "The Changing Demands of Society."

service to the sovereign prince was not eliminated by this new emphasis upon royal administration; rather, it was transformed by administrative public- ity into the *honnête* conception of the institution as an honorable company of learned men beholden to the ethics and values of the wider Republic of Letters. Here one sees vividly how the Republic of Letters in the decades around 1700 served as the institutional bridge between the courtly publicity of the classi- cal age and the more inclusive and demotic publicity of the eighteenth-century Enlightenment.[92]

Administrative monarchy encouraged rather than retarded this transforma- tion. Royal administrators very often valued a more talent-based conception of rank and status while bristling at the older, dynastic conceptions of the same thing defended by traditional elites. Many also saw the reformed Academy of Sciences as both a model and a catalyst for effecting a wider transformation of French society and politics through the application of these values. Bignon spoke for many, both inside and outside the monarchy, when he wrote in 1717 that "when science and *politesse* walk together, they make a forceful impression on people who conduct themselves according to the light of reason."[93] To wed science with manners, therefore, was to make a new political culture, and in its broadest sense administrative publicity at the royal academies served these larger political agendas. Moreover, because this program had broad appeal, royal administrators and Republicans of Letters more often became partners than rivals in France as a result of these initiatives. Knowledge (especially sci- entific knowledge), publicity, and vigorous communication were highly valued by each constituency, and for this reason the initiatives of the French adminis- trative monarchy usually harmonized with the aspirations of the international learned community. This was especially true in France, where the boundary separating savants from royal officeholders was often impossible to draw and included a revolving door linking the Republic of Letters with the French state. Fontenelle, for example, served for a brief period as a writer of royal propa- ganda for the regent Philippe d'Orléans, and other academicians also held rel- evant positions within the French royal administration.[94]

The signs of a reinvigorated royal administration exerting an increasing in- fluence on the academy after 1713 are thus many, even if the continuity between

92. On the courtly orientation of the early Royal Academy, see Shank, "Before Vol- taire," chaps. 2–3.

93. L'abbé Jean-Paul Bignon to Jean-Pierre de Crousaz, 8 August 1717, Fonds Jean- Pierre de Crousaz , 6: 167.

94. On Fontenelle's royal service, see Niderst, *Fontenelle*, 267.

these and earlier efforts mark them as anything but innovations. The Royal Observatory, long a site of lavish state support, was especially revitalized by these efforts even if it had never ceased to be a vital center of official French science after its founding in 1671. In 1700, Fontenelle announced the resumption of the Crown's map-making efforts that had been shelved after the death of Colbert in 1683 and by the financial burdens of the War of the League of Augsburg (1688–1697).[95] Geodesy was an astronomical science rooted in triangulations based on stellar observations. This geodesic revival therefore sent Cassini I and his young son, along with an arsenal of complex instruments, on a set of surveying operations designed to produce more accurate geodesic measurements of the French kingdom. The War of Spanish Succession (1702–1713) again derailed the effort, but it was revived again in 1718, this time under the direction of Cassini II. Overall, these surveying operations were heavily funded and highly visible, and they further encouraged the shift within the academy's culture away from the more abstract work of the analytical mathematicians and toward the more empirical astronomers and *mécaniciens*.[96]

The overall connection, central to Phélypeaux-style royal administration, between the Crown, the Navy, and the maritime/colonial commerce of the kingdom also continued to shape the culture of the Royal Academy after 1715. In 1701, Pontchartrain announced the new state-sponsored academic initiative to collect and disseminate more accurate tide data, and in the same year he also charged the academy with the development of better gunpowder technology.[97] These and other similar efforts to use the academy for utilitarian state work would continue under Maurepas as well. In 1710 the Phélypeaux further centralized the administration of colonial and maritime policy into a new Bureau of Maritime Affairs. Navigation science, including the astronomy that was central to it, were major concerns for this new bureau, as were ship technology, precision instrumentation, cartography, and the acquisition and circulation of all manner of maritime knowledge.[98] The "Long Peace" that began in 1713 per-

---

95. PVARS, July 28, 1701. The announcement was made at the public assembly as well. See PVARS, November 12, 1701; *Mercure galant* (November 1701): 173–174.

96. See Josef Konvitz, *Cartography in France, 1660–1848: Science, Engineering and Statecraft* (Chicago, 1987); Monique Pelletier, *Les cartes de Cassini: La science au service de l'état et des régions* (Paris, 2000).

97. *HARS-Hist.* (1701): 11–13.

98. See Kenneth Banks, *Chasing Empire across the Sea: Communications and the State in the French Atlantic, 1713–1763* (Montreal, 2002); Dale Miquelon, "Envisioning the French Empire: Utrecht, 1711–1713," *French Historical Studies* 24 (2001): 653–677.

mitted the conversion of these administrative concerns into new imperatives governing the scientific work of the Royal Academy. Similarly, new efforts to promote maritime exploration, expansion, and commerce in this period sent many savants overseas on expeditions. These voyages further linked the state, both formally and informally, with the scientific work of the Royal Academy.[99]

The Regency period only intensified these trends. A year after assuming the throne, the regent Duc d'Orléans announced that he would award a prize of one hundred thousand livres (a staggering sum at the time) to the first person capable of demonstrating a reliable method for determining longitude at sea. Academicians were eligible to compete for the prize, but the institution more importantly became the official judge of any proposed solution. In this way, the new prize both clarified and reasserted the academy's status as the official scientific tribunal for the Crown.[100] A private gift to the academy made in 1714 with the explicit aim of "advancing the knowledge of those sciences most useful to the public" also reinforced this identity. It endowed a fund that allowed for a set of annual prizes to be awarded by the institution.[101] In these contests, academicians would be prohibited from competing, but they would select the problems to be explored and act as the ultimate judge in the contests. Reflecting the nature of the academic culture at the time, the academy also decided to create two prizes with the money, one awarded every two years for work on more theoretically conceived questions, and another awarded in the alternative years for winning solutions to more practical and utilitarian problems. In each case, the establishment of these prizes translated into a renewed importance for the French Royal Academy as a site where publicly useful knowledge was collected and adjudicated according to the interests of the state.[102]

The Crown could also advance its interests with respect to the academy through appointment and promotion decisions, and here Bignon's connections with Réaumur, along with their bias toward more directly utilitarian mechanics, appear to have been very influential. The academy had the right to nominate replacements for academic seats in order of preference, and the Crown almost always acceded to the academy's wishes even if it was in no way constrained to do so. This meant that the nomination process and the voting within the acad-

99. See François Regourd and James McClellan III, "French Science and Colonization in the Ancien Régime: The 'Machine coloniale,'" *Osiris* 15 (2000): 31–50.

100. PVARS, March 21, 1716. See also Ernest Maindron, *Les fondations de prix à l'Académie des Sciences* (Paris, 1881), 23.

101. PVARS, December 24, 1718. See also Maindron, *Les fondations de prix*, 3–16.

102. A list of the prize questions and winners is found in Maindron, *Les fondations de prix*, 17–22.

emy were crucial in the admission and promotion of new members. Bignon and Réaumur appear to have used their influence to shape outcomes that served their interests and those of the state. With the notable exception of Maupertuis, who presents a special case, only one person devoted to the analytical approach to mechanics was appointed to the academy between 1715 and 1731. However, even this figure, Pierre Rémond de Montmort, proved insignificant since he was appointed very early in the century and died in 1719, less than two years after joining the institution.[103] Similarly, only one member of the original analyst community, Nicole, was promoted during this period.[104] By contrast, those with a more utilitarian and empirical bent were advanced, and this indicates a new potency in the bias toward direct utility and practicality in the academy that was not present in the decades after 1690.

In many cases these same shifts were also encouraged by the academicians themselves. Writing in his 1720 *mémoire* on "The Utility of the Academy for the Kingdom," Réaumur argued for even closer ties between the academy and the ministries of the administrative monarchy. He especially encouraged the placement of scientific academicians directly in the state bureaucracies and royal councils.[105] The abbé Saint-Pierre, Varignon's one-time financial benefactor and his Parisian roommate during the 1680s, also articulated many of the same goals. The abbé was never a member of the Royal Academy of Sciences, yet he circulated actively in the cafés, salons, and public lecture courses that also defined Parisian scientific sociability during these years. Réaumur described him as "one of my friends, and a man of the world most interested in the public good."[106] He was also recognized as a serious savant by others as well, a reputation that earned him admission into France's older and more literary royal academy, the Académie française. As a young man, Saint-Pierre had actively pursued scientific study, and his views about the proper relationship between science and society in Old Regime France were representative of the intellectual views of many in France, including those who inhabited the Royal Academy and the *bureaux* of the administrative monarchy that supported it.[107]

103. See Fontenelle, "Éloge de M. de Montmort," *Oeuvres de Fontenelle*, 6: 465–479.

104. He was named an *associé mécanicien* in 1718 and a *pensionnaire mécanicien* in 1724. Ac. Sci., dossier Nicole.

105. Ac. Sci., dossier Réaumur. See also Hahn, *Anatomy*, 68; and Maindron, *Les fondations de prix*, 103–110.

106. René Antoine de Réaumur to Jean-Pierre de Crousaz, 26 February 1729, in Fonds Jean-Pierre de Crousaz, 14: 121.

107. On Saint-Pierre, see Joseph Drouet, *L'abbé de Saint-Pierre: L'homme et l'oeuvre* (Paris, 1912); Merle L. Perkins, *The Moral and Political Philosophy of the Abbe de Saint-*

At the center of Saint-Pierre's vision was a renewed importance for scientific knowledge in the governance of France. Equally important was the reorganization of the French state in ways that would channel scientific knowledge toward the improved governance of the kingdom. The academy system constituted for him the most effective vehicle for realizing these goals, and thus in his many speculative essays written in the early years of the eighteenth century, as well as in his notorious *Discours sur la polysynodie* first published to scandal in 1719, the abbé laid out a program for reforming France by reforming its royal academies.[108]

At the center of his system was a master academy that Saint-Pierre called the *Académie politique*. Akin to a central ministry of information and culture, this academy would coordinate the affairs of all the other academic institutions in the kingdom, insuring that their work served the interests of the public good and thus the state. A set of specialized, individual academies would also be created for each discipline, including a new institution devoted exclusively to the applied arts and industry. Academies of medicine, naval and military science, and engineering would also be founded to go along with the existing academy of sciences.[109] Utility was to be the watchword in all of these institutions. Even Saint-Pierre's reformed academy of literature and belles lettres would focus its energies exclusively on utilitarian pursuits since he imagined only two projects for its members: the production of morally uplifting plays and historical narratives or the pursuit of antiquarian scholarship that produced reliable historical knowledge.[110]

Pierre (Geneva, 1959); Nannerl O. Keohane, *Philosophy and the State in France: The Renaissance to the Enlightenment* (Princeton, 1980), 361–376; Thomas E. Kaiser, "The Abbé de Saint-Pierre, Public Opinion, and the Reconstitution of the French Monarchy," *Journal of Modern History* 55, no. 4 (December, 1983): 618–643; and J. B. Shank, "The Abbé de Saint-Pierre and the Emergence of the 'Quantifying Spirit' in French Enlightenment Thought," in *A Vast and Useful Art: The Gustave Gimon Collection on French Political Economy,* ed. Mary Jane Parrine (Stanford, 2004), 29–47.

108. The most complete collection of Saint-Pierre's published writings is Saint-Pierre, *Ouvrajes politiques de Mr. L'Abbé de St. Pierre, Charles Irenée Castel,* 10 vols. (Rotterdam, 1734). My citations will be from this edition.

109. The creation of a new *Académie politique* is a component of a number of the abbé's concrete political proposals. Perhaps the most comprehensive treatment, however, is found in "Sur le Ministère de l'Intérieur de l'État," in Saint-Pierre, *Ouvrajes,* 7: 1–279.

110. See "Projet pour rendre l'Académie des bons Écrivains plus utile," in Saint-Pierre, *Ouvrajes,* 4: 165–195.

Overall, there was little room for creative speculation or idle innovation in Saint-Pierre's highly centralized scheme of state-sponsored learning. Furthermore, once this useful knowledge was produced, publicity and communication would play an essential role in disseminating it to the public and throughout the state. The abbé imagined even more public outreach for his academies than had been created in 1699, including the establishment of newly created public lecture courses supervised by the academies, and a new set of ever more specialized journals, gazettes, and periodicals.[111] He also imagined the creation of formal mechanisms that would tie scientific savants to the decision makers inside the French state. In one representative example, he imagined an academy of population studies that would join together mathematicians and other scientific practitioners with royal *intendants* and others in charge of fiscal and administrative governance.[112] In this way, Saint-Pierre's utopian vision of a reformed monarchy using publicly-oriented academies to produce and disseminate useful knowledge for the greater good of France captured in ideal form the motivations informing the administrative approach to academic science in this period. Indeed, his work can be read as the admittedly utopian mission statement of royal administrators like the Phélypeaux and their intellectual allies such as Bignon and Réaumur.

Saint-Pierre found Louis XIV's monarchy deficient on these counts, especially in its preference for traditional dynastic warfare and Baroque Catholicism more than these innovating tendencies. Yet when Saint-Pierre publicly declared as much in print in his *Discours sur la polysynodie,* he crossed a line that could not be crossed in absolutist France. Many in the Regency government were sympathetic to Saint-Pierre's ideas, including the regent himself, who was perceived by many to be a knowledgeable partisan of modern, scientific thinking.[113] As the ruler of France, however, the regent could not tolerate open criticism of the Sun King's reign, and thus he sanctioned the abbé harshly for his transgressions. Saint-Pierre was removed from the Académie française, an unprecedented act at the time, and one that would not again be repeated.

---

111. "Projet pour des conferences de Fizique," in Saint-Pierre, *Ouvrajes,* 5: 317–344. For discussion of how these institutions were related to political publicity and transparency, see also "Conferences politiques tres avantageuses aux particuliers et au service du roi et de l'état," in Saint-Pierre, *Ouvrajes,* 4: 88–101.

112. Saint-Pierre, *Ouvrajes,* 4: 255–267.

113. Bignon praised the "good fortune of having a prince who takes an interest in the sciences and knows the value of a company of savants." Jean-Paul Bignon to Jean-Pierre de Crousaz, 20 November, 1714, Fonds Jean-Pierre de Crousaz , 6: 159.

His case also generated a scandal that rocked the Republic of Letters even if the savant ultimately emerged with his reputation intact.[114] Less than the scandal, however, it was the influence of Saint-Pierre's utopian vision of reformed monarchy that was more important in this context. His writings, as well as the widespread discussion of them, added significantly to the convergence that occurred after 1715 pushing for a reinvigorated alliance between academic science and royal administration in the improvement of the French commonweal.

This alignment also reinforced the gradual marginalization of abstract mathematical mechanics within the life and work of the Royal Academy of Sciences. With more and more of the academy's time being taken up with state-oriented utilitarian work of this nature, analytical mathematicians like Varignon and Nicole found it increasingly difficult to find support and justification for their abstract mathematical work. In this way as well, the analytical program in mechanics was forced into retreat. But this is not to say that this science died altogether. Varignon remained an active practitioner until 1722, and Nicole carried the mantle after him until he was joined by the next generation of analysts in the 1730s. Saurin likewise remained a source of support for these efforts until his retirement in 1731, even if his interests were increasingly directed elsewhere. Fontenelle too offered consistent advocacy even if the secretary increasingly found himself called upon to use his voice for other causes after 1715. Accordingly the centrality of this science and the constituency that pursued it was displaced within the academy while those oriented toward Bignon, Réaumur, and their agendas grew in its place.

French Academic Science after 1715 Personified: Jean-Jacques Dortous de Mairan

More than anything else, the emergence of Jean-Jacques Dortous de Mairan at the center of French academic science after 1715 illustrates well the character that French academic science had acquired on the eve of the eruption of the Newton wars after 1728. Since Dortous de Mairan was also a crucial player

---

114. Saint-Pierre, *Discours sur la polysynodie: où l'on démontre que la polysynodie, ou pluralité des conseils, est la forme de ministère plus avantageuse pour un roi, & pour son royaume* (Amsterdam, 1719). The details of the scandal are offered in Drouet, *L'abbé de Saint-Pierre*. For a sampling of the reaction within the Republic of Letters, see *L'Europe savante* 6 (September 1719): 43–88; *Nouvelles littéraires contenant ce qui se passé de plus considerable dans la République des lettres*, 11 vols. (The Hague, 1715–1720): 8: 408–416.

FIGURE 13. *Jean-Jacques Dortous de Mairan.*
Courtesy of the Bibliothèque nationale de France.

in a number of important developments, it is worth pausing in this context to consider his ascent to prominence in some detail.

Born to a minor noble family in Béziers, Dortous de Mairan went to Paris in 1698 at the age of twenty to further his education.[115] Once in the capital, he followed the intellectual path of so many by making the acquaintance of Malebranche who, in Dortous de Mairan's words, "was so kind as to explain to me M. de l'Hôpital's [treatise on infinitesimal analysis] and to give [me] some further instruction in mathematics and in physics."[116] These encounters launched Dortous de Mairan on his scientific career. Although he returned to Béziers in 1704 to begin a career in the clergy, he also began participating in the scientific activities of the Bordeaux Academy of Sciences. The significance of Dortous de Mairan's Malebranchian training was revealed in his first scientific paper, an analysis of the so-called "roue d'Aristote." This classical problem involves accounting for the difference in length between two curves traced in equal time by discrete concentric circles that revolve on the same uniformly rotating turntable.[117] The problem was tailor-made for the new infinitesimal calculus, and Dortous de Mairan applied this mathematics in offering his solution. Satisfied by his effort, he also sent the solution to the Paris Academy, which recorded the receipt of a solution from "M. Ortous de Meyran de Béziers" on June 19, 1715.[118] Following protocol, they then appointed a committee to study it. The report was favorable, and given the topic and the author's method of approach, Fontenelle presented the solution to the public in his *histoire* for 1715 even though "Ortous de Meyran" was not yet a member of the company.

As always, the secretary used the opportunity to further his larger goals. "Physics," he began, "which contains many unknown principles and many others that are known but still very difficult to apply, offers us a great many obscurities and mysteries. Geometry, however, has only clear principles and it should therefore never present any."[119] Fontenelle then explained "M. d'Ortous

115. The most extensive scholarly treatment of Dortous de Mairan's life and work is found in Abby Rose Kleinbaum, "Jean Jacques Dortous de Mairan (1678–1771): A Study of an Enlightenment Scientist" (Ph.D. diss., Columbia University, 1970). See also "Éloge de M. de Mairan," *HARS-Hist.* (1771): 89–104.

116. Dortous de Mairan to Malebranche, September 17, 1713, in *Malebranche's First and Last Critics: Simon Foucher and Dortous de Mairan*, ed. and trans. by Richard A. Watson and Marjorie Grene (Carbondale and Edwardsville, 1995), 68.

117. For a discussion of the problem and Dortous de Mairan's solution, see Kleinbaum, "Jean Jacques Dortous de Mairan," 9–10.

118. PVARS, June 19, 1715.

119. *HARS-Hist.* (1715): 30.

de Meyran's" solution, drawing connections between this mathematical paper and the mechanics of moving bodies as well. Declaring it a definitive answer, Fontenelle then drew the necessary conclusion. "Voilà, a false marvel absolutely exposed. M. d'Ortous de Meyran has successfully revealed a truth that was not only hidden but also contradicted by powerful prejudices."[120]

The secretary clearly recognized in "Meyran" a kindred spirit, but unfortunately for the legacy of Malebranchian mathematicism in France, Dortous de Mairan's early analytical orientation, like that of Réaumur, was destined to be short-lived. The disintegration was already underway by 1715. Between 1712 and 1713, Dortous de Mairan studied philosophy and theology, and sometime during that year he came across a copy of Spinoza's *Ethics*. The work changed his life. He knew that the *Ethics* was a dangerous work and that its conclusions led directly to atheism and other morally repugnant conclusions. However, as he read the text he could not find any error in its demonstrations. In his mind, Spinoza presented an ironclad geometric demonstration, and this sent the young savant into despair.

Looking for guidance, he decided to seek the advice of Malebranche. Describing his dilemma, he explained in a letter dated September 17, 1713, that "the character of this author, so different from all that I had previously seen—the abstract, concise, and geometrical form of his work, the rigor of his arguments—seemed to me worthy of attention. So I read him attentively, and he impressed me. I have since reread him, and I have meditated [about him] in solitude, and in what you call the silence of the passions; but the more I read him, the more I find him sound and full of good sense. In a word, I do not know where to break the chain of demonstrations. . . . I wanted to forget him. But when one is deeply moved by the desire to know the truth, can one forget what has seemed evident? On the one hand, I cannot envisage without compassion for humanity, and without sorrow, the consequences that follow from his principles; on the other, I cannot resist his demonstrations. It is, Reverend Father, in order to extricate myself from so distressing a state that I have the honor to write to you. Be so kind as to develop for me the paralogisms of this author, or what would suffice to indicate to me the first step that has led him to the precipice. . . . Point it out to me concisely, I beg you, and in the manner of the geometers. It is the method that he has adopted and the least apt for concealing error; let us attack him at his strong point, and with his own weapons."[121]

120. Ibid., 35.

121. Dortous de Mairan to Malebranche, September 1713, in *Malebranche's First and Last Critics*, 68–69.

Dortous de Mairan's missive reached Malebranche in the country, and at first the philosopher did not seem to appreciate the depth of its author's study. He claimed not to own Spinoza's work, and only to have read parts of it before becoming "disgusted" with its "horrifying consequences" and its "alleged demonstrations." "Take the trouble, sir, to read the definitions etc. that he cites in his demonstrations, and you will discover, if I am not mistaken, the equivocation that invalidates his proof."[122] This response frustrated Dortous de Mairan. Ignoring the reverend father's intimations of illness, he penned a lengthy response a month later in which he laid out the problem in greater detail.

In this second letter to Malebranche, Dortous de Mairan revealed his full philosophical arsenal. He targeted in particular Malebranche's claim that Spinoza blurred the distinction between intelligible creation and actual or material creation. Where Malebranche saw a clear contradiction in Spinoza's position, Dortous de Mairan saw a clarity not found in other accounts of the same problem. As he explained, "it is true that, once one has reduced *created or material extension* and *intelligible extension* to these clear and exact notions of substance or mode, . . . one sees arising from this several consequences embarrassing for the theological system. For either that extension is in God and constitutes his essence as an attribute, or it is not in God." Drawing out the implications of this choice, he continued: "If it is in God, then the whole universe and all bodies are only affections of one of the divine attributes, or are only God modified in such and such a manner insofar as he is extended—which is the pure doctrine of [Spinoza]. Or, if extension does not belong to God, then there is something which is not God and does not constitute his essence which exists necessarily, which is infinite, eternal, indivisible, etc." Neither choice was acceptable, yet the rigor of Spinoza's demonstration seemed to demand that one or the other be admitted. Moreover, Dortous de Marain alleged that Malebranche's distinction between *created extension* and *intelligible extension* offered little more than a specious escape from this dilemma, one that has been "imagined only to cover these difficulties and to explain, thanks to the equivocation it contains, things in fact inexplicable by the ordinary system."[123]

Two equally provocative charges are contained in Dortous de Mairan's assertion. First, he claims that rigorous geometrical reasoning supports only two possible conclusions given the Cartesian definition of matter as mere extension. Either this matter is God, and Spinoza's pantheism is thus justified, or this matter is something completely alien from God and the world is nothing more than a blind, purposeless machine. Second, Dortous de Mairan also con-

122. Malebranche to Dortous de Mairan, October 1713, in ibid., 69.
123. Dortous de Mairan to Malebranche, November 1713, in ibid., 75.

tends that Malebranche's solution to this dilemma—his distinction between intelligible and material extension, which is the linchpin of his entire system of metaphysics—is less geometrically rigorous than Spinoza's formulation. Thus, those who feel compelled to let the "cold rigor of geometry" lead them to the truth are lead to Spinoza's conclusion and not Malebranche's.

Capturing this dilemma in an analogy relevant to the wider discussion here, Dortous de Mairan likened the predicament posed by a careful reading of Spinoza to the one confronted by the early students of infinitesimal analysis. "I admit," he wrote, "that . . . in general [Spinoza's] system is difficult to understand if one is not suited to his method, that the method is rigid and abstract, that it demands a great deal of attention, and much practice in accuracy. But what matter, provided that one can fully succeed in understanding it? The subject is worth the trouble. The same difficulties, though in a different subject matter, were found in the new geometry of the infinite and in the system of differential calculus. The famous book of M. de l'Hôpital, which contains its principles and analysis, was at first understood in France only by a very small number of mathematicians; several wrote against it, or neglected to study it because of the alleged absurdities it seemed to them to contain, and you know, Reverend Father, that not long ago one of the members of the Academy of Sciences, though a great geometer and arithmetist, was still unable to bring himself to subscribe to it, and will perhaps never do so. However, the difficulties die away and the system remains victorious. I will not believe it impossible for the same thing to happen to that of our author, until I have been shown quite precisely the paralogism of his reasoning."[124]

Malebranche's subsequent replies frustrated Dortous de Mairan, and the correspondence only reached a resolution when Malebranche died in October 1715. Yet viewed from the perspective of Dortous de Mairan's subsequent scientific work, two aspects of this philosophical exchange are worth emphasizing. First, the future academician reveals in these letters his deep commitment to the Cartesian epistemology of clear and distinct reasoning and *évident* demonstration. This tendency would become characteristic of his scientific work throughout his career. More important, he also reveals his precise orientation within this otherwise pervasive French philosophical outlook. Dortous de Mairan was not the only savant to challenge Malebranche on the precise relationship between intelligible and material extension. Leibniz's attacks on occasionalism pursued a similar line of reasoning, and the German's own complex balance between mathematical phenomenalism and the ontology of active forces is in certain respects a different solution to the same Spinozist

124. Ibid., 73. The "great geometer and arithmetist" is Michel Rolle.

problem.[125] Similarly, the French Jesuits also criticized Malebranche on these grounds, and their general hostility toward abstract, geometric philosophy was largely driven by their belief that this was at the root of both Malebranche's and Spinoza's errors. Finally, the Jansenist philosopher Arnauld, one of the authors of the influential Cartesian treatise *The Port-Royal Logic*, likewise engaged Malebranche on precisely this point. His solution was a more pragmatic and empirical approach to the Cartesian definition of matter that drew sharp lines between metaphysics, theology, and physics.[126]

A variation on Arnauld's approach was further pursued by "empirical Cartesians" such as Fontenelle and Pierre-Sylvain Régis who had no patience whatsoever for metaphysico-theological subtleties.[127] Whatever rigorous, geometric reasoning demanded was acceptable to them so long as it agreed with sensate experience as well. Dortous de Mairan's philosophy ultimately adhered most closely to this latter stance. It is for this reason that his exchange with Malebranche was important in shaping his scientific career. For one, his despair with Malebranche's metaphysico-theology may have triggered his departure from the clergy, a decision that he made soon after his exchange with the reverend father. Whatever its influence on this choice, his dialogue with Malebranche certainly helped Dortous de Mairan to crystallize his own philosophical orientation and to launch him on his particular scientific trajectory. After 1690, the wide influence of Malebranche's mathematical Cartesianism combined with his deep phenomenalism about causality helped to inspire the development of analytical mechanics in France. By contrast, Dortous de Mairan's frustration with these same mathematical and metaphysical abstractions oriented him toward the more empirical, causal, and mechanistic approach to science that was increasingly becoming ascendant in France.

"Mairan," writes Abby Rose Kleinbaum, "was incapable of accepting . . . abstraction as a method in science; he saw it as a dangerous crack in the dam erected against the dark and murky occult qualities medieval philosophers had ascribed to nature. Calculation, Dortous de Mairan insisted, was no substi-

125. Leibniz's exchanges with Malebranche on these and other questions are chronicled conveniently in André Robinet, *Malebranche et Leibniz relations personelles* (Paris, 1955). See also idem, *Système et existence dans l'oeuvre de Malebranche* (Paris, 1965); and Robert Adams, *Leibniz: Determinist, Theist, Idealist* (Oxford, 1994).

126. For a good discussion of Arnauld's exchanges with Malebranche, see Steven J. Nadler, *Arnauld and the Cartesian Philosophy of Ideas* (Princeton, 1989).

127. I borrow the term "empirical Cartesians" from Roger Ariew, "Radical Cartesians: Empiricism, Metaphysics, and the Method of Doubt," unpublished manuscript.

tute for explanation."[128] In place of abstract mathematics, Dortous de Mairan believed science needed to offer clear and distinct mechanical explanations grounded in evident, geometric demonstrations. This made him among other things a strong advocate of Cartesian vortical mechanics throughout his life. As he summed up: "Whenever pure and inanimate geometry is found to be noticeably insufficient for the production of certain effects, it is absolutely essential to have recourse to these subtle and active fluids."[129]

His earliest work in physics applied this mechanistic approach to a number of problems, most importantly the formation of ice. The resulting papers earned him the Bordeaux Academy's prizes for the years 1716, 1717, and 1718, and this success emboldened him. Having surrendered his ecclesiastical positions once and for all and armed with an arsenal of "extremely good recommendations," he set off for Paris in 1718 in order to "look, listen, and learn—in a word, to make myself a bit more polished, and to familiarize myself with the affairs of the Republic of Letters."[130] His recommendations were clearly potent, for in only a few months he had made the acquaintance of Fontenelle, Bignon, and even the regent Duc d'Orléans. In December 1718 he was offered an *associé géomètre* position in the academy, significantly succeeding the recently deceased analyst Guisnée in this post. By entering the academy as an *associé*, Dortous de Mairan became one of only two academicians ever to enter the academy above the lowest rank. He attributed this exceptionalism to favor rather than talent.[131] He continued, moreover, to use his connections to good effect after 1719. In little more than a year he ascended to the rank of *pensionnaire,* and in January 1721, only two years after his *entrée* into the academy, he was made a director of the institution. He served in this capacity on a number of other occasions, and in 1728 the Crown offered him a suite of apartments in the Louvre, a rare honor for an academician. The former royal palace in Paris was home to all the royal academies after the reforms of 1699, and Réaumur also had apartments there. The two men accordingly became close associates in this period, forging a friendship that would remain strong throughout their lifetimes. Dortous de Mairan also received other appointments and honors, and his apotheosis was completed in 1740 when he was appointed to succeed Fontenelle as the perpetual secretary of the Royal Academy.

128. Kleinbaum, "Jean Jacques Dortous de Mairan," 38.

129. Ibid., 37.

130. Cited in Kleinbaum, "Jean Jacques Dortous de Mairan," 10.

131. He expressed this belief in a letter to Bouillet in March 1734. Cited in Kleinbaum, "Jean Jacques Dortous de Mairan," 11.

Dortous de Mairan's power in the academy after 1720 was thus great, and this alone placed him alongside Réaumur and Bignon. His scientific outlook, moreover, was also exceedingly compatible with theirs. Like Réaumur, Dortous de Mairan was a devoted empiricist who believed that sound science began with the collection and interpretation of empirical data. He was perhaps more attached to mechanistic system building than Réaumur, for as Kleinbaum states, "He could not single out any problem about the cosmos without presenting an entire cosmology."[132] Nevertheless, despite his particular inclination toward grand structures, Dortous de Mairan's science was compatible with Réaumur's, especially in its concrete and empirical approach to nature.

Dortous de Mairan's work on the aurora borealis, analyzed thoroughly by J. Morton Briggs, illustrates well these similarities.[133] Briggs calls work on the northern lights in this period "a Baconian orgy of detailed description of particular sightings."[134] Dortous de Mairan's contribution epitomized this approach. He made an extensive study of all references to the aurora both in recent times and throughout history, and found reports of the phenomena that dated from as early as 400 C.E. He also conducted observations of his own and corresponded with others asking that they send reported sightings to him. Once the data were amassed, Dortous de Mairan synthesized them and used the results to develop a mechanical explanation of the observed phenomenon.[135] Briggs stresses that Dortous de Mairan was in step with the other leading students of the aurora in this period, Edmund Halley and Euler, in privileging clear and distinct mechanical pictures over quantified mathematical analysis: "In all

132. Ibid., 1.

133. Dortous de Mairan's work on the aurora borealis culminated in a treatise, *Traité physique et l'historique de l'Aurore boréale,* that was given the exceptional honor of being published as an addendum to the academy's *mémoires* of 1731. Only Cassini II's *Traité sur la figure de la terre* (1720) and Fontenelle's *Eléments de la géometrie de l'infini* (1727) had been published in a similar fashion, and Dortous de Mairan's work received wide attention and acclaim as a result. Dortous de Mairan's *Dissertation sur la glace, ou explication physique de la formation de la glace et de ses divers phénomènes* was also published as an addendum to the academy volume of 1749. See J. Morton Briggs, "Aurora and Enlightenment," *Isis* 58 (1967): 491–503.

134. Briggs, "Aurora and Enlightenment," 501.

135. The cause of the aurora, Dortous de Mairan contended, was the "zodiacal light," a part of the solar atmosphere that forms a pyramid in the northern skies. In the second edition of his work, which appeared in 1754, Dortous de Mairan explained this effect according to the principle of attractive matter. In 1731, however, Dortous de Mairan offered a more mechanistic account, arguing that the solar atmosphere produces the zodiacal light by acting physically and mechanically upon our own atmosphere.

cases, mathematical analysis was missing."[136] Dortous de Mairan's work did employ mathematics and calculations quite often, but mathematics was in no way essential it. Rather, his science was rooted in empirical descriptions explained by causal mechanisms. Whatever mathematics was employed derived from these mechanistic models and served them as mere routines of calculation. Varignon's science, which conceptualized physical problems in terms of abstract mathematical categories, could not have been more different, and while Dortous de Mairan's approach to the mathematical sciences reflects his lifelong alienation from Malebranchian mathematicism, it also shows his persistent sympathy for the kind of empirical and applied mechanics supported by Bignon and Réaumur.

Dortous de Mairan's commitment to mechanical causality also brought him together with Réaumur in one other way as well. His nonmathematized approach to physics very often led him to the construction of elaborate explanatory pictures, which, as Kleinbaum notes, "sometimes contained analogies that seem more akin to poetry than geometry."[137] In his *Traité de glace*, Dortous de Mairan likened the subtle matter in the interior of a fluid to "the cool winds which penetrate the center of a forest."[138] In his other scientific work, he often used similar imagery. Rhetoric such as this was an essential requirement for academic and scientific success given the academy's attachments to Parisian public science after 1699, and while Dortous de Mairan never possessed the literary panache of a Réaumur or a Buffon, he was a far better stylist than most. He became a member of the Académie française in 1743, and as his eventual appointment to succeed Fontenelle as academy secretary suggests, these literary talents were as important as the content of his scientific works to the establishment of his intellectual reputation.

Unlike Réaumur, Dortous de Mairan also reinforced his public persona with an active presence in Parisian society. He joined Fontenelle, Pierre Carlet de Marivaux, and Baron de Montesquieu as one of the "Seven Sages" that frequented the salon of Mme. Tencin. He was also a frequent guest at the salons of Mme. Lambert and Mme. Dupin.[139] Comparing Dortous de Mairan to d'Alembert late in the life of both academicians, Marmontel described the former as having "a gentle and laughing disposition." "Age has done for him

136. Briggs, "Aurora and Enlightenment," 501.

137. Kleinbaum, "Jean Jacques Dortous de Mairan," 42.

138. Ibid.

139. See Pierre-Maurice Masson, *Madame de Tencin (1682–1749: Une vie de femme de XVIIIe siècle* (Paris, 1910), 194; and Roger Marchal, *Madame Lambert et son milieu* (Oxford, 1991), esp. 241–243.

what nature has done for d'Alembert," the writer explained. "It has tempered all the movements of his soul; and what he has left of heat . . . is composed, impartial, and wise, of an original turn, and a gentle and fine saltiness."[140] Sociable virtues such as these were more than mere advantages for an academician in the new public academy after 1699; they were potent requirements for asserting and establishing intellectual authority. Dortous de Mairan's career illustrated the centrality of these new cultural imperatives, for possessed of the right combination of intellect and sociability he moved quickly to the center of public and academic science after 1720.

Fontenelle was already at the center of French public science when Dortous de Mairan arrived. He had been there since the late 1680s, and Réaumur joined him there after 1715. Dortous de Mairan completed the trio when he ascended to academic prominence in the early 1720s. Once forged, moreover, the Fontenelle–Réaumur–Dortous de Mairan triad quickly became the dominant force in French academic science over the next two decades. Overall, this triumvirate exerted an enormous influence over the actual work done inside the institution. Just as important, it also shaped the wider public perception of the academy and its work. Abstract mathematical mechanics did not fare well in this new climate despite Fontenelle's continuing centrality and advocacy of it. The geodetic and astronomical work of the Cassinis and their astronomical colleagues; Dortous de Mairan's elegant mechanistic models of ice and the aurora borealis; the latest natural curiosity unearthed by Réaumur; the advances being made in metallurgy, clock making, and the other publicly useful utilitarian arts: these were the topics that dominated the public discourse of the academy after 1715. The shift to these new priorities reflected the new prominence of Bignon's agendas within the institution as well as the ascent of his supreme triumvirate—Fontenelle, Réaumur, and Dortous de Mairan—to the center of French public science. They also paved the way for the new understanding of Newtonian science that began to take hold in France after 1715.

140. Marmontel, *Mémoires*, 1: 162, cited in *Malebranche's First and Last Critics*, 66.

CHAPTER 2

Sources of Enlightenment
Newtonianism Toward a
New Climate of Science in
France after 1715

Between the publication of Newton's *Principia* in 1687 and the death of its author in 1727, the tenor of intellectual life within the Republic of Letters underwent a transformation. The change resulted in a new comfort with openly partisan criticism, one that made possible the critical energies of the Newton wars and those in turn of the French Enlightenment. Before the Enlightenment, Ann Goldgar writes, "the most common concern of the members of the Republic of Letters was their own conduct. Ideology, religion, political philosophy, scientific strategy, or any other intellectual or philosophical framework were not as important as their own identity as a community."[1] This was because Republicans of Letters saw themselves first and foremost as "a scholarly community . . . tied together by certain shared values which were created by the interactions among its members."[2] What this meant in practice was that when Republicans of Letters were confronted with differences of opinion rooted in politics, religion, philosophy, or something else, their first priority was to resolve these differences collegially by choosing "the way of moderation" over critical debate and assertive contestation. This is not to say that arguments never occurred or that they were neither heated nor bitter when they did erupt. Quite the contrary, the Republic of Letters was as contentious as any other intellectual community, and very often its members engaged in heated debate. For Republicans of Letters, however, argument, even moderate argument, had a negative rather than a positive valence. Intellectual *commerce* implied above all else honorable dialogue governed by a deeply internalized ethic of civility.[3]

This ethic prized courtesy and harmonious discussion over the pursuit of individual convictions at all costs. It also produced a culture where friendship and gentlemanly agreement were esteemed while the aggressive assertion and argumentative defense of private positions were viewed with suspicion. As Goldgar notes with reference to Gisbert Cuper, a representative man of letters

---

1. Goldgar, *Impolite Learning*, 6–7.
2. Ibid., 4.
3. See also Shapin, *Social History of Truth*, esp. chaps. 1 and 5.

of the decades around 1700, "the goal for him was not to publish; rather it was to know things and to have something to say to people when writing letters. Communication, not the thing communicated, was his focus. The community, not any project, was the goal."[4] Goldgar is right to emphasize the bias toward civility and sociable community in the pre-Enlightenment Republic of Letters. The *pax analytica,* produced by the protocols of French public science as they were instituted after 1699, was a further illustration of it. European science in the decades around 1700 was indeed *honnête* science, yet Goldgar is also right that a change occurred in this respect in the early eighteenth century. By 1750, a new and more overtly argumentative style had become widespread even if this new orientation remained in perpetual contest with the older values out of which it was born. Enlightenment may in fact find its definition in the emergence of this new individualistic, critical, and argumentative *esprit* within the otherwise consensual, sociable, and communitarian culture of the Republic of Letters. But if so, how does one account for the shift?

Jonathan Israel offers one model by arguing for the unfolding of a new critical energy out of the philosophy of Spinoza.[5] For him, Spinoza's materialist metaphysics, his disenchanted skepticism about revealed religion and clerical authority, his radical republicanism, and his rationalist critical style all mark a rupture in the history of European culture. Enlightenment for Israel thus follows from the increasing diffusion and acceptance of Spinozist radicalism throughout Europe. Margaret Jacob, however, rightly sees this view as too philosophically determinist.[6] The radicalism that Spinoza authorized, she contends, only developed punch when it became a vehicle for fighting the perceived triangular oppression of absolute monarchy, hereditary aristocracy, and authoritarian religion in Old Regime society. For her, the Huguenot expulsion from France in 1685, driven by a reactionary and Catholic absolutism, combined with the progressive political upheavals of England and the Low Countries, especially the "Glorious Revolution" of 1688–1689, offer a more concrete social and political source for these new critical energies.[7]

To the extent that Spinozism, critical natural religion, anticlericalism, and Anglo-Dutch political culture were inseparably entwined after 1689, the dif-

---

4. Goldgar, *Impolite Learning,* 11.

5. Israel, *Radical Enlightenment.*

6. See Jacob's review of Israel's *Radical Enlightenment* in *Journal of Modern History* 75, no. 2 (June 2003): 387–389.

7. Jacob has articulated her views in a number of books, articles, and unpublished lectures; however, an economical synopsis of them is found in her *The Enlightenment: A Brief History with Documents* (Boston, 2001).

ferences between Israel and Jacob are somewhat moot. What is more important is the entanglement of all these forces in the production of the new critical discourses associated with Enlightenment. Also part of this scenario was a set of simultaneous institutional changes in the nature and structures of European intellectual life. In a pioneering study, Jürgen Habermas theorized the birth of what he called a "bourgeois public sphere" in eighteenth-century Europe, one that made possible a new criticism rooted in private individuals coming together as a public to openly debate their intellectual views. Recent historians have improved upon this Habermasian model by showing the many ways in which new media and new practices of sociability generated new possibilities for public criticism after 1700. From this neo-Habermasian perspective, Enlightenment criticism is less the sociologically determined result of changing class relations as Habermas would have it, and more a historically produced effect of the new forms, practices, and institutions of intellectual life that became available in Europe in the eighteenth century. "Such practices and institutions might include philosophical argument," Jonathan Sheehan writes in an incisive presentation of this viewpoint, but they would also "encompass such diverse elements as salons, reading circles, erudition, scholarship and scholarly techniques, translations, book reviews, academies, new communication tools including journals and newspapers, and so on." [8] Israel's radical philosophy and Jacob's radical politics also came to life at the center of this new eighteenth-century media, and at the center of it as well was Newtonian science. Therefore, whether it served as yet another impetus toward these changes or as one more effect of them, the new discourse of Newtonianism that developed within the Republic of Letters after 1715 participated directly in production of this new Enlightenment criticism.

The dynamics that led to these shifts, as well as their effect in producing the new critical discourses of Newtonianism, are the subject of this chapter. The career of the Swiss savant Jean-Pierre de Crousaz can serve to tie together its arguments.

Newtonian Physics in the Republic of Letters circa 1715

Crousaz personified the values of the Republic of Letters as Goldgar describes them. [9] He became a public presence in the Republic fairly late in his life,

8. Jonathan Sheehan, "Enlightenment, Religion, and the Enigma of Secularization: A Review Essay," *American Historical Review* 108, no. 4 (October 2003): 1075–1076.

9. Jacqueline de La Harpe offers a thorough intellectual biography of Crousaz in *Jean-Pierre Crousaz et le conflit des idées au siècle des Lumières* (Geneva, 1955). Also revealing are

at the age of forty-nine, after he published his first book, *Nouvel essai de logique*. The text, which appeared in 1712, offered "reflections suitable for contributing to the precision and scope of our understanding," and before its debut Crousaz had already become an established figure in the city of Lausanne through his work as a cleric, a teacher at the Lausanne Academy, and a city notable.[10] Educated at Leiden and Paris, he also served as a bridge between the wider learned world of the Republic of Letters and the provinces of Switzerland.[11]

His provincial location made him especially attached to the codes of gentlemanly conduct that defined the Republic of Letters in this period. For him, membership in this community meant inclusion in a meritocratic republic of peers that elevated him beyond the smallness of his immediate home. It also allowed him to contribute beneficially to the advancement of knowledge and the public good while also affording him the opportunity to participate in the learned conversations that raised his own status within this estimable community. His *Logique* of 1712 marked his public debut, and the manner of his intervention says much about him and the values of the period. Bignon was one of many who admired Crousaz's work, and the two men began a correspondence in 1713 that would continue until the minister's retirement in the 1740s. "The public will justly recognize your talents," he wrote to Crousaz in August 1713. "It will also recognize your moderation and your love of truth. Nothing is more estimable in a savant."[12] Crousaz, in fact, received nothing but warm praise from Bignon throughout his life, for as the abbé wrote, "what honor for the sciences if those who cultivate them also bring that spirit of gentleness and modesty that is so rare today."[13] Crousaz combined learning and intellect with civility and *politesse* in precisely this way, and for this reason he was a model Republican of Letters in the minds of Bignon and others.

Crousaz also pursued an intellectual career after 1712 that confirmed this identity. Rather than focusing on a single topic or a specialized body of work,

---

Crousaz's voluminous personal papers currently housed at the Bibliothèque cantonale et universitaire, Lausanne, Switzerland, and especially his vast correspondence, also housed there, which is conveniently cataloged in Marianne Perrenoud, *Inventaire des archives Jean-Pierre Crousaz, 1663–1750 (IS 2024)* (Lausanne, 1969). The correspondence is collected in the fourteen-volume archival collection Fonds Jean-Pierre Crousaz (referred to as FJPC). All of my citations will be to this collection.

10. Jean-Pierre de Crousaz to l'abbé Jean-Paul Bignon, 23 October 1712, FJPC, 1: 5–6.

11. In addition to La Harpe, an account of Crousaz's life is found in "Éloge de M. de Crousaz," *HARS-Hist.* (1750): 258–274.

12. Bignon to Crousaz, 21 August 1713, FJPC, 6: 135.

13. Ibid.

FIGURE 14. *Jean-Pierre de Crousaz.*
Courtesy of the Bibliothèque cantonale et universitaire, Lausanne, Switzerland.

Crousaz cultivated a learned and gentlemanly polymathy. In 1714, his *Traité du beau* appeared to great acclaim, and in the subsequent decades he published works on mathematics, including an important treatise on the infinitesimal calculus; mechanics, including a contribution to the ongoing *vis viva* debate about the correct measure of forces in moving bodies; philosophy, including a widely discussed critique of Anthony Collins's deist tracts *A Discourse on Free-thinking* and *A Philosophical Inquiry concerning Human Liberty* along with a critical examination of ancient and modern Pyrhonnianism; theology, especially an edition of sermons and critical works on Calvinist doctrine; and education, including pedagogical works on teaching arithmetic, geometry, rhetoric, and philosophy to children.[14] His publications were further supplemented by an active correspondence that brought him into contact with many of the leading luminaries of the Republic of Letters. Johann Bernoulli was an early correspondent, as was Jean-Baptiste Rousseau after he was exiled from France for his libelous suit against Saurin. He also formed especially close ties with Bignon and Réaumur, and through them Dortous de Mairan and Fontenelle. In 1720, these ties helped him to win the first-ever Paris Academy prize on the nature of motion, and in 1725 he was named one of the academy's official foreign correspondents.

The many letters that Crousaz exchanged with Bignon and Réaumur after 1712 reveal the depth of their shared vision of appropriate scientific practice. Responding to Crousaz in October 1716, Bignon agreed that "metaphysicians are not sufficiently careful." "In matters of metaphysics," he continued, "it is dangerous to want to engage oneself too far. Much better to simply hold back," especially since "the speculations that form the basis of their study often lead directly to paralogisms."[15] Crousaz often expressed similar sentiments. "With the exception of minds of a superior order," he wrote to Bignon in December 1718, "men almost never contain their reflections or their taste within just bounds."[16] Connecting these reckless tendencies to the sciences in a letter of August 1718, he further added that "making systems is the most dangerous temptation of the physicist." The foundations of a system are usually "obscure, even incomprehensible" yet the system itself appears to "explain things ingeniously." In this way, "one gets more stubbornly attached to it than to the desire to remove incomprehension in the first place." This brings an illusion of intelligibility when critical skepticism is in fact warranted.[17] Crousaz explained the same fallacies to an approving Réaumur with respect to the selection of the upcoming academy prize questions. "It is best to suggest simple subjects that can little by little be built gradually into more complicated ones," he wrote.

14. La Harpe provides an appendix with a convenient chronological list of all of Crousaz's many publications in *Crousaz et le conflit des idées*, 268–271.

15. Bignon to Crousaz, 7 October 1716, FJPC, 6: 163.

16. Crousaz to Bignon, 10 December 1718, FJCP, 1: 255.

17. Crousaz to Bignon, 19 August 1718, FJCP, 1: 105.

"Otherwise, the questions that it poses will push today's physicists even more toward that taste for systems that is already too widespread." [18]

Crousaz's view of proper science, which combined careful empiricism, judicious philosophical reasoning, and a critical skepticism toward passionate, sectarian theorizing, was representative of the wider Republic of Letters of the period. These shared values cemented his close union with the leading figures at the Paris Academy. They also constituted the nub of a dispute that erupted between him and the English defenders of Newton sometime after 1714. In his *Traité du beau,* Crousaz used the Newtonian theory of universal gravitation to illustrate what he called "false beauty" in the sciences. "It might be true that bodies, by their very nature, tend essentially toward one another, and if it is true, then the effects of their *pesanteur* (the French term that describes the weightedness of bodies) will follow the laws established by Monsieur Newton. Furthermore, the most beautiful phenomena of the universe will be demonstrated by these laws." However, the beauty thus achieved will only be a "true beauty," Crousaz explained, if "the great principle itself is real, and this reciprocal *pesanteur* is not a chimera." If it is instead a "dream" or "a supposition in air," then "the only beauty that this wise Author will have achieved" is the beauty derived from "the ingenious application of this principle." There is a sort of elegance in this kind of intellectual coherence, Crousaz continued, but its aesthetic appeal is false and should not be confused with the true beauty of a real science. Indeed, an ingenious but false system does possess a certain kind of beauty, but one more like that of "a fiction" or a "novel" [*roman*] than that of a true philosophical explanation. Yet even with "false beauty," Crousaz added, "one still prefers to see these fictions approach reality and unite with it in its own class called verisimilitude [*vraisemblance*]." [19]

Newton's theory of universal gravitation, Crousaz implied, achieved at best a beautiful *vraisemblance* with nature, yet he offered no other commentary about Newton in this treatise, nor did he argue that Newton's theory of universal gravitation was indeed akin to a philosophical fiction. Instead, he let his readers draw their own conclusions. Other works by Crousaz also engaged with topics disagreeable to him in this decorously critical manner, and in this way he personified the values of the *honnête hommes* who dominated the Republic of Letters. His suspicions about Newtonian gravitational attraction were equally representative of the opinions of the period. In the decades around 1700, when Crousaz's intellectual views were formed, virtually no one took Newtonian

---

18. Crousaz to Réaumur, 29 November 1718, FJCP, 1:243.

19. Jean-Pierre Crousaz, *Traité du beau* (Amsterdam, 1714), 112–114.

physics seriously. One reason for this was because Newton's argument for universal gravitation in the *Principia*, rooted as it was in a complicated use of mathematics, experiments, and empirical inductions to defend physical claims about the world, was simply at odds with the prevailing epistemological assumptions in Europe at the time. Since few rose in this period to defend Newtonian physics, there was no conception of Newtonianism available, and no men or women who called themselves (or were called by others) Newtonians.

Throughout the Republic of Letters in the decades around 1700, in fact, the mathematical reading of the *Principia* that was influential in France remained the most common interpretation for everyone. This was true even in England, even if Newtonian physics found its first advocates here.[20] An indicator of this fact is offered by Samuel Clarke, who mirrored the changing climate of opinion in 1710 when he added a set of footnotes to the third edition of his English translation of Jacques Rohault's Cartesian physics textbook that challenged Rohault's principles on Newtonian grounds. His translation was destined for use in the classrooms of Oxford and Cambridge, and accordingly his footnotes, which explicitly called the celestial vortices a "chimerical fiction," helped to define the approach to Newtonian physics shared by Englishmen who came of age after 1710. In preparing the first edition of this work in 1697, however, the same Clarke felt no need to add any comparable editorial commentary, nor even to invoke the Newtonian theory of gravitational attraction in any way. Instead, he let Rohault's Cartesian physical account of celestial mechanics and terrestrial gravity pass into English university classrooms without further commentary. This anecdote illustrates well the absence of any self-conscious rivalry in physics between Newtonians and Cartesians before 1700.[21]

Few if any defenders of Newtonian gravitational attraction existed anywhere in Europe before 1710, and for this reason, the label "Newtonianism," defined as a set of physical arguments about matter, space, motion, and physical causality, simply had no meaning. Even Newton attested to this consensus, for as he wrote to Richard Bentley in 1698, summarizing the sentiments of virtually everyone within the European scientific community at the time: " 'Tis inconceivable that inanimate brute matter should, without mediation of something else which is not material, operate and affect other matter without mutual contact,

---

20. The research of Niccolò Guicciardini reinforces this point. See his *Reading the Principia: The Debate on Newton's Mathematical Methods for Natural Philosophy from 1687–1736* (Cambridge, 1999), esp. chap. 7, and idem, "Dot-age: Newton's Mathematical Legacy in the Eighteenth Century," *Early Science and Medicine* 9 (August 2004): 218–256.

21. Aiton, *Vortex Theory*, 71–72.

as it must be if gravitation, in the sense of Epicurus, be essential and inherent in it. That gravity should be innate, inherent, and essential to matter so that one body may act upon another at a distance through a vacuum without the medium of anything else by and through which their action or force may be conveyed from one to another is to me so great an absurdity that I believe no man who has in philosophical matters any competent faculty of thinking can ever fall into it. Gravity must be caused by an agent acting constantly according to certain laws, but whether this agent be material or immaterial is a question I have left to the consideration of my readers." [22]

This deep incredulity with respect to the physics of universal attraction through empty space was widespread in Europe around 1700, and Crousaz's expression of it in his 1714 *Traité du beau* was in no way rare. Yet the sentiments provoked the ire of a newly self-conscious group of English Newtonians when they read Crousaz's book in London. What had brought about this transformation? A number of convergent factors, and perhaps the first catalyst for change was the publication in 1706 of Newton's *Opticks*.[23] The theory of gravitational attraction was articulated much more emphatically in the *Opticks* than in the *Principia*. Newton also employed common language and experimental observations to articulate and defend his views rather than complex mathematics. The publication of the *Opticks*, therefore, cast a new light upon the physical dimension of the *Principia*, the dimension so often dismissed by European thinkers prior to this date.

Viewed now as a couple, the *Opticks* and the *Principia* together forced savants to confront one overarching philosophical dilemma. As the *Journal de Trévoux* stated in its February 1709 review of the *Opticks:* "Physics is divided into two parts: natural history and the study of causes. The new system of light invented by M. Newton and proposed here renders the first more complete and the second more difficult." [24] The *Opticks* made natural history more

22. Cited in Alexandre Koyré, *From the Closed World to the Infinite Universe* (Baltimore, 1968), 178–179.

23. The first, English edition of Newton's *Opticks* appeared in London in 1704. However, given the barrier separating France from the Anglophone public sphere, this edition attracted little attention in France. In 1706, Newton published a Latin version of the same text, and this edition attracted much attention, at least in scholarly circles. Widespread study of Newton's *Opticks* in France was further made possible in 1720 by the appearance of Pierre Coste's French translation. This edition had a wide influence in France. All of my references will be to the modern edition of the *Opticks* edited by I. B. Cohen (New York, 1979), which is based on the fourth and final London edition of 1730.

24. Review of Isaac Newton, *Opticks*, in *Journal de Trévoux* (February 1709): 186.

complete because the work demonstrated experimentally the polychromatic nature of white light, an empirical fact that the French had little problem accepting once the evidence was made clear.[25] The same was true of Newton's inverse-square law of planetary motion offered in the *Principia,* which few in France challenged as an empirically demonstrated fact. However, Newton's *Opticks,* and by extension his *Principia,* rendered the search for causes more difficult because it also argued that light was composed of material corpuscles containing forces subject to a law of universal gravitational attraction. This idea challenged the deepest foundations of European science. Newton, for example, explained the bending of light when passing through dense media in terms of the extra gravitation exerted by the denser substance as the particles of light passed through it. This theory explicitly connected the particular physics of light in the *Opticks* with the more general and mathematical laws of moving bodies in resisting media demonstrated in the *Principia.* Accordingly, the book forced a new confrontation with attractionist physics in general.

With the *Opticks* now available as the referent, Newton's name began to be associated more and more after 1706 with the physics of gravitational attraction. This occurred even though many also retained their preexisting understanding of his science formed from reading the *Principia* alone. The *Opticks* clearly complicated the purely mathematical reading of the *Principia,* yet savants very often dismissed these more disturbing implications while absorbing those features compatible with existing scientific assumptions. This was certainly the case with the practitioners of analytical mechanics in France, and the reception of Newton's theories of light followed a similar pattern. However, not everyone could sustain this hybrid stance in the face of Newton's actual writings, and this led some to an acceptance of his more provocative views about universal gravitation and the physics of attraction while others were provoked to challenge them outright.

Perhaps the earliest Frenchman to take the idea of attractive matter seriously was the chemist and academician Étienne François Geoffroy. Geoffroy's reasons for embracing Newtonian attraction were tied closely to his particular location within the field of eighteenth-century science. As Arnold Thackray has discussed, Newton's conception of an interparticle force of material attraction

25. For a discussion of the history of French acceptance of Newton's polychromatic theory of light, see Henry Guerlac, "Where the Statue Stood: Divergent Loyalties to Newton in the Eighteenth Century," in *Essays and Papers in the History of Modern Science* (Baltimore, 1977), 137–139, and idem, *Newton on the Continent* (Ithaca, 1980), 78–163.

found its earliest and most vigorous defenders among the chemical community in England.[26] Chemists such as John Freind and doctors such as George Cheyne were at the leading edge of the movement to apply Newton's theory of attractive matter to the study of matter in general. Astronomers and physicists such as David Gregory and John Keill soon joined these doctors and chemists in pursuing this view, and by 1706 these Englishmen had begun to define Newton's achievement in terms of an alleged discovery of a material force of attraction operative in all bodies whose laws could be established quantitatively. The *Opticks* catalyzed this movement, becoming the scientific Bible for this subset of the broader Newtonian movement. The *Journal de Trévoux* captured this trend in 1709, writing of the Latin edition of Newton's *Opticks* that "there is no chemical operation which is not to him a good proof of the efficacy of attraction."[27] By this date, attractionist Newtonianism was growing in importance in Britain, and Geoffroy was an important conduit for these ideas into France.[28]

Taking advantage of the brief peace of 1697–1702 between the War of the League of Augsburg and the War of Spanish Succession, Geoffroy spent most of 1698 in England. He was elected a fellow of the Royal Society at this time, and he also studied with Freind and Keill while there. Once back in Paris, he continued to correspond with the Royal Society through its secretary Hans Sloane. Given these ties, it was Geoffroy who presented the first English edi-

26. Arnold Thackray, *Atoms and Powers: An Essay on Newtonian Matter-Theory and the Development of Chemistry* (Cambridge, 1970). Geoffroy's connection to English Newtonianism is discussed in detail (85–95).

27. Review of *Opticks*, *Journal de Trévoux* (February 1709): 199.

28. The literature on the culture of Newtonianism in England is vast, but my understanding of the topic has been most informed by the following: Thackray, *Atoms and Powers*, esp. chaps. 2–3; Robert E. Schofield, *Mechanism and Materialism: British Natural Philosophy in the Age of Reason* (Princeton, 1970), chaps. 1–3; Westfall, *Never at Rest*; Margaret Jacob, *The Newtonians and the English Revolution* (Ithaca, 1974), idem, *The Radical Enlightenment: Pantheists, Freemasons, and Republicans* (London, 1981), chaps. 1, 3; Margaret Jacob and Betty Jo Teeter Dobbs, *Newton and the Culture of Newtonianism* (Atlantic Highlands, 1995); Simon Schaffer, "Natural Philosophy," in *The Ferment of Knowledge: Studies in the Historiography of Eighteenth-Century Science*, ed. G.S. Rousseau and Roy Porter (Cambridge, 1980), idem, "Natural Philosophy and Public Spectacle in the Eighteenth Century," *History of Science* 21 (1983): 1–43, idem, "Newtonianism," in *Companion to the History of Modern Science*, ed. R. C. Olby et al. (London, 1990); and Larry Stewart, *The Rise of Public Science: Rhetoric, Technology, and Natural Philosophy in Newtonian Britain, 1660–1750* (Cambridge, 1992).

tion of Newton's *Opticks* to the academy in 1706.[29] As one of the only English speakers in the company during these years, Geoffroy produced a hastily prepared translation of Newton's work. Between August 1706 and June 1707, ten academic sessions were devoted to a reading of Geoffroy's translation before the appearance of the Latin edition of the *Opticks* made such presentations unnecessary.[30] The topic of attraction was no doubt broached in these sessions, but as is customary no record of the discussions appears in the academy registers.[31] Geoffroy, however, was clearly influenced by Newton's ideas, and his chemical work after 1707 began to show the influence of attractionist Newtonianism.

Developments in Britain and France further encouraged Geoffroy's turn. Freind's deeply attractionist *Praelectiones Chymicae* appeared in 1709, and Geoffroy received on behalf of the academy a copy of Francis Hauksbee's *Physico-Mechanical Experiments* in 1710.[32] Both works emphasized the role of attractive forces in the explanation of natural phenomena. Freind's work was reviewed in the *Journal des sçavans* in 1711, and the reviewer noted that "the design of the author, as he explains it to the illustrious M. Newton, to whom he dedicates the work, is to explain the operations of chemistry by the attractive virtue."[33] The *Journal de Trévoux* pointed to the same orientation in its review of Freind's work in 1712, and it was this strand of Newtonian chemistry that Geoffroy pursued.[34] Hauksbee's work also supported Geoffroy's thinking, since it offered among other things an experimental treatment of the phenomenon of fluid rise in capillary tubes and an attractionist explanation for it. Interest in the "capillary effect" was great at the Paris Academy, and Geoffroy served as the intermediary between Hauksbee and the French on this question.[35]

29. Newton was a corresponding member of the Paris Academy at this time and the presentation of gift copies of published works was an established protocol of corresponding membership.

30. This episode is discussed at length in I. B. Cohen, "Isaac Newton, Hans Sloane, and the Académie Royale des Sciences," in Alexandre Koyré, *Mélanges Alexandre Koyré*, ed. I. B. Cohen and René Taton, 2 vols. (Paris, 1964), 1: 61–116.

31. The academy's records during this period often contain only the following entry: "Discussion of M. Newton's *Opticks* continued."

32. John Freind, *Praelectiones Chymicae, In quibus omnes fere Operationes Chymicae* (London, 1709); Francis Hauksbee, *Physical-Mechanical Experiments on Various Subjects* (London, 1709).

33. Review of John Friend, *Praelectiones Chymicae*, in *Journal des savants* (1711): 10.

34. Review of John Friend, *Praelectiones Chymicae*, in *Journal de Trévoux* (October 1712): 1780–1786.

35. Work on capillary rise by French academicians was published in 1705. See *HARS-Hist.* (1705): 21–25 and *HARS-Mem.* (1705): 241–254. But this work is hard to connect with

By 1712, Geoffroy had expressed his sympathy for the Newtonian theory of gravitational attraction unequivocally in a letter to Hans Sloane.[36] Likewise, in early 1718, Geoffroy began to reveal the scientific results of his own attractionist chemical work to the academy. During this year, he first presented his "Table of Different Relationships [*rapports*] Observed between Different Substances" to the company. He also appeared at the spring public assembly of 1718 to talk about this work.[37] As the title suggests, his contribution was an elaborate and rather Hermetic-looking table that offered a systematic analysis of the affinities between different chemical substances.[38] Newton's *Opticks*, which appeared in a second English edition in 1717 and in Pierre Coste's influential French translation in 1720, offered a similar list of relative affinities. Geoffroy's table was clearly influenced by Newton's work, while rooted in his own empirical investigations. No record of the reactions to Geoffroy's table is found in the academy registers, but in his "Éloge de M. Geoffroy" delivered in 1731, Fontenelle noted that "these affinities upset some who believed they were only disguised, and therefore more dangerous, versions of attraction (for clever men had already succeeded in giving attraction seductive forms)."[39] Fontenelle's *éloge* is suspect, however, since it was written after 1730, when discussions of attraction had become more heated and the intellectual climate overall more polemicized.

Less anachronistic, and likewise more interesting, was Fontenelle's treatment of Geoffroy's work in the academy's *histoires* of 1718 and 1720. In his second report on Geoffroy's table in 1720, Fontenelle indicated that the initial table had generated controversy. But, he added, "such criticism is inevitable given a work that is so new."[40] He also indicated that the complaints were "empirical and practical rather than conceptual," further asserting by way of conclusion that "if mathematics had the same difficulties that chemistry does, it would be even more difficult than it is. This is because of the extreme complexity of

---

the explicitly Newtonian tenor of discussions about capillary rise that followed Hauksbee's work. The academy records, however, make it virtually impossible to discern what the views of the academicians as a whole were toward Hauksbee's work.

36. Cited in Cohen, "Newton, Sloane, and the Académie royale des sciences," 100–102.

37. Étienne François Geoffroy, "Table des différents rapports observé en chimie entre différentes substances," PVARS, January 7, 1718; April 27, 1718 and *HARS-Mem.* (1718): 202–212.

38. The table itself was published as a foldout leaf in *HARS-Mem.* (1718): plate 8, 212. A modern reproduction of the table is found in Thackray, *Atoms and Powers*, 91.

39. *Oeuvres de Fontenelle*, 7: 220.

40. "Éloge de M. Geoffroy," *HARS-Hist.* (1720): 32–33.

the latter and the inexactitude in the meaning of the terms employed." [41] This assessment resonates with others by Fontenelle that also stressed the difficulty of attaching pure geometry to mechanics and physics. Placed in the context of Fontenelle's public academic discourse as whole, it also worked to connect Geoffroy's chemistry with other programs important to the secretary.

Fontenelle's previous and more extensive account of Geoffroy's table in 1718 explains the deeper assumptions underlying the secretary's presentation. He began this earlier first report by noting that many substances appear to display a natural attraction and repulsion for one another. Yet chemists do not understand why this behavior occurs. "What principle of action can one conceive to account for this affinity?" Fontenelle asked. "It is here that the sympathies and attractions would come to the rescue, if only there were such things." "Leaving unknown that which is, and holding only to certain facts," Fontenelle asserted that chemical experiments [*experiences*] make it clear "that one body has a greater tendency to unite with some bodies and not others and that these different dispositions have different degrees." [42] Geoffroy's table is "in some respects prophetic," the report continued, because it accounts for these empirical facts and thus allows one to predict what will happen when certain substances are mixed. But, he added, "[Geoffroy] does not consider here the different forces acting between these bodies, but only their *rapports*." [43] In other words, Fontenelle suggested, Geoffroy was eschewing claims about physical causes and treating instead the descriptive relationships between observable effects. "The more chemistry is perfected," the reported concluded, "the more M. Geoffroy's table will be perfected as well." For "if physics can never achieve the exactitude of mathematics, it can at least imitate the order of it. And a chemical table is by itself an agreeable spectacle to the human mind, just like a table of numbers ordered according to certain properties or *rapports*." [44]

This account resonates strongly with Fontenelle's Malebranchian discourse about analytical mechanics. Indeed, it has the effect of assimilating Geoffroy's work to the program in Malebranchian mathematical phenomenalism despite its thoroughly unmathematical and unphenomenalist character. Geoffroy's chemistry appeared to Fontenelle as an analytical description of empirical *rapports*, and not as a physical and experimental science of forces in nature. Of course, the actual chemistry that Geoffroy offered was anything but this, yet framed in terms of the Malebranchian language of *rapports* it could be made

41. Ibid., 36.
42. *HARS-Hist.* (1718): 35–36.
43. Ibid., 36.
44. Ibid., 37.

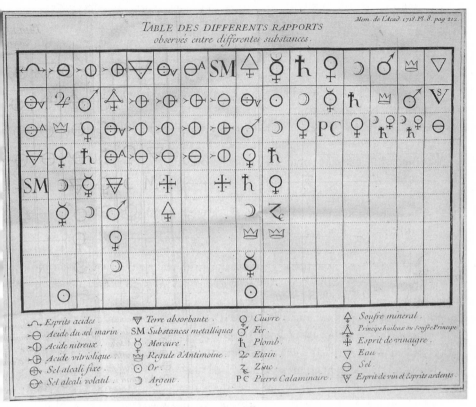

FIGURE 15. *Étienne François Geoffroy, "Table of the different* rapports between
the chemical elements,"from Histoire de l'Académie royale des sciences *(1718).
Courtesy of the Herzog August Bibliothek, Wolfenbüttel, Germany.*

compatible with an approach to science that eschewed direct claims about the
existence and action of interparticle forces in the name of predictive descriptions. Positioning Geoffroy's science this way obviously made it compatible
with Fontenelle's scientific assumptions while also serving many of the latter's larger agendas, and for this reason the secretary saw no need to criticize
Geoffroy's work despite its attachments to Newtonian attraction. For his part,
Geoffroy appears to have handled the politics of attraction in the academy in
an equally adroit manner. He neither emphasized the physical dimension of
his chemistry in print or in private, nor did he effect any strong association
between his work and that of the English Newtonians. In this way, Geoffroy
became an early defender of Newtonian attraction in France while the wider
culture both in and out of the academy remained unmoved by such a view.

Yet if Fontenelle and Geoffroy agreed (in public at least) to collegially paper over their scientific differences, Crousaz found no such collegial toleration among the English Newtonians who read his *Traité du beau*. They saw a criticism they disagreed with in Crousaz's text, and they, accordingly, responded with criticisms of their own. In doing so, they showed that it was not only a new willingness to believe in Newton's attractionist physics that gave birth to Enlightenment Newtonianism, but a new eagerness to critically defend it in the public sphere as well. The latter was a product of a host of institutional and cultural changes, and key to them was the new practice of critical learned journalism that also began to grow more prominent after 1700.

Newtonian Science, Eighteenth-Century Journalism, and the New Radical Philosophy

Jean Le Clerc was at the forefront of these developments. His *Bibliothèque universelle,* along with Bayle's *Nouvelles de la republique des lettres,* pioneered a new and distinctively Dutch form of learned journalism after 1675 that was deeply influential throughout Europe.[45] Grounded like the earliest scholarly journals in the authoritative book review, the personal letter, and the occasional *mémoire,* Le Clerc's *journaux* found their distinctive character in the intellectual style of their editor and the content that his editorial style produced. Le Clerc was an ardent Protestant, and his *journaux* reflected this bias in the attention they gave to the latest works in Calvinist theology.[46] He was also a student of English philosophy, particularly the metaphysics of John Locke and the natural philosophy of Isaac Newton. His *journaux* devoted much attention to these and related philosophical discussions as well. In 1703, he even began offering English-style experimental philosophy classes in Amsterdam, a trademark Newtonian practice as we will soon see. Le Clerc was likewise a republi-

---

45. On Le Clerc, see Annie Barnes, *Jean Le Clerc (1657–1736) et la république des lettres* (Paris, 1938). See also Goldgar, *Impolite Learning.*

46. An interesting side note to this story is that the brother of the French academic mathematician Joseph Saurin was a scholar and minister in The Hague and one of Le Clerc's favorite theologians (Saurin of Paris was also a Calvinist minister himself before turning to mathematics after his conversion to Catholicism in the 1690s. See Fontenelle, "Éloge de M. Saurin," *Oeuvres de Fontenelle,* 7: 271–284). The Dutch Saurin's writings were reviewed frequently in Le Clerc's journals. See, for example, the long review of "Discours de M. Saurin sur le Pentateuque," in *Bibliothèque ancienne et moderne* 12 (1719): 237–320; and the review of "Catéchisme de M. Saurin," *Bibliothèque ancienne et moderne* 18 (1722): 424–445.

can and a bitter critic of French and Spanish absolutism. These political views also shaped his writing.

Le Clerc's self-fashioning as a learned journalist and man of letters, however, was just as important to his widespread cultural influence. An exiled French Huguenot, Le Clerc was part of the expanding cadre of would-be public savants who used the religious toleration and intellectual liberty of the Netherlands to foster a new identity as citizens of the international Republic of Letters.[47] Le Clerc's journalism was one of several crucial practices constitutive of this increasingly self-conscious community of savants. Furthermore, as an ambitious and influential citizen of the cosmopolitan learned world, Le Clerc also aspired, in ways very much like Fontenelle, to shape the discourse within the Republic of Letters according to his assumptions.

Each was an esteemed member of the Republic of Letters, but after 1697 Fontenelle was also beholden to the discipline of the official French intellectual establishment. This social position constrained the public discourse he produced. Le Clerc, by contrast, adhered only to the discipline of the Republic of Letters. His public discourse, therefore, while no less disciplined than Fontenelle's, acquired a different character. In particular his *journaux* cultivated a more openly critical and opinionated form of intellectual dialogue than was possible within the official organs of the French public science. Nevertheless, Le Clerc also strove to maintain the appropriate decorum expected among Republicans of Letters, and in this respect he and Fontenelle were more alike than different. Both also worked hard to channel controversies and disagreements into a reasoned consensus agreeable to their particular points of view. For this reason as well, each was widely respected within the wider learned community.

Le Clerc's initial periodical, *Bibliothèque universelle,* ceased publication in 1693 after a seven-year run, but it was revived in 1703 under a new title, *Bibliothèque choisie.*[48] This journal ran until 1713. Then, after a brief hiatus of only a few months, Le Clerc launched his third and final journal, the *Bibliothèque an-*

47. This development is explored most fully in two recent books: Goldgar, *Impolite Learning;* and Dena Goodman, *The Republic of Letters: A Cultural History of the French Enlightenment* (Cornell, 1992). Goodman's book deals primarily with the second half of the eighteenth century but it opens with an account of the transformations in the Republic of Letters around 1700. Goldgar's book, by contrast, studies the early century developments in detail. I am grateful to both Goldgar and Goodman for participating with me in a very instructive e-mail discussion of this topic on the SECSF listserv in June 1999.

48. Publication details about these and all other Francophone journals in this period can be found in Sgard, *Dictionnaire des journaux.*

cienne et moderne. All three periodicals were identical in form and content, and the last ran until Le Clerc's retirement in 1727 at the age of seventy. The success of his periodicals insured successors, and by 1730 several other Le Clerc–style journals were already in circulation. The *Journal littéraire de la Haye* was began in 1713, and produced a spin-off, *L'Europe savante,* which ran from 1718 to 1720. The first *Journal littéraire* was reformed in 1718, this time under the direction of a new group of editors, including the self-proclaimed Newtonian Willem 'sGravesande. Under various imprimaturs, this journal ran until 1737, becoming an even more explicitly scientific and philosophical periodical during these years. The explicitly philosophical orientation of these journals was also carried on by a further spin-off, the *Bibliothèque raisonée des ouvrages des sçavans de l'Europe,* which included many of the same editors and ran uninterruptedly from 1728 until 1753.[49] These journals were modeled after Le Clerc's, and the editors responsible for them often knew him personally. Together, they established the Francophone Dutch journal as a widely influential organ of European thought and criticism after 1700.

The discourses sustained by the learned Dutch periodicals were also supported by an equally active culture of book publication. Not only did the editors of these journals, including Le Clerc and 'sGravesande, produce works of their own that were reviewed in the press, the Dutch printing houses that published each also helped to expand the wider print culture of the period by producing first editions of books, reissues of hard-to-find texts, and pirated editions of volumes that were either scandalous or restricted. The second edition of Newton's *Principia,* for example, was produced in such limited numbers in England that a Dutch printer in Amsterdam produced a rival edition designed to satisfy this demand.[50] It was this Amsterdam edition rather than the London edition that the editors of both the *Journal des savants* and the *Journal de Trévoux* reviewed in 1715 and 1718, respectively.[51] Pierre Coste's important French translation of Newton's *Opticks* published in 1720 was also a product of a Dutch publishing house, and these are but two examples of the widespread expansion of print culture in this period catalyzed by the Francophone publishing structures of the Netherlands.

49. On this journal, see Bruno Lagarrigue, *Un temple de la culture européenne (1728–1753): L'histoire externe de la Bibliothèque raisonée des ouvrages des sçavans de l'Europe* (Nijmegen, 1993).

50. On this, see I. B. Cohen, *Introduction to Newton's Principia* (Cambridge, 1978), 256–258. The first English edition had a print run of only 750 copies.

51. Review of Newton, *Principia,* in *Journal des savants* (1715): 157–60; review of Newton, *Principia,* in *Journal de Trévoux* (February 1718): 466–475.

How widely these Dutch-produced texts circulated in France before 1730 is not clear. Technically, books produced without explicit French approval were forbidden, and book smuggling was prosecuted in the Old Regime.[52] On the other hand, ample evidence indicates that this literature was widely available even if it sometimes had to pass under the cloak. In the private library inventories studied by Daniel Mornet, Le Clerc's journals, along with Bayle's *Nouvelles de la république des lettres,* were the most widely prevalent of all journals in the collections.[53] Mornet admittedly studied libraries from a later period (1750–1780), and thus it is hard to draw firm conclusions about the early part of the century from his research. But it is nevertheless clear that the works reviewed in these Dutch journals and the discussions that they triggered increasingly found their way into the French public sphere even if an aura of dangerous illegality often remained attached to them as well.

The rise of this non-native, Francophone publishing culture also triggered a set of important changes in the French press. By law, French periodicals, like other Old Regime institutions, held monopoly rights over their intellectual terrain. Thus, founding a new periodical in France was not always easy given the official privileges accorded to both the *Journal des savants* and, after 1724, the *Mercure de France.*[54] Exemptions could be won, however, as the Jesuits had achieved with the *Journal de Trévoux.* Yet they required political backing like that which the Duc de Maine had offered in this case. Another option was the clandestine circulation of newssheets, but this was a risky and shady affair even if dozens of such works are extant from the period.[55] The clandestine circula-

52. On the history of books in Old Regime France, see Henri-Jean Martin, *Livre, pouvoirs et société à Paris au XVIIe siècle,* 2 vols. (Geneva, 1969); Roger Chartier and Henri-Jean Martin, eds., *L'histoire de l'édition française,* 4 vols. (Paris, 1983–1986), esp. volume 3, *Le livre triumphant, 1660–1830*; and David T. Pottinger, *The French Book Trade in the Ancien Régime, 1500–1791* (Cambridge, 1958);

53. Daniel Mornet, "Les enseignements des bibliothèques privées (1750–1780)," *Revue d'histoire littéraire de la France* 17 (1910): 453. A useful table summarizing Mornet's findings regarding periodicals is found in Goldgar, *Impolite Learning,* 60.

54. Malesherbe's papers, housed in the Bibliothèque nationale, offer a fascinating window into the difficulties associated with journalism in Enlightenment France. As director of the book trade from 1744 to 1757, Malesherbe collected dozens of proposals for new journals since he had the authority to grant a new journal royal approval. Yet virtually none of these journals were sanctioned, and overwhelmingly it was the monopoly rights of the preexisting periodicals in France that he invoked in rejecting these initiatives. See BN, MSS f. fr 22133.

55. Sgard's *Dictionnaire des journaux* is a rich source for tracing the ephemeral periodicals of this period.

tion of books, journals, and manuscripts was important, and we will return to it in a moment, but on the whole the printed word remained remarkably controlled in France before 1720.This control helped to limit the intellectual discourse of the period, especially any discourse challenging to the official consensus prevalent in the public sphere.

Gradually after 1715, however, this containment began to weaken. One early development was the rise of a new class of French journals rooted in the reporting of foreign literary news. The first of these was the *Bibliothèque anglaise*, which ran from 1717 to 1728. It was succeeded by the *Bibliothèque britannique*, which ran from 1733 to 1747. Both works offered an important conduit for English ideas into France. Following the lead of the *Bibliothèque anglaise*, the *Bibliothèque germanique* appeared in 1720. The *Nouvelles littéraires de la Suisse*, the *Bibliothèque française*, and the *Bibliothèque italique* soon joined them, appearing in 1722, 1726, and 1728, respectively. Not all of these works were produced in France, and few of these were authentically "foreign" in their editorial and intellectual sensibilities. But all received the royal privileges that their Dutch counterparts lacked.[56] Taken as a group, they also greatly expanded the size of the literate public sphere in France between 1715 and 1730 while also reinforcing the perception that "national" literatures and cultures existed and were coherent groupings. This expansion was supported by a parallel expansion in the availability of books in France, and viewed retrospectively it is clear that the French public sphere after 1730 was a vastly larger space than it had been two decades earlier.[57]

This overall expansion in the range and quantity of texts accompanied a related transformation in the tone and style of the public discourse found in them. Both made possible the new consciousness about attractionist Newtonianism that took shape in France after 1715. Perhaps the first aggressive attractionist to appear in the newly expanding French public sphere was David Gregory. In 1702, Gregory published at Oxford his explicitly attractionist *Astronomiae physicae et Geometricae Elementa*. Gregory's work, in its stress upon the physical fact of universal gravitation, was representative of the attractionist approach to Newton's *Principia* then growing in importance in Britain and Holland.[58] In

56. Sgard's *Dictionnaire des journaux* offers an encapsulated history of each of these publications.

57. See Roger Chartier, "Do Books Make Revolutions?" in *The Cultural Origins of the French Revolution*, trans. by Lydia G. Cochrane (Durham, 1991), 67–91. Goldgar, *Impolite Learning*, also notes the importance of the physical expansion of the public discourse in the first decades of the eighteenth century.

58. David Gregory, *Astronomiae Physicae et Geometricae Elementa* (Oxford, 1704).

February 1710, less than a year after Saurin had broken the official silence surrounding this topic in France by refuting the Newtonian theory of gravitational attraction publicly at the academy's spring public assembly,[59] the *Journal de Trévoux* ran a lengthy review of Gregory's treatise. As a Latin tome intended primarily for university students and scholars, Gregory's work would not otherwise have found many readers in France or elsewhere had not the French Jesuits chosen to discuss it. Thus, the lengthy and exceedingly readable review that the journal offered was crucial in disseminating Gregory's attractionist views to a wider audience.

"Occult qualities, especially those associated with attraction and *pesanteur*, were once thought to be so banished from physics and from nature by M. Descartes, that greatest of geometers. One only talked about them when looking for an example of false reasoning founded upon infantile prejudices," the reviewer began. "Able English mathematicians, however, have returned them to the world and anyone who has read the *Principia* and the *Opticks* of M. Newton has no doubt observed his credulity on this topic." Here one sees clearly the new conception of Newtonianism that emerged once one could read the argument for universal gravitation in the *Principia* in light of the related arguments about gravitational attraction in the *Opticks*. The reviewer also echoed the old understanding of Newton's work as well, lamenting that "if only one could ignore [the attractionist] principle, the subtlety of the geometry that is used throughout would make it an excellent work." It was no longer possible to affect this distinction, however, for what Gregory had done in the name of Newton was to build "a system of the world founded upon attraction, or the mutual *pesanteur* operative between bodies by which they tend according to an inclination natural and occult to unite with one another."[60]

A number of features of this precise presentation are revealing. First, the review describes an explicitly English community of mathematicians centered around Newton who are united in their belief in attractive matter. No such awareness can be found in earlier French writings. The reviewer also reverses the traditional Jesuit hostility toward Descartes by making him "the greatest of geometers" and the defender of true natural philosophy through his anni-

59. Saurin's presentation is noted in PVARS April 10, 1709. The presentation was reviewed in *Mercure galant* (April 1709): 176–184. The paper was subsequently published as "Examen d'une difficulté considerable proposé par M. Huygens contre le système Cartesien sur la cause de la pesanteur," *HARS-Mem.* (1709): 131–149. On Saurin's vortical physics, see Aiton, *Vortex Theory*, 170–177; and Brunet, *L'introduction des idées de Newton*, 25–29.

60. Review of David Gregory, *Astronomiae Physicae*, *Journal de Trévoux* (February 1710): 252–253.

hilation of "infantile" attractions. Before this date, the Jesuits typically associated Descartes with atheistic and materialistic mechanism, while making his mathematical approach to natural philosophy the source of these dangerous errors. As George Healy, who has studied the scientific discourse in the *Journal de Trévoux* in great depth, notes: "when Newton published the *Principia*, Descartes remained the main problem of the French Jesuits even as he was the chief hope of other French intellectuals. But increasingly Newton replaced Descartes in this role." [61] The journal's review of Gregory's treatise indicates that this shift was already well underway by 1710.

The rest of the journal's discussion of Gregory also helps to explain the reasons for this reversal. After first attempting to disconnect the mathematical dimension of Gregory's treatise from its attractionist physics—a by now classic French rhetorical strategy—the reviewer returned to the principle of attraction in his account of Gregory's first book. Here the journal declared: "Since attraction or *pesanteur* is, to state it simply, the soul of this universe—the entity that creates the order between all the principal parts—it is here that one must begin." [62] The reviewer continued by lamenting the difficulties of this assignment since Gregory offered no "clear and distinct explanation" of the principal of attraction at the heart of his work. The Jesuit also expressed little surprise with this result since no such explanation of attraction is possible. Countering Gregory, he instead deployed the by then classic rationalist arguments to demonstrate the absurdity of the claim that matter acts at a distance across empty space without any intervening, mechanical medium. [63]

This use of the Cartesian language of *évidence* against the "occult mystery" of attraction is significant because it reveals a new willingness on the part of the Jesuits to embrace Cartesian rationalism in this way. More innovative and more significant, however, is the language found in the opening passage. No philosopher was more dangerous in the mind of French Jesuits than Benedict de Spinoza. To call attraction, therefore, the "soul of this universe" and the source of "order between all the principal parts," as the reviewer did, was to connect Gregory's work, and Newtonian attractionism more generally, with the most threatening of all early eighteenth-century heretical philosophies: Spinozist pantheism. John Toland's *Pantheisticon*, first published in 1720, was

61. George R. Healy, "Mechanistic Science and the French Jesuits: A Study of the Responses of the Journal de Trevoux (1701–1762) to Descartes and Newton" (Ph.D. diss., University of Minnesota, 1956), 105. A similar argument is made in Alfred R. Desautels, *Les Mémoires de Trévoux et le mouvement des idées au XVIIIe siècle* (Rome, 1956), 48–59.

62. *Journal de Trévoux* (February 1710): 254.

63. Ibid.

the manifesto of eighteenth-century pantheism, and in this work, he summed up the pantheist worldview:

"[Pantheists] say *All things are from the whole and the whole is from all things. . . . They assert that the Universe (of which this world we behold with our eyes is but a small portion) is infinite both in Extension and in Virtue, but one in the continuation of the Whole. . . . [its] integrant parts are always the same, and constituent parts always in motion. . . .* From that motion and Intellect that constitute the Force and harmony of the infinite Whole, innumerable species of things arise, every individual of which is both a Matter and a Form, Form being nothing else than a disposition of Parts in each Body. From whence therefore we may conclude that the best Reason, and most perfect order regulate all things in the *Universe.*"[64]

The review of Gregory's treatise in the *Journal de Trévoux* implicated the theory of gravitational attraction with this pantheist philosophy. This association marks an important development in the emergence of eighteenth-century French attitudes about Newtonianism. Toland, who first used the term "pantheism" in his exceedingly heterodox *Origines Judaicae* published at The Hague in 1709, was a radical Irishman who turned his devotion to Lockean sensationalism and Newtonian physics into a self-consciously subversive political and religious philosophy.[65] The *Pantheisticon* was the culmination of this project, but Toland began articulating his pantheist program much earlier. In his *Letters to Serena* first published in 1704, he devoted an entire letter to a defense of the notion that motion is essential to matter. In this letter, Toland invoked the authority of Locke and Newton directly to argue that matter is necessarily active, that the activity derives from inherent natural forces, and that the action of the entire universe can be reduced to these innate, material principles.[66] Fur-

64. John Toland, *Pantheisticon, or The Form of Celebrating the Socratic Society* (London, 1751), emphasis in the original. This citation is drawn from *Pantheisticon,* facsimile edition, ed. René Wellek (New York, 1976), 15–16.

65. Much of the best scholarship on Toland's life and thought is in Italian. See Chiara Giuntini, *Pantheismo e idelogia republicana: John Toland (1670–1722)* (Bologna, 1979); and Giancarlo Carabelli, *Tolandiana* (Florence, 1975; supplement 1978). My understanding of Toland has been most informed by the work of Margaret C. Jacob, who uses these Italian sources. See esp. *Radical Enlightenment.* Also useful to me in situating Toland in a wider context has been Justin Champion, *Republican Learning: John Toland and the Crisis of Christian Culture* (Manchester, 2003); and John Y. Yolton, *Thinking Matter: Materialism in Eighteenth-Century Britain* (Minneapolis, 1983).

66. John Toland, *Letters to Serena* (London, 1704), facsimile edition, ed. René Wellek (New York, 1976), 163–239.

thermore, while claiming to refute Spinoza through his reasoning, he actually linked Newton to him by arguing that "Newton's words are capable of receiving an interpretation favorable to my opinion."[67]

Toland believed in particular that Newton's conception of space, along with his understanding of the relationship between matter and motion in the universe, was akin to his own Spinoza-like understanding of the same things. In this conception, matter, motion, and space are but three manifestations of one universal, monist Nature-God. Toland's other writings attached this materialist natural philosophy to a defense of paganism, pantheism, and radically democratic politics as well. The result was a philosophy very close in spirit and substance to Spinoza's, if not literally consonant with it, especially in its support of monism, materialism, rationalism, republicanism, and an open hostility to the mysteries of revealed religion.[68]

Spinozism was without question the great "other" against which all pious philosophy was measured in the eighteenth century. Toland's affinities with it, therefore, made him a dangerous thinker. However, by connecting Spinoza's dangerous metaphysics and theology explicitly with the theory of gravitational attraction, Toland also added something new: he made it possible to link Newtonian physics with these same subversive currents as well. It was David Gregory who first taught Toland about Newton's theory of universal gravitation during the latter's studies at the University of Edinburgh in the 1690s.[69] After this introduction, Toland never wavered in his conviction that his philosophy was grounded solidly in Newton's own thought. Toland's works nevertheless became scandalous throughout Europe, and in England they were the focus of much consternation, especially among Newton and his circle.

As Margaret Jacob has shown most fully in a number of works, Newtonians such as Richard Bentley, William Whiston, and Samuel Clarke, often guided by Newton himself, crystallized their own understanding of the Newtonian creed by fighting to distinguish it from the materialist and deist interpretations of it made by Toland and others. In particular, they used the public forum of

---

67. Ibid., 182–183.

68. I am grateful to Jeffrey R. Wigelsworth for discussing the Newtonianism of John Toland with me and for sharing the fruits of his then unpublished research. For Wigelsworth's views, see "Lockean Essences, Political Posturing, and John Toland's Reading of Issac Newton's *Principia,*" *Canadian Journal of History* 38 (December 2003): 521–535.

69. Jacob, *Newtonians and the English Revolution,* 211; Alexander Grant, *The Story of the University of Edinburgh during Its First Three Hundred Years,* 2 vols. (London, 1884), 2: 296; and John Friesen, "Archibald Pitcairne, David Gregory, and the Scottish Origins of English Tory Newtonianism, 1688–1715," *History of Science* 41 (2003): 174.

the Boyle Lectures, an annual program of ten public sermons delivered at different London churches and devoted to the relationship between natural philosophy and religion, to define an orthodox Newtonianism hostile to Toland's pantheism.[70] Samuel Clarke in fact used these very lectures in 1705 to explicitly and publicly refute the materialism articulated in Toland's *Letters to Serena*.[71] Central to the Newtonianism that Clarke and others developed through the Boyle Lectures was an ardent belief in the validity and scientific superiority of Newton's attractionist physics. Thus, while Clarke railed against Toland's pantheist interpretation of innate forces, he did so while distinguishing it from the Newtonian conception of gravitational attraction for which he had nothing but praise. This distinction, however, was often difficult to maintain in practice, and in this way, despite constant efforts to exorcise the demon, the specter of subversive radicalism, irreligion, and Spinozism continually haunted discussions of Newtonian attraction throughout the eighteenth century.

In France, where skepticism toward attractionist Newtonianism was the conventional wisdom, it was easy to use Toland's inflammatory heresies to subvert the position of the more orthodox Newtonians as well. This is clearly what the *Journal de Trévoux* was attempting to do in its review of Gregory. Yet by adopting this of all discursive strategies, the journal was also revealing an important truth about the intellectual climate of early eighteenth-century France. In a pioneering work that has not generated the subsequent research that it deserves, Ira Wade outlined in 1938 the existence of a clandestine libertine community active in France in the early decades of the eighteenth century.[72] This community, which operated via a secret practice of manuscript circulation and illicit sociability, served as a conduit for a host of radical and subversive ideas into and out of France. Toland's works figure prominently in the list of clandestine texts complied by Wade, and overall works that discuss Spinoza and Spinozist themes are omnipresent. Prevalent also are other texts devoted to favorite pantheist topics such as the materialist nature of the soul; the natural religion of Bruno and Vanini; rational Biblical criticism; and the pagan materialism of Pythagoras, Epicurus, and Democritus.[73]

Wade's greatest achievement rests in revealing the existence of this subversive intellectual culture in France from as early as the 1690s. His work has also

70. This argument is made most thoroughly in Jacob, *Newtonians and the English Revolution*. See also Jacob, *Radical Enlightenment*, 90–93.

71. See Jacob, *Radical Enlightenment*, 105.

72. Ira Owen Wade, *The Clandestine Organization and Diffusion of Philosophic Ideas in France from 1700 to 1750* (Princeton, 1938).

73. For a list of the 102 extant manuscripts found by Wade, see ibid., 11–18.

been advanced recently by that of Jonathan Israel, who has traced the growth and maturation of a Spinozistic underground throughout Europe in the decades after Spinoza's death in 1675.[74] France figures in Israel's account, as does Toland, and read alongside Wade a broad outline of the French radical community, its practices, and its intellectual concerns emerges. At the center of this culture in France in the first two decades of the eighteenth century was the Comte de Boulainvilliers.[75] Educated at the Oratorian college of Juilly outside Paris, where he was taught Cartesian philosophy, Boulainvillers began to congregate late in life at the home of the Duc de Noailles with "a small society of free thinkers [*libres chercheurs*]." Here, according to a contemporary observer, "ideas that anywhere else would have been prosecuted were exchanged."[76] Boulainvilliers also wrote tracts of his own, including several efforts to explicate Spinoza's philosophy.

In these ways, Boulainvilliers helped to foster the circulation of radical and heretical ideas throughout France.[77] He also possessed ancient noble lineage and the friendship of powerful aristocrats such as Noailles, and this allowed him to preside over what was obviously an illicit intellectual network. When he died in 1722, dozens of subversive tracts reflecting the entire range of radical heretical thought in this period were found at his estate. Among the savants known to have associated with him were the Marquis d'Argenson, an honorary member of the Academy of Sciences and a close friend of the abbé Saint-Pierre with whom he founded the Club de l'Entresol, an English-style political club, in 1723; Duclos, a member of the Académie française who, together with Dortous de Mairan and Fontenelle, was a leading light at the salons of Mme. Lambert and Tencin and a regular at the Café Gradot; Nicolas Lenglet Dufresnoy, a critical antiquarian, historian, and man of letters who linked the Boulainvilliers circle

74. Israel, *Radical Enlightenment*.

75. Wade devotes an entire chapter to Boulainvilliers, *Clandestine Organization*, 97–123. See also Israel, *Radical Enlightenment*, 565–590.

76. Bougainville, "Éloge de M. Fréret" (Paris, 1792), cited in Wade, *Clandestine Organization*, 98.

77. These works are analyzed in detail by Wade, *Clandestine Organization*, 99–123. See also Israel, *Radical Enlightenment*, 565–575; Paul Vernière, *Spinoza et la pensée française avant la révolution*, 2 vols. (Paris, 1954), 2: 306–325; and J. S. Spink, *French Free Thought from Gassendi to Voltaire* (London, 1960), 267–272. Spink sees a strong affinity between the French Spinozism articulated by Boulainvilliers and the writings of John Toland, and he lists the evidence that Toland was a source for Boulainvilliers (272, n.3). Spink also notes that the influential French cleric Huet spoke out publicly against Toland in 1714, an attack that was reported in the Dutch *Journal littéraire de La Haye* (vol. 4, 252). Overall, the works of Israel, Spink, and Vernière help fill in some of the gaps left by Wade's study.

to the world of Dutch publishing and to the Académie des inscriptions et belles lettres in Paris; Nicolas Fréret, an important luminary of the early Enlightenment, famous for his libertine philosophical tracts; Jean Levesque de Burigny, a cosmopolitan man of letters who, together with his brothers, founded *L'Europe savante* in 1718; and the scientific academician and astronomer Joseph-Nicolas Delisle, who professed a more mercenary if no less occult relationship with the count and his circle: he cast Boulainvillier's astrological charts.[78]

Through these networks it is very likely that savants, particularly those that circulated in Parisian society, became aware of the radical, Spinozist interpretation of Newtonian attraction articulated by men like Toland. Reinforcing this understanding was the appearance of this same constellation in the Dutch Francophone periodicals that became increasingly numerous after 1715. A survey of one representative volume from 1720—Michel de La Roche's *Mémoires littérai-res de la Grand-Bretagne* published in The Hague—reveals discussions of the *mémoires* of the Paris Royal Academy integrated together with summaries of the English *Philosophical Transactions of the Royal Society* for that year. Works by Toland are also discussed alongside those of his opponents, like John Clarke, the like-minded brother of Samuel. Similarly discussions of Newtonian science and method are also interspersed as well. Given this manner of presentation, it is likely that French readers of such journals thus grew accustomed to associating Newtonianism with debates about radical metaphysics and theology.[79]

The *Journal de Trévoux,* in its evocation of Spinozistic pantheism in the context of Gregory's Newtonian *Astronomiae physicae,* alerts us to the powerful resonance of these themes in the French public sphere by 1710. After this date the pantheist discourse about Newtonianism only became more widespread. It was catalyzed further by the appearance of a variety of different works that pointed to the same heretical conclusions. In 1711, the *Journal de Trévoux* reviewed William Whiston's *Praelectiones astronomicae* published at Cambridge in 1707.[80] After noting that Whiston succeeded Barrow and Newton as the Lucasian Professor of Mathematics at Cambridge, the reviewer quickly went on the attack. He claimed to hear Epicurus singing throughout Whiston's treatise and he cited a Latin passage celebrating the all-pervasive *vis animus* as a repre-

78. Wade, *Clandestine Organization,* 98. Voltaire used the pseudonym St. Hyacinth to immortalize this community in his *Dîner du Comte de Boulainvillers* (Paris, 1728).

79. *Mémoires littéraires de la Grand Bretagne,* 8 vols. (The Hague, 1720–1724). Toland's *Tetradymus* is discussed in 6: 361–421. On this journal, see Sgard, *Dictionnaire des journaux,* 823–824.

80. William Whiston, *Praelectiones astronomicae Cantabrigae in Scholios publicis habitae* (Cambridge, 1707).

sentative synopsis of the work.[81] Other overtly attractionist works of Newtonian science published before 1720 were treated by the journal in the same way, and through these means a wider public was exposed to attractionist Newtonianism and the dangerous thought increasingly associated with it.

Importantly, however, the *Journal de Trévoux* never let their critical opposition to these ideas stop them from giving a thorough exposition of the scientific claims themselves. Gregory's treatise, for example, received two long reviews and Whiston's one. In each presentation, the scientific ideas of these English Newtonians were given a full airing, even if the editorial slant in each case was antagonistic. Thus, while the French public learned through these reports of the existence of English Newtonians possessing suspect philosophical views, they also learned that these same men produced serious and often esteemed scientific works. More restrained in its reporting, the *Journal des savants* reinforced this perception. It too ran reviews of works by avowed attractionists, yet not motivated by Jesuit religious imperatives the editors at this journal rarely invoked the specter of dangerous religion in their accounts. Instead, restricted to a few hundred words, and beholden to a stern policy of objective neutrality, reviewers for the official French journal merely reported the contents and significance of the text in a matter-of-fact way. Such reporting nevertheless exposed the French public to these controversial ideas, and by implicitly conferring a quasi authority upon them through the very act of their restrained, judicious presentation, the journal contributed directly to the growing understanding of attractionist Newtonianism in France.

The bias against polemical discourse at the *Journal des savants* reflected its continuing attachments to the seventeenth-century learned world that gave birth to the journal. More in step with the new eighteenth-century journalism of Le Clerc and his followers was the *Journal de Trévoux,* which grew increasingly comfortable with a more stridently critical tone as the century progressed. The arrival of Father Louis Bertrand Castel as the new science editor of the journal in 1720 especially catalyzed this trend, and for this reason his case will be discussed separately. But in this context it is important to note how journalism as whole, given its precise position in the wider field of European learning in the early eighteenth century, contributed in a singular way to the rise of a new culture of criticism in Europe after 1700. It also contributed directly to the rise of critical Enlightenment Newtonianism, as the exchange between Pierre Rémond de Montmort and Brook Taylor in the pages of the newly created journal *L'Europe savante* reveals.

81. Review of William Whiston, *Praelectiones astronomicae,* in *Journal de Trévoux* (July 1711): 1263–1264.

Prelude to the Newton Wars:
The Rémond de Montmort–Taylor Debate

With the signing of the Treaty of Utrecht ending the War of Spanish Succession in 1713, a noticeable increase in Franco-British and Franco-Dutch intellectual exchange occurred. One of the first French savants to take advantage of this new accessibility was Pierre Rémond de Montmort. He went to London with an international group of savants at the invitation of the astronomer Edmund Halley to observe the total eclipse of the sun, which Halley had predicted for 1715. Rémond de Montmort's invitation was an acknowledgment of his strong ties to England formed prior to this date. According to Fontenelle, the young savant had fled across the Channel in 1697 at the age of twenty in order to escape the tyranny of his oppressive father. There he learned English and developed an affection for English society, but he also read Malebranche's *Recherche,* falling under its spell in the manner so typical of philosophically minded Frenchmen in the 1690s.[82] When Rémond de Montmort returned to France in 1699, he thus began to implicate himself within the Malebranche circle, studying infinitesimal analysis with Carré and Guisnée, and making the acquaintance of Varignon, who in turn put him in touch with Johann Bernoulli.[83]

Rémond de Montmort also maintained his ties with England, however, returning there briefly in 1700 and making the acquaintance of Newton during this visit. His first and most important mathematical work, *Essai d'analyse sur les jeux de hazard,* published in 1708, reflected both sides of his identity.[84] This pioneering work in mathematical probability spoke simultaneously to the interest in algorithmic and algebraic analysis characteristic of French mathematics in this period and to the interest in utilitarian "statistics," or state mathematics, that figures like William Petty and Edmund Halley had made important in England.[85] In his cosmopolitan wedding of English and French scientific

82. "Éloge de M. de Montmort," *Oeuvres de Fontenelle,* 6: 465–478.

83. Jean Bernoulli to Pierre Varignon, 20 March 1703, in *Der Briefwechsel von Johann I Bernoulli. Band 3,* 75; Pierre Varignon to Jean Bernoulli, 7 May 1703, ibid., 87.

84. Pierre Rémond de Montmort, *Essay d'analyse sur les jeux de hazard* (Paris, 1708).

85. See Mary Poovey, *A History of the Modern Fact: Problems of Knowledge in the Sciences of Wealth and Society* (Chicago, 1998), esp. 141–165; Patricia Kline Cohen, *A Calculating People* (Chicago, 1982), 28–30; and John Brewer, *The Sinews of Power: War, Money, and the English State* (Cambridge, 1988), 223–225. The history of political arithmetic in France has received less study, but its importance is noted in Simone Meyssonier, *La balance et l'horloge: La genèse de la pensée liberale en France au XVIIe siècle* (Paris, 1989), 66, 71, 141–142; and Joseph Klaits, *Printed Propaganda under Louis XIV: Absolute Monarchy and Pub-

attitudes within a learned culture that had not yet become conscious of any perceived national differences between English and French thought, Rémond de Montmort and his work personified the ties that bound these two communities in the early decades of the eighteenth century.

Rémond de Montmort's trip to England in 1715 rekindled these connections. He was made a fellow of the Royal Society during this visit, and he also made the acquaintance of several mathematicians there. These included Abraham de Moivre, a French Huguenot exile who was also doing pioneering work in mathematical probability, and Brook Taylor, a young mathematician who, like de Moivre, had close ties to Newton. Rémond de Montmort's relations with all of these English mathematicians were extremely friendly.[86] Even de Moivre, who Rémond de Montmort suspected of plagiarism, remained a cordial colleague throughout his life.[87] Rémond de Montmort also became particularly close with Taylor, beginning a vigorous correspondence with him when he returned to France in the fall of 1715.[88]

The Rémond de Montmort–Taylor correspondence reflects an intimate friendship that was only ended when the Frenchman caught smallpox and died in 1719 at the age of forty-one. The Frenchman's love affair with his future wife features prominently in the early letters, and discussions of mathematics and natural philosophy are interspersed throughout with anecdotal accounts of family matters and the intimacies of private life. Included in many letters was news about the latest shipment of champagne that Rémond de Montmort was sending to his English colleagues from his estate near Epérnay. Overall, then, Rémond de Montmort's relationship with Taylor was more than just an *honnête* acquaintance among savants. As Fontenelle described it, "they shared a tenderness somewhat like that of brothers."[89] The depth of their friendship made

---

lic Opinion (Princeton, 1975), 107, 178–179. A discussion of the European-wide phenomena of political arithmetic in the Enlightenment is also found in Andrea A. Rusnock, "Biopolitics: Political Arithmetic in the Enlightenment," in *The Sciences in Enlightened Europe*, ed. William Clark, Jan Golinski, and Simon Schaffer (Chicago, 1999), 49–68.

86. Further discussion of this relationship is found in W. Young, ed., *Contemplatio philosophica: A posthumous work of the late Brook Taylor. To which is prefixed a life of the author... With an appendix containing sundry original papers* (London, 1793); and D. Brewster, *Memoirs of the life, writings, and discoveries of Sir Isaac Newton*, 2 vols. (London, 1855).

87. See Rémond de Montmort's preface to *Essay d'analyse sur les jeux de hazard*, 2nd ed. (Paris, 1713).

88. Rémond de Montmort's letters to Taylor, including photocopies of those originals found in the Archives of the Royal Society of London, are found in Ac. Sci., Dossier Rémond de Montmort. All of my citations will be drawn from this dossier.

89. *Oeuvres de Fontenelle*, 6: 475.

the fact of their public debate about Newtonian natural philosophy in 1718 and 1719 all the more intriguing.

In the first extant letter from Rémond de Montmort to Taylor dated September 6, 1715, the Frenchman expressed frustration that, unlike Nicole, he had not received copies of Taylor's new book (most likely *Methodus incrementorum directa & inversa* published in London in 1715) or the letter that contained "several objections against the physics of M. Descartes and Father Malebranche." "I hope it is not lost," he wrote to Taylor, "for I have only received one letter from you since my departure and it contained no discussion of science."[90] In the intervening months, Rémond de Montmort's frustration was alleviated. In January 1716 he wrote to Taylor promising to "respond to all the philosophical difficulties that you proposed to me."[91] But only in March did Rémond de Montmort again broach the topic of natural philosophy. "You must be shocked not to have received my response to your objections against our properties [*propriétés*] and our manner of philosophizing in physics," he wrote. "I hope that it will soon be in a state where I can send it to you. However, since it is destined for judges as sharp as you and your colleagues, I cannot take enough precautions in writing it. We are as divided on physics as theologians, but at least we can see a bit more clearly in science than in religion. You deserve good proofs. I hope to find them for you, and since nothing pleases the mind more than natural and methodical order I will attend not only to things but also to the manner of expressing them [*non seulement aux choses mais aussi de la manière de les dire*]."[92]

More letters passed between Rémond de Montmort and Taylor over the next eighteen months, but only in August 1718 does the topic of philosophy reappear. By this date a new context for it has emerged. Thanking his friend for the "extreme pleasure" that his letter of July 14 had given him, Rémond de Montmort expressed surprise that Taylor had not yet received a copy of the manuscript in which "I undertake to defend our properties [*propriétés*] of physics." "M. de Rostaine promised to send it to you," he wrote, and he ended his letter by promising to follow up with this matter and with a new shipment of "vin de bourgogne" which is "especially good this year."[93] Yet by August Taylor had not received Rémond de Montmort's "dissertation on the properties [*propriétés*] of physics of the Cartesians compared to those of the English," which the Frenchman had entrusted to "a mathematician and philosophe"

90. Rémond de Montmort to Taylor, 6 September 1715, Ac. Sci., dossier Rémond de Montmort.

91. Rémond de Montmort to Taylor, 2 January 1716, in ibid.

92. Rémond de Montmort to Taylor, 31 March 1716, in ibid.

93. Rémond de Montmort to Taylor, 1 August 1718, in ibid.

named Rostaine to deliver.[94] Two further letters were sent to Taylor over the next year without any mention of the dissertation, and then in the summer of 1719 Rémond de Montmort succumbed to smallpox, dying in October before offering any further clarification to Taylor about his piece.

The reference to Rostaine, however, points to the posthumous history of Rémond de Montmort's work. In addition to being a mathematician and a philosophe, Rostaine was most likely a would-be man of letters with ties to the publishing houses of Holland. His precise place within the Republic of Letters is not at all clear, but what is evident is that he served as one link between Rémond de Montmort and the world of Dutch publishing.[95] Rostaine's circle, which must have included among others Thémiseul de Saint-Hyacinthe and the three Levesque brothers, was one of many in this period who took inspiration from the careers of Bayle and Le Clerc in formulating their own ambitions as men of letters. Accordingly, the group became involved in journalism, participating in the creation of the *Journal littéraire de la Haye* in 1713, and then launching their own successor, *L'Europe savante*, when the former foundered in 1718.[96] Rostaine was most likely on the fringes of each of these initiatives, and whatever his actual tie to them, Rémond de Montmort's "Dissertation on the principles of physics of M. Descartes compared to those of the English philosophers" that Rostaine delivered to Taylor at the Frenchman's request appeared in *L'Europe savante* in October 1718.[97] It was followed immediately by an anonymous response, most likely from the editors, and then, in May 1719, just months before Rémond de Montmort's death, by a longer response from Taylor.[98]

The journal shaped the meaning of Rémond de Montmort's text in important ways. Published in The Hague, *L'Europe savante* was one of several new

94. Rémond de Montmort to Taylor, 24 April 1718, in ibid.

95. I have not been able to unearth any documentation regarding "M. de Rostaine." Jean Sgard does not include him in his *Dictionnaire des journalistes, 1600–1789* (Oxford, 1999), and no other biographical dictionary of which I am aware mentions him either. Accordingly, my portrait of him is speculative and is drawn principally from Rémond de Montmort's descriptions and the history of the text with which the French academician entrusted him.

96. See Sgard, *Dictionnaire des journalistes,* 513–517, 729–730, idem, *Dictionnaire des journaux,* 395–396, 693–695.

97. Rémond de Montmort, "Dissertation sur les principes de physiques de M. Descartes comparés à ceux des philosophes anglais." "Réflexions de M. . . . sur la dissertation de M. de Montmort," *L'Europe savante* 5 (1718): 209–294.

98. Ibid., 5 (1718): 294–303; and Brook Taylor, "Response à la dissertation de M. de Montmort," *L'Europe savante* 9 (1719): 83–134.

Francophone periodicals to appear in the early years of the "Long Peace." In the first volume the editors situated their journalistic agendas within the longer history of the European learned journal. They saw Denis de Sallo's *Journal des savants*, founded in 1665, as the mother of the genre, and their initial "Preface" narrated a history of journalism that emphasized the crucial role that learned periodicals played in unifying "all the *honnête gens* of the Republic of Letters, be they in England, Italy, Germany, Holland, or France." The Jesuit *Journal de Trévoux* made an appearance in their narrative, as did Otto Mencke's Latin *Acta Erdutorum,* and the Francophone works of Bayle and Le Clerc. Newer works, like the *Journal littéraire de la Haye,* revived by Willem 'sGravesande and Prosper Marchand in 1718, were also mentioned, as was de la Roche's *Bibliotheque anglaise* begun in 1716. Overall, the history spoke to the new self-consciousness that made learned journalism a preferred occupation for many aspiring Republicans of Letters after 1700.[99]

However, the text also exposed a problem. In the wake of the overall expansion of the learned periodical in the early eighteenth century, what justification existed for creating another new monthly? The editors of *L'Europe savante* offered the usual pledge of comprehensive coverage, originality, balance, and erudition, but they also added a new justification. "We will do more than this," the editors wrote, "we will judge." This willingness to criticize, they argued, was the principal difference between their journal and others, and in their "Preface" the editors promised to "join criticism with factual reporting [*le critique à l'historique*]" so as to "create a journal that is useful even to those that have already read the books that we review."[100]

This avowedly critical perspective was not yet the norm in the Republic of Letters, which instead aspired to an ethic of disinterested and judicious neutrality. Yet while *L'Europe savante* emphasized its attachment to these older values too, it also articulated a new and more overtly critical conception of intellectual discourse that marked it as part of a new trend. From the perspective of Rémond de Montmort's debate with Taylor, which appeared in the journal's ninth and eleventh issues, the critical slant of *L'Europe savante* exerted an important effect on its reception. Earlier issues of the journal had included everything from discussions of Whig and Tory politics to debates about the theology of the Eucharist and the fad for the commedia dell'arte in Paris. Furthermore, the journal discussed these polemical topics in a style that Pierre Burger calls one of "sagacious vivacity."[101] The simple fact that Rémond de Montmort and

99. Ibid., 1 (1718): iii–xix.

100. Ibid., xiv.

101. Sgard, *Dictionnaire des journaux,* 395.

Taylor felt compelled to debate the physics of universal gravitation was an important marker of the changing intellectual climate of the time, yet the appearance of their thoroughly moderate and judicious exchange within a journal self-consciously devoted to promoting and provoking public, critical debate of philosophical matters was even more catalyzing.

There was in fact an important tension between the critical agendas of the journal's editors and the reasoned cordiality of the debate on natural philosophy that they printed. True to his promises, Rémond de Montmort constructed a carefully reasoned defense of Malebranchian physics that was rooted, in his mind, in the fundamental similarity between it and a correct reading of Newton's *Principia*. He began, therefore, neither with a defense of vortical cosmology, nor even with a critique of Newtonian attraction even if this indictment appeared later in the paper. Instead, he built his argument upon Malebranchian philosophical skepticism and phenomenalism as the proper foundation for a well-constituted physics. "One must judge things by the ideas that represent them and not by the thoughts [*sentiments*] we have about them at each instance," wrote Rémond de Montmort. "This is the famous Cartesian principle that delivered us from all the obscurity of vague words like quality, appetite, affinity, tendencies, intentionalities, substantial forms, etc., that were the basis of ancient philosophy."[102] By recognizing that we have no idea about what things are in themselves, we see the regular laws that nature obeys. "It is clear," the essay argued, "that God in his wisdom regulates everything toward certain ends through the execution of certain wise laws." It is furthermore obvious that recognizing this "simple, clear, and well-accepted principle leads one naturally toward the most sublime and important truths."[103] Proper physics follows, then, from the use of sensate appearances, properly disguised, to reveal the laws operating in God's creation.

This was no doubt a sincere statement of Rémond de Montmort's Malebranchian views, but it was also a very clever argumentative strategy. Newton had also insisted in his *Principia* that physics should be practiced in terms of the quantitative treatment of observed phenomena. As he had written in his "General Scholium": "As a blind man has no idea of colors, so we have no idea of the ways in which the most wise God senses and understands all things. . . . We have ideas of his attributes, but we certainly do not know what is the substance of any thing." Natural philosophy, therefore, should restrict itself to a "treatment of God from the phenomena."[104] On many occasions, Newton likewise

102. *L'Europe savante* 5 (1718): 209–210.
103. Ibid., 212.
104. *Princ.*, 942–943.

resisted the strong, physicalist interpretation of gravitational forces that others attributed to him. His 1698 letter to Richard Bentley made this position clear. Newton's later statements in the "Queries," added to the 1718 edition of his *Opticks,* however, suggested a different understanding,[105] yet Rémond de Montmort's strategy involved collapsing the distinctions that allegedly separated "English Newtonians" from "French Cartesians." Given the Frenchman's familiarity with Newton's work, not to mention his intimacy with the English mathematicians who admired it, he might also have been pointing to a shared opposition among all the mathematical Newtonians, including those in England, to the more overtly physicalist interpretation of Newtonianism found in the works of John Freind, George Cheyne, John Keill, and even Newton in the *Opticks.* It is also worth emphasizing that Rémond de Montmort was debating Taylor, and not Newton directly, and that he had directed his piece at "the physics of English philosophers" and not at Newton or the Newtonians per se. Framed this way, it is likely that Rémond de Montmort's real agenda was not to refute Newtonian physics itself but rather to defend the mathematical phenomenal interpretation of it that he felt united savants on both sides of the Channel.[106]

Rémond de Montmort's appropriation of Newton's "Rules of Reasoning" in book 3 of the *Principia* further supports this understanding. Rule 1 insists that no cause is to be admitted into philosophy beyond that which is sufficient to explain the phenomena. Rule 2 stipulates that we must, as much as possible, attribute the same cause for the same effects.[107] From this, Rémond de Montmort concluded that since all matter must be uniform, and motion obeys a single set of regular laws ordained by God, the most natural way to account for this motion according to Newton's own rules is to posit the existence of "certain fluids in nature" that are the cause of the motions we see.[108] This of course challenged Newton's argument in the *Principia* at its core, even if Newton's speculations about a mechanical ether in the "General Scholium," and then again in the

---

105. In addition to the attractionist physics that informs Newton's theories about the nature and propagation of light throughout the text, his queries were also used to advance his attractionist views quite forcefully. Especially pointed was query 31, which opened by asking whether "the small particles of bodies have not certain powers, virtues, or forces by which they act at a distance, not only upon the rays of light . . ., but also upon one another for producing a great part of the phenomena of nature?" *Opticks,* 375–376.

106. On the commonalities among English and Continental mathematicians such as Taylor and Rémond de Montmort before 1720, see Guicciardini, *Reading the Principia.*

107. *Principia,* 794–795.

108. "Réflexions de M. . . . sur la dissertation de M. de Montmort," 221–224.

"Queries" to the *Opticks* could be deployed to make Newton a defender of fluid mechanisms. Rémond de Montmort did not pursue this strategy. He rather argued for the fundamental similarity between his own Malebranchian approach to physics and the one offered in the *Principia*. Against Roger Côtes, whose "Preface" to the 1713 edition of the *Principia* he referenced directly, Rémond de Montmort argued that the existence of the vortices was not a "gratuitous and unproven hypothesis." [109] Rather, he reasoned, the hypothesis follows naturally from the correct application of Newton's stated and well-conceived method. "Given known truths one tries to find those that are hidden," Rémond de Montmort explained. "This is the method of analysis used by mathematicians, and the art of pulling the unknown out of the known is what renders this science the most beautiful and the most useful production of the human mind. Its usage is not limited to discovering the properties of numbers and curved lines; it extends itself to everything that is susceptible to *rapports*, and in particular to the discovery of truths in physics." [110]

Côtes had made Newton's "experimental philosophy" the champion of analytical thinking of this sort while the Cartesians were caricatured as fabulous builders of hypothetico-deductive systems. Rémond de Montmort turned this conception on its head. He made the Cartesians the real masters of analysis through their reasoned discovery of the necessity of the vortices while the "attractionist Newtonians" were positioned as the misguided *romanciers* who invented physical principles without sound reasons for them. He also offered book 6 of Malebranche's *Récherche* as the real model of the correct application of analysis and synthesis in direct contrast to Côtes' claims for the *Principia*. [111] Crucially, however, Rémond de Montmort also asserted French scientific superiority in a way that preserved the fundamental harmony between Newton and Malebranche, thus leaving the door open to a possible a reconciliation between French and English philosophy were one to be desired.

Rémond de Montmort's critique of the theory of universal gravitation was a case in point. True to his Malebranchianism, Rémond de Montmort neither insisted dogmatically that the vortices existed nor even that gravitational attraction was an obvious fiction. Rather, he argued, we can never know what the real nature of bodies is, and we are thus restricted to rational descriptions of empirical phenomena. What made attraction untenable and the vortices in turn preferable for him was the disregard that Newtonian physics seemed to have

109. Ibid., 254.
110. Ibid., 257.
111. Ibid., 259.

for these fundamental philosophical principles. Much of Rémond de Mont-mort's essay, in fact, was spent exposing these philosophical problems. Writing about space, he stressed the paradoxes that emerge when one treats space as something distinct from the extension inherent to bodies. He also noted the impossibility of understanding the interaction of bodies through means other than direct contact. "When explaining phenomena," he wrote in summary, "is it necessary to prefer abstract ideas over the true laws of nature that hold that a body only moves another body when it touches it directly and immediately? It is upon this point, Sir, that our entire controversy rests, as well as the controversy that places the English against almost every other philosopher in Europe."[112]

The rest of Rémond de Montmort's critique pursued a similar approach. He never challenged the physics of universal gravitation per se, but rather argued that there were no reasonable foundations for accepting this theory and many irrefutable arguments against it. "It is to cast ourselves into the shadows to admit as principles forces and virtues of which we have no idea."[113] Similarly, with regard to Newtonian empty space, Rémond de Montmort reasoned through all the arguments supporting skepticism about the vacuum, concluding ultimately that "after so many arguments that prove the existence of the plenum, to accept it is not to make a hypothesis; it is rather to refuse to make one."[114] He also stressed that his conclusions were sustained by a careful reading of Newton's *Principia*. Ignoring altogether Newton's empirical and physical attacks on the system of the vortices in book 2 and the "General Scholium," Rémond de Montmort stressed Newton's own indeterminancy about the physics of attraction. "Never has anyone carried the exactitude and the inventiveness of mathematics and experiment further," he wrote, and on this count he agreed wholeheartedly with all the praise that the English had lavished upon Newton. "But I stop there," he continued, "and I know M. Newton is too equitable and enlightened to demand more." Others, however, were not so judicious. "They want the law of universal gravitation to be more than just a complete account of the movement of the heavens; they also want it to be a physical account of these movements and many other motions as well."[115] Such a deduction was

112. Ibid., 290.

113. Ibid., 233.

114. Ibid., 267.

115. Ibid., 290. Rémond de Montmort cited in particular the chemical work of Cheyne and Freind, recently published in the *Philosophical Transactions of the Royal Society*, as illustrative of this overextension of the theory of universal gravitation.

wholly ungrounded in Rémond de Montmort's estimation, and by deploying "his Newton" against the Newton of the attractionist Newtonians, he was both defending his authentic Malebranchianism and arguing for a unity between his views and those of a properly conceived Newtonianism.

For his part, Taylor seemed anything but uncomfortable with the moderated consensus that Rémond de Montmort was suggesting, so long as his own revisions to the Frenchman's arguments were accepted too. His tone was equally cordial throughout, and like Rémond de Montmort he also emphasized the differences that existed between his position and those held by the more stridently attractionist Newtonians. Against his opponents' argument that reason demanded a mechanical point-contact understanding of bodily interaction, Taylor responded with his own skeptical argument. "It is not proven that material principles alone account for the phenomena of nature," he asserted.[116] We have no idea how bodies in fact interact, he argued, echoing Malebranche verbatim, and thus to attribute bodily action to material contact is just as gratuitous as attributing it to attractive forces operating across empty space. Furthermore, agreeing with Rémond de Montmort that it was wrong to make attraction a fundamental property of bodies, he then challenged the Frenchman by saying the same thing about impulsion. "We know that motion is caused somehow, but how it is caused we cannot say for certain."[117]

His own defense of Newtonian attraction was built on the same skeptical foundation. "It is fallacious to claim that impulsion is the only possible cause of motion," he wrote, "because this is asserting with certainty that which we do not know."[118] Viewed this way, gravitational attraction was no less philosophically suspect than impulsion, and it could, therefore, ground a true physics. Nature, in fact, should remain the ultimate judge, Taylor argued, and in making this point the real difference between him and Rémond de Montmort emerged. "You say that neither reason nor experience offer any proof that attraction exists, but experience in facts proves the contrary. . . . Reason and experience show us that there are in fact attractions that are not caused by impulsions even if this is not the same thing as saying these motions are caused by attraction." Explaining further, he added: "It is true that we have no idea what attraction is, . . . which is to say that we do not know in what manner it is caused. But we do have an idea of it in another sense. For it consists in the tendency of one body toward another and we have an idea of this tendency even if we do not understand it at all. We cannot reject something just because the

---

116. "Réponse à M. de Montmort," *L'Europe savante* 9 (1719): 85.
117. Ibid., 88.
118. Ibid., 90.

cause of it is unknown to us. Therefore, gravitation is not a hypothesis, it is a fact that experience obliges us to accept."[119]

Here was the difference that would forever disrupt any unity among the mathematical Newtonians. Whereas Rémond de Montmort and his French colleagues always stressed the philosophical irrationality of attraction and its corollary conceptions of space and physical action, Taylor and his English compatriots always stressed the validity of the theory as a possible empirical description of nature while rejecting that nature had to act in philosophically rational ways. In the rest of his presentation, therefore, Taylor offered what would become the standard arguments leveled against rationalist Cartesian physics in the eighteenth century. On the question of space, Taylor acknowledged that the distinction between space and bodily extension was difficult to comprehend philosophically. But, he argued, "the entire dispute reduces to a question of fact. It is entirely a question of whether the space that we both conceive is actually full or not; this cannot be decided by abstract reasoning, but only by experience."[120] Similarly, against Rémond de Montmort's argument that God could not allow his particular will to be the cause of action in nature because to do so was to violate reason, Taylor countered: "we have no idea how God works," and "therefore you cannot exclude the will of the creator as a cause."[121] He made it clear that this was not the same thing as saying that the will of the creator was *in fact* the cause of gravity; rather, he insisted, it *could be* the cause, and, therefore, cannot be ruled out as such on rational grounds alone.

This pragmatic empiricism laid the foundations for Taylor's final arguments. Insisting that Rémond de Montmort was erroneous in ruling out the theory of gravitation on philosophical grounds alone, Taylor argued that the theory in fact accounted for the empirical facts much more completely than the system of vortices. He hit particularly hard on the recent research about resisting fluids, arguing in detail that this work demonstrated the devastating inadequacies of the vortical system.[122] He similarly challenged Malebranche's theory of light and color, finding his explanation "far less satisfactory" than Newton's.[123] In these passages, Taylor rebutted Rémond de Montmort's Cartesianism directly, making it clear that the English possessed a superior natural philosophy, and one fundamentally at odds with the prevailing assumptions

119. Ibid., 96–97.
120. Ibid., 108–109.
121. Ibid., 105.
122. Ibid., 120–126.
123. Ibid., 113–115.

in France. Yet even here the opposition between the two antagonists was less irreconcilable than it would later become. Each shared a common devotion to philosophical skepticism and an eagerness to let nature and reason stand as the ultimate arbiters of their dispute. They also adhered strictly to the canons of *honnête* decorum prized by Republicans of Letters. Later exchanges between Cartesians and Newtonians would not be so judicious, and the appearance of the Rémond de Montmort–Taylor exchange among the critically edged content typical of *L'Europe savante* reveals some of the factors responsible for this transformation.

Writing to each other, Rémond de Montmort and Taylor adhered strictly to their ethic as savants and academicians and to their personal ties of friendship. In this way, their actual published debate stood as a paragon of the reasoned intercourse prized by academicians and Republicans of Letters more generally. The editors of *L'Europe savante*, however, had different commitments. They spoke to a wider public, including those not beholden to these stringent codes of intellectual *honnêteté*. They also sought to participate themselves in the philosophical discussions of the day by injecting a critical voice into them. As journalistic entrepreneurs, they also wanted to offer a compelling journalistic product to their readers, one that was likely to be more publicly appealing to the extent that it attracted interest through the spectacle of debate. Seeking to satisfy these multiple agendas and true to their pledge to act as a critical voice, the editors thus appended a brief set of "Reflections" on Rémond de Montmort's initial dissertation that cast the debate in very different terms.

The text started by framing the dialogue in terms of a discourse of revolution and upheaval. "Descartes imagined a happy system by which he alleged to explain all the phenomena of physics," the writer explained. "In it, he equated the certitude of his ideas with the certitude of geometric demonstrations. Philosophers applauded almost all the ingenious speculations of this great man, but some asked whether it would be advisable to consult nature through experiments in order to learn from it the manner of its action." These critics also discovered "the true laws of motion, which are very different than Descartes'."[124] Soon after this discovery, "a book bristling with geometry appeared in England." This was Newton's *Principia*, and it claimed to "destroy the Cartesian system at its foundation and to establish another in its place." To the English, the writer continued, "the demonstrations on which this new system was built appeared to be as exact as those of Pappus and Archimedes." Beyond Albion, however, they did not have the same reception. "M. Newton, who made all of England his servant, could not find a single partisan in France. Is it possible

124. Ibid., 294–295.

that Fortune, who intervenes to limit the conquests of princes, is also meddling in the conquests of philosophers? If so, then one is pleased to see M. de Montmort rallying thinkers to end a division that is so prejudicial to the progress of the sciences. Here, however, we offer several reflections that will raise doubts about whether he has perfectly fulfilled our hopes."[125]

This formulation frames the French-English dialogue about Newton in completely different terms than those used by Rémond de Montmort and Taylor themselves. Neither Rémond de Montmort nor Taylor thought in terms of an all-or-nothing choice between Descartes or Newton, nor did they see the work of Newton as a "revolutionary overturning" of the Cartesian system. Instead, they looked at the work of each philosopher in terms of a unified understanding of the development of physics. The opposition they saw between Descartes and Newton, therefore, was not couched in terms of absolute conceptions of right and wrong, or French versus English philosophy, and especially not in terms of parties and revolutions. What Rémond de Montmort and Taylor saw were different approaches to the same intellectual goal. Similarly, while the commentator stressed the irreconcilable divide between English and French attitudes on these questions and suggested a powerful role for national prejudice in sustaining them, Rémond de Montmort and Taylor personified the cosmopolitan culture of the Republic of Letters that saw such conceptions as petty, partisan, and unworthy of honorable savants. The editors of *L'Europe savante,* by contrast, demonstrated the potential that caricatures like "English Newtonian" and "French Cartesian" held for sensationalizing a complex intellectual debate.

The remainder of the response itself, which was directed entirely toward exposing the errors of Rémond de Montmort's reasoning, also sensationalized the debate in other ways. Rémond de Montmort had invoked theological notions of a rational God when arguing for the impossibility of gravitational attraction, for as he saw it the theory of attraction could only be justified by invoking the arbitrary will of God as its ultimate cause. Yet since this conception deprived God of reasoned foresight, while invoking his will as a cause for physical action in the world, for the Frenchman it was tantamount to offering no justification at all. English Newtonians, however, had been defending this precise claim for over a decade. Samuel Clarke asserted it from a London pulpit in 1705 as part of his Boyle Lecture, arguing that the course of nature is "nothing else but the will of God producing certain effects in a continued, regular, constant and uniform manner."[126] Rémond de Montmort was likely aware of such arguments as he

---

125. Ibid., 5: 296.

126. E. Valiati ed., *Samuel Clarke. A Demonstration of the Being and Attributes of God and Other Writings* (Cambridge, 1998), 149.

wrote. The text of Clarke's 1705 Boyle Lecture circulated widely and was especially well known among the Newtonians with whom he associated after 1715. A French translation of the text published in Amsterdam was also reviewed by the *Journal de Trévoux* in February 1718.[127] Taylor also used a similar argument in his response to Rémond de Montmort even if their discussion stopped short of articulating this precise disagreement. Since this issue raised larger theological issues that that the editors of *L'Europe savante* were more eager to exploit, and since they had already devoted the pages of their journal to discussions of the new Spinozist, deist, and pantheist philosophies that were beginning to circulate within the Republic of Letters, discourses in which the question of God's will and its rationality were central, the respondent to Rémond de Montmort seized on these possibilities, linking the Frenchman's conception of a rational God to Spinozistic determinism. He also made a point of challenging Rémond de Montmort's argument about the inseparability of space and matter since this too smacked of Spinozistic pantheism.[128]

The agenda of Rémond de Montmort and Taylor was of course very different. Yet because their debate appeared in a journal framed by a critical editorial commentary, many readers took a different message from their exchange. In the Paris Academy, where Rémond de Montmort had been a member since 1716, it is most likely that the subtle philosophical nuances of the debate were appreciated while the crass and unseemly polemicism of the journalists was ignored. In the wider public sphere, however, different outcomes were possible, and here the expansion in the number of critical learned periodicals in this period finds its ultimate significance. These new journals effectively opened up the Republic of Letters to a more widely inclusive and boisterous discourse about natural philosophy than previously existed. They also challenged the control over public scientific discourse exercised by institutions like the Paris Academy and the older, more official periodicals like the *Journal des savants*. None of Rémond de Montmort's arguments, in fact, appear to have been aired in an academic session, nor was there any noticeable discussion of Taylor's Newtonian positions in the assemblies. Certainly, the absence of any self-proclaimed Newtonians in the academy goes a long way toward explaining this outcome, but equally important was the way that partisan disputation remained something inappropriate for a gentlemanly learned society.

127. Review of Samuel Clarke, *De l'existence et des attributs de Dieu, des devoirs de la réligion naturelle, et de la verité de la réligion chréstienne*, in *Journal de Trévoux* (February 1718): 247–283.

128. "Réflexions de M. . . . sur la dissertation de M. de Montmort," 297.

Fontenelle did mention the exchange with Taylor in his "Éloge de M. Mont-mort" delivered only a month after Taylor's response was printed.[129] He failed to mention the journal where it appeared by name, however, and he reduced his discussion to but a few sentences. Nevertheless, the secretary did use the opportunity to exert his authority in shaping the reception of the Rémond de Montmort–Taylor debate. Pointing to Rémond de Montmort's strong affection for England, he wrote that these associations in no way led his academic col-league to "accept the attractions, abolished, one had thought, by Cartesian-ism, yet revived by the English who nevertheless sometimes hide the love they feel for them."[130] Fontenelle continued in this vein when discussing the de-bate itself. "[Rémond de Montmort] had a great quarrel with M. Taylor about [gravitational attraction], and even composed, with care, a rather long disserta-tion in which he returned the attractions to the nothing from which they had escaped."[131] The secretary further noted Taylor's "long reply" but made no mention of the role that the journal and its respondent played in the exchange. Instead, he summed up the conclusions of the debate judiciously, writing that "if one wants to understand the discussion, then there are only impulsions. But if one does not care about understanding it, then there are attractions and anything else one would like. But since Nature remains so incomprehensible to us, maybe it is wiser to just leave it where it is."[132]

This *honnête* management of the dispute was typical of Fontenelle, yet in its overt and public invocation of the opposition between French Cartesian-ism and English attractionist Newtonianism it was also a significant innova-tion as well. Among other things, it indicates that the attractionist hypothesis was becoming more important in France by 1719 and that the academy was increasingly being forced to acknowledge its existence and influence. Crousaz learned a similar lesson when he tried to engage in a debate about experimen-tal physics with a leading English Newtonian, John Theophilus Desaguliers. Larry Stewart has studied the rise of the English experimental Newtonians in Britain that Crousaz confronted in 1718.[133] Their science was deeply rooted in the practice of public display and urban sociability, but it also associated itself quickly and explicitly with the authority of Newton. Crousaz found him-self at odds with the leading experimental philosophers of England in the final

129. *Oeuvres de Fontenelle*, 6: 465–478.
130. Ibid., 6: 474.
131. Ibid.
132. Ibid.
133. Stewart, *Rise of Public Science*.

years of his life, despite his devotion to experimental natural philosophy as he understood it. The reasons for this friction reveal another dimension of the changing culture of the Republic of Letters that gave birth to Enlightenment Newtonianism.

Critical Newtonianism and the New Experimental Philosophy

Perhaps the first of the Newtonian experimental philosophers was John Harris, who began giving public demonstrations of the "true Mechanick philosophy" in London in 1698. Harris was succeeded by John Hodgson in 1707, and by this date Newtonian experimental physics was well established in England. Francis Hauksbee, an instrument maker by trade, began giving public demonstrations in this period, and he was joined by university professors such as William Whiston and John Keill and public impresarios and entrepreneurs such as Humphry Ditton and Benjamin Martin. These public experimental demonstrators wore their Newtonian colors as a mark of honor. And since the first decade of the eighteenth century also witnessed the first public attacks on Newton, especially those made by Leibniz and Bernoulli in the context of the famous calculus priority dispute to be discussed in the next chapter, the Newtonianism of the experimental physicists quickly became polemical.[134] These polemics often turned the new experimental philosophy, which these individuals believed they had learned from Newton himself, into rhetorical vehicles for defending their master. Accordingly a vigorous and polemical discourse in support of Newtonian experimental philosophy was forged, one that evoked the coffeehouses and other public spaces from which it was born.

Francis Hauksbee captured the new esprit of the Newtonian experimental philosopher in his 1709 *Physico-Mechanical Experiments on Various Subjects*. "The learned world is now almost generally convinced," he wrote, "that instead of amusing themselves with vain hypotheses, which seem to differ little from romances, there's no other way to improving natural philosophy but by demonstrations and conclusions founded upon experiments judiciously and accurately made."[135] Pronouncements such as these helped to define an assertively experimental approach to Newtonian mechanics and philosophy. Côtes reinforced this development in his 1713 "Preface" to Newton's *Principia*, of-

---

134. On the priority dispute, see A. R. Hall, *Philosophers at War: The Quarrel between Newton and Leibniz* (Cambridge, 1980).

135. Cited in Stewart, *Rise of Public Science,* 118. I have modernized the spelling and punctuation in the original.

fering a systematic, philosophical defense of Newton's "experimental philosophy" couched within the same partisan rhetoric. After dismissing, with Hauksbee, those philosophers "who take a liberty of imagining at pleasure unknown
figures and magnitudes" and who thus "run out into dreams and chimeras,"
Côtes defined the more prudent "experimental philosophy" as that one which
"derives the causes of all things from the most simple principles possible." Experimental philosophers "contrive no hypotheses, nor do they admit them into
natural science otherwise than as questions whose truth may be discussed," he
wrote. Instead, they proceed by a "twofold method, analytic and synthetic,"
by first deducing forces from phenomena through analysis, and then, through
synthesis, showing the consolidation of the rest.[136]

The new partisan polemicism of the Newtonian experimental philosophers
is clear in Côtes' preface. Less evident is the role of experimental practice itself
in this philosophy.[137] Newton's work filled this gap, and the Newtonian experimentalists were quick to draw upon it when constructing their ideal image of the
experimental philosopher. Newton's *Opticks* was easily read as a report of truths
derived and demonstrated experimentally, yet the *Principia,* with all of its complex mathematics and mathematico-physical arguments, was a tougher sell as
experimental physics. Eager to make Newton one of their own, however, the
English experimental Newtonians transformed the complex mathematical and
philosophical architecture of the *Principia* into a defense of this very thing.

Three aspects of the *Principia* supported its status as a manual of experimental philosophy. First was the role of Johannes Kepler in Newton's work. In
his widely influential *A View of Sir Isaac Newton's Philosophy* first published in
1728, Henry Pemberton, a public, experimental demonstrator himself, stated
the essential claim. "Now since each planet moves in an ellipsis and the sun is
placed in one focus, Sir Isaac Newton deduces from hence, that the strength
of this power (the centripetal force towards the sun) is reciprocally in the du-
plicate proportion of the distance from the sun."[138] Kepler had empirically established the elliptical character of planetary motion over a half-century before
the publication of the *Principia,* and to link Newton's work to Kelper's as Pemberton did was to make the theory of universal gravitation a physical induction
from these preexisting empirical facts. Interpreting Newton's achievement this
way made the *Principia* essentially a book of inductive, empirical physics in

136. *Principia,* 386.
137. On this relationship, see Alan E. Shapiro, "Newton's 'Experimental Philosophy,'"
Early Science and Medicine 9, no. 3 (September 2004): 185–217.
138. Cited in Curtis A. Wilson, "From Kepler's Laws, So-called, to Universal Gravitation: Empirical Factors," *Archive for History of Exact Sciences* 6 (1969–1970): 89.

contradistinction to the mathematical and phenomenalist reading of the text that first took hold in France and elsewhere in the 1690s.[139]

Diligent experimental practice was also crucial to this particular understanding of Newton's achievement. At the heart of the central argument of the *Principia* is Newton's demonstration that mass and weight have a reciprocal, quantitative relationship. The significance of this demonstration in the treatise cannot be overstated. As Dana Densmore explains: "That these two considerations counter-balance is not, perhaps, surprising when both are brought to the mind together. But that matter's resistance to being set in motion should exactly equal the gravitational attraction between it and the earth is mysterious."[140] Newton proved that they are equal, establishing the physical foundations of classical mechanics as a result, and experiment figured conspicuously in his demonstration of this truth.

Newton described the process of discovery that led to the mass/weight relationship in book 3, proposition 6 of the *Principia*. "Others have long since observed that the falling of all heavy bodies toward the earth . . . takes place in equal times, to a very high degree of accuracy." Based on this understanding, he then set out to experimentally confirm the mass/weight correlation. Using pendulums, he tried experiments whereby bobs of equal weight, hung at equal heights, were filled with different substances and then swung to determine the comparative times of their oscillation. "The weight of the one to the weight of the other" was correlated, he found, "and the like happened in the other bodies. In these experiments, in bodies of the same weight, a difference of matter that would be less than the thousandth part of the whole could have clearly been noticed, had any such been."[141] In fact, no such difference exists.

In this demonstration, Newton offers the mass/weight relationship as a truth derived through experimental practice with pendulums. Such stories offered powerful evidence that Newton was primarily an experimentalist in his approach to physics despite the deeply mathematical character of so much of his work. Yet even Newton's mathematics, when understood carefully, could

---

139. The idea that Newton took Kepler's laws as empirical givens and then derived his celestial mechanics from them was long taken for granted in Newton scholarship. For a representative example, see E. J. Dijksterhuis, *The Mechanization of the World Picture* (Oxford, 1961), 477. This idea has now been significantly undermined by more recent scholarship. Most important is Wilson, "From Kepler's Laws, So-called, to Universal Gravitation." See also Richard Westfall, *Never at Rest*, chap. 10; and Cohen, *Introduction to Newton's* Principia, 47–54.

140. Dana Densmore, *Newton's* Principia: *The Central Argument* (Santa Fe, 1995), 346.

141. *Principia*, 806–807.

also be used to claim that the term "experimental philosophy" described his science as a whole. Writing in the *Principia*'s "General Preface," Newton described geometry as founded in mechanics. "Geometry does not teach us to describe these straight lines and circles, but postulates such a description," he wrote. "Therefore geometry is founded in mechanical practice, and is nothing but that part of *universal mechanics* which accurately reduces the art of measuring to exact propositions and demonstrations."[142] Stated another way, Newton is claiming that the objects of geometry—lines, circles, triangles—originate in the physical world and that "geometry does not teach us to describe these straight lines and circles, but postulates such a description."[143] By framing the foundations of geometry this way, Newton claims that mathematics, while abstracted, is nevertheless an empirical science rooted in objects that exist in the physical world. Such a view helped to make Newton's complex mathematics an adjunct to his experimentalism in the mind of those seeking such a hierarchy. It also helped to reinforce the idea that Newton was literally a "mechanical philosopher," a laborer who tinkered with machines, despite the abstruse mathematical thinking and philosophy that was obviously of deep concern to him as well.[144]

Such a view resonated powerfully with the experimental demonstrators who wanted to place the *Principia* alongside the *Opticks* in canonizing Newton as a hero of experimental physics. Out of this fusion, experimental Newtonianism was born. Côtes' more philosophical approach to these matters (he was a professor at Cambridge) and Newton's embrace of them gave this experimental philosophy a much needed intellectual cachet, yet the public, experimental demonstrators themselves remained the heart and soul of the movement. It is here that the influence of John Theophilus Desaguliers in this period looms so large. Born of a Huguenot family in La Rochelle, Desaguliers fled to England with his family after the revocation of the Edict of Nantes in 1685.[145] As a young

---

142. *Principia*, 381–382.

143. Ibid.

144. D. T. Whiteside supports this reading in "Patterns of Mathematical Thought in the Later Seventeenth Century," *Archive for History of Exact Sciences* 1 (1960–1962): 179–388. Also supportive is the work of Niccolò Guicciardini. His forthcoming book on Newton's "philosophy" of mathematics, which he generously allowed me to read in manuscript, illustrates the numerous ways that Newton was an empiricist, and even a kind of mechanical experimentalist, as a mathematician.

145. On Desaguliers's career, see Stewart, *Rise of Public Science*, 119–133; Jacob, *Radical Enlightenment*, idem, *Living the Enlightenment: Freemasonry and Politics in Eighteenth-Century Europe* (Oxford, 1991); Schofield, *Mechanism and Materialism;* Jean Torlais, *Un Ro-*

man he studied philosophy at Oxford under John Keill, and in 1709 he succeeded his master in his university post as physics demonstrator. He also made
the acquaintance of Newton in this period, and after proving himself a loyal
and talented ally, he was made a fellow of the Royal Society in 1714.

By this date, Desaguliers was living in London and offering public courses
in experimental physics in coffeehouses and other public venues. His courses
were perhaps the most popular and widely attended of all the public lecture
courses in London. In 1717, he added publishing success to his list of triumphs,
bringing out a textual accompaniment to his public courses, *Physico-mechanical
lectures,* to great public acclaim.[146] As a member of the Royal Society, Desaguliers also participated in a number of official scientific activities. In 1715, he
was chosen to demonstrate Newton's prism experiments before Rémond de
Montmort and the rest of the international delegation in England to observe the
solar eclipse. In the same year, he also performed a more complete set of physical experiments for the members of the Dutch Embassy, a group that included
Willem 'sGravesande. Desaguliers staged similar displays for dignitaries from
Spain, Sicily, Venice, and Russia in subsequent years.[147] He was also involved
in the emerging culture of Freemasonry, helping to establish some of the first
Masonic lodges on the Continent in the 1720s and 1730s.[148]

Through these and other networks Desaguliers defined an important standard for experimental Newtonian philosophy. He combined serious philosophy of a strongly Newtonian sort with an attention to those topics that amused
and entertained. In his *Physico-mechanical Lectures,* Desaguliers defined experimental Newtonianism as a "mathematical method, but only analogically
so." "Like geometers," he wrote, "I offer a chain of propositions proving each
other. But instead of definitions, axioms, and postulates, purely geometrical,
the experiments made at the first lecture prove the precepts of the second, and
so on."[149] In Desaguliers's hands this approach to physics became both an influential example of the emerging experimental understanding of Newtonian
science and a tremendous popular success. British experimental Newtonianism also spread to Holland, and by 1705 Jean Le Clerc was giving English-style

chelais grand-maître de la Franc-maçonnerie et physicien au XVIIIe siècle: Le Révérend J. T.
Desaguliers (La Rochelle, 1937).

146. John Theophilus Desaguliers, *Physico-mechanical lectures, or An account of what is
explain'd and demonstrated in the course of mechanical and experimental philosophy, given
by J. J. Desaguliers, M.A. F.R.S.* (London, 1717).

147. Stewart, *Rise of Public Science,* 121–122.

148. Jacob, *Living the Enlightenment,* esp. 88–89.

149. Cited in Stewart, *Rise of Public Science,* 123.

experimental physics classes in Amsterdam,[150] while 'sGravesande, encouraged by Desaguliers, began doing the same in The Hague by 1715.[151] Other Dutch savants soon joined them, and the culture of public experimental science became as popular in Holland as it was in Britain.

There was much in the method and philosophy of experimental Newtonianism that Crousaz could agree with. He too believed that sound natural philosophy began with clear empirical principles and ended with general conclusions demonstrated through rigorous, mathematical reasoning. He also worried about those who moved too quickly to synthetic demonstrations, producing as a result imaginative chimeras in the guise of philosophical systems. He had in fact applied the same reasoning in his implicit critique of Newtonianism in his *Traité du beau*. Thus if Crousaz did not share Desaguliers's infatuation with Newtonian physics, it was not a difference rooted in disagreements about philosophical method.

So what triggered the feud between Crousaz, Desaguliers, and the other English Newtonians that began after 1714? Social and cultural differences rooted in rival conceptions of the place of experimental Newtonianism within the Republic of Letters were more central than any particular methodological or philosophical difference. After Desaguliers had demonstrated Newton's prism experiments to Rémond de Montmort in London in 1715, similar demonstrations were also performed in Paris before the members of the Paris Academy. Dortous de Mairan's first academic presentation in 1718, in fact, was a performance of Newton's prism experiment to his new colleagues in Paris. It triggered virtually unanimous acceptance afterwards of Newton's claims about the polychromatic nature of white light as a fact of nature proven by experiment. In the discourse that followed, however, Dortous de Mairan rejected Newton's theories about the propagation and refraction of light, arguing instead for a theory of his own demonstrated according to the rigorous methods of synthetic, mechanical reasoning.[152] Dortous de Mairan was representative of many in his eagerness to reason experimentally while drawing anti-Newtonian conclusions from such practices, and Crousaz likewise ran afoul of the English Newtonians for practicing what they perceived as exactly this kind of anti-Newtonian and Cartesian mechanist science.

Prodded perhaps by the reports of Dortous de Mairan's demonstration in Paris or by the wider discussion of Newton's optical theories in the Republic of

150. Barnes, *Jean Le Clerc*, 144.

151. Stewart, *Rise of Public Science*, 122.

152. On this demonstration, see Kleinbaum, "Jean Jacques Dortous de Mairan," 150–152; and Guerlac, *Newton on the Continent*, 65–113.

Letters, Crousaz began to experiment with prisms himself in 1718. In August, he wrote to Bignon of his work. He stressed the complexities that must be attended to before experiments can be performed correctly and the errors that emerge from careless experimental practice. Summing up, he wrote that "my experiments seem to perfectly conform with the hypothesis of M. Newton on color that is today so fashionable." Yet, he also concluded cautiously that it was prudent to resist drawing any overly hasty conclusions from such "simple, even puerile, experiments."[153] Such reticence resonated with his other cautions against overly zealous theorizing, and it also echoed Fontenelle's caution about experimental philosophy, which he expressed in the pages of the academy's *histoires*.

The work of a young French academician named Saulmon can illustrate Fontenelle's stance vis-à-vis the new emphasis on experiment in natural philosophy that became more pronounced after 1715. In 1707, Saulmon entered the academy as an *éleve*, and soon after he began working on the Cartesian vortical theory of weight. In one of his first recorded interventions in the academy, he rose in 1712 to defend Cassini II's vortical theory of the tides against the criticisms launched against it by Varignon.[154] After this intervention, Saulmon began to study more closely the forces created by swirling fluids and their mechanical effect on material bodies. This work culminated in his first appearance at an academy public assembly in the fall of 1714.[155] At this session, he presented a discourse to the public on the vortical foundations of *pesanteur*, but one with a rare twist: he further offered the public an experimental demonstration supporting his arguments.

Saulmon was aware of Huygens's 1669 bucket experiment and the difficulties with the vortical theory that it raised. He, therefore, set out to repeat the experiment in a way that would resolve its problems. His solution was a newly shaped cylindrical vortex, which he used to model the mechanical effect. In November 1714 he demonstrated the experiment and the theories he drew from it to the assembled public at the academy's fall session.

The decision to offer the public an experimental demonstration of *pesanteur* was a brilliant ploy on Saulmon's part. The *Mercure galant* in particular attests to the widespread interest in experimental demonstrations among elite society in this period.[156] The Paris Academy, however, was not at the center

153. Crousaz to Bignon, 19 August 1718, FJPC, 1: 105.
154. PVARS, May 28, 1712.
155. PVARS, November 14, 1712. See also the correspondence between Bignon and Saulmon in preparation for this presentation in Ac. Sci., Dossier Saulmon.
156. See, for example, *Mercure galant* (August 1701): 281–285.

of this culture. Unlike England, where public displays of experimental physics had become part of the fabric of urban coffeehouse culture and fixtures of official science at the Royal Society, experimental demonstrations in Paris remained largely the preserve of university professors who used these displays to illustrate physical principles so as to instruct their students. Such demonstrations had nevertheless become public events by 1700, attracting audiences that far exceeded the number of students enrolled at the university. Sorbonne professors such as Guillaume Dagoumer and Pierre Polinière attracted large audiences to the University of Paris after 1690 to witness experimental demonstrations, and in 1709 Polinière issued a textual accompaniment to his course, *Experiences de physique,* that went through many editions while attracting a wide audience.[157] The new public academy had yet not succeeded by 1714 in tapping into this reservoir of interest in public experimental science, however. As a result, when Saulmon connected the academy to this emerging culture, he was initiating an important innovation.

The arrival of public experimental physics at the Royal Academy also forced Fontenelle into a new rhetorical position. The secretary was anything but uncomfortable with the vortical physics that Saulmon demonstrated, yet his work forced the secretary to deal publicly with something new: an overtly experimental justification for it. For this reason, his account of Saulmon's work is revealing. "Nothing would be more glorious for physics aided by geometry than to discover how the mechanical laws that we know produce celestial motions, and give every appearance of being the general cause of *pesanteur* as well, do indeed produce both," he began. Such a discovery "would give us an ex-

---

157. The *Mercure galant* attests to the vogue for public experimental demonstrations in turn-of-the-seventeenth-century Paris. See *Mercure galant* (August 1701): 281–285. The connection between the University of Paris and public physics demonstrations is explored in Laurence Brockliss, *French Higher Education in the Seventeenth and Eighteenth Centuries: A Cultural History* (Oxford, 1987). The best work on public experimental demonstrations in eighteenth-century France is Michael R. Lynn, *Popular Science and Public Opinion in Eighteenth-century France* (Manchester, 2006). On Polinière, see Gad Freudenthal, "Littérature et sciences de la nature en France au début du XVIIIe siècle: Pierre Polinière, l'introduction de l'enseignement de la physique expérimentale à l'Université de Paris et *l'Arrêt burlesque* de Boileau," *Revue de synthèse* 99–100 (1979): 267–295, idem, "Electricity between Chemistry and Physics: The Simultaneous Itineraries of Francis Hauksbee, Samuel Wall, and Pierre Polinière," *Historical Studies in the Physical Sciences* 11 (1981): 203–229; Blake T. Hanna, "Polinière and the Teaching of Physics at Paris: 1700–1730," in *Eighteenth-Century Studies Presented to Arthur M. Wilson,* ed. Peter Gay (Hanover, 1972), 13–39; Geoffrey Sutton, *Science for a Polite Society: Gender, Culture, and the Demonstration of Enlightenment* (Boulder, 1995), 195–197, 201–204.

planation of a phenomenon that is both exceedingly common and exceedingly difficult to explain." Moreover, since the a priori reasons for these motions are often too hidden from our view, it is necessary to climb to them little by little by experiments. For this reason, Saulmon gives us his "artificial heavens," or his "rotating fluids that carry objects in the manner of planets."[158]

Saulmon justified his experimental approach in similar terms. He stated in his *mémoire* that "the movements of circular fluids produce a tremendous range of effects in nature including the periodic movements of the planets and *pesanteur*. Experiments are one of the surest means of recognizing these causes."[159] Fontenelle, however, was less assertive in his conclusion. He stressed the plethora of complicating factors that separated an experimental model from nature itself, while cautioning against any overly strong conclusions from experimental demonstrations of this sort. Furthermore, as Saulmon himself admitted, the demonstration left a number of key problems unresolved. "Some of the results produced by the experiments are precisely the opposite of those to be expected given the system of *pesanteur*," the secretary warned. "But this should not be a source of despair for those accustomed to the practice of research. Sometimes the thing that seems to lead you most astray ends up taking you to your goal in the end."[160]

In this description, Fontenelle revealed his deep suspicion of inductive, experimental approaches to natural philosophy. His caution was not exaggerated, for writing about Jakob Bernoulli in his 1705 *éloge*, he noted that Bernoulli had "arranged assemblies and created a sort of academy where he performed experiments that were either the foundation or the proof of his geometrical calculations." Bernoulli was "the first to establish in the city of Basel this manner of philosophizing," he continued, "which is the only reasonable one despite the fact that it has arisen very slowly."[161] Nevertheless, the academy secretary only found experimental demonstrations valid so long as they were used in either one of two ways. Experiment could be employed after the fact, as a tool of clarification or an instrument of confirmation of truths previously demonstrated by other, more rigorous means. Or experiments could be employed as aids in discovery, so long as the results produced in this way were then subjected to reliable, synthetic demonstrations before they were claimed as scientific truths. Saulmon's approach appeared to invert this hierarchy at times because he

158. *HARS-Hist.* (1714): 102.

159. Saulmon, "Experiences sur des corps plongées dans un tourbillon," *HARS-Mem.* (1714): 381.

160. *HARS-Hist.* (1714): 106.

161. *Oeuvres de Fontenelle*, 6: 111–112.

claimed to demonstrate the existence of the mechanisms themselves through analogies with experimental models. This was a highly suspect approach in Fontenelle's estimation, and for this reason he was careful to caution readers about it.

Crousaz shared Fontenelle's understanding of the values and pitfalls of experimental reasoning in natural philosophy. "I am entirely for mechanical principles, and in this respect I am a Cartesian," Crousaz wrote to Réaumur in 1718.[162] Several months later, having read Rémond de Montmort's critique of Newtonian philosophy in *L'Europe savante,* he added that "at Lausanne one would have thought that this piece was by me if the author's name had not been printed since our ideas could not be more in conformity." Crousaz was quick to distance himself from any sectarian allegiances, however, adding that he had strong disagreements with Descartes on the nature of creation and that he disagreed with Rémond de Montmort's Malebranchian conception of occasional causes.[163] In this complex and independent-minded relationship to the emerging philosophical sects of the period, he and Fontenelle were also of like mind.

However, when Crousaz attempted to address English savants such as Desaguliers with this same combination of reasoned independence and gentlemanly collegiality, he was met with a very different response. Sometime in 1717 a Bern savant named Schurer, having just returned from England, reported to Crousaz that English savants routinely spoke disparagingly of him and his work. In particular, he said, the passages in Crousaz's *Traité du beau* critical of Newtonian attraction were commonly invoked to mock and discredit their author.[164] This report outraged Crousaz, and since he had recently received a copy of Desaguliers's *Physico-mechanical lectures* and admired the work, he wrote a letter to its author attempting to clear the air and start a correspondence. In his letter, Crousaz strictly followed the protocols of courtesy dear to the Republic of Letters. He opened with praise, writing that "there is not an academy that does not envy the good fortune of your disciples." He also situated both his correspondent and himself within the world of cosmopolitan learning and politesse as a way of introducing his concerns. "I have several friends who have traveled in England full of my philosophical ideas," he wrote. "They tell me that when a foreigner badly pronounces a phrase in front of the English, he is made to feel his fault by a chorus of laughter. It is the same, they

162. Crousaz to Réaumur, 29 November 1718, FJPC, 1: 243.

163. Crousaz to Réaumur, 2 April 1719, FJPC, 1: 366.

164. Crousaz recounts this episode in a letter to Bignon, 10 September 1717, FJPC, 1: 59.

say, whenever someone advances a physical or mathematical position that contradicts the method and principles of M. Newton."[165]

In this passage, Crousaz makes manners the central obstacle separating him from the Newtonians. He further reinforces this point by also noting the real nature of his relationship to English science. "You must know, sir, that in all sincerity I have the most ardent admiration for M. Newton. It is just that I make it a principle to adhere to a certain timidity and circumspection in my thoughts, and this prevents me from adopting his principles." He also worried that devotion to Newton had become "an affair of nation" in England, and he addressed Desaguliers as a fellow cosmopolitan savant by asking him whether "the English are really so bellicose as to entertain such campaigns." "A wise man, it seems to me, must abandon such a thought." Concluding, he politely and matter-of-factly suggested some corrections and criticisms of Newton's theory of color based on his own experiments and then ended with a cordial invitation for future commerce.[166]

No response arrived. Even after a second and similarly cordial letter was sent two years later, Desaguliers remained silent.[167] Instead, Crousaz received a letter from François de Pillonière, a client traveling in England, that explained the silence while further angering the Swiss savant about it. Pillonière reported that Crousaz's reputation in England was only getting worse. "You know that you have the entire nation of bigots, pedants and radicals [*esprits forts*] against you," he wrote, a reference to the reception of Crousaz's 1718 critique of Anthony Collins's deist tracts. "But you should also know that the friends of M. Newton, whose status as figures of superior ability, especially in matters of physics, makes them the recognized arbiters of reputation here, are not too favorable to you, either." Like Schurer, Pillonière pointed to the passages in the *Traité du beau* critical of Newtonian attraction as a source for this animosity. For the Newtonians, he explained, the theory of universal gravitation is simply beyond question. "They believe it to be applied in the most solid manner and to be proved demonstratively, or at least as much as is necessary, by its wonderful explanation of the phenomena." "With them speaking in terms of the world and heavens of Aristotle, or of substantial forms is no longer a crime. Only the defense of the hypothesis of the plenum or other similar Cartesian ideas is suspect." Pillonière worried about the fate of Crousaz's non-Newtonian color theory in this climate. He also lamented how "these prejudices, so recently rejuvenated and intensified, make it impossible for [the Newtonians] to

165. Crousaz to J. T. Desaguliers, 1 September 1718, FJPC, 1: 123.
166. Ibid.
167. Crousaz to Desaguliers, 26 March 1720, FJPC, 2: 137.

see any of the other good and useful things in your works." He ended by suggesting that independent arbiters might be needed to resolve the dispute.[168]

Crousaz was outraged by each of these reports, and he expressed his frustration to virtually all of his correspondents over the next three years. Bignon responded by saying that "the rest of the world does not resemble you, Sir. To serve without homeland and without interest is a rare virtue in this century." He also pointed to the peculiar nature of English society as a further palliative. "The spirit of party reigns there more than anywhere else, and I know of no illness more contagious than that one."[169] Réaumur responded in similar terms, writing that "the objections of those who take sides without meditating are among the worst. . . . This spirit of contestation accomplishes nothing except inspiring a partisan zeal for one's compatriots."[170]

Crousaz was heartened by this sympathy, but a second letter from Pillonière in late 1721 only fueled his despair. "The unmitigated intolerance of the philosophers and mathematicians of England astounds me," Pillonière wrote. "I find your judgment regarding this species of philosophical Papism, and the mysterious manners that it inspires, as just as it is moderate. I also applaud you for retaining a wise Pyrrhonianism in the face of this extreme and despotic dogmatism. What a dishonor to see [physics], a beautiful and peaceful profession, endowed with politeness and grateful to all who pursue it, invaded by those who lack equity and express ingratitude toward the most considerable and useful people."[171] Crousaz channeled these same sentiments toward Réaumur soon after. "The spirit of party," he wrote with respect to England, "is the most general characteristic of this nation. . . . As for me, I will always make it a law to separate the certain from the uncertain, for it is precisely this that these gentlemen do not do regularly enough." After a long defense of his own position, he further summed up the dispute itself by saying that "since I was not sufficiently devoted to M. Newton and did not become his disciple at all cost, M. Desaguliers concluded that I was not worthy of his commerce. That is how much the spirit of party dominates there."[172]

Desaguliers's actual attitude toward Crousaz is not at all clear, but from the perspective of Bignon and Réaumur, savants who adhered first and foremost to the honorable decorum of the Republic of Letters, Desaguliers's conduct was clearly disgraceful. Nevertheless, not all English savants, nor even English

168. François de la Pillonière to Crousaz, 30 June 1720, FJPC, 2: 174.
169. Bignon to Crousaz, 15 January 1718, FJPC, 6: 171.
170. Réamur to Crousaz, 9 March 1719, FJPC, 1: 353.
171. François de la Pillonière to Crousaz, 1 November 1721, FJPC, 5: 145.
172. Crousaz to Réaumur, 21 November 1721, FJPC, 2: 188.

Newtonians, were to be tarred with this same judgment. Thus despite these insults, Crousaz continued to seek out a cordial English colleague. Having failed to build a bridge with Desaguliers, he tried others, and in the end he found one who adhered, like him, to the appropriate code of gentlemanly reason. Colin Maclaurin was a Scottish mathematician with impeccable Newtonian credentials. Writing to Réaumur in late 1721, Crousaz gushed about the "infinitely obliging letter" he had just received from him. "It would be ungrateful if I did not acknowledge him as a true friend," he wrote, further adding that having now proposed five or six experiments for discussion, it was clear that Maclaurin was someone with whom he could engage scientifically. "I am in dialogue with a man of perfect politeness," he declared, "something that is rare in that country."[173]

Crousaz's success with Maclaurin reveals that it was not English Newtonianism per se that created fissures within the Republic of Letters after 1715, but the particular entanglement of certain English Newtonians with certain discursive tendencies of the period. English society as a whole played a part in this shift, for after 1689, a new and more openly partisan and contentious political culture did take hold there. One of the consequences of the "Glorious Revolution" was the establishment of permanent electoral and party politics in England, and this laid the foundation for a new political culture. Public venues like coffeehouses, taverns, and bookstores were particularly transformed into sites of open and vigorous political contestation.[174] This in turn fostered the sense, echoed by Bignon and Crousaz, that England was a land where party ruled and sectarian strife was the norm rather than the exception.

Louis François Dubois de Saint-Gelais, a Parisian journalist, mirrored this increasingly widespread perception when he wrote in his 1717 *Rémarques sur l'Angleterre* that "the English claim to be the freest nation in the world." English coffeehouses, he continued, offered the most visible manifestation of this liberty, for "there is one for every estate and every party, and they are the ordinary meeting point for everyone from lords to common people. Major business is conducted there, and inside one does whatever one wants: talks, gambles, smokes, dreams, or mingles. Politics and government are the most common

---

173. Crousaz to Réaumur, 23 December 1721, FJPC, 4: 321.

174. Linda Colley, *In Defiance of Oligarchy: The Tory Party, 1714–1760* (Cambridge, 1985); Geoffrey Holmes, *British Politics in the Age of Anne* (London, 1987); Jacob, *Living the Enlightenment;* Roy Porter, *English Society in the Eighteenth Century* (London, 1990); and James O. Richards, *Party Propaganda under Queen Anne: The General Elections of 1702–1713* (Athens, 1993).

topics of conversation for that is the genius of this nation." Dubois de Saint-Gelais also noted the role of the press in fostering this same political culture. At cafes, he explained, "one finds the gazettes from France, Holland, and England along with abundant other printed matter, and in these texts the spirit of the different parties and sects do not hide themselves."[175]

To the extent that experimental Newtonianism developed within this same coffeehouse and journalistic milieu, it too acquired this orientation toward sectarian affiliation and partisan polemicism. Yet England was more than coffeehouse politics, and coffeehouse discourse found a home in sites other than London. In the first volume of his short-lived *Histoire journalière de Paris*, Dubois de Saint-Gelais captured this complexity perfectly while conveying a different image of English culture. "Academies of sciences and the arts are like neutral republics," he explained. "Only those who apply themselves without regard for distinctions of nation or even religion are admitted. The English, who understand what this is worth, think the same way about these matters, even though they do not all think alike. Thus, the members of the famous Academy of England, known by the name of the Royal Society, are neither all Englishmen nor all Anglicans."[176]

Dubois de Saint-Gelais was describing an important perception here. Like the Paris Academy, the Royal Society of London was a gentlemanly society committed to republican ideals of *honnête*, cosmopolitan science.[177] The inclusion in the society of coffeehouse demonstrators like Desaguliers and Hauksbee or journalists like those at *L'Europe savante* certainly created tensions within this ideal. Often these tensions erupted into controversies that needed to be policed. Yet the French academies were no different, as the calculus wars of the early 1700s or the career of Joseph Saurin attests.[178] No monolithic contrast between English and French manners emerged in this period, therefore, even if a discourse distinguishing aggressive English Newtonians from mannered continental Cartesians did arise. The exceedingly cordial interaction

---

175. Louis François Dubois de S. Gelais, *Rémarques sur l'Angleterre*, in *Pièces échapées du feu* (n.p., 1717), 5. J. Macary argues that this piece was actually written by André-François Deslandes. See *Masque et Lumières au XVIIIème siècle* (La Haye, 1975), 63–64. On Dubois de Saint-Gelais, see Sgard, *Dictionnaire des journalistes*, 1: 330.

176. Louis François Dubois de Saint-Gelais, *Histoire journalière de Paris*, 2 vols. (Paris, 1716–1717): 1: 58–59. On the history of this journal, see Sgard, *Dictionnaire des journaux*, 1: 549–550.

177. See Shapin, *Social History of Truth*, esp. 122–123.

178. See Shank, "Before Voltaire," chap. 3.

between Rémond de Montmort and Brook Taylor during exactly these years, in fact, reveals the deeper complexities that were still prevalent within the Republic of Letters during this period. And if the Crousaz-Desaguliers confrontation was, to be sure, a harbinger of things to come, the Rémond de Montmort–Taylor debate shows that these later developments were anything but overdetermined.

Indeed, savants far removed from the cauldron of English coffeehouse politics also developed their own inclination toward strident partisanship. Johann Bernoulli spent most of his life in the quiet backwater of Basel, yet he was arguably the most contentious and disruptive savant of his generation.[179] Always quick to detect an affront to his honor and reputation, Bernoulli clashed with dozens of savants throughout his lifetime, including many of the leading luminaries of the Republic of Letters. His public battles with his brother Jakob were legendary, and they almost kept him from becoming a foreign correspondent of the Paris Academy.[180] Bernoulli likewise turned hostile toward l'Hôpital and Crousaz after each behaved in ways that he deemed inappropriate. In the case of Crousaz, it was his victory in the Paris Academy prize contest of 1720, a prize that Bernoulli thought he should have won, that triggered Bernoulli's ire.[181] Other savants were treated similarly, and overall Bernoulli's conduct reveals the friction that individual, contentious personalities always created within the otherwise harmonious ideals of the Republic of Letters.

Bernoulli, however, was the exception that proved the rule, for his peers in the Republic of Letters routinely described his conduct as demeaning to his status as a great savant. Harder to classify and police were the critical journalists since their assertive, opinionated approach to learning was often praised even as it provoked contentious disputes that challenged the journalist's status as *honnête hommes*. A case in point is the career of the French Jesuit Louis-Bertrand Castel. Born in Montpellier, Castel found his way into the Society of Jesus by distinguishing himself at the Jesuit College of Toulouse. By 1719, he had become a full-fledged member of the order, teaching philosophy at the College of Cahors. When some of his writings caught the attention of Fontenelle

179. What little secondary literature on Bernoulli exists is highly technical in nature and mostly in German. For a readable summary of it in English, see E. A. Kellmann and J. O. Fleckstein, "Johann Bernoulli I," in Charles Coulston Gillispie, ed., *Dictionary of Scientific Biography* (New York, 1970–1980), 2: 51–55.

180. See the letter from Pierre Varignon to Johann Bernoulli, 31 January 1699, in Costabel and Pfeiffer, *Der Briefwechsel von Johann I Bernoulli. Band 2*, 215–218.

181. See Bernoulli's letter of 28 August, 1720, which was critical of Crousaz, in FJPC, 12: 297.

and Father Tournemine, then an editor at the *Journal de Trévoux,* Castel soon found himself in Paris with a prestigious teaching post at the Collège Louis-le-Grand and an editorial position at the Jesuit periodical. From the moment of his arrival, he began to use the journal as an organ for his particular brand of Jesuit philosophy and science. Castel became in effect the journal's de facto science editor, and given his critical proclivities, he in turn injected a new polemical energy into the periodical. He in fact helped to make the journal into something like the French Catholic counterpoint to the critical Francophone journals of Holland. In this way, Castel became a major player in the French public sphere after 1720.[182]

Castel's precise views will be explored through his many interventions into the scientific debates of the period; however, one theme relevant to the discussion here can be noted now: controversy. Castel rarely entered a discussion without enflaming it, and he therefore made the Jesuit journal a much more partisan and openly polemical voice on questions of philosophy and science than it ever was before. He and his Jesuit supporters had their own, peculiarly confessional reasons for making their journal more opinionated, but overall their attitude also reveals how the simple expansion of journalism itself directly generated new critical energies. Since this expansion and the new critical possibilities that it spawned were part of a European-wide phenomenon, they likewise locate the source of Enlightenment criticism within the Republic of Letters as a whole and not exclusively in post-1688 England.

Nevertheless, in the same way that a thoroughly *honnête* debate about natural philosophy between a French academician and his mathematical colleague in England could be turned into a vehicle for partisan polemicism by a feisty Dutch journalist, the gentlemanly civility that dominated intellectual relations within the Republic of Letters before 1715 was transformed by the more boisterous and openly opinionated culture of the new journalism. Moreover, since England and Holland often provided a more welcoming home to this journalism (Castel in France being the exception that proves the rule), these two places often did more to catalyze these changes than other locales. Enlightenment Newtonianism was a product of all these changes. It emerged after 1715 as an entanglement of all of the hot-button topics of the day—science, nature, experiment, materialism, Spinozism, radical religion, publicity, and politics, among others. It also came to life as one of many new critical zones activated

182. On Castel, see Desautels, *Les Mémoires de Trévoux et le mouvement des idées au XVIIIe siècle;* Healy, "Mechanistic Science and the French Jesuits"; Catherine M. Northeast, *The Parisian Jesuits and the Enlightenment, 1700–1762* (Oxford, 1991); and Donald S. Schier, *Louis Bertrand Castel: Anti-Newtonian Scientist* (Cedar Rapids, 1941).

by the new social practices and media structures that created and catalyzed these new developments. The Newton wars in France erupted out of this same cauldron when French men and women began to openly contest this new critical object—Newtonianism—for the first time. It is therefore to the beginnings of those battles that we now turn.

Preparing the Battlefield

Fighting For and Against

Newton after 1715

In 1727 Fontenelle celebrated two anniversaries. One marked his birth in February seventy years earlier; the second his appointment at forty years of age as the first-ever perpetual secretary of the Royal Academy of Sciences. The historical record does not relate what, if any, festivities occurred to honor these milestones but what is clear is that Fontenelle was not showing any urges toward retirement. He would in fact serve another thirteen years as academy secretary and live another two decades after that (he died a month shy of his one hundredth birthday), and he thus began his fourth decade in the academy with as much vigor as his first. The year 1727 was nevertheless crucial for Fontenelle even if it was not marked by any significant turn into retirement. Rather, when viewed retrospectively, this year marked a moment of transition in his academic career. During his first three decades as academy secretary, he had defined himself and his new office by working vigorously to defend an image of *honnête* public science that came to define official French science as whole. In his final decades in this post, this effort did not so much change as find itself confronted with a host of new challenges produced over the previous decades. Fontenelle's intellectual trajectories were similarly challenged anew after 1727. Having developed in early adulthood a complex philosophical stance rooted in a mutually reinforcing combination of Cartesian physical rationalism and Malebranchian mathematical phenomenalism, he was increasingly forced as an older man to defend this view against the new scientific discourses of Newtonianism that had emerged during the previous decades.

Compounding these struggles was the way that Fontenelle's position in French public science had changed since his first battles on behalf of infinitesimal analysis in the 1690s. During the calculus wars, Fontenelle and Bignon had been forced to negotiate a settlement between a dozen or so academicians, three journals (the *Mercure galant, Journal des savants,* and *Journal de Trévoux*), and a handful of other interested parties, all of whom were susceptible to the powerful pulls of honor and civility that were then central to the wider ethic of public science. This remained a difficult challenge, yet their success produced a peace that cemented a structure for public science that defined Fontenelle's

FIGURE 16. *Bernard le Bovier de Fontenelle.*
Courtesy of the Archives de l'Académie des sciences, Paris, France.

work over the next three decades. By contrast, when in 1728 Privat de Molières began to position the Paris Academy squarely on the Cartesian side of the new cosmological debate that had opened during the previous decades, Fontenelle found himself in a very different position. Rather than three journals, he had dozens to contend with, including many produced outside of France by savants with a dizzying array of agendas. He also spoke for an academy that had had grown more diverse, specialized, and imbedded within the organs of the French state than ever before, and one that had a different center of intellectual gravity as well. Finally, and perhaps most importantly, he confronted a new intellectual climate, one that had grown larger and more diverse over the last three decades, and one that was more comfortably contentious, polarized, and rambunctious than the one within which he had come of age.

The old discourses and regulating structures of academic public science in France proved no match for these new developments and as these changes established themselves Fontenelle found himself in a new position. In hindsight, 1727 appears as a turning point in this respect because in this year events conspired to place him in a renewed position of authority on the very eve of the Newton wars, which would later place the academy and the public into open contestation in a completely new way. The fall public assembly, held in November 1727, concentrated this power into a single event. At this assembly, Fontenelle delivered two orations, one a familiar *éloge*, but the second a very uncharacteristic piece of mathematical science. Each represented a summation of Fontenelle's prior work, and each also pointed toward the new struggles that he and the scientific culture he personified would confront in the decades to come. To trace the passage into the Newton wars in France, therefore, let us focus on these two Fontenellian efforts.

Fontenelle's Crucial Year: 1727

The first to be delivered was the *éloge*, and the deceased academician that it honored was none other than Isaac Newton, a foreign associate of the Paris Academy, who died in London in March 1727. Fontenelle began his *éloge* by briefly tracing Newton's family genealogy before quickly turning to his early mathematical education. "In order to learn mathematics," Fontenelle wrote, "[Newton] never studied Euclid, which seemed to him too clear, too simple, and not worthy of his time. He knew it almost before having read it, for a quick glance at the theorems was sufficient to demonstrate everything to him. He jumped right away to other works such as Descartes' *Geometry* and Kepler's *Optics*. One can thus apply to him what Lucan said of the Nile: 'The ancients did not know its source because it was not permitted for them to see the Nile weak and fledgling.'"[1] This characterization of Newton's mathematical development accomplished two things at once. On the one hand, it initiated the discourse of high esteem and glorious praise that would characterize Fontenelle's *éloge* throughout. On the other hand, it also prepared the ground for what would become a central argument of Fontenelle's presentation. By describing Newton's pedagogical leap beyond Euclid and straight to the analysis of Descartes, the secretary not only illustrated Newton's superhuman genius but also situated Newton's mathematical thought in a very precise way.

Throughout his other *éloges*, Fontenelle made the passage from Euclid to more complex mathematics the normal intellectual trajectory for mathematicians. In his eulogy for Varignon, for example (read, not coincidentally, only four years before his *éloge* to Newton) Fontenelle described the young Norman's first discovery of Euclid in a Caen bookstore while a student of the Jesuits. He became "charmed by the book," Fontenelle explained, "not only because of its order and coherence [*enchainement*] but also because of the ease with which he could enter into the book's arguments." In Varignon's case, Euclid also helped him "to taste the clarity, the linkages, and the surety of geometric truths." It further aided his adolescent fight against "the eternal incertitude, the embarrassing sophistry, and the useless and often affected obscurity of the Philosophy of the Schools." Descartes' writings, particularly his *Geometry,* were then revealed to him, and building upon these Euclidean foundations the young Norman then went on to experience "the new light that spread throughout the thinking world from them."[2]

This trajectory from Euclid into Cartesian analysis was the standard one in Fontenelle's *éloges*.[3] It also reflects the more pervasive understanding of mathematical education in France, which saw analytical mathematics as a more advanced field of study appropriate only for those students who had first mastered the classical methods of rigorous Euclidean geometry.[4] The Oratorians and the Jesuits also accepted such a view, and framed by this context Fontenelle was

1. *Oeuvres de Fontenelle,* 6: 111.

2. Ibid., 6: 19–20.

3. See also the *éloges* of Carré (*Oeuvres de Fontenelle,* 6: 249–256), Rémond de Montmort (*Oeuvres de Fontenelle,* 6: 465–478), and the Chevalier de Louville (*Oeuvres de Fontenelle,* 6: 459–464) for a similar presentation.

4. Discussions about appropriate mathematical education in France can be followed by studying reviews of the leading mathematics text books of the period. For a representative example, see the discussion of Clairaut's *Elemens de géometrie* and *Elemens d'algebre* in *Journal des savants* (1741): 574–581 and *Journal des savants* (1747): 96–103, respectively.

making Newton's mathematical work, which skipped Euclid entirely, the end result of a pedagogical *sonderweg*. Furthermore, since pedagogical discourses such as these were closely tied to debates about proper philosophical method, this precise framing was not without significance.

Having thus introduced Newton's mathematics in this particular way, Fontenelle continued by surveying Newton's mathematical achievements. He asserted that Newton had made "great discoveries in geometry" by the age of twenty-four and that these discoveries "served as the foundations for his two most famous works, the *Principia* and the *Optics*."[5] Situating these achievements more precisely, Fontenelle claimed that Newton's first great accomplishment rested in offering a more general solution to the problem of the quadrature of curves than any yet achieved. "The solution of M. Newton was not limited to hyperbolas alone but extended to general formulas applicable to all sorts of curves, even mechanical ones, and to the determination of the quadratures, their rectifications, their centers of gravity, and the solids formed by their revolution. . . . It was a great reward for geometry to possess a theory so fecund and general, and it was a glory to M. Newton to have invented a theory so surprising and ingenious."[6] The secretary further noted Newton's invention of the calculus, adopting the "required neutrality of the historian" in relating the priority controversy with Leibniz that this work triggered. He also used this episode to stress the modesty and propriety of Newton's character, describing him as a model Republican of Letters. As he summed up: "Newton was content with the riches [of his mathematical discoveries], and he was never agitated about glory."[7]

Having laid out Newton's mathematical achievements, Fontenelle then introduced the *Principia* as the public demonstration of Newton's mathematical genius. The secretary described the treatise as a "revelation" of all of Newton's previous mathematical work. He also characterized the book as a work in which "the most profound geometry serves as the foundation for an entirely new physics." Acknowledging the complex reception of the text, he further noted that "it did not at first attract the attention that it merited and that it deserves to merit someday." Fontenelle offered nothing but praise in all of these descriptions; however, by introducing the *Principia* in this precise way he also framed the book's intellectual contributions quite specifically. Positioned as the culmination of two decades of pure mathematical work and described as the product of Newton's desire to demonstrate the power of his new mathemati-

5. *Oeuvres de Fontenelle*, 6: 112.
6. Ibid.
7. Ibid., 6: 114.

cal innovations, the *Principia* became in Fontenelle's description essentially a mathematical treatise and little else. Nowhere, for example, did Fontenelle mention Newton's experimental work with pendulums, his efforts to collect astronomical and geodesic data, his work on the physics of fluids, or any of the other more overtly empirical and experimental work that was equally crucial to the construction of the treatise. Instead, Fontenelle declared, "two theories dominate in the *Principia:* that of central forces and that of the resistance of media to movement, both almost entirely new and both treated according to the sublime geometry of the author." [8]

This was less a complete description of Newton's actual work than a description of the French interpretation of the text that was developed in the decade after 1687. Varignon's work in particular was essentially concerned with these two features of Newton's treatise, and by defining Newton's achievement this way Fontenelle was once again situating Newton at the center of French analytical mechanics. "One cannot touch on either of these topics without having Newton in front of one's eyes," the secretary enthused. "Either one repeats what he did or one follows him. He cannot be ignored, for he appears wherever one looks!" [9] Fontenelle went on to explain the mechanics of elliptical planetary orbits and the role of central forces within it, using language that precisely echoed his descriptions of French central force mechanics over the previous three decades. In this way, Newton's work and that of French analysts like Varignon were once again joined as they had been, at least in Fontenelle's mind, since 1692.

However, having used the early paragraphs of the *éloge* to reconnect French analytical mechanics with Newton's *Principia* in the traditional manner, the secretary also worked to redefine the autonomy of French science in the remainder of his text. Raising the question of Newton's physics, Fontenelle declared that "no one knows what the nature of *pesanteur* is, and neither does M. Newton." [10] The secretary then traced the alternatives that Newton presented. Either *pesanteur* is a product of mechanical impulsion, such that a block of marble is pushed toward the earth without the earth in any way acting upon it. Or heaviness in bodies results from the material attraction of the bodies themselves, such that both the marble block and the earth act upon one another despite the absence of any material connection between them. Claiming to reveal the real views of Newton, Fontenelle wrote that he "always speaks in terms of the reciprocal *pesanteur* of bodies determined by the relationship of

8. Ibid., 6: 115.
9. Ibid.
10. Ibid.

their masses, and in this way he appears to suggest that *pesanteur* is really just attraction." "He always uses [attraction] to express the active force of bodies," he continued, yet in truth "the nature of this force is unknown and Newton does not attempt to define it. If impulsion could be the agent behind this action, then why not use the clearer term? For one will agree that the terms cannot be used interchangeably since they are diametrically opposed." [11]

With Newton's tendency to insinuate a principle of universal gravitational attraction duly noted, Fontenelle then drew the essential conclusion. "M. Newton's constant use of the word *attraction,* supported by such great authority and, perhaps, also by the sympathy that one feels toward him and the idea itself, has made readers comfortable with an idea once banned by the Cartesians and ratified by all the other philosophers as well. One must today be on guard not to imagine any reality to this idea, for one is exposed to the peril of belief as soon as one hears it." This shift in focus from mathematics to physics, and from a mathematical conception of central forces to one rooted in material and metaphysical powers of attraction, allowed Fontenelle to redefine the meaning of Newtonian science. Despite Newton's complex stance on the status of gravity in the *Principia,* a stance shared by all judicious and careful savants, Fontenelle suggested that his language misled people into attributing physicality to an idea—central forces—that was really only a mathematical concept. Furthermore, invoking the label Cartesian to describe the opponents of such thinking, he then positioned Newton's conception of central force mechanics, steeped in talk about attraction, against the approach of French mathematicians (and implicitly all other responsible philosophers as well) who speak only in terms of clear and distinct mechanical impulsions.

In the next passages Fontenelle furthered this redefinition. He admitted that Newton's system agreed quite well with the observed phenomena, noting in particular the movements of Jupiter and Saturn. But is this empirical agreement the same thing as a proof of gravitational attraction? Fontenelle clearly thought not, and he further pointed to the source of Newton's confusion in this area. "What a display of geometry was necessary to uncover this chaos of *rapports*! It appears foolhardy to even have attempted it. One can only view with astonishment that from a theory so abstract, formed from a variety of particular theories that are each difficult to manipulate, a set of facts in absolute agreement with those established by astronomy are born." [12]

11. Ibid., 6: 118.

12. Ibid., 6: 118–119. This characterization echoes Huygens's sentiments expressed to Leibniz in a letter dated November 1690: "I am by no means satisfied . . . by all the other theories that he builds upon his principle of Attraction, which seem to me absurd, as I have

Fontenelle's description ultimately celebrates the mathematical brilliance of Newton's work while nevertheless suggesting caution about any physical induction from it. He continued in this vein in the next passages when examining book 2 of the *Principia*. Here, Fontenelle declared, "one enters the principal phenomena of Nature such as the movements of the celestial bodies, light, and sound." As always, Newton established his theories "upon the most profound geometry," but he also drew conclusions from them that were unwarranted. In particular, Newton claimed to "destroy the vortices of Descartes and to overturn this grand celestial edifice that one thought indestructible," but was this reversal justified? Summing up, the secretary noted that "bodies [in Newton's system] move in a great void with only their emissions [*exhailaisons*] and the rays of light from the sun . . . mixing to fill the almost infinite reaches of empty space. Attraction and the void, banished from physics by Descartes, and banished forever, one had thought, is returned by M. Newton, armed now with an entirely new force that one did not think was possible even if it was perhaps a little disguised."[13]

By this point, Fontenelle has articulated two very different images of Newton. On the one hand, he has offered Newton the brilliant mathematician who had pioneered the mathematical approach to mechanics that culminated in the work of people such as Varignon. On the other hand, Newton is the deceptive *physicien* who had disguised illegitimate natural philosophy within brilliant *math*ematics. This was the interpretation of Newton offered by the *Journal des savants* in 1688, and at one level Fontenelle was merely repeating this description in his *éloge*, even if he also framed it according to the new polarized language of Newtonians against Cartesians. At the same time, Fontenelle was also offering a more complex assessment of this emerging philosophical divide. Viewed mathematically, Newton's work and Varignon's could not have been more alike. Both deserved the highest esteem, Fontenelle argued, since each understood the power of mathematics in unmasking physical truths. Newton's mathematical project, however, had somehow gone astray while Varignon's remained on the right path. Explaining this fissure was crucial to Fontenelle if he was to provide a vision of the Newtonian *sonderweg* in physics that harmonized with what he saw as the parallel triumph of French analytical mechanics. Fon-

---

already mentioned in the addition to the *Discourse on Gravity*. And I have often wondered how he could have given himself all the trouble of making such a number of investigations and difficult calculations that have no other foundation than this very principle." Cited in I. B. Cohen, *The Newtonian Revolution* (Cambridge, 1980), 80.

13. *Oeuvres de Fontenelle*, 6: 120–121.

tenelle found the solution he needed in a difference of method, and in his final paragraph describing the *Principia* he asserted his central argument.

The issue was framed as a methodological clash between Descartes and Newton. "These two great men, who now find themselves so opposed to one another, actually had much in common," Fontenelle explained. "Both were geniuses of the first rank, born to rule over other minds and to found empires. Both were excellent geometers who saw the necessity of transporting geometry into physics. Both founded their physics upon a geometry that was at the very heart of their manner of thinking as a whole. But the first [Descartes], taking the more ambitious route, wanted to place himself at the source of everything and so made himself master of the clear and fundamental first principles. He then descended through a set of necessary steps to the particular truths of nature. The other [Newton], more timid or more modest, began his journey by applying himself to phenomena in order to ascend to the unknown principles. He resolved to admit as truths only those things that could give him a systematic understanding of their consequences. The first left behind what he understood immediately in order to discern the cause of what he saw; the other parted from what was clear in order to find the cause, be it clear or obscure. The evident principles of the first did not always lead to phenomena as they are, but the phenomena of the other did not always lead to sufficiently evident principles. The barriers in these two contradictory routes hindered the progress of two men of great stature. These were thus not the barriers of their own minds, but the barriers of the human mind itself." [14]

In the complexities of this brilliantly rich description are found all of the key transitions that Fontenelle was struggling to reconcile in 1727. One involved the changing character of the Republic of Letters. For over half a century, Fontenelle had lived according to the collegial ethic of this community, one that despised partisan dispute and valued intellectual harmony. He had also applied these same values to his work as academy secretary, attempting to use his public discourse as a vehicle for harmony and consensus within French science and the Republic of Letters as a whole. Newtonian philosophy had recently become a site of unprecedented divisiveness within this community, yet in his *éloge* Fontenelle demonstrated his continuing attachment to the traditional *honnête* conception of intellectual dialogue by attempting to collapse rather than inflame these differences. "Both [Descartes and Newton] were geniuses of the first rank," the secretary declared, and the differences between them, therefore, should not be reduced to mere differences of philosophical opinion.

14. Ibid., 6: 121.

It was, in fact, the complexities of the human mind itself that produced this opposition, or so Fontenelle said.

To frame the Cartesian/Newtonian rivalry this way was not to make it the basis for a partisan choice but rather to imply that Newton and Descartes presented an intellectual pinnacle upon which all reasonable savants should gaze together in wonder. Nevertheless, a bitter battle was emerging throughout the Republic of Letters over their relative values. Even if Fontenelle wanted to remain detached from these new battles and to position himself above the fray in the manner typical of *honnête* savants, his own absorption within the very same currents he sought to contain was in evidence as well. His *éloge*, in fact, constituted a final attempt to restitch the fraying fabric of republican collegiality within the wider learned world, but it also marked his own entrance into this newly partisan environment as a defender of Cartesianism and a critic of Newtonianism. Despite the *honnête* equity of his comparison of Descartes and Newton, Fontenelle made clear in other places where his own partisan sympathies lay. They were never expressed with open polemicism since had he done so he would have violated his core ethical convictions. He also would have violated his responsibilities as the official and royally sanctioned eulogist of an esteemed academician. His convictions appeared nevertheless, and this made the text at once Fontenelle's last stand as an *honnête*, nonpartisan savant and his first volley in the French Newton wars that were soon to erupt.

Philosophical method served as the foundation for Fontenelle's partisan critique of Newtonianism. Drawing upon Newton's own pronouncements about method in describing his divergence from Cartesian reasoning, the secretary focused on the key relationship between analysis and synthesis in the philosophical approach of each. In the second edition of his *Opticks,* which appeared in English in 1717 and in Pierre Coste's French translation in 1720, Newton appended a statement of method that resonated strongly with Fontenelle's description in his *éloge*.[15] "As in Mathematicks, so in Natural Philosophy, the Investigation of difficult things by the Method of Analysis ought ever to precede the Method of Composition," Newton wrote. Composition as used here was a synonym for synthesis, and in order to show why the sciences should avoid proceeding by this method, Newton defined his terms more carefully. "Analysis consists in Making Experiments and Observations, and in drawing general Conclusions from them by Induction," he explained. "Although the arguing from Experiments and Observations by Induction be no Demonstration of general Conclusions, yet it is the best way of arguing which the Nature of things admits of. . . . By this way of Analysis we may proceed from Compounds to

15. Isaac Newton, *Opticks* (New York, 1979), 404.

Ingredients, from Motions to the Forces producing them, and in general from Effects to their Causes, and from particular Causes to more general ones 'til the Argument end in the most general." Roger Côtes' "Preface" to the 1713 edition of the *Principia* used the same hierarchy in framing Newton's "experimental" approach to natural philosophy, and in his *éloge* Fontenelle intervened at the center of these discussions, trying to reposition this debate in ways that supported his thinking while challenging the Newtonian position.

Fontenelle's strategy was centered on inverting the normative hierarchy between analysis and synthesis that Newton and Côtes argued for. Fontenelle argued that the method of analysis was not the foundation of sound scientific method but only a precursor to true scientific thinking, which must follow the method of synthesis. In making this inversion, his presentation closely followed the very different description of the analytical and synthetical methods offered in Antoine Arnauld and Pierre Nicole's widely influential *Art de la penser,* or *Logique de Port-Royal* as it commonly came to be called, published in its definitive edition in 1682.[16] Part 4 of the *Logic* treats scientific method, and the second chapter of this section, subtitled "Two kinds of method, analysis and synthesis," draws heavily on Descartes' *Rules for the Direction of the Mind* in distinguishing the two methods most appropriate for proper scientific inquiry. Analysis, they state, is concerned with discovering truth and can also be called the method of resolution or the method of discovery. Synthesis, by contrast, is the method "for making truth understood by others once it is found." This method can also be called the method of composition or the method of instruction.[17] Both philosophical methods have their proper role in the pursuit of properly scientific knowledge, but a hierarchy between them nevertheless exists.

Proper physical science, they claim, only emerges when the necessary hierarchy between these two methods is maintained. In particular, while English "experimental philosophers" saw analysis as the foundational scientific method,

16. This was the final edition published under the direction of the authors and the last to include revisions. The book was continually reissued throughout the eighteenth and nineteenth centuries, however, and became arguably the most influential introductory philosophy text of its day. As Jill Vance Buroker states in the introduction to her new English translation of the text: "The *Port-Royal Logic* was the most influential logic from Aristotle to the end of the nineteenth century. The 1981 critical edition. . . . lists sixty-three French editions and ten English editions, one of which (1818) served as a text in courses at the Universities of Oxford and Cambridge." Antoine Arnauld and Pierre Nicole, *Logic, or The Art of Thinking,* trans. and ed. by Jill Vance Buroker (Cambridge, 1996), xxiii. All citations will be from this edition.

17. Arnauld and Nicole, *Logic,* 233–234.

the *Port-Royal Logic* asserted the opposite. Physical reasoning from particular known effects to general causes constituted speculative, and hence uncertain, science, and this same reasoning also supported Arnauld and Nicole's very un-Newtonian suspicions about induction and empiricism. Inferring conclusions inductively from sensate experience yielded only probable knowledge, the authors stated. Moreover, while this knowledge could in fact agree with true or scientific knowledge, one could not be certain that it did based on the evidence of our senses alone. "Our certainty that (the senses) are not deceiving us comes not from the senses, but from a reflection of the mind by which we distinguish when we should and when we should not believe the senses."[18] Analysis alone, therefore, cannot provide a scientific understanding of nature. It was certainly useful in the discovery of new ideas, and as such it served an important function. However, the knowledge it provided was unreliable and thus properly unscientific. Only after a synthetic demonstration of truth had been attained could one claim to possess more than hypothetical conjectures.

Fontenelle was clearly using this understanding of proper scientific method in his critique of Newton's work in his *éloge*. While Descartes begins with clear and distinct principles and then derives truths synthetically from them, Newton begins with observations and tries to induce truths through analysis. Newton also claims that general causes can be inferred from what Arnauld and Nicole would describe as a mere coincidence between empirical events. Viewed in light of these methodological assumptions, which are Fontenelle's as well, this is a flawed method that produces tentative results at best. Cartesian science produces reliable scientific knowledge because it operates synthetically; Newtonianism, by contrast, produces only analytical conjectures that have no demonstrative (and thus scientific) validity.

Fontenelle's previous account of Newton's mathematical education also finds its full significance when viewed through the lens of this methodological critique. By noting Newton's avoidance of Euclid as a youth, the secretary fore-shadowed Newton's later descent into pure, undisciplined analysis by pointing to the absence of any sustained engagement with the rigor of Euclidean synthetic reasoning. Similarly, Newton's analytical approach to science also explained Fontenelle's ability to downplay the empirical correspondence between Newton's theories and the observed facts. Since synthetic demonstration defined proper scientific method for Fontenelle, a correspondence between a phenomenon and theory was not demonstrative unless a synthetic account of causality was attached to it. This is precisely what Newton's theory of universal gravitation refused to offer, however, and as a result Fontenelle and like-minded

18. Ibid., 228–229.

Cartesians treated the theory as a provocative conjecture only and not yet as a synthetically demonstrated scientific explanation.

This argument from method was to prove powerful in France, where it would ultimately become one of the leading Cartesian arguments against Newtonianism after 1728. It was also important, if less obviously so, for Fontenelle's other academic address delivered before the assembled public in November 1727. This was the preface to a massive philosophical treatise on infinitesimal mathematics that Fontenelle published in the same year.[19] *Eléments de la géometrie de l'infini* appeared as an official volume of the Paris Academy, and in many respects the treatise was the culmination of Fontenelle's one authentic and continuously pursued scientific project throughout his career: the philosophy of infinitesimal mathematics. When the calculus wars erupted, Fontenelle delivered his one and only academic *mémoire* on this topic in 1702. The paper was not recorded in the academy registers and is now lost, but the academician Bourdelin noted in his diary that it was *"fort beau"* and was "applauded by everyone."[20] The *mémoire* also was not printed because rather than publish it in the academy's annual volume, the secretary chose to develop it instead into a formal treatise on this topic. In 1714, he delivered the rough draft of a treatise on the foundations of infinitesimal analysis to Rémond de Montmort, who wrote favorably of the work to Nicolas Bernoulli, and the manuscript also appears to have circulated within the wider Republic of Letters and to have and generated further positive commentary.[21] *L'Europe savante* confirmed this assessment, writing in its very first issue of January 1718 that Fontenelle was preparing "an ample treatise on the metaphysics of infinitesimals" that was currently being examined by "two able mathematicians" and would "appear very soon."[22] By 1719, however, the work had still not appeared. Varignon wrote to Johann Bernoulli that he had devoted his entire fall vacation of 1719 to reading and evaluating the manuscript, and Bignon wrote to Crousaz in the same year saying that "Fontenelle's text has not yet gone to press, but it might shortly."[23] Despite these efforts, however, the text was still not printed, and Fontenelle does not appear to have pursued the work again with any intensity until 1724.

19. For a thorough account of the genesis of Fontenelle's *Eléments de la géometrie de l'infini,* see Michel Blay, introduction to Fontenelle, *Eléments de la géometrie de l'infini,* ed. Michel Blay and Alain Niderst (Paris, 1995), 7–32.

20. Claude Bourdelin, BN, N.A. f. fr. 5148.

21. See Blay, introduction, 10–11.

22. *L'Europe savante* 1 (1718): 106–107.

23. L'abbé Jean-Paul Bignon to Jean-Pierre Crousaz, 1 April 1719, in FJPC, 6: 179.

The changing mathematical culture of the 1720s prompted Fontenelle to return to the work, and since these same changes also colored the work and its reception, they must be considered here. Together with the larger shifts in which they participated, the new mathematical disputes that erupted after 1715 further helped to generate the intellectual and institutional field that produced both Fontenelle's treatise and the emerging Newton wars into which it entered. These precise developments in the culture of eighteenth-century mathematics may in fact have a very good claim on being the central triggers of the French Newton wars themselves.

All the shifts noted in the last chapter were necessary for the production of the critical Enlightenment Newtonianism that drove the Newton wars. Nevertheless, the precise battle itself really began out of the calculus priority dispute that erupted in earnest in 1713. This battle pitted Newton against Leibniz over authorship of the infinitesimal calculus, and yet out of this narrow mathematical struggle emerged many of the categories and discourses, not to mention much of the polemical heat, that came to define the Newton wars later on. This dispute also activated in a singularly intense way all the new media essential to Enlightenment criticism—journals, books, correspondence networks, urban sociability, etc.—and in this way too it was a catalyst for change. It also led to the production of arguably the first full-blown bomb of the Newton wars—the Leibniz-Clarke correspondence—a work that created the conceptual and rhetorical battlefield of the Newtonian/Cartesian dispute in more influential ways than any other single text. This correspondence was not about mathematics in the strict sense, but since it never let go of its connection to the priority dispute out of which it was born, the mathematical dispute was a key trigger for these larger eruptions as well.

Finally, and more particular to France, the mathematical disputes of the priority controversy also provided the context for the arrival of the Jesuit Father Louis Bertrand Castel in the French public sphere. He arrived at the *Journal de Trévoux* in 1720 as an innovative new teacher of mathematics at the Parisian Jesuit Collège-le-Grand. At the journal, a new tendency toward energetic criticism was his great contribution to public scientific discourse, yet it was in mathematics more than any other subject where his critical energies were most exercised. Fontenelle's treatise on infinitesimal analysis was in fact an intervention in a newly contested mathematical public sphere, one made more contentious by Castel's work in the 1720s. Castel's response to Fontenelle's work, and to French academic mathematics more generally, also played a singularly important role in defining the institutional dynamics that made possible the Newton wars over the subsequent decades. For all these reasons, then, Fontenelle's mathematical efforts in 1727 need to be contextualized in terms of the arrival of Castel and the institutional and intellectual dynamics that his debut further catalyzed.

The First Shots of the Newton Wars: The Calculus Priority Dispute

The calculus priority dispute was the trigger of all triggers in the Newton wars, and since A. R. Hall has admirably accounted for the historical details of this dispute, there is no reason to rehearse them here.[24] Instead, consider how the expanding struggle illustrated the changing culture of the Republic of Letters in this period while also contributing significantly to its transformation into a more openly polemical and critical space. The first shot in the dispute was fired in 1699 by Newton's protégé Nicolas Fatio de Duillier, an expatriate Swiss savant known to Crousaz. Fatio de Duillier asserted in print that Newton was "the first and by many years the most senior inventor of the calculus" and that Leibniz's claim to be its inventor was rooted in little more than "eager zeal."[25] Hall argues persuasively that Fatio de Duillier was acting alone in this effort and that his motives were jointly rooted in bitterness against Leibniz and an eagerness to see his master's reputation elevated in the eyes of others. By channeling his anger and ambitions this way, however, Fatio de Duillier violated the core values of the Republic of Letters by inciting its members to challenge the honor of a venerable savant. The Swiss mathematician also had far less stature within the Republic of Letters than Leibniz, and he possessed nowhere near the credibility necessary to sustain such a charge. Accordingly his "*demarche*," to quote Hall, became little more than "a damp squib."[26]

Johann Bernoulli, who learned of Fatio de Duillier's charge from Varignon, was nevertheless quick to react with hostility, as was his wont. He suggested offering Fatio de Duillier a humiliating public test that would expose the limited nature of his mathematical abilities.[27] Leibniz, however, reacted more calmly. He was informed of the charge by the Marquis de l'Hôpital, and he responded that a gentle rebuke would suffice "for it would be a ridiculous spectacle if learned

24. A. R. Hall, *Philosophers at War: The Quarrel between Newton and Leibniz* (Cambridge, 1980).

25. Cited in ibid., 106–107.

26. Ibid., 120.

27. See Varignon's letter to Bernoulli, 10 December 1699, in Bernoulli, *Briefwechsel von Johann I Bernoulli. Band 2*, 233. See also Hall, *Philosophers at War*, 121.

men who profess higher standards than others should exchange insults like fish-wives."[28] Prodded by Bernoulli, he did pen a very restrained, if devastatingly critical, rebuttal, yet he waited a full year before publishing it in the *Acta Eruditorum*.[29] In the interim, Leibniz learned from John Wallis that Hans Sloane, the secretary of the Royal Society, attributed the society's printing of Fatio de Duillier's charge to trickery and that Newton was angry at his protégé for his impudent behavior. Wallis also expressed regret at Fatio de Duillier's conduct, and speaking with a courtesy that Leibniz recognized as exceptional, he reassured him that the incident was nothing more than a personal aberration.[30]

Nothing therefore came of Fatio de Duillier's charge, at least until the next decade, and this illustrates the gentlemanly cordiality that still dominated mathematical and philosophical discussions within the Republic of Letters in this period. It also illustrates the quiet that still surrounded Newton and his work in the *Principia*. Samuel Clarke, who would later become a leading polemicist for the Newtonian cause, confirmed the character of this climate of opinion in his English translation of Rohault's Cartesian *Traité de physique* for use at Cambridge University in 1697. In this first edition, it will be recalled, Clarke offered neither criticism of Rohault's vortical physics nor any mention of Newton's rival theory of universal gravitation in the text. In the third edition of 1710, by contrast, Clarke added his famous footnotes that used Newtonian principles to undermine the core Cartesian arguments of the text.[31] The priority dispute, which erupted again with new vehemence in 1709, played a key role in bringing about this new polemical urgency.

Clarke's intellectual transformation during these years also mirrored larger trends, for the same period likewise witnessed the rise of the public experimental physics demonstrators and their own vigorous advocacy of Newtonianism within the coffeehouses of London. It also saw the emergence of the radical materialist and deist reading of Newton pioneered by Toland and others, along with the "orthodox Newtonian" response to it that Clarke and his allies began to preach from the pulpits of the Boyle Lectures and to disseminate in print. This too encouraged an overall increase in the polemical energy surrounding Newton after 1700.[32] Flowing throughout and stimulating this polemical turn

28. Cited in Hall, *Philosophers at War,* 121. See also Gottfried Wilhelm Leibniz, *Leibnizens mathematische schriften,* 7 vols., ed. C. I. Gerhardt (Halle, 1849–1863), 2: 596–620.

29. *Acta Eruditorum* (May 1700): 198–202.

30. Hall, *Philosophers at War,* 121–122.

31. Eric J. Aiton, *The Vortex Theory of Planetary Motion* (New York, 1972), 71–72.

32. See especially Margaret C. Jacob, *The Newtonians and the English Revolution* (Ithaca, 1974).

as well was the ongoing political turmoil of England, including the debates around the Act of Settlement that secured a Protestant successor to the English throne in 1702 and the resumption of war to thwart the claims of Catholic, absolutist France to the throne of Spain in 1703. The coffeehouses, taverns, bookstores, and other public spaces of England became absorbed in partisan political disputation around these issues, and this discursive climate shaped mathematical discussions as deeply as any other.[33]

Out of this crucible emerged the new polemics on behalf of Newton that were increasingly commonplace after 1710. In this new climate, passionate advocacy for the master's positions became not merely acceptable but a required marker of true philosophical convictions. Hostility toward those opposed to Newton's theories and especially toward Cartesians on the Continent likewise became commonplace in England and throughout the Republic of Letters. National categories also became entangled in this new partisanship as defending Newton and Newtonianism increasingly became synonymous with the defense of England itself. Cartesianism and Leibnizianism similarly came to be associated with the national tendencies, be they positive or negative, of France and Germany, respectively. Here the birth of a more widespread and vigorous English nationalism in precisely this period was clearly one source for this new philosophical jingoism.[34] Yet English parochialism alone cannot account for this new climate. Republicans of Letters in England as elsewhere still valued the disinterested, cosmopolitan pursuit of knowledge more highly than nation and party. Consequently, had there not been a strong conviction in England that Newton was right and that his philosophy was being unjustly ignored and disparaged, he would not have become such a powerful focus of nationalist energies.

Yet even here, one suspects that English insecurities about their perceived inferiority within the international Republic of Letters played a role in their newfound zeal to defend England's one recognized genius at all costs. Class politics also contributed its part to the turmoil as experimental demonstrators, anxious to wash away the taint of their nongentlemanly status, latched onto Newton as a vehicle for their own social elevation. For such men, moderate and complex discussion about Newtonian philosophy was unacceptable

33. See Holmes, *British Politics in the Age of Anne;* and Richards, *Party Propaganda under Queen Anne.*

34. On the rise of nationalist sentiment in eighteenth-century England, see Linda Colley, *Britons: Forging the Nation, 1707–1837* (New Haven, 1992). Patricia Fara also discusses the entanglement of nationalism with English natural philosophy in *Sympathetic Attractions: Magnetic Practices, Beliefs, and Symbolism in Eighteenth-Century England* (Princeton, 1996).

since it smacked of the haughty pedantry of the learned men that excluded them from elite intellectual status.[35] Politics exerted an influence as well, for as Steven Shapin has shown in a brilliant article, the roots of the Leibniz-Clarke dispute can also be found in the politics of the 1714 Hanoverian succession to the English throne.[36] Overall, the only thing that one can say with certainty is that a new, nationally defined crusade on behalf of Newton did take shape in England after 1710 even if the reasons for this new polemicism were complex and dispersed.

The resumption of the calculus priority dispute in 1709, after it had lain dormant for almost a decade, proved to be the flashpoint for these changing realities. The trigger man this time was John Keill, a Scottish born savant who learned his Newtonianism, like John Toland, at the foot of David Gregory at the University of Edinburgh. Keill began his career in the late 1690s as the experimental "deputy" to the Oxford philosophy professor John Millington. Keill's job, which he ceded to Desaguliers in 1707, was to give experimental physics demonstrations that illustrated and explained Millington's philosophical lectures. Parisian professors such as Guillaume Dagoumer and Pierre Polinière also attracted large audiences to the University of Paris after 1690 to witness similar displays, and in 1709 Polinière issued a textual companion to his demonstrations, *Experiences de physique,* that went through many editions while attracting a wide audience.[37] Keill did the same in 1700, publishing two texts, *True Physics* and *True Astronomy.* Each offered a textual accompaniment to Keill's physical demonstrations, yet unlike Polinière's similarly conceived works, his included something different: a partisan defense of a single natural philosophical system—Newton's.

Keill's stance as a Newtonian advocate made his books arguably the first popular expositions of Newtonian philosophy ever published. They were not, however, a huge success. Keill would soon join the Royal Society, and in 1712 he was named the Savillian Professor of Astronomy at Oxford. Nevertheless, the reception of *True Physics* and *True Astronomy* contributed less to this career ascent than the other activities he pursued after 1708. In 1710, he published his first original scientific paper in the *Philosophical Transactions of the Royal Society,* a mathematical treatment of the laws of centripetal force. In the paper, without really needing to do so, Keill also launched a bombshell. He asserted that his work was based on "the highly celebrated arithmetic of fluxions, which

35. This point is drawn out most fully in Stewart, *Rise of Public Science.*

36. Steven Shapin, "'Of Gods and Kings': Natural Philosophy and Politics in the Leibniz-Clarke Disputes," *Isis* 72 (1981): 187–215.

37. See chapter 2, note 154 above.

Mr. Newton beyond any shadow of a doubt first discovered." He further added, again gratuitously, that this same mathematics "was afterwards published by Mr. Leibniz in the *Acta Eruditorum* having changed the name and symbolism." Here was not only a claim of priority for Newton in the discovery of the calculus but also a charge that Leibniz had plagiarized his version of the calculus from Newton's earlier efforts.[38]

The entire priority dispute, which quickly became the most acrimonious intellectual struggle anyone had ever seen, unfolded out of the reaction to this initial charge by Keill. Yet who was this upstart physics demonstrator, and what was he up to in acting so belligerently in 1710?[39] Hall judiciously weighs all the possible motives, including an alleged pugnacity that he attributes to Keill's "Scottish blood." In the end, he concludes that the best explanation comes from situating him among a group of English Newtonian "revolutionaries fighting for recognition within a Cartesian universe."[40] This judgment is on the mark, so long as one qualifies it by noting that the "Cartesian universe" against which Keill and his fellow revolutionaries began to fight existed far more in their own minds than anywhere within the Republic of Letters. A widespread skepticism about the plausibility and rigor of attractionist Newtonian physics certainly existed in 1710. A commonplace belief in the methodological superiority of Cartesian mechanistic physics was also present. Yet little partisan devotion to Descartes' precise theories of celestial or terrestrial mechanics was evident, save in the widely held assumption that sound physics meant rigorous, mechanical explanations of physical causes. Working from these premises, public dismissals of Newtonian attraction had recently been issued by the Paris Academy (Saurin in his public assembly presentation of 1709), the French Jesuits (their reviews of Gregory and Freind in the *Journal de Trévoux* in 1710 and 1712, respectively), and Leibniz (in the *Acta Eruditorum* and in Leibniz's *Theodicy* that appeared in 1710). In response to these public rejections of Newtonian physics, argues Hall, Keill launched his war as one personal initiative within an increasingly collective English effort to secure Newton's name and reputation against the perceived slights of his critics.[41]

Whatever the originating motives, the dispute itself turned bitter quickly. Invoking his authority as a fellow of the Royal Society, Leibniz protested Keill's

38. Cited in Hall, *Philosophers at War*, 145.

39. Very little modern historical scholarship about Keill exists, but what is available is ably summarized in David Kubrin, "John Keill," in Gillispie, *Dictionary of Scientific Biography*, 7: 275–277.

40. Hall, *Philosophers at War*, 143.

41. Ibid., 144–145.

use of the *Philosophical Transactions,* the society's official organ, to slander his honor. He further appealed to the society's officers asking for an appropriate response. Hans Sloane, the secretary, turned to Newton, who in turn turned to Keill. Keill proceeded to justify his argument to Newton while also showing him some of the recent published critiques of his natural philosophy, especially those written by Leibniz. Based on this demonstration and some amount of personal animosity toward Leibniz, Newton then agreed to let Keill construct a more systematic defense of his original charges. In January 1713, a prosecutorial document against Leibniz articulating and defending the original charge of plagiarism was published by the Royal Society and distributed free of charge to the public. In the years that followed, Keill led the public defense of Newton, rallying numerous other Newtonians to his cause, while Leibniz and his close ally Johann Bernoulli mustered a vigorous defense of their own and a critique of the Newtonian position.

The overall result was an intensely bitter battle that touched all intellectual activity in its vicinity. This included debates about natural philosophy tangential to the technical mathematical issues at stake. Indeed, observers at the time (including Fontenelle) and historians ever since have agreed that there was nothing of substance in the mathematical dispute itself since the calculus probably had no single inventor, and the important achievements of Leibniz and Newton were realized independently at more or less the same time. Nevertheless, the years 1713–1718 were marked by a titanic struggle, unprecedented in its breadth and bitterness, over the invention of the calculus. This battle turned an emerging polemicism about Newton and Newtonianism in England into an all-out national crusade to defend Newton against all perceived attacks, especially those launched by foreigners in France and Germany. It also forever transformed the courteous world of the Republic of Letters into a place where partisan disputation about natural philosophy became more and more the norm.

Crousaz's *Traité du beau,* with its decorous challenges to the epistemological foundations of Newtonian physics, entered this environment in 1714. Consequently, the newly activist English Newtonians turned the book into an exemplar of the obdurate Cartesian hostility to true natural philosophy and not, as the author hoped, into an exemplar of judicious and cordially reasoned science. The appearance in London around the same time of a broadsheet screed against Newton authored by a writer named "Crucius" only made matters worse. Many of the Newtonians believed that Crousaz wrote the pamphlet, even though he in fact had no connection to the work. In the heated climate of the priority dispute, however, the mere suggestion that Crousaz harbored anti-Newtonian sentiments was enough to tar him as an unreasonable Cartesian hell-bent on undermining the true physics.[42] In this way, Crousaz's efforts to interact with the English Newtonians in the courteous manner of honorable savants were sabotaged from the outset by the polemical environment of the priority dispute.

Crousaz's experience was emblematic, for throughout the Republic of Letters the polemical energies unleashed by the priority dispute transformed philosophical dialogue. The 1713 edition of the *Principia* was one outgrowth of this upheaval since Newton's motivation for a new edition was triggered in no small part by Bernoulli's relentless attacks on the mathematics of the first edition, attacks that only became vigorous after he allied himself with Leibniz in the calculus struggle.[43] Similarly, the polemical character of Côtes' new preface was motivated by a parallel desire to aggressively articulate the newly polemical Newtonian ideology and to associate it emphatically with the *Principia*. In fact, while the priority controversy was rooted in a set of narrow mathematical and historical technicalities, one of its interesting consequences was that it focused new critical attention on Newton's natural philosophy as a whole.

The 1713 edition of the *Principia* fueled this development since it included not only the revised mathematical proofs that responded to Bernoulli's criticisms and Côtes' new polemical manifesto on behalf of Newtonian experimental philosophy, but also the new "General Scholium" that added Newton's own voice to the debates about God, nature, and matter triggered by Toland, Collins, and the other Spinozists and deists. In the "Scholium," Newton defined his own understanding of space, matter, and the deity in a way that directly challenged Toland's Spinozistic understanding of these principles. His arguments, moreover, agreed perfectly with the Newtonianism that Clarke and others had developed in the Boyle Lectures. In this way, Newton's actual science, at least as revealed in his foundational texts, was also transformed by the new polemical climate of the years after 1710.

Yet as much as the mathematical issues of the priority dispute mostly provided the spark necessary to set an externally prepared tinderbox aflame, they were not entirely tangential to the disputes themselves. For one, they reopened the topic of infinitesimal analysis for intellectual dispute, an opening that made available vast reservoirs of polemical fuel for the fire. The narrow mathematical differences that defined the priority dispute as an actual intellectual struggle

42. Crousaz was informed of this confusion by Pillonière in June 1720. See François de Pillonière to Jean-Pierre Crousaz, FJPC, 2: 174. Crousaz then spread the story to others, including Réaumur. See Crousaz to René Antoine de Réaumur, FJPC, 2: 188.

43. On the origins of the second edition of the *Principia*, see Richard Westfall, *Never at Rest: A Biography of Isaac Newton* (Cambridge, 1980), 729–751.

were also central rather than tangential to the deeper intellectual issues at the heart of debates about Newtonianism and Cartesianism. Accordingly, the linkages that quickly formed between the mathematical terms of this dispute and the larger philosophical ones that became the basis of more widespread argument were not accidental. It actually made sense to connect the mathematical differences between Newton and Leibniz to the larger differences that increasingly isolated Newtonians and Cartesians, and this linkage made possible Castel's seemingly odd entry into the Newton wars through a set of critiques of advanced analytical mathematics. The same explains Fontenelle's apparently odd choice of entering the emerging Newton wars through a treatise on infinitesimal geometry.

But what exactly were these mathematical differences, and how exactly did the priority dispute polarize and polemicize them? To fully answer this question, a detailed survey of the history of mathematics after 1650 would be required, but since several highly skilled historians of mathematics have already written much of this history, a summary of their conclusions can be offered instead.[44] To be brief, Newton and Leibniz employed contrasting mathematical approaches in their work, and each conceived of their calculus differently

44. Central to my understanding of these topics are Michel Blay, *La naissance de la mécanique analytique: La science du mouvement au tournant des XVIIe et XVIIIe siècles* (Paris, 1992), idem, *Les raisons de l'infini. Du monde clos à l'univers mathématique* (Paris, 1993); H. J. M. Bos, "Differentials, Higher-Order Differentials and the Derivative in the Leibnizian Calculus," *Archive for the History of the Exact Sciences* 14 (1974–1975): 1–90; Carl B. Boyer, *The History of the Calculus and Its Conceptual Development* (New York, 1959); Niccolò Guicciardini, *The Development of Newtonian Calculus in Britain, 1700–1800* (Cambridge, 1989), idem, *Reading the* Principia*: The Debate on Newton's Mathematical Methods for Natural Philosophy from 1687 to 1736* (Cambridge, 1999), idem, "Dot-Age: Newton's Mathematical Legacy in Eighteenth-century Mathematics," *Early Science and Medicine* 9, no. 3 (September 2004): 218–256; Thomas Hankins, *Jean d'Alembert: Science and the Enlightenment* (Oxford, 1970); Morris Kline, *Mathematics: The Loss of Certainty* (Oxford, 1980), idem, *Mathematical Thought from Ancient to Modern Times* (Oxford, 1990); Michael S. Mahoney, "Changing Canons of Mathematical and Physical Intelligibility in the Late Seventeenth Century," *Historia Mathematica* 11 (1984): 417–423, idem, "On Differential Calculuses," *Isis* 75 (1984): 366–372, idem, "Infintessimals and Transcendant Relations: The Mathematics of Motion in the Late Seventeenth Century," in *Reappraisals of the Scientific Revolution*, ed. David C. Lindberg and Robert S. Westman (Cambridge, 1990), 461–491; Helena Pycior, *Symbols, Impossible Numbers, and Geometric Entanglements: British Algebra through the Commentaries on Newton's Universal Arithmetick* (Cambridge, 2000); and D. T. Whiteside, "Patterns of Mathematical Thought in the Later Seventeenth Century," *Archive for the History of the Exact Sciences* 1 (1961): 179–388.

even if they ultimately developed equivalent systems of infinitesimal analysis independently and largely simultaneously. The equivalency of the two systems meant that savants trained in one approach or inclined for whatever reason toward one system as opposed to the other, could easily recognize the similar work being done within the other system. For this reason, Varignon, l'Hôpital, Fontenelle, and others had no problem recognizing Newton's calculus in the *Principia* and in turn conceiving of the work as "all about this new calculus" even though Newton did not employ his method of fluxions in the treatise. He used instead an idiosyncratic mathematics that adhered more strictly to classical geometry, and this meant that at the level of mathematical style (if not always substance) the science of the *Principia* and the Leibniz-inspired analytical mechanics practiced by the French were distinct.

Mathematicians were also conscious of these differences, even if they rarely invested great intellectual significance in them before 1710. Indeed, in the first fifteen years after the appearance of the *Principia,* the talk was largely about reconciliation as the correspondence networks that linked the leading mathematicians of Europe buzzed with rumors that a new edition of the treatise was imminent, one that would be authored by Newton but revised in accordance with the new calculus. Interestingly, in light of his early advocacy of Newtonian attractionist physics, the Scottish mathematician David Gregory, who taught Keill and Toland their Newtonianism at the University of Edinburgh, was a leading proponent of such a revised edition. Roger Côtes wanted one as well, as did Rémond de Montmort's English colleagues Brook Taylor and Abraham de Moivre. That so many Englishmen after 1690 were eager to see a new, "analytical" *Principia* developed using the newest infinitesimalist mathematics reveals the extent to which the "Continental" understanding of Newtonian science was not so exclusively "Continental" after all.[45]

Nevertheless, within the cosmopolitan correspondence networks that actually constituted the mathematics "profession" around 1700, Newton stood out as especially idiosyncratic. Leibniz too was no ordinary mathematician, and since contemporaries perceived each to be a sublime mathematical genius, their mathematical differences, which were indeed substantial, formed one basis for the future rivalry between them. Their disagreements hardly made the bitter acrimony of the priority dispute inevitable; however, several fundamental differences did mark Newton's approach to mathematics as distinct from Leibniz's. One had to do with how each conceived of the "content" of his mathematics. For Newton, it was crucial that mathematical objects be "real things," even if reality here could be construed in a variety of ways. This allied him strongly

45. See especially Guicciardini, *Reading the* Principia, chap. 7.

with English thinkers such as Barrow (Newton's first mathematical mentor) and Hobbes who argued that mathematical symbols were only justified if they could be referred to real, meaningful objects.[46] In his "Preface" to the *Principia,* Newton offered one illustration of his thinking by describing mathematics as simply a more abstract rendering of the real shapes and magnitudes found in nature. Likewise, in his calculus, he anchored the epistemologically slippery concept of the infinitesimal fluxion by representing it through references to the empirical phenomenon of continuous motion in nature.[47]

In each of these cases, Newton's mathematics was disciplined by his conviction that mathematical concepts must have clear meaning according to evident standards of intelligibility. A comparison he offered in the late 1670s between Cartesian mathematical reasoning and that offered by the ancient sources illustrates well this epistemological orientation. According to Newton, Descartes achieved his result "by an algebraic calculus which, when transformed into words (following the practice of the ancients in their writings) would prove to be so tedious and entangled as to provoke nausea." While correct, the solution might not even be understood at all, he continued, and the solution of the ancients was better since it used "certain, simple propositions . . . concealing the analysis by which they found their constructions." Newton also agreed with the ancients that "nothing written in a different style was worthy to be read," a pronouncement that illustrated well his general practice of holding mathematics to a higher epistemological standard than mere correctness. He wanted mathematics to be intelligible as well, and for him this meant achieving symbolic clarity and linguistic legibility along with accuracy.[48]

Empiricism offered a particularly strong way to guarantee this intelligibility, and for this reason it became one of Newton's most important epistemological principles. His differential calculus illustrated this tendency since his epistemological confidence about the evanescent fluxion, the core mathematical concept of his system, derived from his belief that it merely captured in mathematical terms the empirically common-sense phenomenon of continuous motion in nature.[49] Not all mathematical objects could find their meaning em-

---

46. See Mahoney, "Canons of Intelligibility"; Pycior, *Impossible Numbers*; and Whiteside, "Changing Patterns of Mathematical Thought." Also useful on this topic is David Sepkoski, *Nominalism and Constructivism in Seventeenth-Century Mathematical Philosophy* (London, 2007).

47. See Boyer, *History of the Calculus,* 187–223.

48. Cited in Westfall, *Never at Rest,* 379.

49. See Boyer, *History of the Calculus,* 213.

pirically, however, and here Newton's conservatism about mathematical legibility was also reinforced by a corresponding vigilance about mathematical rigor. He recognized the value of all kinds of mathematical methods, at least as tools of discovery, and he employed a variety of mathematical methods himself, including the Cartesian analysis criticized above. Nevertheless, mathematics proper for him was demonstrative mathematics that followed the strict canons of deductive, synthetic proof.

Newton expressed his convictions about rigor when choosing his mathematics for the *Principia*. He opted not to use his more economical and conceptually potent fluxional calculus to develop his arguments since he believed that this mathematics was not sufficiently rigorous for a formal, demonstrative treatise. Instead, he developed a hybrid, yet synthetically developed, form of quantitative geometry for his text, a choice that further reflected his general preference for geometry over algebra since the former was, for him, more legible in its concepts and more rigorous in its demonstrative methods.[50] Newton also viewed with suspicion the new, "modern" mathematics developed in the wake of Cartesian analysis even if he nevertheless contributed in important ways to its development. In his mind, mathematical analysis was a powerful tool that was especially useful in the search for new discoveries. It was also a great convenience since it offered mathematicians a shorthand for quickly economizing complex mathematical arguments. Thus, as either an aid to discovery or as a method of convenience, analysis posed no problems for Newton.

However, when analytical solutions were offered as definitive solutions or when economy and convenience were offered as epistemological substitutes for rigorous, synthetic demonstration, then Newton was quick to call foul. He in fact cultivated a self-conscious antimodernism in this respect as he celebrated the superiority of ancient mathematical methods over those used by the moderns. He even allied himself to some degree with the ancients themselves by describing his own innovations as recoveries of ancient techniques since lost to history. Scholarly research into ancient mathematical texts and methods was in fact an important part of Newton's regular mathematical practice, and in this respect his understanding of the "progressive" advance of mathematical knowledge was at odds with those with a more aggressively "modern" orientation.[51]

Leibniz's relationship to the same mathematical developments was very different. Arriving at his differential calculus through a program of more general philosophical research into the relationship between thought and symbols in

---

50. See especially Guicciardini, "Dot-Age"; and *Reading the* Principia, chap. 3.
51. Guicciardini, "Dot-Age."

human reasoning, he never shared Newton's concern about the precise relationship between mathematical symbols and their content.[52] For Leibniz, as for Malebranche, who shared much in common with him on this point, mathematical concepts were not necessarily mirrors of nature but rather instrumental tools of human reasoning. They allowed for a scientific comprehension of nature's fundamental relationships, and as such their meaning derived not from their intrinsic intelligibility but rather from their power of simplification and rational clarification. Understood this way, mathematical symbols did not themselves need to be meaningful or legible to be legitimate. They simply needed to cohere into rational systems that effectively accounted for the general relationships being studied.[53]

Within such a framework, mathematical discovery also became an epistemological end in itself for Leibniz, one whose attainment was confirmed through the achievement of increased generality or economy of mathematical articulation. Synthetic demonstration, by contrast, became less important since the innovative mathematical achievements themselves, measured in terms of their universality, utility, and economy, became their own demonstrations. Leibniz, for example, never worried about the epistemological validity of his new calculus or its symbols, writing to Varignon on one occasion that "even if someone refuses to admit infinite and infinitesimal lines in a rigorous metaphysical sense as real things, he can still use them with confidence as ideal concepts that shorten his reasoning, similar to what we call imaginary roots in ordinary algebra."[54] Newton, by contrast, rejected this "modern" conceptualization, insisting that mathematics was meaningless if its concepts were merely imaginary. Newton also believed that rigorous, synthetic geometry constituted the real foundation of mathematics, not the symbolic analysis of numerical relationships.

The irony here is that each mathematician nevertheless developed an equivalent, if distinct, system of differential calculus despite these philosophical

52. For an incisive recent analysis of the beginnings of Leibniz's conception of infinitesimal analysis, see Matthew L. Jones, *The Good Life in the Scientific Revolution: Descartes, Pascal, Leibniz and the Cultivation of Virtue* (Chicago, 2006), chaps. 5–6.

53. In addition to Boyer, *History of the Calculus*, 187–223; Blay, *Raisons de l'infini*, 138–159; and Bos, "Differentials in the Leibnizian Calculus"; see Louis Couturat, *La Logique de Leibniz* (Paris, 1901); and Robert M. Adams, *Leibniz: Determinist, Theist, Idealist* (Oxford, 1994), esp. 219–234.

54. Leibniz to Varignon, 2 February 1702, in Leibniz, *Leibnizens mathematische schriften*, 4: 92–93.

differences. Moreover, while the contrasts between the two systems are clear if one studies the full documentary record behind each in detail, no one at the time had full access to this record, and few mathematicians at the time troubled themselves with these philosophical issues in any case. Furthermore, since it was the results themselves that most often circulated, usually in the form of discrete presentations offered in letters, manuscripts, and journal articles (i.e., not formal, deductive proofs), it was easy for contemporaries to miss the formal differences that had driven the origination of the mathematics in the first place. The priority dispute would eventually give people a reason to search for these differences, but before it erupted there was no reason to look for them. There were also few pressures preventing mathematicians from interpreting the results however they liked. Accordingly, it is anything but surprising that the two mathematical systems coexisted for decades without provoking any animosity or rivalry between them.

The priority dispute changed this situation. It compelled mathematicians to make a choice between these two mathematicians. The trial over the invention of the calculus also shaped this new rivalry in important ways. Most directly, it gave a new impetus for inquiring into the relative character of Newton's and Leibniz's mathematical work, along with a new motive for critically scrutinizing the differences that this inquiry revealed. To counter the charge of plagiarism, Leibniz and his supporters (notably Johann Bernoulli) rigorously attacked Newton's mathematics. To further this attack, Bernoulli also persuaded Leibniz to issue some challenge problems that posed, he believed, difficulties of resolution from within Newton's system. The idea was to expose in this public way the inferiority of Newton's mathematical methods. However, since these methodological differences were more perceived than real, these "ill-regulated mathematical contests," to quote Hall, "did nothing to clear up the points at issue" while enhancing greatly "irritation and ill-feeling."[55]

These challenge problems did clarify the Leibnizian position, however, since they reinforced the central point of his defense: namely, that the Leibnizian calculus was original and superior to the one developed by Newton. The Newtonians found it difficult to respond to this claim directly, for when Bernoulli and Leibniz exposed the weakness of Newtonian mathematics in solving difficult problems, they were in fact pointing to real differences between the two systems. In the face of these difficulties, the Newtonians were forced to construct a coherent defense of their own approach. This resulted in the creation of a new Newtonian mathematical ideology that had not existed before.

55. Hall, *Philosophers at War*, 216.

Rigor was a central theme in this new discourse. The Newtonians claimed that Leibnizian mathematics was reckless and misleading because it failed to respect the appropriate hierarchy between analysis and synthesis. As the Royal Society's original report authorizing the calculus dispute declared in 1713, "Mr. Leibniz's [method] is only for finding out. . . . [It is not for] demonstrating, but only investigating a Proposition, [or] for making dispatch."[56] The Newtonians thus conceded that their method was less convenient. But what it surrendered in economy, it more than recuperated in its superior rigor, or so they argued. Indeed, since only Newton's method was grounded in rigorous synthetic geometry, his supporters claimed, the method of fluxions was the only truly scientific calculus while Leibniz's infinitesimal analysis was little more than an economical knockoff of the Newtonian original.

The calculus of fluxions was in fact no more rigorous than Leibnizian infinitesimal analysis since the conceptual tools necessary to provide such demonstrative rigor were not available in the eighteenth century.[57] In practice, this meant that the bitter arguments about rigor and method that raged throughout the 1710s really amounted to a debate between two distinct but equally valid mathematical epistemologies. What gave the Newtonian method added rigor in the mind of its supporters was its strict adherence to ancient canons of synthetic deduction. Also crucial was its refusal to offer analytical solutions as definitive solutions. The superior legibility of its core concept, the evanescent fluxion, was also praised, while Leibniz's infinitesimals were mocked as either meaningless fictions or metaphysical monsters. For his part, Leibniz never wavered in his defense of infinitesimals, arguing that the Newtonians misunderstood and misrepresented the nature of his core concept. Infinitesimals indeed have no discernible reality, he argued, and they are little more than "well-founded fictions."[58] But this did not make them any less rigorous as a result. He likewise rebutted Newton's bias toward geometry and synthetic proof by continually demonstrating the superior explanatory power of his "less rigorous" analytical methods. As a result, no reconciliation occurred, nor was any ever likely, while the battle itself produced a new clarity about the mathematical differences at stake. This clarity, when combined with the acrimony that produced it, ultimately supported a new polarization among mathematicians.

56. *Account of the Book Entitled* Commercium Epistolicum (London, 1713), 206. This citation is drawn from the facsimile copy of the original publication printed in Hall, *Philosophers at War*, 296.

57. See especially Kline, *Loss of Certainty*.

58. Leibniz to Varignon, 2 February 1702, in Leibniz, *Leibnizens mathematische schriften*, 4: 94–95.

At the same time, the focus on rigor and method in these mathematical disputes contributed in important ways to the emerging debates about Newtonian natural philosophy. Since the alleged rigor of Newton's mathematics came in part from the apparently empirical foundations of his evanescent fluxion, and since Leibniz's infinitesimals were chastised as imaginary and chimerical fictions, the centrality of empiricism to Newton's mathematical philosophy allowed the defenders of Newtonian experimentalism to apply many of their favorite arguments to Newtonian mathematics as well. In fact, "Newtonian experimental philosophy," with its core belief in the "analytical-synthetic method" as the only basis for sound philosophy, emerged in precisely this period through an odd alliance between Newtonian experimentalists and Newtonian mathematicians. Mathematics contributed one key component since the terms "analysis" and "synthesis" were first used in mathematical discussions, where they had precise methodological meaning. They were applied later to natural science, where they became the watchwords for a new debate about appropriate philosophical method. Thus, when Côtes confirmed their applicability to eighteenth-century methodological debates about physics and natural philosophy, as he did in his 1713 "Preface" to the *Principia's* second edition, he was at one level appropriating a precise mathematical terminology for other uses.

In the context of the polemics of the priority dispute, however, these precise differences were less important than the cause of Newton's reputation that they jointly supported. Furthermore, since Newton encouraged an epistemological link between his mathematical thinking and his empiricism, it was easy to join these two dimensions of the Newtonian program into one unified conception.

In this way, through a complex alchemy of scientific, political, and social factors, a newly aggressive discourse about Newtonianism was created, one that stressed Newton's superior appreciation for the proper relationship between analysis and synthesis. The errors of Newton's alleged opponents were also increasingly located in their methodological confusion. English mathematicians found this discourse reassuring to their belief that Newton's stress upon conceptual legibility, the rigor of synthetic demonstrations, and empiricism offered the right foundations for sound mathematical science. It also joined them conveniently with the English experimental physicists who had their own reasons for defending this very same triad of epistemological principles. This marriage was highly ironic since many of the same experimental physicists who rallied with the Newtonian mathematicians in the larger Newtonian cause often downplayed and even disparaged mathematics in their conception of Newtonian science. Yet it was accidental marriages such as these that gave birth to the hybrid outcome called Enlightenment Newtonianism.

Father Louis Bertrand Castel and
Critical Mathematical Discourse in France

Another odd set of partnerships involved Father Louis Bertrand Castel's position within these same disputes. From the moment he arrived in Paris in 1720, Castel made his critical voice felt in the French public sphere. At the center of his critical agendas, moreover, was a very particular and highly idiosyncratic set of mathematical views that he nevertheless defended vigorously and often polemically throughout his life. These critical interventions shaped French discussions of Newtonianism in important ways. Castel used the pages of the *Journal de Trévoux* in 1722 to began articulating his positions. The occasion was the publication of Crousaz's *Commentaire sur l'analyse des infiniment petits*, a work that offered itself as an improvement upon l'Hôpital's pioneering calculus textbook of 1696. As the *Journal des savants* noted in its 1722 review, Crousaz was motivated to write his commentary by a belief that l'Hôpital's treatise assumed too much to be understood by a wide audience.[59] In response, the Swiss savant proposed a more accessible introduction to the differential calculus that would follow l'Hôpital closely but also clarify his work at every step. Castel set out to review Crousaz's work, and in doing so he set the tone for all of his subsequent mathematical writings.[60]

Criticism was the pervasive theme, a slant that he drew from both the polemical tone characteristic of the priority dispute and from the new trends in eighteenth-century learned journalism. Castel mirrored each, devoting twenty-seven pages to a fully developed polemic about the current state of mathematics. His prime target was the "Récents," his term for the practitioners of the new algebraic analysis, especially the infinitesimal calculus. The priority dispute also figured centrally in Castel's analysis, for it had revealed that mathematics, "once recognized for its solidity, righteousness, equity, accessibility, and spirit of peace and concord," had become "a battlefield, a court of justice, and even a death trap, the theater of a thousand riots."[61] What had caused this turn of events was the "*nouveaux calculs*" that the ancients had humbly shunned. "[This mathematics]," Castel argued, "inspires a spirit of softness, dissension, chicanery, and even warfare in its partisans." It is also the reason that mathematics has recently become "an affair of nation and state." Castel ultimately called for greater rigor in mathematics as a solution to these difficul-

59. *Journal des savants* (April 1722): 214–219.
60. Review of Jean-Pierre de Crousaz, *Commentaire sur l'analyse des infiniment petits*, in *Journal de Trévoux* (July 1722): 1189–1215.
61. Ibid., 1191.

ties, for as he summed up: "Enigmas incapable of divination are proposed; a science is then made through the use of new expressions and arbitrary words; everything is then transformed into words, symbols, and letters. Is this a mathematical science? Hardly, for it is not even a grammatical science. Yet *voilà* (as we know all too well) the Republic of Letters falls into disorder. Enlightenment goes into hiding, pushed into the shadows by the spirit of contradiction. Meanwhile, contentiousness, born of this, proceeds to render all knowledge suspect." [62]

Castel eventually settled into a discussion of Crousaz's book, yet even here his account served mostly as a platform for his own critical agendas. The result, therefore, was an original, argumentative essay in the guise of a book review—an approach that would become standard in Castel's journalism. His analysis also revealed many of the mathematical positions that eventually made him attractive to the English Newtonians. The latter had found their identity within the crucible of the priority dispute, a battle that ostensibly focused on the question of who first discovered the calculus, but one that quickly transformed itself into a debate over rival mathematical philosophies. Castel's review was not really an intervention in this debate, yet his frustrations with modern mathematics nevertheless echoed many of the views expressed by the English Newtonians in their struggles with Leibniz. Each, for example, lamented the perceived departure of ancient, synthetic rigor, and each worried about the illusions created by the abstract and ill-defined symbolism of the new mathematics. The illegibility of the new analysis was also scorned, while the symbolic clarity of geometry was praised. The English also claimed that Newton's fluxional calculus was more rigorous than Leibniz's because it was grounded in nature and synthetically demonstrated, and Castel seemed to agree. For as he wrote, "it is to risk fortifying the very obscurity that needs dissipating" to attempt to clarify, as Crousaz proposed to do, the new calculus from within its own terms. "One must part ways with calculus and analysis and rejoin synthetic geometry if one wants to seize the luminous side of the analytical calculus. For the first speaks only to the imagination while the second speaks directly to the eyes." [63] Formulations such as these mirrored the Newtonian insistence on clear symbolic legibility in mathematics, and Castel also seemed to agree with the Newtonian insistence upon empirical clarity and synthetic rigor in mathematical epistemology. Both convictions were in fact characteristic of his mathematical philosophy, but it was his Jesuit affiliation more than any direct or indirect attachment to the English Newtonians that explains this tendency in his thought.

62. Ibid., 1195–1196.
63. Ibid., 1201.

Throughout its history, the *Journal de Trévoux* was a consistent critic of abstract and abstruse mathematical thought since its Jesuit editors believed that such reasoning constituted the slippery slope to metaphysical and theological disaster. Consonant with their modernized Aristotelian approach to natural philosophy, the Jesuits celebrated pragmatic, common-sense empiricism in science while valuing a judicious balance between the eye and the mind when reasoning about nature's mysteries. They also celebrated humility and wonder as the stance most appropriate to natural philosophical inquiry.[64] As Father Tournemine, Castel's predecessor at the *Journal de Trévoux* and one of the leading Jesuit intellectuals of the early eighteenth century, once declared: "There is not in life an inclination more dangerous, or more ridiculous, than to conduct oneself by means of geometry, unless one can be entirely sure that the idea, or the principle of demonstration, is conformable to reality."[65]

Descartes and Spinoza were particularly "dangerous" and "ridiculous" in this respect, and the Jesuit hostility toward the perceived mathematical excesses of these philosophers made them keen to criticize mathematical rationalism in whatever form it took. Castel's diatribes against the new analytical mathematics, therefore, were typical of the Society of Jesus, even if he offered the public a newly critical and journalistic variation on this familiar Jesuit theme. That Castel's writings also resonated so strongly with the discourse of English Newtonianism was a further coincidence produced by the changing intellectual tides of the period. Without intending to produce this harmony, French Jesuits and English Newtonians ultimately sounded a lot alike when they linked the reckless mathematical abstraction of Cartesian and Leibnizian mathematics to the dangerous physical and metaphysical errors of their natural philosophy. The Scottish Newtonian Colin Maclaurin, in fact, made a similar point, without the Jesuit motivation, when he wrote with respect to Leibnizianism in 1742 that "an absurd philosophy is the natural product of a vitiated geometry."[66] The priority dispute only heightened the charge of such rhetoric, and given the new polarization of the Republic of Letters into national philosophic parties, French Jesuits often found cultural opportunities in their intellectual affinities with the English.

64. On the general philosophical inclinations of French Jesuits, see Healy, "Mechanistic Science and the French Jesuits"; Desautels, *Les Mémoires de Trévoux;* Gaston Sortais, *Le cartésianisme chez les Jésuites français au XVIIe et au XVIIIe siècle* (Paris, 1929); and Catherine M. Northeast, *The Parisian Jesuits and the Enlightenment, 1700–1762* (Oxford, 1991).

65. Cited in Healy, "Mechanistic Science and the French Jesuits," 72–73.

66. Colin Maclaurin, *A Treatise on Fluxions,* 2 vols. (London, 1801), 1: 47.

Castel was a case in point. He used his Angloid mathematical sensibilities to cultivate a public identity as a mathematical Anglophile while also positioning himself critically with respect to the dominant mathematical discourse in France.[67] At the *Journal de Trévoux*, he began the practice of reviewing the *Philosophical Transactions of the Royal Society*, a decision that gave French readers new access to this largely Anglophone publication. He also used his monthly reviews to make clear his knowledge of and affection for English mathematics. As early as 1726, he wrote privately to John Woolhouse of his desire to become a critical French advocate for English mathematics, and over the next decade he put this ambition into practice.[68] One climax was reached in 1735 when Castel supervised the French translation of Englishman Edmund Stone's treatise on the differential calculus. Stone's text was largely a first-ever English translation of l'Hôpital's 1696 *Analyse des infiniment petits*, but he called his work *The Method of Fluxions* and offered his readers a version of l'Hôpital's text recast in terms of Newton's fluxional calculus.[69] At one level, the differences were merely symbolic, since the two systems of differential calculus were largely equivalent. But since the priority dispute had charged these symbolic differences with deep cultural meaning, Stone gave many English readers what they wanted: a calculus textbook tailored toward their native Newtonian affinities. Castel, meanwhile, presented French readers with an even stranger result. He gave them a French translation of Stone's English translation of l'Hôpital's original French text, a project that only makes sense once one understands the cultural and mathematical translation that was involved as well. He also penned the work's preface, using his characteristic polemical style to situate the book within the ongoing mathematical debates of the period.[70] Johann Bernoulli, Leibniz's great ally in his battles with the British, was one mathematician who penned a critical response to Castel's work.[71]

67. On Castel's "Anglophilia," see Healy, "Mechanistic Science and the French Jesuits," 105; Desautels, *Les Mémoires de Trévoux*, 51–54; and John Greenberg, *The Problem of the Figure of the Earth*, 241–243, 260–265.

68. Father Louis Bertrand Castel to John T. Woolhouse, May 1726, Archives of the Royal Society of London, letterbook 18: 333–337.

69. Edmund Stone, *A Treatise on Fluxions* (London, 1730).

70. Edmund Stone, *L'analyse des infiniment petits comprenant le calcul intégral dans toute son étenduë*, trans. by Fr. Louis Bertrand Castel (Paris, 1735).

71. Johann Bernoulli, "Remarques sur le livre intitulé *Analyse des infiniment petits comprenant le calcul intégral dans toute son étenduë, etc.*" in Johann Bernoulli, *Opera omnia*, 4 vols. (Lausanne, 1742), 4: 169–192.

Castel also found himself more and more in conflict with the mathematicians at the French Royal Academy as a result of his work, even if he was supported in his efforts by his colleagues in the French wing of the Society of Jesus. Castel's inclination toward empiricism, for example, was shared by many French Jesuits, and this supported the impression that the society as a whole shared his affection for English science. French Jesuit colleges were in fact ahead of the curve in adding experimental physics courses to their science curriculum, and since their courses, like their counterparts in England, were public spectacles, open to a wide audience, they further publicized the association between the Society of Jesus and this kind of science.[72] Given their wider philosophical agendas, the Jesuits also used these channels to promote their particular fusion of empiricism with sound natural philosophy. The sixteenth general congregation of the order confirmed in 1731 that Aristotelian doctrine would remain the core of Jesuit scientific teaching. However, within these constraints, "the most agreeable manner of teaching" was permitted. Teachers of physics were particularly encouraged to "use mathematical principles and the experiments of savants to explain and illustrate the most remarkable phenomena of nature." The French embraced this pedagogical model earlier than most, for as Jesuit Father Croiset described the goal: "Ours is not at all an abstract and unintelligible physics, but an agreeable science of natural causes and their effects."[73]

Jesuit Father Noël Regnault's widely read *Les entretiens physiques d'Ariste et d'Eudoxe* further confirmed this orientation when it appeared in 1729.[74] The book went through seven editions, becoming one of the biggest bestsellers of the period, and overall it popularized the characteristically Jesuit approach to experimental natural philosophy within the wider Francophone world.[75] In the text, Eudoxe performs experiments and Ariste explains them, an approach that mirrored Jesuit science teaching more generally. Furthermore, despite the book's precise orientation toward Aristotelianism, it echoed the methods and epistemological assumptions of English experimental philosophy. Especially resonant was the book's suspicion of mathematical abstraction and its insis-

72. François de Danville, S.J., "L'enseignement scientifique dans les collèges des Jésuites," in *Enseignement et diffusion des sciences en France au XVIIIe siècle,* ed. René Taton (Paris, 1964), 27–65.

73. Ibid., 37.

74. Père Regnault, *Les entretiens physiques d'Ariste et d'Eudoxe* (Paris, 1729).

75. Daniel Mornet offers an assessment of the popularity of various eighteenth-century French publications in "Les enseignements des bibliothèques privées," *Revue d'histoire littéraire de France* 18 (July–September 1910): 449–492. Regnault's work ranks high on Mornet's list of bestsellers.

tence upon the epistemological validity of common-sense empiricism. These were widely held Jesuit values, and their agreement with English experimental philosophy made the society an equally important conduit for them within France as a whole.

Regnault also stressed the value of the easy, the unaffected, and the agreeable in scientific practice, and in this respect his empiricism also had a particular worldly French flavor that cut against the grain of eighteenth-century English experimentalism. Yet the same emphasis was characteristic of Castel's work as well, and it further placed him alongside Regnault and other scientifically oriented Jesuits at the center of the urbane discourse of science that more secular works like Fontenelle's *Entretiens sur la pluralité des mondes habités* had made central to French public science after 1690. Other French Jesuits preceded Castel and Regnault in this sphere, including Father Gabriel Daniel, whose witty satire of the "world of Descartes" was a popular hit in the 1690s.[76] Non-Jesuits also contributed to this literature, including the abbé Noële-Antoine Pluche, whose wildly popular *La spectacle de la nature* began to appear in 1732, and William Derham, whose popular works of English natural theology appeared in a set of popular French translations in the 1720s.[77] Together, these texts and others like them sustained the characteristic French fusion of public science with urbane sociability that was its hallmark in the early decades of the eighteenth century.[78]

76. Gabriel Daniel, *Voyage du monde de Descartes* (Paris, 1690).

77. Noël-Antoine Pluche, *La spectacle de la nature, ou Entretiens sur les particularités de l'histoire naturelle* (Paris, 1732); and William Derham, *La théologie physique, ou Demonstrations de l'existence de Dieu*, trans. by Jean Daniel Beman (Rotterdam, 1726), idem, *La théologie astronomique*, trans. by l'abbé Bellanger (Rotterdam, 1729).

78. It is worth noting that among the most influential conduct manuals of the seventeenth century was a set of dialogues written by Jesuit Father Dominique Bouhours. Bouhours's works in particular appealed strongly to the emerging noncourtly sensibilities of salon society. Bouhours was the author of many works, but two of the more influential in this genre were *Les entretiens d'Ariste et Eugène* (Paris, 1671) and *La manière de bien penser dans les ouvrages d'esprit* (Paris, 1687). Each of these works was reprinted several times in the eighteenth century. On the noncourtly bias of Bouhours, see Daniel Gordon, *Citizens without Sovereignty: Equality and Sociability in French Thought, 1670–1789* (Princeton, 1994), 86–88. One way of viewing the rise of these explicitly *mondain* Jesuit scientific dialogues in the eighteenth century is to see them as the result of a marriage between the Jesuit preoccupation with right-minded conduct as exemplified by Bouhours and the Jesuit interest in the popular dissemination of correct philosophy as exemplified by their teaching, their work as journalists, and their publication of popular philosophical works such as Father Daniel's *Voyage du monde de Descartes* (1690).

Castel was particularly influential in taking this wider *mondainité* and injecting it into the mathematical discourse of the period. At the Collège Louis-le-Grand, he instituted a new curriculum, his "smiling method," as the historian François de Danville has called it, that involved replacing the cold abstraction of traditional mathematics with more pleasurable mathematical pursuits.[79] The thinking that informed these reforms is well illustrated in a dialogue that Castel published in early 1735 entitled *La géometrie naturelle en dialogues*. The work was published first as a serial publication in the literary journal *Nouveaux amusemens du coeur et d'esprit* and then later as a book published in Amsterdam. When it appeared, it joined other Jesuit scientific dialogues in illustrating how these *mondain* idioms had infiltrated both Jesuit and French mathematical discourse.[80]

Castel's text employs the standard Jesuit interlocutors, Ariste and Eudoxe, to explain the author's worldly approach to mathematical instruction. Ariste opens by challenging the idea that geometry is a truly natural science, but Eudoxe counters with a classic justification of natural mathematics, asserting that "the sensible world is entirely governed by Geometry, or by Mathematics in general. Almost everything can be reduced to counting, weighing, and measuring."[81] Such views were widely accepted, but at this juncture the dialogue takes a very striking turn. Attempting to sum up Eudoxe's argument, Ariste declares: "The object of mathematics is thus natural and the pure work of nature. Yet we render it completely artificial by making it the pure work of our hands and our imagination?" "Correct," Eudoxe declares. "Everyone can become a geometer because nature is entirely geometrical." Mathematicians, however, corrupt this natural approach by making it "overly intellectual and overly abstract." Eudoxe particularly despises the use of unnatural figures and problems in geometry, for as he explains: "As soon as one assumes an air of science by

79. Danville, "L'enseignement scientifique dans les collèges des Jésuites," 59.

80. My citations will refer to the version that appeared in *Nouveaux amusemens du coeur et d'esprit* 1 (741): 80–93, 185–202. A preface to this journal, which promises "a collection consecrated exclusively to the pleasures of the heart and mind," defines well the intended audience for this work. I consulted the copies found at BN, Z 24508–24524. This appears to be a reprint edition of the actual journal edited by l'abbé Philippe and printed in Paris, which Sgard dates from 1734–1735 and 1737–1745. See Sgard, *Dictionnaire des journaux, 1600–1789*, 2: 1182. Castel's dialogue was printed in the first volume indicating that it appeared in late 1734 or early 1735. For a review of Castel's dialogue, see *Journal des savants* (1739): 238–243.

81. Castel, *Géometrie naturelle en dialogues*, 1: 86–87.

adopting an apparatus of mysterious characters and terms, . . . things change from their natural state."[82] This especially occurs when mathematical figures and symbols are detached from nature through abstract symbolism or unnatural complexity. "According to this view," Ariste asks in conclusion, "you must hardly approve of the cryptic [*grimoire*] calculus and algebra that the mathematicians of recent times have used to cloud the most simple geometry?" "This is all that it takes to close the door of natural geometry to everyone," Eudoxe replies.[83]

Castel's hedonistic mathematical philosophy informed the pedagogy that he attempted to institute at Louis-le-Grand after 1720. It was also the philosophy that informed his many efforts at critiquing contemporary mathematics. There was of course nothing English or Newtonian about these efforts. They rather derived directly from his Jesuit, and especially French Jesuit, commitments. Nevertheless, his emphasis on naturalizing mathematics by bringing them more in accord with the facts of nature appealed to the sensibilities of the English Newtonians. So did his aversion to the vain and misleading abstraction of modern, symbolic mathematics. When, for example, he explained that the fundamental concepts of Apollonian geometry were not hard to conceive since "our cakes [*pains de sucre*] are cones, our yew trees conic sections, our dice cubes, and our eggs ellipsoids," the Newtonians had no problem recognizing their own empirical conception of mathematics in Castel's discourse even if they may have found his precise idioms alien to their own. Similarly, when Castel took Crousaz to task for recklessly connecting the infinitude of God with l'Hôpital's notion of an infinitesimal magnitude, the English had no problem echoing his views even if the philosophical and theological reasons that led them to this position could not have been more different.

In 1730, the *Mercure de France* confirmed these connections, reporting that "the mathematicians in London" found Castel's work "marvelous and extraordinary."[84] Four months earlier, Castel had been made a Fellow of the Royal Society. He is still the only French Jesuit ever made a fellow of the Royal Society, and one of only a handful of Jesuits ever admitted into this bastion of Britishness from anywhere in the world. His reputation in England was thus great, and his perceived Anglicism also shaped his reputation in France. Especially noteworthy was his rapid establishment as vociferous critic of French academic

82. Ibid.

83. Ibid., 1: 88–89.

84. *Mercure de France* (September 1730): 1974, cited in Danville, "L'enseignement scientifique dans les collèges des Jésuites," 59.

mathematics, a stance that allowed his mathematical Anglophilia to become a discourse of opposition directed against the Paris Academy.

As Castel rose to public prominence after 1720, a group of publications by members of the French Academy appeared that defended in one way or another the methods and assumptions of the Leibnizian calculus. Among them were Crousaz's commentary mentioned above (he was a foreign associate); a posthumous treatise on the infinitesimal calculus by Varignon; and Fontenelle's 1727 treatise, which he revived in 1724 as a response in some measure to the new climate of mathematical opinion triggered by the priority dispute and catalyzed by Castel. During the winter of 1724–1725, Fontenelle engaged in a serious intellectual exchange with Johann Bernoulli about his dormant treatise. The letters suggest that Fontenelle had resisted publication earlier because of fears of the work's inadequacy, but he also expressed a new eagerness to clear up any errors and get the work into print.[85] In August 1726, *Eléments de la géometrie de l'infini* was presented to the academy for review and authorization. It was approved in February 1727, and since the academy's volume of *mémoires* for 1725 was then in press, the members further decided to publish the treatise as an addendum to the academy's official *Histoire de l'Académie royale des Sciences* for that year.[86] In November, just prior to the work's appearance, Fontenelle read the "Preface" to his *Eléments* at the academy's fall public assembly. The public reading was covered in the *Mercure,* which noted the elegance and eloquence of Fontenelle's presentation, but also the difficulty of the subject matter.[87]

Reviews of the treatise were mixed. The *Journal des savants* devoted two long, thorough, and typically neutral articles to the text, while the *Journal littéraire de la Haye* also ran a review, written by 'sGravesande, that offered some criticisms of Fontenelle's reasonings couched within general praise for the work and its author.[88] Fontenelle responded to 'sGravesande immediately, publishing a set of clarifications in a subsequent issue of the same journal.[89] Overall, the exchange between Fontenelle and 'sGravesande was marked by cordiality and mutual respect, and since the points of debate were extremely esoteric it did little to transform the reception of Fontenelle's treatise within the wider public sphere. Castel's review, which ran in July 1728, was also a

85. Blay, introduction, 11.

86. PVARS, August 14, 1726; February 27, 1727.

87. *Mercure de France* (November 1727).

88. Review of Bernard le Bovier de Fontenelle, *Eléments de la géometrie de l'infini,* in *Journal des savants* (1728); and in *Journal littéraire de la Haye* 14 (1728): 363–384.

89. *Journal littéraire* 16 (1730): 1–9. A facsimile of this text also appears in Blay and Niderst, *Eléments,* 615–619.

model of judiciousness.[90] He opened by praising Fontenelle and by expressing the public's gratitude for this long-expected work. He also situated the treatise squarely within Fontenelle's public work as academy secretary. "Those who judge with balance and wisdom have already discerned in the *Histoire de l'Académie des Sciences* the fortunate alliance that this author has made between stylistic elegance and geometrical knowledge," he wrote. "They have come to expect [from M. Fontenelle] precise and ingenious perspectives, an ordered spirit, luminous expressions: in a word, manners. One sees this and more in these new *Eléments*."[91]

Setting the stage for the rest of his review, however, Castel also added that "the passage from the beautiful mind that produced the best works of M. Fontenelle to the geometrical genius that gave birth to this text is no easier to explain than the passage from the finite to the infinite."[92] With this introduction, Castel then undertook a systematic and subtly critical examination of Fontenelle's work. Castel's strategy was ingenious given his ambitions and the character of Fontenelle's text. Rather than challenge directly Fontenelle's discourse of clarification, Castel went Fontenelle one better. He reviewed each of Fontenelle's claims in great detail, pointing out the ambiguities and problems of each. He further deployed Fontenelle's own language of neutral presentation in his discussion. This in effect turned Fontenelle's characteristic discourse of objective clarity back against him.

Castel built his review upon the "two great paradoxes of infinitesimal analysis," the same paradoxes that had plagued analysis from its inception. Castel emphasized again and again that Fontenelle's work did nothing to resolve these conundrums. By exposing these inadequacies with Fontenellian clarity and eloquence, Castel thus countered Fontenelle's effort to demonstrate the rationality of infinitesimal analysis with a clear and distinct exposition of its irrationality. Given the nature of Fontenelle's treatise, this was a devastating strategy. Fontenelle's book, while masterful in many respects, was anything but an accessible and easily comprehended explanation of infinitesimal analysis. At its best, it established Fontenelle as a serious student of these issues and demonstrated that a rational argument in defense of infinitesimal analysis could be made. However, the mathematical community soon came to see the work as a failure, and without the authority of experts behind it, the wider public, prevented from understanding the work directly because of its abstruse

90. Review of Bernard le Bovier de Fontenelle, *Eléments de la géometrie de l'infini*, in *Journal de Trévoux* (July 1728): 1233–1263.

91. Ibid., 1234.

92. Ibid., 1235.

character, remained unmoved.[93] Castel's clear and readable critique of the treatise reinforced this impotence.

But this was only one half of Castel's response. In 1728, little more than a year after the publication of Fontenelle's *Eléments,* Castel published *Mathématique universelle, abregée à la portée de tout le monde,* his own popular exposition of mathematical philosophy. This was an attempt to systematize his easy, naturalist approach to mathematics and present it to a wide audience.[94] As he summed up in his "Preface": "My sole aim is to demonstrate that one can (and I have been able to) restore the most unnatural geometry to its proper nature; that one can reduce it to ordinary ideas, to link it with the most common notions, and thus to pass from practice to theory in a natural way."[95] In the review in the *Journal de Trévoux,* which Castel penned himself, he struck the same chord. "Geometry is the principal, even the unique, object of this work. However, not wanting to treat only dry and purely speculative geometry, I have neither restricted myself to this science alone nor presented its truths from the abstract and detached perspective common in the other sciences."[96] Overall,

93. Blay is the most careful modern student of Fontenelle's work, and he offers a conveniently brief assessment of the historical significance of the *Eléments* in the introduction cited above. See also idem, *Les raisons de l'infini: Du monde clos à l'univers mathématique* (Paris, 1993), 175–213. Blay is largely a defender of Fontenelle, but modern historians of mathematical thought remain as divided as Fontenelle's contemporaries about the significance and quality of his treatise. Lagrange called the *Eléments* and Fontenelle's mathematical writings more generally "unintelligible gibberish," and some eighteenth-century critics even wondered whether Fontenelle knew any mathematics at all. These negative assessments of Fontenelle's work are collected by John L. Greenberg and used to present a much more critical assessment of his achievement. For a typical example, see Greenberg, *The Problem of the Earth's Shape from Newton to Clairaut: The Rise of Mathematical Science in Eighteenth-Century Paris and the Fall of "Normal" Science* (Cambridge, 1995), 676–677, n. 51. A more neutral assessment that surveys contemporary reactions to Fontenelle's work is offered in Suzanne Delorme, "La géometrie de l'infini et ses commentateurs de Jean Bernoulli à M. de Cury," *Revue d'histoire des sciences* (1957): 339–359.

94. Louis Bertrand Castel, *Mathématique universelle, abregée à la porté de tout le monde* (Paris, 1728).

95. Ibid., iii.

96. Review of Louis Bertrand Castel, *Mathématique universelle, abregée à la porté de tout le monde,* in *Journal de Trévoux* (May 1729): 855. Castel's full review was published in three installments: *Journal de Trévoux* (April 1729): 695–721; *Journal de Trévoux* (May 1729): 855–875, and *Journal de Trévoux* (September 1729): 1587–1612. Taken together, the reviews really constitute a set of essays about mathematics in their own right rather than a book review narrowly defined.

the work offered a cornucopia of different ideas about mathematics couched in the *mondain* discourse that was quickly becoming Castel's trademark.

Castel also combined his *mondain* response to Fontenelle's abstruse treatise with a more pointed critique of French academic mathematics. As science editor of the *Journal de Trévoux*, he was responsible for reviewing the academy's *Histoires et mémoires* whenever they appeared. Early in his tenure, he carried out these responsibilities attentively, writing reviews that were marked by judicious criticism tempered by expressions of sincere intellectual respect. His 1722 review of Saurin's *mémoire* on parabolic curves in the academy's volume for 1718 was illustrative of his general approach. He noted that Saurin "finds and demonstrates his solution like an able analyst," but he also "asked permission" to point out that the solution could have been found more simply and clearly by using traditional geometric methods.[97] This manner of academic critique was the norm for the Jesuit throughout the 1720s, but in 1730 he broke with it, writing a very polemical review of the academy's *mémoires* for 1725.[98] This was the volume that carried Fontenelle's *Eléments* as an addendum, and it also appeared in the immediate wake of Castel's rival *Mathématique universelle*. After reviewing the academy's other published papers in the November 1729 issue of the journal, Castel used a second review, published in January 1730, to focus exclusively and critically on the academy's mathematical work in this volume.

His bias against modern algebraic analysis was in full evidence in this review, as was his affection for the English mathematicians that positioned themselves against it. François Nicole, the last member of the original Malebranchian cohort still practicing mathematics in the academy, received the lion's share of Castel's attack. He was accused not only of plagiarizing work already done by the mathematicians at the Royal Society, but of making their results more muddled and confused as well. Castel also offered a general critique of French academic mathematicians, complaining of their unwillingness to bring their work down to the level of ordinary readers and of their preference for algebra and analysis over clear, demonstrative geometry. Dortous de Mairan's thoroughly nonanalytical mathematics was also criticized, and thus the Jesuit's critique of the French academy was general.[99] Read alongside the escalating mathematical provocations of his earlier writings and the wider polarities trig-

97. Review of *Histoire de l'Académie royale des Sciences* (1718), in *Journal de Trévoux* (June 1722): 989–990.

98. Review of *Histoire de l'Académie royale des Sciences* (1725), in *Journal de Trévoux* (January 1730): 103–120.

99. Ibid.

gered by the priority dispute, this review constituted a direct challenge to the authority of academic mathematics in France.

Not surprisingly, therefore, Fontenelle and his allies were quick to respond in kind. The secretary had been a key player in the maneuverings that brought Castel to Paris in 1720, and after his arrival he maintained a cordial correspondence with the reverend father.[100] Later in the decade, however, Fontenelle broke off this relationship, writing to Castel in an undated letter that he regretted the "unfortunate dispute" that had arisen between them and then terminating all correspondence with him.[101] In 1729, the *Journal des savants* also published an uncharacteristically hostile review of Castel's *Mathématique universelle*, stating that the treatise might be appropriate for small children but that Castel's cavalier dismissal of traditional geometric reasoning was misguided.[102] This review was followed a few months later by an even more hostile treatment of Castel in the guise of an anonymous pamphlet.[103] This scathing critique was authored by Saurin. Perhaps because of Saurin's close ties with the editors at the *Journal des savants* (he had once been a member of Bignon's editorial team), the journal chose to review the pamphlet even though it rarely discussed pieces of polemical ephemera of this sort. Also apologizing at the outset for breaking its avowed policy of neutrality, the journal published a lengthy excerpt from Saurin's screed, claiming that the text was "simply too interesting to be refused." [104]

Saurin's pamphlet assumed the form of a letter to a friend about the mathematical education of his son. It offered the following assessment of Castel's *Mathématique universelle:* "It is a work of systematic confusion, a poorly conceived chaos of definitions, divisions, and subdivisions that resists even the most patient reader." [105] Saurin also took Castel to task for calling the geometric method "the most defective of all the didactic methods because of its rigidity."

100. See Healy, "Mechanistic Science and the French Jesuits," 117.

101. The correspondence between Fontenelle and Castel is found in *Oeuvres de Fontenelle*, 3: 348–364. Many of the letters, however, carry abbreviated dates such as "lundi 6" or no date at all, thus dating the precise chronology of the dispute is difficult.

102. Review of Louis Bertrand Castel, *Mathématique universelle, abregée à la porté de tout le monde*, in *Journal des savants* (1729): 351–354.

103. *Lettre sur le Traité de mathématiques du P.C. et les extraits qu'il fait dans les Journaux de Trévoux des mémoires de l'Académie des Sciences* (Paris, 1730).

104. Review of *Lettre sur le Traité de mathématiques du P.C. et les extraits qu'il fait dans les Journaux de Trévoux des mémoires de l'Académie des Sciences*, in *Journal des savants* (1730): 603–611.

105. Ibid., 604.

Saurin, by contrast, praised the strict rigor of geometric thinking, calling it the source of certainty in the sciences and the appropriate expression of "a judicious mind."[106] Making clear the precise audience at stake in this struggle, Saurin also lampooned Castel's appeal to the female audience so important to elite taste in *le beau monde*. Castel had suggested that women would appreciate the virtues of his approach to geometry because they were better equipped than men to judge matters of taste. This was an argument, ironically, that Perrault and Fontenelle had used to great effect in their battles with traditionalists during the "ancients and moderns" battles of the seventeenth century.[107] But to it Saurin responded: "Changing fashions inspired by geometry? So now we have female geometers and women more geometrical than men. What a surprise for them!"[108] Addressing directly Castel's pointed critiques of academic mathematicians, Saurin ended his letter by calling Castel's manner of critique both "vague" and "injurious." "He stains the respectable memory of those who merit our *éloges*."[109]

While Saurin, albeit anonymously, took the academy's response to the public sphere, the members also arranged for an official response. In April 1730 the following declaration was read to the members of the Royal Academy: "M. le President de Maisons [an honorary member of the institution] reports that M. le Duc de Maine, under whose authority the *Journal de Trévoux* is printed, having learned that in several of the most recent issues of this journal the works of the Academy have been treated in a manner entirely contrary to propriety, has ordered that the journal make a formal apology [*une satisfaction authentique*] to the Academy in the next volume. He also declared that those who have produced these negative pieces will be denied employment at the journal. Father Castel is said to be the source of these attacks."[110]

In May, the journal published its promised apology, without mentioning Castel's name or the precise criticisms that had generated the uproar directly. Instead it simply apologized for suggesting that the Royal Academy was anything other than an esteemed institution while promising to "take the measures necessary to prevent such complaints in the future."[111] Castel was not

106. Ibid., 605.

107. See Joan DeJean, *Ancients against Moderns: Culture Wars and the Making of a Fin de Siècle* (Chicago, 1997).

108. Review of *Lettre sur le Traité de mathématiques du P.C.*, in *Journal des savants* (1730): 606.

109. Ibid., 610–611.

110. PVARS, April 1, 1730.

111. *Journal de Trévoux* (May 1730): 893–894.

dismissed from the journal, however, and in 1732 he resumed his crusade, publishing a review in the *Journal de Trévoux* of Edmund Stone's *The Method of the Fluxions*.[112] The review noted that Stone's treatise was largely a translation of l'Hôpital's 1696 work on the differential calculus, but one recast in the Newtonian symbolism of the fluxional calculus. He also noted and praised the largely English techniques of integration that Stone included. But after briefly summarizing the book's technical mathematical contents in a neutral if praiseworthy way, Castel spent most of the review narrating the story of Stone's "discovery" as a mathematician. He was the self-taught servant of a distinguished aristocrat, Castel explained, whose mathematical genius had been discovered by accident. His mathematical knowledge was achieved naturally, therefore, without the taint of prejudice that so often inflicts others. By implication, therefore, his particular synthesis of the differential and integral calculus also possessed an exceptional naturalness not found in other works. Castel continued his advocacy of Stone by also arranging for a French translation of his work in 1735, and by contributing his own preface to the text. Castel's witty and urbane *La géometrie naturelle en dialogues* appeared in 1735 as well, and thus whatever retribution Castel suffered in 1730 contributed little to disrupting his ongoing mathematical campaigns.

Furthermore, even if Castel was chastised publicly for his polemics, the outcome ultimately served his agendas far more than those of Fontenelle and the academy. An official declaration of censure against one of the academy's most outspoken and widely read critics had to have been among the least desirable outcomes Fontenelle could have imagined. He had spent his entire life trying to build bridges between the academy and the elite public, yet Saurin's attack on Castel suggested that an irreconcilable opposition existed between serious, academic, masculine geometry and Castel's frivolous, *mondain,* feminine alternative. This was exactly the divide that Castel had worked to open, and the academy's authoritarian response only reinforced Castel's perspective in the public mind. Furthermore, while Castel was silenced in this particular context, he remained free to develop his discourse against infinitesimal analysis in other ways. The academy, by contrast, had its options constrained since the controversy confirmed its public image as either a bastion of curmudgeonly officialdom or as an authoritarian policeman oppressive to contrary viewpoints. Neither image agreed with the conception of the academy that Fontenelle had worked hard to create after 1699, and consequently his battles with Castel were a sign that a new era was dawning.

112. Review of Edmund Stone, *The Method of Fluxions*, in *Journal de Trévoux* (January 1732): 103–113.

The new public perception catalyzed by Castel of an increasingly inward-looking and intellectually homogenous Royal Academy using its authority to fend off the wider scientific views circulating in the public sphere was a key component of the Newton wars that erupted after 1730. Fueling this perception as well was the growing sense that a great philosophical divide had emerged and that the Paris Royal Academy of Sciences had positioned its institutional authority squarely on one side of it. Privat de Molières' work in vortical celestial mechanics, begun in 1728, combined with Fontenelle's public presentations of it that appeared in print after 1730, clarified this perception. And as it clarified, it allowed the discourses that had come to define this philosophical divide over the preceding decades to become weapons, akin to Castel's mathematical Anglophilia, in a struggle over the institutional authority of official French science. More than any other single source, the Leibniz-Clarke correspondence provide these would-be scientific warriors with the discursive weapons appropriate for fighting these battles. The origination of this philosophical exchange out of the calculus priority dispute, therefore, and then its particular passage into the French public sphere after 1720, completed the necessary preparation for the subsequent Newton wars in France.

Framing a Philosophical War: The Leibniz-Clarke Correspondence

The correspondence began in November 1715, when Leibniz penned a short letter to Caroline, Princess of Wales, responding to her inquiries about his *Theodicy*.[113] This text, published in 1710, contained Leibniz's first public criticisms of the Newtonian theory of universal gravitation, along with an important intervention into the pressing debate about freedom and determinism in nature.[114] Leibniz's correspondence with Caroline was begun in the context of the erupting priority dispute, and since the princess was also residing in London at the time as part of the entourage of the newly arrived Hanoverian court of George I, there was a significant political background to the corre-

---

113. H. G. Alexander offers a very concise introduction to the correspondence in his introduction to the modern, critical edition of the letters. Introduction, to H. G. Alexander, ed., *The Leibniz-Clarke Correspondence* (Manchester, 1956), ix–lvi. All of my citations from the correspondence itself will be drawn from this edition. See also Westfall, *Never at Rest*, 760–780; Ezio Vailati, *Leibniz and Clarke: A Study of Their Correspondence* (Oxford, 1997); and Hall, *Philosophers at War*.

114. Leibniz, *Essais de théodicée, sur la banté de Dieu, la liberté de l'homme, et l'origine du mal* (Amsterdam, 1710).

spondence as well. Many believed that Caroline was sympathetic to Leibniz's philosophical views, and since he was the king's official librarian and historian back in Hanover, her interest raised the specter of a further "Germanification" of England. Would Leibniz be invited to London as the court philosopher of the new Hanoverian monarch? And if so, what would this mean for England, and for those who had come to define their own cultural identity, stridently so after 1713, through Newtonian natural philosophy? These politics further fueled the debates around Newton that erupted in England at this time, and accordingly when Leibniz's first letter to Caroline arrived in London in the fall of 1715 it was greeted with intense anxiety, agitation, and polemical fervor by Newton and his supporters.[115]

The letter itself was brief, offering only four numbered assertions, but each was a bombshell. First, Leibniz declared, "natural religion itself seems to decay (in England) very much. Many will have human souls be material; others make God himself a corporeal being."[116] This opening salvo effectively linked a general decline in natural religion with the rise of English materialism. In the subsequent lines of the letter, Leibniz isolated the roots of this contagion. One was the influence of John Locke, since "his followers are uncertain at least whether the soul be not material and naturally perishable."[117] The other was Newton, whom Leibniz first accused of making space an organ of God's being. Stressing the Spinozist implications of such a view, he further suggested that Newtonian natural philosophy made God a material presence in the world, not an immaterial creator separate from it. In his fourth and final assertion, Leibniz further criticized the weaknesses of the Newtonian conception of God. "According to their doctrine," Leibniz mocked, "God Almighty wants to wind up his watch from time to time; otherwise it would cease to move. He had not, it seems, sufficient foresight to make it a perpetual motion. Nay, the machine of God's making is so imperfect, according to these gentleman, that he is obliged to clean it now and then by an extraordinary concourse, and even to mend it as a clockmaker mends his work."[118]

The pointed rhetoric of these charges was a product of the priority dispute, which was two years old when Leibniz penned his first letter to Caroline. Taken as a whole, his letter also amounted to a salvo in this battle, one directed at a powerful figure and charging that Newtonian philosophy supported the most dangerous sorts of irreligion. Not surprisingly, the English Newtonians were

115. See Shapin, "Of Gods and Kings."
116. Alexander, *Leibniz-Clarke Correspondence*, 11.
117. Ibid.
118. Ibid.

quick to respond. Samuel Clarke took up the pen, but it is likely that Newton was behind every stroke of the quill.[119] In November, Clarke sent his first reply to Leibniz offering a correspondingly numbered point-by-point refutation of Leibniz's charges. The response triggered a further reply from Leibniz, and the debate continued in this manner until Leibniz's death in November 1716. In all, Leibniz wrote five letters, and Clarke responded to each with a letter of his own. Since neither antagonist conceded anything throughout these exchanges, the resulting correspondence constituted nothing less than a philosophical "clash of the Titans" over some of the most vexing and controversial intellectual issues of the day. Once published, it further offered readers throughout the Republic of Letters unrivalled access to the positions and justifications informing these contending approaches to natural philosophy.

Several features of the Leibniz-Clarke correspondence were especially influential in framing the Newton wars that followed in its wake. One important division centered on the philosophical roots of irreligion. In his first reply to Clarke, Leibniz reiterated the charge that Newtonian philosophy supported materialism. "I believe the author had no reason to add that the mathematical principles of philosophy are opposite to that of the materialists," Leibniz wrote. "On the contrary, they are the same only with this difference, that the materialists in imitation of Democritus, Epicurus and Hobbes, confine themselves altogether to mathematical principles and admit only bodies; whereas Christian mathematicians also admit immaterial substances."[120] Clarke's reply, however, inverted this formulation. Leibniz isolated active matter atomism as the source of materialism, an argument that was already gaining strength in France by 1720. Clarke, by contrast, found it in rationalism and deterministic metaphysics instead. "Whereas the materialists suppose the frame of nature to be such as could have arisen from mere mechanical principles of matter and motion, of necessity and fate," he wrote, "the mathematical principles of philosophy show on the contrary that the state of things (the constitution of the sun and planets) is such as could not arise from any thing but an intelligent and free cause."[121] This divide between a fear of materialism rooted in rationalist necessity (the Newtonian position) and one rooted in reckless and groundless physi-

---

119. The most forceful case for Newton's participation in the Leibniz-Clarke correspondence is made by Alexandre Koyré and I. B. Cohen, "Newton and the Leibniz-Clarke Correspondence," *Archives Internationales d'Histoire des Sciences* 15 (1962): 63–126. Newton's role is seen as less central, however, by Vailati, *Leibniz and Clarke*, 4; and Hall, *Philosophers at War*, 219–220.

120. Alexander, *Leibniz-Clarke Correspondence*, 15.

121. Ibid., 20.

cal theorizing about forces in matter (the Leibnizian alternative) would become a commonplace in the wider European discussions of materialism after 1720.

The subsequent exchanges between Leibniz and Clarke only made the irreconcilable nature of this opposition more precise. Throughout, Leibniz insisted that his principle of sufficient reason was the real measure of pious and truthful natural philosophy. According to him, there must be a reason for everything in God's creation, and Clarke was mistaken in using God's freedom to account for things that could not be explained rationally. Indeed, Leibniz insisted that Clarke's views supported a Socinian conception of God's total and unrestricted freedom combined with its Spinozist corollary: the immediate, material omnipresence of the divinity. Clarke, for his part, admitted that God never acts without a reason. However, turning Leibniz's principle on its head, he also stated that "this sufficient reason is oft-times no other than the mere will of God." "Why this particular system of matter should be created in one particular place, and that in another particular place, when (all place being absolutely indifferent to all matter) it would have been exactly the same thing *vice versa* . . . there can be no other reason but the mere will of God. Which if it could in no case act without a predetermining cause . . . this would tend to take away all power of choosing and to introduce fatality." [122]

In this way, the Leibniz-Clarke correspondence defined a philosophical and theological debate that would remain a source of contestation for at least half a century. This philosophico-theologic dispute also informed the more explicitly scientific dimension of the exchange. In his first letter, Leibniz had likened a strong belief in God's thoroughly free will to substituting the miraculous for rational explanations. In his second letter, he made the point more even polemically, stating that "if God is oblig'd to mend the course of nature from time to time, it must be done either supernaturally or naturally." If supernatural causes are at work, "we must have recourse to miracles in order to explain natural things, which is reducing an hypothesis *ad absurdum,* for everything may easily be accounted for by miracles." But if the causes are natural, then "God will not be *intelligentia supramunda.* He will be comprehended under the nature of things; that is, he will be the soul of the world." [123] This and other formulations explicitly linked Clarke's Newtonian position to Spinozism, further implicating it with the discourse of pantheist materialism then becoming widespread thanks to Toland and others. Clarke countered this challenge by again invoking the omnipresence and freedom of God. "To cause the sun (or the earth) to move regularly, is a thing we call natural," he explained. "To stop its motion for

122. Ibid., 21–22.
123. Ibid., 20.

a day we call supernatural. But the one is the effect of no greater power than the other; nor is the one with respect to God more or less natural or supernatural." Furthermore, "God's being present in or to the world does not make him to be the soul of the world. . . . [He] is present to the world not as a part, but as a governor, acting upon all things, himself acted upon by nothing.[124]

Newton had made an identical argument in his "General Scholium," and Leibniz also invoked the arguments of the *Principia* in his reply. "If God would cause a body to move free in the ether round about a certain fixed center, without any other creature acting upon it, I say it could not be done without a miracle, since it cannot be explained by the nature of bodies." He therefore maintained that the "attraction of bodies, properly so called, is a miraculous thing since it cannot be explained by the nature of bodies."[125] This was the first appearance of Newton's theory of universal gravitation in the correspondence (literally, at least, since the whole correspondence was at one level about this), and it prompted Clarke to offer an explicit defense of Newtonian gravitation. "The question is not what it is that divines or philosophers usually allow or not allow," he reasoned, "but what reasons men allege for their opinions . . . For a body to move in a circle round a center *in vacuo* if it be usual (as the planets moving about the sun) 'tis no miracle, whether it be effected immediately by God himself or mediately by any created power; but if it be unusual (as for a heavy body to be suspended and move so in the air) 'tis a miracle whether it be effected by God himself or mediately by any created power."[126] Leibniz would have none of this distinction, for if it were used then "monsters would be miracles."[127] Thus, he reiterated his earlier charge: "'tis a supernatural thing, that bodies should attract one another at a distance without any intermediate means; and that a body should move around without receding in the tangent though nothing hinder it from so receding. For these effects cannot be explained by the nature of things."[128]

In the final exchange of letters, the bitterness of the dispute over Newtonian attraction became even more pronounced. Asking at the outset whether Clarke was a "lover of truth or whether he will only cavil,"[129] Leibniz reasserted the principle of sufficient reason as "the foundation of all true philosophy." "Reasonable and impartial men will grant me that having forced an adversary to

124. Ibid., 24.
125. Ibid.
126. Ibid.
127. Ibid., 43
128. Ibid.
129. Ibid., 55.

deny that principle is reducing him *ad absurdum.*[130] He also reiterated the Spinozist nature of Newtonian attraction. Assuming this theory, "everything will then be equally natural or equally miraculous." The doctrine will "tend to make God the soul of the world" and "God will be a part of nature."[131] Further responding to Clarke's contention that there might be a mechanism of gravity that is "invisible, intangible, and perhaps not mechanical," Leibniz quipped, "He might well have added inexplicable, unintelligible, precarious, ground-less, and unexampled."[132] Clarke had also conceded that "an attraction without any means intervening would indeed be a contradiction." "Very well!" Leibniz replied. "But then what does he mean when he will have the sun to attract the globe of the earth through an empty space? Is it God himself who performs it? But this would be a miracle, if ever there was any. Or perhaps some immaterial substances, or some spiritual rays, or some accident without a substance, or some kind of *species intentionalis,* or some other I know not what?" "The author seems to have still a good stock [of such things] in his head," Leibniz concluded, but he reasons "without explaining himself sufficiently."[133]

Despite the increased vigor of Leibniz's final critique, Clarke refused to budge. He reiterated that God was free to create any world he saw fit. There was nothing, therefore, irrational or miraculous about the Newtonian theory of gravitational attraction in the void. Likewise, he stated again that denying God a direct place in the action of nature, as he believed Leibniz's system did, was to make the world a godless machine. He also reiterated that space is neither an attribute of God nor Leibniz's divine "soul of the world" but instead an infinite *immensum* created by God yet separate from him.[134] Clarke also used the privilege of having the last word to give his arguments a slightly different spin in his final letter. Most important, he drew out more explicitly than ever before the connections between his own views and those of the experimental Newtonians. "It is very unreasonable to call attraction a miracle and an unphilosophical term," he declared. "[For] it has so often been distinctly declared that by that term we do not mean to express the cause of bodies tending toward each other, but barely the effect, or the phenomenon itself."[135] Indeed, Clarke continued, "that the sun attracts the earth, through the intermediate void space, . . . with a force that is in direct proportion to their masses . . . and that the space

130. Ibid., 96–97.
131. Ibid., 91.
132. Ibid., 94.
133. Ibid., 92–94.
134. Ibid., 41.
135. Ibid., 115.

betwixt them is void . . . all this is nothing but a phenomenon, or actual matter of fact, found by experience."[136]

Here Clarke explicitly employs the experimental philosophy as articulated by Côtes in his "Preface" to the *Principia* and by the experimental physicists such as Hauksbee and Desaguliers. In the remainder of his letter, he cemented this connection. In response to Leibniz's request that Clarke explain just what attraction is, he responded, "it is nothing but a phenomenon, or actual matter of fact, found by experience."[137] "That this phenomenon is not produced . . . without some cause capable of producing such an effect is undoubtedly true," he declared. "Philosophers therefore may search after and discover that cause, if they can, be it mechanical or not mechanical. But if they cannot discover the cause, is therefore the effect itself, the phenomenon, or the matter of fact discovered by experience (which is all that is meant by the words attraction and gravitation) ever the less true? Or is a manifest quality to be called *occult* because the immediate efficient cause of it (perhaps) is occult or not yet discovered? . . . This is very singular reasoning indeed."[138]

This letter ended the debate, or rather launched it into the public sphere. In the end, what the Leibniz-Clarke correspondence offered was a clearly defined battleground for debate about Newtonianism. One theater of combat concerned the relationship between natural philosophy and pious natural religion. Clarke and Leibniz were eloquent and persuasive in defending two largely irreconcilable positions on this question. Another zone of combat was the proper epistemology for natural philosophy itself. Here the theory of Newtonian gravitation was the real focus of the argument, but since this debate occurred alongside the parallel religious debate as well, the letters worked to intensify the theological concerns already hovering around discussions of Newtonian physics. They also revealed two persuasive, yet irreconcilable epistemologies for physical science. Clarke continually insisted that regular, empirically documented facts were true irrespective of their philosophical rationality. Leibniz, by contrast, continually retorted that true physics meant rational, mechanical explanations of causes. Thus, in each of these ways, the Leibniz-Clarke correspondence worked to place the defenders and opponents of Newtonianism on opposite sides of an irreducible intellectual divide.

In France, where this chasm was already beginning to form for other reasons, the results of its publication were provocative. By early 1717, Clarke was working to have the correspondence published, complete with annotations

136. Ibid., 118.
137. Ibid., 119.
138. Ibid., 118–119.

that clarified his own positions. He also added citations from Newton's own works to reinforce his claims and an appendix offering damaging citations from Leibniz's writings. Clarke's edition of the letters appeared in London sometime in 1717, elevating the ongoing mathematical priority dispute into a full-fledged philosophical and religious war.[139] In 1713, the new *Journal littéraire de la Haye* published a vitriolic attack on Leibniz written by John Keill that advanced many of the same critiques used by Clarke. Keill, however, framed his attack as further support for the Newtonian position in the calculus dispute, and accordingly his larger philosophical, theological, and scientific critiques were less potent than they might otherwise have been.[140]

Clarke's publication changed this situation. It avoided the calculus question entirely, transforming the dispute between Newton and Leibniz into a straightforward philosophical confrontation, and one with enormous stakes. Clarke's initiative was furthered by others. In 1716, a new Dutch Francophone journal began appearing devoted specifically to the intellectual life of England. Called *Bibliothèque Anglaise*, it reviewed the latest English books, and in its first volume it offered a brief, if typically neutral, account of Clarke's "Collection of Papers."[141] In subsequent issues, the journal also discussed other relevant works in English philosophy and natural religion, including Collins's *Philosophical Enquiry concerning Human Liberty*, explicitly refuted by Clarke in his edition of the correspondence with Leibniz; William Whiston's *Astronomical principles natural and revealed;* Dr. Thomas Bennett's discourse on the "everblessed trinity" that challenged Clarke's theology directly; and several scientific works by leading Newtonians, including Keill's *Introductio ad veram Astronomiam* and Brook Taylor's *Methodus Incremetorum*.[142] These discussions connected the Leibniz-Clarke correspondence directly to the wider debates over natural philosophy and theology then occurring in the Republic of Letters. In fact, since the same journal also published extended discussions of John Toland's *Pantheisticon* (published at "Cosmopolis" in 1720) and his *Tetradymus* in later issues, readers of this periodical, like many other literate observers, would have seen the entanglement between all these issues directly.[143]

139. *A Collection of Papers which passed between the late Learned Mr. Leibnitz and Dr. Clarke in the years 1715 and 1716 relating to the Principles of Natural Philosophy and Religion* (London, 1717).

140. *Journal littéraire de la Haye* 1 (1713): 206–214, 445–453. On this article, see Alexander, introduction to *Leibniz-Clarke Correspondence*, ix–x; Westfall, *Never at Rest*, 763–764.

141. *Bibliothèque anglaise*, 25 vols. (La Haye, 1716–1727), 1: 456–459.

142. Ibid., 1: 414–447, 447–456; 3: 1–65, 226–247; 4: 336–365, 523–539.

143. Ibid., 8: 285–311; 9: 235–265.

Jean Le Clerc furthered this same understanding in 1722 when he published a critical treatment of Leibniz's system of the preestablished harmony in his *Bibliothèque ancien et moderne*.[144] Le Clerc also reported in 1726 that Leibniz's students, including his heir apparent, Christian Wolff, had accused him of harboring Spinozist impulses.[145] These interventions were part of the larger effort, triggered by the priority dispute, to promote Newton and disparage Leibniz within the Republic of Letters. Clarke's English edition of the Leibniz-Clarke correspondence was conceived similarly; however, since Le Clerc spoke to the far larger and more influential Francophone public sphere with his work, while Clarke's English texts presented communication problems in this space, thoughts of translating the Leibniz-Clarke correspondence into French began percolating as soon as Clarke's edition appeared in London.

Pierre Desmaizeaux realized these ambitions. He was another of the exiled French Huguenots that did so much to foster Enlightenment criticism within the Republic of Letters after 1715. He lived the life of an independent man of letters in London, corresponded regularly with Bignon in France, keeping the minister and his "petite académie" of editors at the *Journal des savants* abreast of the latest literary news, and he also had important contacts with Dutch publishers and journalists, especially Le Clerc and Pierre Coste. He likewise maintained close ties with French men of letters such as Houdar de la Motte, Saurin's close friend and a leading figure in his and Fontenelle's circle at the Café Gradot. As a result, Desmaizeaux was well positioned socially to introduce French readers to the debate then raging in England.[146]

Clarke personally shaped the reception of his work by producing a particularly constructed edition, and Desmaizeaux would do the same in his French edition of the correspondence. In Clarke's edition, a dedicatory preface to Princess Caroline framed the debate found in the letters themselves, and to them was also added an appendix with a second correspondence entitled "Letters to Dr. Clarke concerning Liberty and Necessity." Reinforcing this framework, a piece on Anthony Collins's *A Philosophical Enquiry concerning Human Liberty* was included, the very work that provoked Crousaz's critical energies in 1718. Collins, like Toland, was a radical deist and a threat to Clarke's more orthodox form of Newtonianism. By including his refutation of Collins's work within this edition, Clarke was therefore working to turn his book into a care-

144. *Bibliothèque ancien et moderne* 23 (1722): 413–428.
145. *Bibliothèque ancien et moderne* 27 (1726): 110–115.
146. Desmaizeaux's position in the Republic of Letters is treated well in Anne Goldgar, *Impolite Learning: Conduct and Community in the Republic of Letters* (New Haven, 1995).

fully constructed Newtonian manifesto targeting a number of different philosophical issues simultaneously.

. Desmaizeaux also put his own stamp on his edition of the correspondence, even if his intellectual agendas are much harder to define. Newton believed that Desmaizeaux was a Leibnizian sympathizer, and there is some documentary evidence of this, even if the case against him is not transparent.[147] Clarke also believed that Desmaizeaux posed a threat, but it is more likely that he held no strong views one way or the other and that he saw in this dispute an interesting (and potentially profitable) intellectual battle and nothing more. Whatever his motivation, Desmaizeaux worked diligently after 1717 to bring out a French edition of the Leibniz-Clarke debate. His work, *Collection of diverse pieces by Mssrs. Newton, Leibniz, and other famous authors treating philosophy, natural religion, history and mathematics,* appeared out of Amsterdam in 1720, virtually simultaneously with Coste's similarly Amsterdam-published French edition of Newton's *Optics*.[148]

The precise contents of the work were significant. In the first of its two volumes, Desmaizeaux offered a faithful French translation of the entire contents of Clarke's 1717 English text. This included the entire correspondence with Leibniz, Clarke's letters regarding liberty and necessity, and the latter's refutation of Collins's *A Philosophical Enquiry concerning Human Liberty.*[149] The only change Desmaizeaux made was to replace Clarke's dedicatory letter to Princess Caroline with his own historical "Preface" accounting for the circumstances that produced the debate.[150] In the second volume, however, Desmaizeaux added a totally original collection of documents. Most important were a number of texts relating to the calculus dispute between Newton and Leibniz.[151] He also included a set of *opuscules* by Leibniz treating such writers as the Earl of Shaftesbury, John Locke, and Malebranche.[152]

This second volume clearly altered the overall meaning of the Leibniz-Clarke exchange, and reactions to the work in France reflected this reorientation. The *Journal de Trévoux* was the first to respond, offering a review of each volume of Desmaizeaux's *Recueil* in its issues of June and July 1721, respective-

147. The key letters are noted in Westfall, *Never at Rest,* 777, n. 262.

148. Pierre Desmaizeaux, *Recueil de diverses pieces, sur la philosophie, la réligion naturelle, l'histoire, les mathématiques, &c par Mrs Leibniz, Clarke, Newton et autres auteurs celèbres,* 2 vols. (Amsterdam, 1720).

149. Ibid., vol. 1.

150. Ibid., 1: i–xxiii.

151. Ibid., 2: 3–126.

152. Ibid., 2: 129–464.

ly.[153] The first review introduced both volumes by defining the collection as a documentary record of the "quarrels that M. Leibniz had at the end of his life with England over the subject of M. Newton, and then with M. Newton himself." The journal also noted that these quarrels "made too much noise for one not to be interested in the subjects at issue and the positions that each side took in the proceedings."[154] The journal distanced Desmaizeaux himself from these battles, praising the "famous" editor, and especially his historical "Preface," for being "disinterested and objectively factual [*fort historien*]."[155] But the journal also claimed to see a bias in Desmaizeaux's presentation. Before launching into his account of the text, the reviewer, who was most likely Castel, invited the "partisans of M. Leibniz" to supplement the work of this illustrious editor with texts that might have escaped his attention.[156] In this way, the journal presented Desmaizeaux's *Recueil* as a piece of pro-Newtonian advocacy, despite the very different, albeit unpublished, views of Newton, Clarke, and others.

Several aspects of the review's opening characterization were important. First, the journal introduced Desmaizeaux's work as a piece of overtly polemical literature. The reviewer also emphasized the scandalous nature of the dispute, preparing the reader in this way for the blow-by-blow account that the journal would then offer. Elsewhere the journal also gave the dispute a particular spin, making the philosophical disagreement between Newton and Leibniz a war of patriotism between England and Germany. "Nothing is more widespread than the spirit of nation, of sect, and, if one can say it, of lineage," the journal declared.[157] By framing the Leibniz-Clarke debate this way, the journal helped to reinforce an emerging nationalist discourse about intellectual matters that was to prove increasingly powerful. It also publicly associated these tendencies with this precise philosophical dispute. Clarke had worked in his edition to frame the Leibniz correspondence very differently, yet the particular

153. Review of Pierre Desmaizeaux, *Recueil de diverses pièces, sur la philosophie, la réligion naturelle, l'histoire, les mathématiques*, in *Journal de Trévoux* (June 1721): 962–989; (July 1721): 1230–1254.

154. Ibid., 962.

155. Ibid., 964.

156. The reviews in the *Journal de Trévoux*, like those in most learned periodicals of the day, were anonymous. It is, therefore, difficult to know with certainty who the authors were. Yet Castel's biographers date his arrival at the journal in 1720, and since Desmaizeaux's text was exactly the kind of material he typically reviewed, it is highly likely that he was the author here.

157. Review of Desmaizeaux, *Recueil de diverses pièces*, in *Journal de Trévoux* (June 1721): 966.

textual architecture adopted by Desmaizeaux, combined with its refraction in the newly critical French press, insured that the French developed their own understanding of the debate and its significance.

In his review, Castel explained that Leibniz had two arguments with "England," one over a "matter of fact"—who invented the calculus—and one over Newton's philosophy. Desmaizeaux, the Jesuit implied, was offering a case against Leibniz in both instances. To treat the two quarrels with sufficient care, the reviewer devoted a separate discussion to each. The first discussion focused on the calculus dispute because, he said, it was the earliest to arise.[158] This in effect inverted Desmaizeaux's presentation in the *Recueil* since he had put the calculus documents in his second volume. The second review, which addressed the actual Leibniz-Clarke correspondence itself, also drew strong connections between this exchange and the calculus controversy, despite the absence of any discussion of these issues in the letters themselves. Castel stated, for example, that Leibniz's offensive against Newtonian philosophy was an effort to end "the poor politics" of defense that had characterized his previous stance in the priority dispute.[159] The overall result was to suggest that the Leibniz-Clarke exchange was an extension of the calculus dispute and perhaps little more than a skirmish within it.

Desmaizeaux had opened this possibility in his *Recueil* by reattaching the Leibniz-Clarke debate to the calculus dispute. His "Historical Preface" also drew connections between the two, even if there is little evidence that he was trying to frame the argument in this precise way. Castel, on the other hand, used his review to stoke the very polemic that Clarke sought to restrain. Whether Castel's interest in fueling the calculus dispute was part of a conscious strategy is not clear, but the results were clearly significant. While Castel was composing his systematic account of the petty, partisan bickering triggered by the invention of the calculus, Fontenelle was trying to deal with Varignon's last public battle with an Italian opponent of his analytical mechanics. In fact, both presentations appeared in print at roughly the same time. Whatever the connection, discussion of the calculus dispute triggered by Desmaizeaux's *Recueil* and Castel's extensive review of it served as the backdrop for Fontenelle's rejuvenated advocacy of advanced, analytical mathematics over the next ten years. It also launched Castel into his characteristic role as a leading public critic of French mathematical work and as a defender of the English.

158. Ibid., 964.
159. Review of Desmaizeaux, *Recueil de diverses pièces,* in *Journal de Trévoux* (July 1721): 1230.

The *Journal des savants*, for its part, also reinforced this particular reception of Desmaizeaux's text.[160] Like Castel, the anonymous reviewer divided his review into two installments, one each for the separate volumes of the text. Furthermore, the reviewer also inverted Desmaizeaux's presentation, presenting his account of the calculus controversy first and then turning the Leibniz-Clarke debate into a consequence of it. However, while the *Journal de Trévoux* used this manner of presentation to accentuate the polemical nature of Desmaizeaux's collection, the *Journal des savants* adopted a different approach. It effectively defused the controversy through its typical strategy of authoritative objectivity. Only four pages were devoted to Desmaizeaux's second volume, an ordinary practice for this journal, but one strikingly different than Castel's lengthy and overtly polemical account. The substance of the two reviews was also different. Castel offered a blow-by-blow account of the calculus dispute itself while the *Journal des savants* merely summarized the barest outlines of the controversy. The reviewer at the official French periodical likewise devoted equal time to the *opuscules* contained in this collection, while Castel quickly left the details of Desmaizeaux's text behind to devote his entire review to the calculus controversy.

Certainly, pointed judgments did break through the reviewer's otherwise objective veneer, such as when M. Leibniz's "affection for systems" was criticized along with his tendency to speak "in a popular manner and without knowledge about the real substance of Malebranche's thought."[161] Short of these brief and undeveloped critiques, however, the *Journal des savants* maintained a stance of neutral objectivity in its reporting, a stance befitting its official commitments to the traditional values of the Republic of Letters. Its review of the Leibniz-Clarke correspondence itself reflected a similar orientation. The text was again quite short, covering little more than four pages, and rather than providing a systematic accounting of each antagonist's arguments as Castel would do, the review restricted itself to defining the essential core of the debate. It called the struggle "a battle between the *Theodicy* and the *Principia*," a concise and powerful characterization, and one that helped to reinforce the conception that the Leibniz-Clarke debate was essentially a trial of strength over Newton's theory of universal gravitation.[162] The remainder of the review reinforced this

160. Review of Pierre Desmaizeaux, *Recueil de diverses pièces, sur la philosophie, la réligion naturelle, l'histoire, les mathématiques,* in *Journal des savants* (1721): 423–427, 474–478.

161. Ibid., 425–426.

162. Ibid., 476.

conception. Restricting its summary to the one question "at the heart of their philosophical battle," the text offered radically condensed synopses of the positions of Leibniz and Clarke on the question of universal gravitation.[163] The review then ended by listing, without comment, the other texts contained in the collection.

This approach in the *Journal des savants* achieved two things simultaneously. First it helped to clarify the reception of the Leibniz-Clarke correspondence by reducing the work essentially to a battle about Newton's theory of universal gravitation. As we have seen, the correspondence itself was about far more than this, and others, especially Castel, would draw different conclusions from it. Nevertheless, the ability of the *Journal des savants* to reduce this complex philosophical exchange to a debate about attraction tout court reveals the emerging sense that it was upon this point that the entire fate of natural philosophy hinged. Furthermore, the review reveals how the theory of attraction could serve as an intellectual rallying point for a host of diverse philosophical, theological, and scientific debates. Despite the range of topics that Leibniz and Clarke had debated—God's freedom, the role of mechanism in nature, the sources of materialism and irreligion, the character of miracles, the nature of force and matter, the limits of human reason, and a host of other complex topics—the theory of universal gravitation increasingly became the critical site for all these discussions.

In this way, the reception of the Leibniz-Clarke correspondence in France solidified the unified and provocative conception of Newtonianism rooted in the theory of universal gravitational attraction that was the object of contestation in the Newton wars. However, the text also added a new provocation: it invited French readers to participate in the fight themselves. Castel's new polemical journalism after 1720 catalyzed these changes, and nowhere was this more evident than in his own use of the Leibniz-Clarke debate to become a participant in these same conflicts. In doing so, he also brought his own, idiosyncratic intellectual agendas to the effort, provoking controversy while also blurring the easy polarities that defined the battle as well.

Castel opened his review by siding squarely with Leibniz. He first noted that Newtonian philosophy offered an easy target for Leibniz's recriminations. "It is, they say, entirely geometrical, and entirely based on experience. But the immense voids that it admits; the nothings of the world at the heart of the world itself; the primitive *pesanteurs* entirely dependent on the quantity of matter and in no way on the structure of the parts in motion; the attractions without impul-

163. Ibid., 476–478.

sions; the mechanisms without contiguous parts and without structures linking them; the actions at a distance without any real intervening medium; the vortices without vortices; the spiritual spaces; a God present at all times to the imagination, and with enormous eyes as well; an extension without parts yet containing indivisible particles: all of this gives an advantage to even the least intelligent adversary, and especially to one whose reasonableness is provoked. M. Leibniz does not run away from the occasion to defend himself." [164]

This presentation placed Castel in a strongly anti-Newtonian position, and his review singled out three aspects of Newtonianism for particular scorn. The first concerned its ties to materialism and irreligion. Castel echoed Leibniz's charge that Newtonian philosophy fostered the decay of religion in England, and in his account of Clarke's rebuttal he left no room for doubt about his own Leibnizian sympathies. "M. Clarke alleges that no philosophy has ever been less suspect than [M. Newton's] on the charge of materialism. He argues, in effect, that the world of M. Newton is so devoid of matter that nothing is more spiritual or more divine. But he pushes this so far as to almost spiritualize everything. With Spinoza, bodies are part of God which is itself material extension. With M. Newton, one must struggle to conceive of how bodies are not part of the spiritual extension, which is nevertheless God. This system wants to appear less injurious to God and men, but does it not really amount to the same thing?" [165] Here Castel pinned the highly charged label of Spinozism squarely upon Newton's breast. Leibniz had implicitly done the same in the correspondence itself, and on this point Castel was comfortable marching with him.

Castel also joined Leibniz in attacking Newton's "mathematical principles of philosophy," calling them only metaphysical principles in disguise. Again invoking the specter of materialism, he supported Leibniz's claim that "the greatest of the materialists and the greatest enemies of the divinity—Democritus, Epicurus, and Hobbes—were entirely beholden to 'mathematical principles' that were really nothing but metaphysical principles in disguise." [166] This assertion reinforced Castel's argument about the materialist implications of Newtonian philosophy while also supporting his more general Jesuit critique of abstract, mathematical systems in general. It also set up his final pronouncement on this topic. "In a word," Castel declared, "it is a fact that the most famous sects of materialists have always obeyed mathematical principles just as

164. Review of Desmaizeaux, *Recueil de diverses pièces*, in *Journal de Trévoux* (July 1721): 1230–1231.
165. Ibid., 1233.
166. Ibid., 1234.

M. Leibniz states. They are also deeply attached to the void as the fundamental principle of their system of chance. They attribute a primary reason to all this, and make of it a divine intelligence."[167]

Castel's diagnosis of the roots of materialism made Clarke's defense of Newtonianism just one more example of it. As a result, Castel's review helped to intensify the materialist fears hovering around Newtonianism in France. It did so, moreover, in ways that extended and intensified Leibniz's own discourse in this direction. Castel ultimately posed a clear choice to his French readers. Newtonianism (understood as a set of physical and metaphysical claims) was the bane of sound natural religion, or so he implied. Thus it needed to be opposed. Castel was less clear about what alternative should be substituted for this dangerous philosophical system, but it was enough that he made the enemy readily apparent. These arguments, moreover, were only a prelude for Castel's even more passionate critique of Newtonianism in the final pages of his review. Here he focused on Clarke's understanding of miracles, and to understand the impact of Castel's discourse one must set his remarks in the context of the wider discussion of miracles as it was evolving in early eighteenth-century France.[168]

By 1720, for reasons that were only obliquely related to the emergence of the miraculous as a key concept within the Leibniz-Clarke dispute, the question of miracles had become a hotly contested topic in France and in the wider Republic of Letters. The category "miracle" had always been bitterly contested in post-Reformation Europe since Protestants had used it in fundamental ways to separate themselves from Catholics. Such confession-defining conceptions as the nature of the ritual of holy communion triggered similarly foundational debates about topics such as the nature of Christ's miracles, the miracle of divine grace, and the ongoing presence of miraculous interventions in nature. Complicating such debates was the rise of a more explicitly rational theology in the seventeenth century, one that held in its most extreme, Spinozistic forms that the concept of the miraculous was a fiction produced by ignorant people who did not understand the natural workings of the universe.[169] More moderate

---

167. Ibid., 1236.

168. Newton's view of miracles is discussed in Peter Harrison, "Newtonian Science, Miracles, and the Laws of Nature," *Journal of the History of Ideas* 56 (October 1995): 531–554.

169. Lorraine Daston and Katherine Park, *Wonder and the Order of Nature* (Boston, 1998); Paul Hazard, *La crise de la conscience européenne (1680–1715)* (Paris, 1935); R. R. Palmer, *Catholics and Unbelievers in Eighteenth-Century France* (Princeton, 1939); and Alan Charles Kors, *Atheism in France, 1650–1729* (Princeton, 1990).

positions were far more prevalent, yet when combined with the ongoing confessional struggles the rise of rational theology further worked to resolve any easy settlement on the question of the miraculous.

The range of opinion about miracles in 1720s France can be sampled by looking at the furor caused by the abbé Houtteville's *The Christian religion proven by the facts.*[170] First published in 1721, Houtteville's work defended one of the important positions to emerge in this period, namely, the claim that the miraculous was merely God's work at a higher level of understanding, one unattainable by human reason. This position allowed authentic miracles to occur while maintaining a thoroughly rational conception of the divinity as a being who acted reasonably and lawfully. Houteville's position thus struck a balance between the more radical position of Spinoza, and the equally radical, if highly Protestant, Socinian view, which claimed that every moment was miraculous as a result of God's total freedom to change the creation at any instant. Given the widespread influence of Cartesian and Malebranchian epistemologies in France, Houtteville's work spoke to a consensus on the issue in the French public sphere. The Jesuits, by contrast, offered a counterweight to this rationalist trend by continually challenging the mechanistic and rationalistic heresies of Cartesian and Malebranchian philosophy. They conceived of nature in highly spiritual terms and resisted the pull toward rational explanations of miracles. They also defended the liberty of God to create and govern the world as he saw fit. This stance pushed the Jesuits toward Socinianism, and they therefore worked hard to distance themselves from this heresy in a manner not unlike Clarke's efforts to distinguish Socinianism from Newtonianism.[171]

Furthering complicating this picture, however, was a third pole: the Jansenists. No religious topic was more controversial in eighteenth-century France than Cornelius Jansen's brand of Catholicism, and likewise no religious group was more attached to particular notions of the miraculous than the followers of Jansen's teaching. As Robert Kreiser writes: "The appeal to, and ideological exploitation of, miracles in times of political adversity had a long history in Jansenist controversy dating back to the mid-seventeenth century. Throughout this stormy period there had been a large number of miraculous cures as well as a variety of other 'supernatural' signs and portents associated with Port-Royal [the key Jansenist institution in Old Regime France], all of

---

170. L'abbé Houtteville, *La réligion chrétienne prouvée par les faits* (Paris, 1721).

171. The English side of this equation is explored in Jacob, *Newtonians and the English Revolution*. For the Jesuits, see, in addition to Healy, "Mechanistic Science and the French Jesuits," and Desautels, *Les Mémoires de Trévoux*, also Gaston Sorais, *Les Cartésiennes chez les jésuites français au XVIIe et au XVIIIe siècle* (Paris, 1929).

which served to sustain the Jansenists' sense of themselves as a specially chosen religious elite. . . . Increasingly the miracle, which bore direct and unequivocal witness to the divine presence, came to constitute perhaps the most important vehicle of expression to the persecuted Jansenist faithful." [172]

This Jansenist attachment to miracles was only strengthened in the eighteenth century by the increasing repression of the movement by the French Crown. In 1709, Louis XIV had the Jansenist convent at Port-Royal razed by royal decree, and in 1713 the king approved the papal bull *Unigenitus,* effectively making Jansenists outlaws in France. The Crown also increased its police efforts in this period, attempting to eradicate Jansenism from the kingdom at the grassroots level. State oppression, however, only added to the aura of martyrdom crucial to the Jansenist cause, and by the 1720s, therefore, their religious fervor was becoming more rather than less intense.[173] The most tangible manifestation of this new religious activism was the appearance in 1728 of the clandestine Jansenist periodical the *Nouvelles ecclésiastiques.* Printed and circulated through entirely underground channels, while subject to constant police persecution throughout its history, this journal helped to crystallize the emerging Jansenist opposition party in France by reporting on religious and political news.[174]

Miracles were a centerpiece of the reporting in the *Nouvelles ecclésiastiques,* and these accounts were framed by an overall editorial slant hostile to all forms of nonsacred learning, especially rationalist natural philosophy.[175] This jour-

---

172. Robert Kreiser, *Miracles, Convulsions, and Ecclesiastical Politics in Early Eighteenth-Century Paris* (Princeton, 1978), 70.

173. The subversive role of the Jansenists in Old Regime France has been argued most forcibly by Dale Van Kley. See *The Religious Origins of the French Revolution: From Calvin to the Civil Constitution, 1560–1791* (New Haven, 1996). Other important work on the role of the Jansenist controversies in eighteenth-century France includes Arlette Farge, *Subversive Words: Public Opinion in Eighteenth-Century France,* trans. Rosemary Morris (University Park, 1994); Catherine-Laurence Maire, *Les convulsionnaires de Saint-Médard: Miracles, convulsions et prophéties à Paris au XVIIIe siècle* (Paris, 1985); and Kreiser, *Miracles.*

174. See Farge, *Subversive Words,* 22–23, 36–37, 41–43, 48–53.

175. The Jansenists opposed modern natural philosophy in a manner largely unique among eighteenth-century French Catholics. The *Nouvelles ecclésiastiques* in particular adhered to a truly antiscience editorial policy, condemning all natural philosophy as an attempt to avoid the real religious duties of mankind. For a representative example, see the discussion of Buffon's *Histoire naturelle* in *Nouvelles ecclésiastiques, ou Mémoires pour servir à l'histoire ecclésiastique* (6 February 1750): 21–24, (13 February 1750): 25–27. On the relationship between Catholic and "modern" thought in the eighteenth century, see Palmer, *Catholics and Unbelievers.*

nal thus supported the rise of a Jansenist culture centered around the public, justificatory miracle. The first such miracle—an inexplicable cure of a young Parisian girl—occurred in 1725. It resulted in an official recognition of authenticity by the French church.[176] More controversial, however, were the miracles associated with the tomb of François de Pâris at the Parisian cemetery of Saint-Médard. Pâris had been a popular priest among the Parisian Jansenists, and after his death in 1727 reports of miracles at his grave began to circulate. The *Nouvelles ecclésiastiques* helped to disseminate these stories, and by 1730 pilgrimages to Saint-Médard, combined with experiences of immediate religious ecstasy, had become a public phenomenon. They soon became a source of great anxiety for the authorities in Paris as well. When the Crown decided to crack down on the practices at Saint-Médard in 1731, Jansenist political consciousness was enflamed anew. A broadsheet posted illegally at the cemetery declared "miracles banned by order of the Crown." Overall, the Jansenists resisted rather than accepted these royal orders, and the public spectacle that ensued thus placed miracles at the center of French public consciousness over the next decade.[177]

Houtteville's book appeared in 1721 amidst the continuing debates across these fault lines. The reactions to it illustrate well the intellectual field within which the Leibniz-Clarke dispute was received. The *Journal des savants* offered a very positive account in 1722, telling readers to "read the book."[178] "The most solid proofs of the truth of the Christian religion are found here, arranged in an order that heightens their convincing force."[179] Furthering this view, the journal also published a letter in the same year from a Jew by the name of R. Ishmael Ben Abraham who claimed to have converted to Christianity after reading Houtteville's treatise.[180] Not surprisingly, the *Journal de Trévoux* was more guarded in its assessment. While it praised Houtteville for his attention to the details of God's creation and for his respect for God's omnipotence, it also cautioned readers about attributing too much rationality to the mystery of God's ways.[181] An anonymous pamphleteer, however, cut across each of these

176. Kreiser, *Miracles,* 74.

177. Ibid. See also Maire, *Les convulsionnaires,* esp. chaps. 2–3; and Kreiser, *Miracles,* chaps. 2–6.

178. Review of l'abbè Houteville, *La rèligion chrétienne prouvée par les faits,* in *Journal des savants* (1722): 102–108, 161–167.

179. Ibid., 167.

180. *Lettre de M. Ismael ben Abraham, Juif, convert, à M. l'abbé de Houtteville,* in *Journal des savants* (1722): 641–647.

181. Review of l'abbè Houteville, *La rèligion chrétienne prouvée par les faits,* in *Journal de Trévoux* (July 1722): 1154–1181.

journalistic accounts in his assessment of Houtteville on miracles. In *Lettre de M. L'Abbé *** à M. L'Abbé Houteville,* discussed in the *Journal des savants,* the pamphleteer charged Houtteville with using the language "familiar to the impious." "The Spinozist always reconciles miracles with the necessary and general laws of nature," he explained.[182]

Arguments such as these were omnipresent in France when the Leibniz-Clarke correspondence appeared. They were especially critical for the French reception of Desmaizeaux's *Recueil,* and especially Castel's review of it. Castel began his discussion of miracles by siding first with Leibniz, calling his approach "the most favorable to truth, to religion, and to the divinity."[183] When Clarke made the distinction between the natural and the supernatural meaningless with respect to God, he was, Castel opined, "losing complete sight of truth and religion."[184] This was a classic Jesuit position, one that preserved the authenticity of God's miraculous intervention in nature while rejecting Clarke's sometimes dangerously Socinian understanding of it. In later passages Castel repeated this position. "According to M. Clarke," he wrote, "it is a miracle that a spider supports itself in air by an invisible thread because it is rarely the case that this thread is visible. Indeed, instead of saying that there would be no miracle at all if the thread were visible, or saying that it would be miraculous if the thread did not exist and nothing material supported it, he says it is supported by the immediate operation of God. After such beautiful reasoning, it is quite charming to see M. Clarke conclude coldly that his wise adversary is *demonstrably* in error regarding the nature of miracles."[185]

Castel's review was full of similarly pointed criticisms of Clarke's position, for as he stressed, "the philosophy that he so vigorously defends offers no understanding of the miraculous at all, unless one wants to believe that the whole world is nothing but a tissue of miracles without any laws of nature whatsoever. What a sad situation for the human mind if this were true." Nevertheless, this was "the unhappy situation that M. Leibniz sees England falling into every day,"[186] and in this instance Castel's Anglophilia took a back seat to his Jesuit theology. Ironically, however, these same theological views also led him to diverge from Leibniz and rejoin the English on other points. In the passages just

182. Review of *Lettre de M. l'abbè *** à M. l'abbè Houteville,* in *Journal des savants* (1723): 207.

183. Review of Desmaizeaux, *Recueil de diverses pièces,* in *Journal de Trévoux* (July 1721): 1240.

184. Ibid., 1241.

185. Ibid., 1243–1244.

186. Ibid., 1245–1246.

cited, Castel joined Leibniz in defending the distinction between natural and supernatural phenomena, asserting in this way his Jesuit belief in the existence of authentic miracles and authentic natural science. However, when Leibniz joined his critique of this Newtonian position with his argument that God's rational perfection necessitated a perfectly rationalist cosmos, Castel ran the other way. On the question of rational determinism in nature, Castel praised Clarke for "reducing the question to its essentials." In fact, after surveying the mechanical and conservationist views of both Descartes and Leibniz together with Clarke's nonmechanical understanding of the same thing, he ended by offering no judgment between them.

Castel's reluctance to defend Leibniz's mechanistic and rationalist stance in these matters was also typical of his Jesuit orientation. In his final assessment of the Leibniz-Clarke debate he also made the anti-Leibnizian dimension of his thinking explicit even if this contradicted his support for Leibniz's other positions elsewhere. "[Leibniz] claims that God does nothing without reason and that his wisdom obliges him to choose what is best."[187] Castel found such ideas misguided. He thus praised Clarke for his "solid refutation of this system." He further praised Clarke for his emphasis on the essential liberty of God, agreeing with him that "the will of God itself can be a sufficient reason, and that liberty itself can produce an optimum. God, like all reasonable creatures, would be a necessary agent, which is to say not an agent at all, if he could not choose the equal or even the least outcome for no other reason than that of his liberty. Without this, one has nothing but destiny, or the fate of the pagans, reigning over everything."[188] This was exactly Clarke's point as well, and here Leibniz is made the enemy of sane religion while Clarke becomes its defender.

Castel offered further attacks along these lines, for overall he found Leibniz's rationalism both unfounded and pernicious. In this spirit, he also offered nothing but praise for Clarke's writings against Collins that were also included in Desmaizeaux's *Receuil*. However, showing his ultimate independence he also praised Leibniz's critiques of Malebranche, which were also included in Desmaizeaux's text. Overall, then, Castel recast the Leibniz-Clarke debate according to his own assumptions. Furthermore, while his views may appear at cross-purposes according to the rationalist/voluntarist, mechanist/empiricist, or Leibnizian/Newtonian dichotomies that framed the debate itself, they fit perfectly with Castel's Jesuit orientation. The important point about this orientation is Castel's vigorous defense of his views in the French public sphere, and the complexity that they added to the French discussion of these issues after 1720.

187. Ibid., 1246.
188. Ibid., 1247.

Jansenists, for example, defended a view of miracles that was very close to Clarke's in substance if not in confessional spirit. Like Clarke, they adhered to a deeply spiritual conception of nature that made God's active presence a source of much that happens in the world. They also shared Clarke's convictions about God's total freedom to effect whatever outcome he saw fit.[189] Given the close affinities that linked Jansenist notions of God's freedom in nature with the theology of the English Newtonians, Clarke's philosophical and theological arguments were, therefore, laden with unintended meaning when they entered France. French Jesuits such as Castel only complicated this situation. They also struggled within this climate to articulate a discourse of natural philosophy and religion that was distinct from Jansenism (a tricky business given the Jesuits's own sympathy for more spiritualized conceptions of nature) while also distant from the excesses of rationalistic, mechanistic materialism. Their work thus clouded even further the theological lenses through which Newtonianism entered France.

On the other hand, Cartesian rationalists like Fontenelle, Dortous de Mairan, and their sympathizers occupied an entirely different intellectual position. They were much more sympathetic with Leibniz's rationalist conception of the miraculous than Clarke, Castel, or the Jansenists. As Réaumur wrote to Crousaz with respect to the *convulsionnaires* of Saint-Médard, "the roots of enthusiasm do not appear to have been eradicated with us. . . . Some of [the *convulsionnaires*] are driven by deceit, and others by a weakness of imagination," but in either case "fear and an absence of reason" is sustaining their fantasies.[190] Réaumur treated alleged miracles as either fictions or as natural phenomena improperly understood, and in this respect he was a Cartesian, one who shared in the emerging critique of Newtonianism that followed, like Leibniz's, from these rationalist convictions.[191] As it developed, moreover, this critique often placed its adherents in opposition to any and all religiously inclined philosophers, both Protestant or Catholic. Indeed, these and other French rationalists positioned themselves in opposition to all of the period's more overtly religious conceptions of God and nature, be it Clarke's latitudinarian Protestantism or the more native Catholicism of the French Jansenists and Jesuits.

189. See Steven Nadler, "Choosing a Theodicy: The Leibniz-Malebranche-Arnauld Connection," *Journal of the History of Ideas* 55 (October, 1994): 573–590.
190. Réne Antoine de Réaumur to Jean-Pierre Crousaz, 29 April 1730, in Fonds Jean-Pierre de Crousaz, 4: 169.
191. On Réaumur's conception of the miraculous, see Daston and Park, *Wonders and the Order of Nature*, 323–324.

Like Leibniz, these increasingly self-proclaimed French Cartesians identified themselves with a deeply rationalist approach to natural philosophy that
pushed the miraculous to the margins of nature's operations. They shared neither the German's metaphysical proclivities in this area, nor his concern with
theological orthodoxy, but they did share his conviction that a scientific explanation rooted in anything other than rational principles and evident mechanical causes was no scientific explanation at all. Furthermore, in making these
claims, the French Cartesians often found French Jesuits such as Castel right
alongside them. When these same Jesuits, however, criticized mechanistic and
mathematical thinking or celebrated the wondrous complexity of God's free
and miraculous intervention in nature, they often parted company with the increasingly secular French Cartesians, even if they could often find themselves
oddly aligned with latitudinarian Protestant Newtonians as a result. The overall outcome, then, was a complex diversity within public scientific discourse in
France, one that belies any simple reduction to coherent binary oppositions.

Amidst this complexity, however, one constant was clear. After 1720, for a
host of different yet interrelated reasons, French scientific discourse became
more overtly combative and contentious. At the center of these new polemics
was the recently clarified conception of Newtonianism rooted in the physics
of universal gravitation. This philosophy had become associated (often unin-
tentionally) with many of the period's most controversial natural philosophical
problems. Some in France nevertheless began to align themselves with these
very ideas anyway, and their numbers began to grow throughout the 1720s.
Yet they remained an exceedingly small minority removed from the public eye
until after 1730. Arrayed against them, by contrast, were a host of individuals
and constituencies who for a dizzying variety of reasons found these views ob-
jectionable. Also situated at the center of these emerging controversies was the
Paris Academy, which entered its sixth decade of authoritative scientific work
in 1727. Its esteemed reputation was implicated in the intellectual changes that
were sweeping through the Republic of Letters, as was that of its septuagenar-
ian perpetual secretary. Fontenelle's activities in 1727 completed the preparation for the Newton wars that erupted after 1730, and it is to these battles that
we now turn.

PART II

The Newton Wars in France

In the late fall of 1732, the French public was introduced to its first self-proclaimed French Newtonian defined according to the criteria that had evolved over the previous decades. The figure was Pierre-Louis Moreau de Maupertuis, and the vehicle for his debut was a work entitled *Discourse on the Different Figures of the Planets*.[1] Less than two years later Maupertuis welcomed a second outspoken Newtonian into the French public sphere: François-Marie d'Arouet, better known through his penname Voltaire. The vehicle for Voltaire's Newtonian debut was his *Lettres philosophiques,* the work that has since been canonized as the first bomb thrown against the ancien régime. These two initiatives were linked historically, biographically, socially, and intellectually, yet the nature of their entanglement was anything but simple. Paying attention to the intricacies of this complex dynamic is nevertheless important, for out of the conjunction between Maupertuis, Voltaire, and their related but distinct Newtonian campaigns emerged both the Newtonianization of French science and the consequent beginning of the French Enlightenment. This dual beginning will be the subject of part 2.

Maupertuis was born in 1698 to an upwardly mobile aristocratic family in St. Malo, and he declared his sympathies for Newtonianism from a position of prominence within official French science and the Parisian society that supported it.[2] He had joined the Royal Academy of Sciences at thirty-five as an

1. Maupertuis, *Discours sur les différentes figures des astres d'ou l'on tire des conjectures sur les étoiles qui paraissent changer de grandeur, et sur l'anneau de Saturne* (Paris, 1732). My citations will be to the edition of the work found in *Oeuvres de Maupertuis,* 4 vols. (Lyon, 1756), 1: 81–170.

2. Mary Terrall's *The Man Who Flattened the Earth: Maupertuis and the Sciences in the Enlightenment* (Chicago, 2002) has superceded all of the previous literature about Maupertuis in print. Accordingly, I will follow her account closely in my own account of Maupertuis's life and work. Still useful as a supplement to Terrall's book, however, is the contemporary account of Maupertuis's life written by his friend and colleague La Beaumelle, *La Vie de Maupertuis* (Paris, 1856). See also David Beeson, *Maupertuis: An*

Le Globe mal connu qu'il a seu mesurer,
Devient un Monument où sa gloire se fonde;
Son sort est de fixer la figure du Monde,
De lui plaire, et de l'éclairer.
par Mr. de Voltaire.

FIGURE 17. *Pierre Louis Moreau de Maupertuis.*
Courtesy of the James Ford, Bell Library, University of Minnesota Libraries.

FIGURE 18. *Voltaire at twenty-four years old. Courtesy of the James Ford Bell Library, University of Minnesota Libraries.*

adjoint géometre, benefitting in his admission from the ties that his father had formed with the Phélypeaux family and the French ministries that they still con-

Intellectual Biography (Oxford, 1992), which complements Terrall's account of Maupertuis's intellectual work, and Pierre Brunet, *Maupertuis: Étude biographique* (Paris, 1929), which adds to her biographical presentation.

trolled. By 1731, Maupertuis had achieved the rank of *pensionnaire*, ascending to the top of the academy hierarchy through a program that directed his massive ambition toward scientific research, academic presentation and publication, and wider intellectual sociability. The details of this ascent will be examined shortly, but the result was that his pronouncements on behalf of Newton after 1732 were backed up by all the prestige and authority that came with esteemed membership in the upper echelons of the French scientific establishment.

Voltaire came to his Newtonianism by a very different route. Born four years earlier than Maupertuis, François-Marie d'Arouet enjoyed neither the wealth nor the social connections that so aided his slightly younger contemporary.[3] He likewise did not share Maupertuis' drive to rise within the official hierarchies that his family position made available to him, even if he was no less ambitious than Maupertuis as a result. Voltaire's chosen vocation—writing—instead created family conflicts, especially with his father, that made his early years fractious in ways that Maupertuis' pursuit of official science was not. His career trajectory after 1718 further confirmed these differences. While Maupertuis was preparing for his admission to the Royal Academy of Sciences in 1723, Voltaire was writing, publishing, and socializing in the manner of many would-be men of letters. His *Oedipe* of 1718 established him as an up-and-coming playwright, and his other writings further helped him to acquire his share of friends and patrons within the international Republic of Letters. His sharp tongue and scathing wit, however, also created enemies, and these included several who were politically powerful. As a result, Voltaire served more than one term in the Bastille before 1724, and on one occasion his incarceration ended with a royal order to leave for England as a political exile.

It was during this forced séjour across the Channel that Voltaire became attached to the details of Newtonian philosophy. Upon his return, he began his project of making his Newtonian predilections public; however, in doing so he did not maneuver within the French scientific establishment in the manner of Maupertuis. Instead, he followed the career path of an aspiring man of letters, pursuing patronage and intellectual esteem within French society and the wider Republic of Letters. It would be convenient to leave the distinction here, marking Maupertuis as the "insider" Newtonian while labeling Voltaire as his more radical and heterodox twin. To do so, however, would be a serious misrepresentation. Certainly Maupertuis' status as a senior member of the

---

3. Throughout this chapter and rest of this book, my account of Voltaire's life and work will be based on two principle sources: his vast correspondence, collected in Besterman, ed., referred to as Voltaire, *Corr.*; and René Pomeau, ed., *Voltaire en son temps*, 5 vols. (Paris, 1985–1994), a biography built primarily out of this correspondence.

French Royal Academy gave his Newtonian pronouncements a character and authority that Voltaire's lacked. Voltaire's position as a gadfly man of letters and budding *esprit fort* also colored his Newtonianism in ways very different than Maupertuis'. The Newtonianism that each man professed was also markedly different in both character and intent. Yet to reduce the early history of French Newtonianism to these social, professional, or political differences would be to seriously misunderstand it.

For one, the boundary between "inside" and "outside," or between establishment science and popular or public science, was anything but firm and established in the 1730s. The Royal Academy of Sciences was at once the great bastion of official science in France and a vital and respected organ of the international Republic of Letters. Publicity was crucial to each identity, as it had been since the 1699 reform, and this meant that academicians continually circulated between the institution and the wider public sphere in a fluid way. Academicians conceived of themselves simultaneously as protoprofessional savants beholden to the increasingly vigorous disciplinary structures of the French academy system and as gentlemanly writers and thinkers who aspired to, and often obtained, public esteem as Republicans of Letters. The wider public, moreover, shared this dual conception. From within these assumptions, Maupertuis' Newtonian campaign launched from within the Royal Academy was easily equated with Voltaire's parallel effort launched from outside its walls. This very conflation was in fact made by many contemporary observers, and since each initiative was ultimately addressed to the same audience—the French (and especially Parisian) wing of the Republic of Letters—and since each advocate spoke from a recognized (if very different) position of authority within this cultural space, the similarities between Maupertuis and Voltaire were very often more important than the differences.

Maupertuis also positioned his own Newtonian interventions in ways that demonstrated the continuing relationship between the French Royal Academy and the larger values of the international Republic of Letters. Having arrived by March 1731 at one of his fundamental Newtonian convictions—namely, the belief that the theory of universal gravitation was a useful conceptual tool for resolving several knotty problems in mechanics—he chose not to present his ideas to his academic colleagues in Paris, but instead to his fellows at the Royal Society of London, to which he had been admitted in 1728. The paper that contained these views directly challenged the published work of Dortous de Mairan, and Maupertuis no doubt concluded rightly that airing these views in Paris would have provoked a confrontation. As he wrote privately to Johann Bernoulli at the time, "I do not have the courage to give [this paper] in a country where it seems they don't think deeply enough and where they do not do justice to M. Newton's system." Instead, as Mary Terrall plausibly suggests, he translated the work into Latin and published it in the *Philosophical Transactions of the Royal Society* so as to secure his good graces at home on the eve of the academy's election for the *pensionnaire* seat to which he aspired.[4]

Even after securing his academic authority by winning appointment to this seat, Maupertuis' reticence about directly confronting his fellow academicians continued. At the time, the academy was embroiled in what had become a decades-long dispute about the correct measurement of force in a moving body. This "*vis viva* controversy" will be discussed more fully shortly, for it pitted the newly self-conscious Newtonians, Cartesians, and Leibnizians into a bitter scientific battle that channeled all the heat of the previous decade's priority dispute into new and influential directions. Maupertuis had privately made his support for the Leibnizian position clear, yet he nevertheless resisted Johann Bernoulli's calls to take up the cause (which was Bernoulli's as well) inside the academy. Here, as with his earliest convictions regarding Newtonian gravitation, Maupertuis preferred to guard his views and to profess an *honnête* neutrality rather than provoke an argument with his fellow academicians.

In each of these cases, Maupertuis demonstrated the continuing power of the gentlemanly norms of conduct that had governed the relationship between the French Royal Academy and the wider Republic of Letters for over half a century. His defense of Newtonianism, which appeared little more than a year after his establishment as a pensioned academician, displayed similar influences. To defend Newtonian gravitation before Fontenelle, Dortous de Mairan, Réaumur, Cassini II, and the other powerful and newly strident Cartesians inside the Paris Academy was indeed to issue a provocation, for as Maupertuis noted at the time the polemics of the previous decade had made academicians "shocked simply to hear the word 'attraction' mentioned in the meetings."[5] Maupertuis accordingly offered his Newtonian challenge with exceeding *honnêteté*. Rather than air his views during an academic session, a move that would have triggered a debate that in turn would have spilled over into the wider public sphere, he instead articulated them outside the institution. He crafted a carefully worded book articulating his position, and then published the text independently with the approval of the academy but in no way under its aegis.[6] In the text itself, he also took pains to avoid enflaming the controversies at the center of the issues he treated. His was not so much a defense of the Newtonian

4. Terrall, *Maupertuis*, 58–60.

5. Maupertuis to Johann Bernoulli, 11 June 1731, cited in ibid., 58.

6. Maupertuis, *Discours sur les différentes figures des astres*.

theory of universal gravitation as a defense of the possibility of such a defense. In this way, he succeeded in becoming France's first declared Newtonian without at first becoming its first Newtonian antagonist.

Voltaire, by contrast, was eager to play the latter role, and he thus read Maupertuis' *Discours* the way that a proselyte reads scripture. Responding to letters from Maupertuis, now lost, that may have contained manuscript versions of the arguments contained in the *Discours*, Voltaire thanked the senior academician for resolving his confusion about Newton's work: "Your first letter baptized me in the Newtonian religion, and your second gave me confirmation. I thank you for your sacraments." [7] Voltaire continued to rely heavily on Maupertuis' guidance as he prepared his own letters on Newtonianism in the *Lettres philosophiques*, and when they were published to scandal in 1734, he approached the academician again, this time to solicit his leadership in the Newtonian crusade that he hoped to launch in France. No record of Maupertuis' response exists, but as Terrall rightly notes it is highly unlikely that Maupertuis seriously considered joining Voltaire's self-styled "Newtonian sect." The author of the *Lettres philosophiques* was, after all, an outlaw. His book had been publicly burned by the royal authorities for "inspiring a license of thought most dangerous to religion and civil order." He was also living in exile as a fugitive from the police. [8] Maupertuis had little need for such rebellion and controversy, and consequently he directed his energies instead toward the academy and those sectors of the Republic of Letters that deferred to its authority.

Yet, here again, it would be misleading to separate Maupertuis and Voltaire too starkly. Voltaire's first letter to Maupertuis was sent before the latter had made his Newtonian views public, and one might wonder, therefore, how Voltaire knew to approach this of all French academicians with his Newtonian inquiries. [9] The most likely answer is that he encountered Maupertuis and his Newtonian views in Parisian society where each was a vigorous participant from 1718 onward. Maupertuis, in fact, found his way into the Royal Academy of Sciences by way of both the official, ministerial channels that mattered and the nonofficial, *mondain* circuits that were just as important. His earliest training as a mathematician occurred under the wing of François Nicole, and

7. Voltaire to Maupertuis, 15 November 1732, in Voltaire, *Corr.*, 86: D537.

8. Cited in René Pomeau, *D'Arouet à Voltaire* (Paris, 1985), and in Pomeau, *Voltaire*, 1: 329.

9. Voltaire's first letter to Maupertuis is dated 30 October, 1732, at least two weeks before the appearance of Maupertuis's *Discours*. On the latter, see Terrall, *Maupertuis*, 67, n. 119. Voltaire's second letter, dated 3 November 1732, indicates an awareness of the book's imminent appearance, but no contact with the book itself.

of all venues it was the Café Gradot, a frequent haunt of Nicole and the other Malebranchian analytical mathematicians in the academy, that cemented this relationship. Fontenelle, Saurin, and other academic analysts were also long-standing Gradot regulars, and when Houdar de la Motte, the acknowledged leader of this cohort, died in December 1731, many reports confirm that Maupertuis succeeded him at the center of this circle.[10]

These sociable connections were anything but marginal to Maupertuis' ascent within the Royal Academy. Saurin had long used the Gradot as a base for securing both his academic and his wider intellectual reputations, and Maupertuis appears to have developed particularly close ties with him during these years. Especially telling was Saurin's support for Maupertuis' position during the *vis viva* controversy. The Café Gradot no doubt provided glue to these alliances, and it is perhaps not coincidental either that Saurin of all people provided the vacant *pensionnaire* seat that Maupertuis filled in 1731. This promotion also occured just weeks before Houdar de la Motte's death allowed for his succession at the Café Gradot as well. Even if Saurin did not facilitate these elevations, the link between the Gradot and the academy evinced here also served Maupertuis.

Like Saurin, Maupertuis also cultivated a dual identity as both an *esprit fort* and a serious academic mathematician. Each role served to situate him alongside the other worldly savants such as Fontenelle and Dortous de Mairan who also occupied prominent places at the top of the academic hierarchy. It also brought him into contact with Voltaire, who circulated with him in the same Parisian social circles, save the Royal Academy. He also shared Maupertuis' ambitiousness as a savant, even if he devoted himself to playwrighting, poetry, and prose composition while Maupertuis occupied himself with serious scientific research. There was, therefore, always a gap between the intellectual identities of the two men rooted in the differences between literature and science in the eighteenth century. Since they shared other, equally potent identities, however, and since the line between literature and science, and especially between academic and nonacademic scientific writing, was blurry and contested, Voltaire could plausibly frame his own intellectual program after 1734 as just one more approach to the common goal that he and Maupertuis were jointly pursuing. Maupertuis for his part also encouraged a blurring of these distinctions since he actively traded upon his status in Parisian society whenever he maneuvered within the Royal Academy. He further evinced no clear sense that he was a professional, academic savant first and foremost and a worldly Republican of Letters second. Rather, he tailored his own intellectual production to

10. See La Beaumelle, *La vie de Maupertuis,* 12; Terrall, *Maupertuis,* 69.

the multifaceted demands of each arena simultaneously. In this way, the cultural changes that he and Voltaire provoked were driven less by the social and vocational differences that kept the two men apart, and more by the complex dynamic that united them.

Nevertheless, it is still useful to trace the contributions of Maupertuis and Voltaire to the commencement of the French Enlightenment by tracing the distinct position of each within the complex web that linked official French science to the wider public sphere, and each to the international Republic of Letters. Inside the Royal Academy, where Maupertuis' ambitions were most strongly focused, his interventions started a series of debates that positioned a new cohort of young mathematicians sympathetic to his views against the newly self-conscious Cartesians who were then dominant within the institution. These Cartesians (Fontenelle, Dortous de Mairan, Réaumur, Cassini II, and Privat de Molières, to name only the most prominent) had moved to the center of French public science by negotiating a set of links between official French science and the wider learned world. To challenge them, Maupertuis needed to cultivate a similar position for his consituency. His early use of the Royal Society of London to articulate his views, and then his decision to publish his *Discours* outside the publication channels of the academy, set the pattern, and after 1734 he engaged in a series of strategic initiatives that effectively played the French scientific establishment and the wider public off of one another in the service of his Newtonian cause, not to mention his own ambitions.

For his part, Voltaire also indirectly aided Maupertuis in these efforts by building his own reputation as an authoritative and critical philosophical thinker who sympathized with the senior academician's Newtonian views. Other ties also brought the two men together even if they never formed an explicit alliance or party in support of their respective Newtonian campaigns. Having formed a close relationship prior to 1734 with Émilie Le Tonnier de Breteuil, the Marquise du Châtelet, a sophisticated scientific thinker in her own right, Voltaire took advantage of the political and material (not to mention intellectual and emotional) resources that she offered by making her provincial home at Cirey the base for his Parisian exile. At Cirey, Voltaire and du Châtelet constructed a laboratory where they performed scientific experiments. They also collaborated on a number of important texts that fueled the first battles of the French Newton wars, while further entertaining numerous Parisian and foreign guests who were drawn to the couple by Voltaire's newfound fame. Among them was Maupertuis and his young, academic ally Alexis Claude Clairaut, who came to Cirey for the sociability but also to work on mathematical problems with the Marquise. Unlike Voltaire, du Châtelet could work with these academicians as a mathematical peer, yet the fledgling philosophe nevertheless benefited from the general discussions of Newtonianism that these meetings generated. They also stimulated his own project, conceived at Cirey, of composing a serious French work exposing the details of Newtonian philosophy in the manner of Henry Pemberton's *A View of Sir Isaac Newton's Philosophy*.

While writing his *Eléments de la philosophie de Newton,* which created a stir when it appeared in 1738, Voltaire also collaborated with du Châtelet on a number of different intellectual projects, including du Châtelet's own, original treatise in natural philosophy that she published in 1740.[11] The couple also remained directly engaged with official French science despite their exile, so much so, in fact, that a politically rehabilitated Voltaire even contemplated replacing Fontenelle as the academy's perpetual secretary when the octogenerian announced his retirement in 1740.[12] Maupertuis also intimated interest in this position, and the fact that each seriously considered assuming this vital public office offers vivid testimony to their parallel ascent during the previous decade. It also confirmed that their career-making campaigns on behalf of Newton were helping to turn the wider intellectual tide. In 1743, Jean le Rond d'Alembert, a recently appointed yet increasingly active member of Maupertuis' Newtonian cohort in the academy, wrote that the Cartesians "are a sect that is very much weakened."[13] Fifteen years later, writing now as a senior member of the French scientific establishment, d'Alembert changed his view, declaring the battle over. "The Cartesian sect barely exists anymore," he opined.[14] Maupertuis was more emphatic in his prediction. Writing to Johann Bernoulli in 1743, he declared "Cartesianism is done for [*foutu*] even in the academy."[15] The years

11. Emilie Le Tonnier de Breteuil, Marquise du Châtelet, *Institutions de physique* (Paris, 1740).

12. The astronomer Jêrome Lalande recorded the following episode in his diary: "When M. de Fouchy was made secretary, M. de Maupertuis and M. de Voltaire both sought the position. M. V. was dismissed after sayign in the halls of the Versailles to a person who asked him what attracted him [to court], "it is not the master of the house." As for M. de Maupertuis, M. de Maurepas kept him well away from the position. Maupertuis returned to the Cardinal his guaranteed pesion of 1,200 livres. He went to sell himself to M. de Réaumur, who lavished him with friendship in return. Nevertheless it all came to nothing." From "Miscellanea notae collectae a J.H.L.F de Lalande, ex geometria potissimum sublimi, et astronomia. ab initio anni 1752," BN, MSS f. fr. 12274, 249. Dortous de Mairan, who served as the interim secretary between Fontenelle and Grandjean de Fouchy, reported something similar. See Kleinbaum, "Jean Jacques Dortous de Mairan," 132.

13. Jean le Rond d'Alembert, *Traité de dynamique* (Paris, 1743), v.

14. D'Alembert, *Traité de dynamique,* 2nd ed. (Paris, 1758), vi.

15. Cited in Terrall, *Maupertuis,* 199.

1734–1758 did witness the conversion of much of French science to the physics of Newton, and it was through the interplay between the internal scientific debates of the Paris Academy triggered by Maupertuis and the wider intellectual discussions instigated by Voltaire, du Châtelet, and others that this change occurred.

The French Enlightenment also commenced in tandem with this shift through a similar entanglement between official science, its institutions, the Republic of Letters, and the discourses of Newtonianism. Maupertuis and Voltaire set the pattern through their parallel use of Newton to open a critical cultural space between establishment French science (i.e., the Royal Academy) and the wider French public. Yet if Maupertuis ultimately fostered and exploited this dynamic in the service of his own ambitions within officialdom, Voltaire used it instead to create a new identity and authority—the persona of the philosophe—that quickly became the avatar of Enlightenment in France. Voltaire's position with respect to official science and especially the Royal Academy was crucial to this formation. By positioning himself as a nonacademic authority on natural philosophy, and by using that position to imply a challenge to the official scientific establishment, Voltaire in effect defined a new conception of the man of letters as an independent, critical thinker beholden only to universal reason and the public that authorized it.

Voltaire, however, did not invent this position out of whole cloth. The expansion of the communication structures of the Republic of Letters—books, journals, letters, travel, sociability, etc.—prepared the social space that made it possible. The new critical climate and provocative philosophical discourses of the early eighteenth-century Republic of Letters also gave Voltaire the critical tools he needed. He in fact drew many of the core features of the philosophe persona from the stridently critical and partisan intellectualism that he encountered in England and Holland during his forced exile of the 1720s. Similarly, without Maupertuis' parallel machinations in and around the Paris Academy, maneuvers that opened the door to Voltaire's public challenge to its authority, his initiatives would not have been possible. As a result, even if he, more than his academic partner, initiated the fusion between Newtonian science and critical, libertine intellectualism that would become the hallmark of the French Enlightenment, his efforts were prepared, however indirectly, by the parallel efforts of Maupertuis.

Consequently, the cultural dynamic that joined each together with Newtonianism after 1732 was the real agent behind these changes. Ultimately the public friction between ideas and institutions that Maupertuis initiated created a battle over Newtonianism that created new possibilities. The philosophe movement, and thus the Enlightenment, was a consequence of these Newton wars, and if Voltaire played the key role in the later stages, his opportunities were prepared by the early efforts of Maupertuis to open up this precise critical space. For this reason, part 2 seeks to understand this cultural dynamic by first examining the early Newtonian initiatives of Voltaire and Maupertuis and then the larger Newton wars they triggered.

CHAPTER 4

The Invention of French
Newtonianism *Maupertuis*
and Voltaire

Maupertuis announced his Newtonianism from the highest levels of the French scientific establishment. Voltaire, by contrast, proclaimed his from within the public space of the Republic of Letters. However, since establishment French science was linked inextricably to the wider public culture of the Republic of Letters, and since the Paris Academy, Maupertuis' prime arena of activity, was especially attached to these wider communities, Maupertuis' proclamations were public addresses as well. The academician in fact came to his Newtonianism through a negotiation between his public and academic identities, and his self-fashioning as the first French Newtonian resulted from an interplay between them. Although never a royal academician of science, Voltaire fashioned himself through a similar negotiation. Public academic science as it had evolved in France since 1699 fostered the authority of the wider public as a legitimating audience for official knowledge. Voltaire laid claim to this authority and used it to position himself as an authoritative public spokesperson for science, one outside of the Royal Academy, and one who challenged the official savants of this body. Maupertuis' maneuverings between the academy and the public nevertheless facilitated Voltaire's initiatives, even if the result ultimately created as many tensions among the new French Newtonians as alliances. From the outset, then, French Newtonianism was a disparate and contentious entity, and to understand these historical complexities let us first consider Maupertuis' contributions to them.

Two Paths to French Newtonianism:
The Early Careers of Maupertuis and Voltaire

Mary Terrall's recent biography admirably recounts the details of Maupertuis' family history, so no summary of them is necessary here.[1] Instead, two threads that proved particularly determinative in shaping Maupertuis' career

1. Terrall, *Man Who Flattened the Earth*, 16–21.

can be emphasized.[2] First was affluence. Since the escapades of Maupertuis' father René as a sea captain and privateer made him extremely wealthy, his son never struggled to secure either patronage or income throughout his life. This material independence should not be forgotten when considering Maupertuis' easy entrance into Parisian society. Nor should it be neglected when accounting for his rapid ascent up the social hierarchies of the day. It should also be remembered when considering Maupertuis' singular ambition to achieve marks of honor and esteem since these were among the few rewards that his vast fortune could not purchase for him.

Second and equally important was the Maupertuis family connections with the French Crown. A seaman by profession, René was not born into the French aristocracy, yet because of his involvement in a number of overseas enterprises that were central to the monarchy, his work facilitated his social and political elevation.[3] René de Maupertuis eventually became a royally sanctioned pirate-warrior, and this career enriched him both materially and politically. His connections with the Phélypeaux family became especially tight, and after he retired from the corsair life to pursue a management role in several maritime businesses, Maupertuis' father also became an important participant in the Pontchartrain's administrative state.[4] In this way, he personified the social elevation that the commercial nobility as a whole experienced under the royal administration of the Pontchartrains.

2. Very little documentation exists about Maupertuis' early years. Most of his surviving correspondence dates from after 1730, and very little else from before this date has survived. Maupertuis' biographers, therefore, have relied heavily on the contemporary account of his life written by his friend and ally La Beaumelle. Maupertuis' life ended amidst scandal, however, and La Beaumelle's work is very much a piece of hagiography addressed to the participants in these disputes. Thus, it is hard to know how reliable his account is. Maupertuis' modern biographers Terrall, Beeson, and Brunet have attempted to fit La Beaumelle's story into a more balanced historical context, and I have relied heavily, therefore, on the revisions that each provides.

3. His move into privateering after 1688 was in fact driven by the larger policies of the Phelypeaux de Pontchartrain ministry, for confronting budget shortfalls, Jerôme de Pontchartrain, then secretary of the navy, turned to the merchant marine as a strategic resource to meet the ongoing military imperatives of the Crown. See Geoffrey Symcox, *The Crisis of French Sea Power, 1688–1697: From the Guerre d'escadre to the Guerre de course* (The Hague, 1974).

4. David K. Smith, "Structuring Politics in Early Eighteenth-Century France: The Political Innovations of the French Council of Commerce," *Journal of Modern History* 74 (2002): 490–537.

Pierre-Louis Moreau de Maupertuis benefited immensely from the political and social advantages that his father's career bequeathed to him. Reaching maturity amidst the political transformations of the Regency period, Maupertuis began his adult career in tandem with the young heir to the Phelypeaux ministerial dynasty, the future Comte de Maurepas. Maurepas, who succeeded his father Jerôme de Pontchartrain in his various posts in 1716, was only fifteen years old at the time of his ascension (Maupertuis was seventeen). He eventually grew to become an able minister who served, like his father, as both the naval secretary and the minister in charge of the academies until his own disgrace in 1749. The Maupertuis family benefited from their ties to this central figure in the French monarchy, and Pierre-Louis in particular formed close ties with Maurepas, who eventually became one of his most loyal supporters and patrons.[5]

During these political transformations, Pierre-Louis also began to fashion his own career. In 1716 he came to Paris for the first time to pursue a degree at the Collège de la Marche. College was a customary rite of passage for an upwardly mobile provincial aristocrat of his stature, but Maupertuis also began to study mathematics privately as well with Varignon's protégé Guisnée.[6] These lessons introduced him to the analytical mathematics that would eventually become central to his scientific work. There is little indication, however, that the young aristocrat was actually contemplating a scientific career at this point. To have done so would in fact have been exceptional given his social rank and the career paths that lay before him. Instead, Maupertuis left Paris in 1718 once his studies were complete, entering the army as a member of the *mousquetaires gris*. He was given a highly desirable officer's position, a fact that attests to the strength of his family's political connections, and this sent him to Lille, where he lived for the better part of the years 1718–1722.[7]

5. On this ministerial history, see James Collins, *The State in Early Modern France* (Cambridge, 1995), 141–143. The bureaucratic "routinization" of administrative monarchy under the Pontchartrains is discussed in Jay M. Smith, *The Culture of Merit: Nobility, Royal Service, and the Making of Absolute Monarchy in France, 1660–1789* (Ann Arbor, 1996), 149–150, 184–185. David K. Smith has also stressed this dimension of administrative monarchy in his work on the Council of Commerce.

6. La Beaumelle, *La vie de Maupertuis*, 7; Brunet, *Maupertuis*, 10–11.

7. David Beeson writes of Maupertuis' military position: "Through a quite extraordinary degree of influence for a first-generation provincial *robin* family, [Maupertuis'] father was able to obtain a lieutenancy for him in the regiment of La Roche-Guyon stationed at Lille, the kind of commission for which young nobles would have given their eye teeth, particularly in peacetime." Beeson, *Maupertuis*, 62.

During the winters, however, he returned to Paris for rest and relaxation, and it was during these sojourns that his career began to turn in a new direction. Maupertuis circulated widely in Parisian society, and he especially enjoyed the company of the savants that gathered each night at the Café Gradot. According to his friend and first biographer La Beaumelle, it was these associations that launched Maupertuis on his scientific career. The Gradot circle was dominated in these years by Houdar de la Motte, and at the center of his circle were Marivaux, Du Bos, and a host of other literary luminaries. Also near its center were the analytical mathematicians Saurin and Nicole, not to mention Fontenelle. According to La Beaumelle, it was the writer and journalist Fréret, another of Houdar de la Motte's allies and a member of Boulainvillier's group of libertine freethinkers as well, who pushed Maupertuis toward mathematics. He believed that this science would help Maupertuis "tame his active and consuming soul." [8] No matter what the motivation, Maupertuis began devoting himself more seriously to mathematical study after 1722.

Nicole in particular became both a tutor and companion during these years, and the two forged a lasting friendship that was prescient of things to come.[9] When Maupertuis met him, Nicole was residing in the desert that was French analytical mathematics in the 1720s. Along with Saurin, who had only joined the original Malebranche circle a decade after its formation, Nicole was the only academic mathematician with strong ties to the original program in analytical mechanics. By 1722, Rémond de Montmort, Guisnée, and Varignon were dead, while Bomie and Bragelogne had become virtual nonentities at the institution. Saurin, moreover, was increasingly devoting his energies elsewhere. Fontenelle was likewise becoming absorbed in the nonmathematical academy culture dominated by Bignon, Réaumur, Dortous de Mairan, and the Observatory-based astronomers. Yet amid these shifts, Nicole remained devoted to the old program in analytical mathematics pioneered by the old Malebranchians, and it was into this program that Maupertuis was initiated after 1720.

As John Greenberg notes, Nicole produced a number of important papers in integral calculus between 1715 and 1725.[10] These papers, however, attracted virtually no attention either inside the institution or elsewhere. As a practitioner of the abstract, analytical mathematics despised by Father Castel, Nicole also became the target of the reverend father's criticisms when he set out to challenge the mathematical work of the French Academy in the *Journal de Trévoux*. It will be remembered that these critiques spurred Fontenelle toward

8. La Beaumelle, *La vie de Maupertuis*, 14.
9. Ibid.
10. Greenberg, *Problem of the Earth's Shape*, 239–240.

his new mathematical initiatives of the 1720s, yet these efforts appear to have done little, at least at first, to advance Nicole's career. For precisely this reason, however, Nicole's new relationship with Maupertuis, which solidified during precisely this period, was deeply significant. Under his wing, the young military officer resumed his study of analytical mathematics begun with Guisnée in 1716, and soon the senior academician had a young and talented mathematical protégé. The sociability of the Café Gradot also united the two men, and since Fontenelle's successor as perpetual secretary of the Academy of Sciences Grandjean de Fouchy further connected Nicole with the Phélypeaux family in his *éloge* of the academician delivered in 1758, they may have shared a set of political ties as well.[11] The alliance further transferred itself into the academy in 1723 when Maupertuis resigned his military command and began to seek admission to the Royal Academy.

Maupertuis' acceptance followed soon after, but the precise character of his appointment was revealing of his early status as a savant. On December 4, 1723, Maupertuis visited the academy to deliver two scientific papers devoted to topics in natural history.[12] The practice of delivering presentation pieces in anticipation of academic admission was by then a common academic practice—Dortous de Mairan, for example, had demonstrated Newton's prism experiments before the members in 1718 as a prelude to his own admission to the institution. Maupertuis' presentation session was nevertheless unique. Both Bignon and Maurepas attended the meeting, each in their capacity as honorary members of the company, and before their guest spoke the ministers announced an administrative change.[13] They ordered the expulsion of a little known *adjoint mécanicien* by the name of François-Joseph Camus for failure to attend academic meetings. They then announced that nominations should be made to replace him.[14] Although not unprecedented, expulsions of this sort were extremely rare, and for Maupertuis, the timing could not have been more fortunate. At the next session, when the academy heard nominations to fill Camus' seat, a plurality of the votes went to Maupertuis.[15] Maurepas notified the academy on December 22 that the king had approved their selection, and he

11. "Éloge de M. Nicole," *HARS-Hist.* (1758): 107–114.

12. PVARS, December 4, 1723.

13. Maupertuis' connections to Maurepas have already been noted, but Beeson sees ties with Bignon as well. He notes, for example, that Maupertuis acted as something of an unofficial secretary for Bignon on a number of matters in the year after his admission to the academy. See Beeson, *Maupertuis*, 63, n. 9.

14. Ibid., 63.

15. PVARS, December 11, 1723.

further announced a switch of titles. Maupertuis was made an *adjoint géomètre* while Beaufort was moved into Camus' vacant *adjoint mécanicien* seat.[16] Thus, by the end of 1723, Maupertuis was an academician and a member of the class of mathematicians.

As his presentation papers in natural history suggested, however, his intellectual identity was not necessarily rooted in mathematics alone despite this affiliation. In fact, the young academician appears to have sampled a number of different intellectual identities in this period, defining an eclectic approach to science that would remain a characteristic feature of his work throughout his lifetime.[17] Early on, Réaumur's public scientific persona appears to have exerted a powerful influence upon him, for Maupertuis' first scientific papers evince a strong desire to balance scientific professionalism with literary *bel-esprit*. In his first public assembly paper delivered in November 1724, Maupertuis spoke on the acoustics and form of musical instruments, a topic tailor-made for the academy's public assembly audience. Since Maupertuis was also an accomplished guitarist—a further outcome of his formation as a cultured and *honnête* aristocrat—his work evoked this experience as well. The paper, which was most likely scrutinized before its presentation by both Bignon and Réaumur, reveals an easy fluency with the literate discourse of science that was still crucial to esteem in the wider public sphere.[18] The *Mercure* called Maupertuis' paper an "exact" account of the "bizarre" affinities that differently shaped musical instruments shared, and praised the discourse overall. Furthermore, since the scientific exposition of interesting curiosities such as these had by this date become Réaumur's stock-in-trade, Maupertuis' presentation positioned him as a possible aspirant to this public reputation.[19] He also devoted himself to Réaumur's preferred natural history in this period, producing an account of salamanders that was published in the academy's *mémoires* for 1727.[20] He

16. PVARS, December 22, 1723.

17. Terrall in particular emphasizes Maupertuis' shifting intellectual identities, seeing in them a response to the different publics that he sought to address. In addition to Terrall, *Maupertuis*, see idem, "Gendered Spaces, Gendered Audiences: Inside and Outside the Paris Academy of Sciences," *Configurations* 2 (1995): 207–232.

18. Maupertuis, "Sur la forme des instruments de musique," PVARS, November 15, 1724; *HARS-Mem.* (1724): 215–226.

19. *Mercure de France* (November 1724): 2429.

20. Maupertuis, "Observations & experiences sur une des especes des salamandres," *HARS- Mem.* (1727): 27–31. Jacques Roger sees a strong affinity between this work and Réaumur's natural history of the same period. See *Les sciences de la vie dans la pensée française du XVIII siècle* (Paris, 1963), 469.

would continue this research throughout his lifetime, producing important work in both the life sciences and natural history, and as his admission to the Académie française in 1743 would confirm, this orientation helped Maupertuis to establish a public intellectual stature not available to pure mathematicians such as Nicole.[21]

In all these ways, then, the young academician showed every indication of moving to the center of public academic science in France by following in the footsteps of the great scientific *hommes des lettres* such as Fontenelle, Dortous de Mairan, and Réaumur. Mathematics, however, remained central to Maupertuis' work, and in the short term at least it was this trajectory that proved most influential for him. In 1726, he delivered his first mathematical paper to the academy, a solution to a problem of maxima and minima.[22] Saurin had presented a related work only weeks before, and in Maupertuis' contribution his debt to his analyst teachers was evident.[23] The paper was published in the academy's *mémoires* of 1726, and both he and Nicole continued this work in 1727, producing mathematical papers that were eventually published as well.[24] Fontenelle further gave this work a brief notice in his *histoires*, a presentation that was dwarfed, if supportively so, by the extensive account of Fontenelle's *Eléments de la géometrie de l'infini* written by another Gradot regular, the abbé Terrasson. Fontenelle's manner of framing this mathematical work, moreover, was revealing of the changing intellectual climate of the time. In 1726, he connected Maupertuis' work explicitly with Newton's *Principia*, an indication that the secretary saw Maupertuis at the center of the Newton/analytical mechanics complex that remained central to his vision of Newton's place within eighteenth-century French science.[25] In his 1727 report, by contrast, he ironically traced the origins of Maupertuis' work back to Réaumur's one and only mathematical paper published in 1707. This was a genealogy that marginalized

21. For a discussion of this work, see Roger, *Les sciences de la vie*, 468–487; Terall, *Maupertuis*, chap. 7; and Brunet, *Maupertuis*, chaps. 7 and 9.

22. Maupertuis, "Sur une question de maximis & minimis," PVARS, January 26, 1726, *HARS-Mem.* (1726): 84–94.

23. Saurin, "Observation sur la question des plus grande & des plus petites quantités," December 18, 1725, *HARS-Mem.* (1725): 238–259.

24. "Quadrature et rectification des figures formées par le roulement des polygones réguliers," in *HARS-Mem.* (1727): 204–214; and "Nouvelle manière de développer les courbes," *HARS-Mem.* (1727): 340–350; and Nicole, "Méthode pour sommer une infinité de suites nouvelles, don't on ne peut trouver les sommes par les méthodes connues," *HARS-Mem.* (1727): 257–269.

25. *HARS-Hist.* (1726): 42–45.

Maupertuis by locating him (and implicitly Nicole as well) in the intellectual hinterlands occupied by French academic mathematics in the 1720s.[26]

Maupertuis did not wait to see the publication of these reviews. In May 1728 he left Paris for an extended set of travels that kept him away from the academy throughout most of the next three years. His first stop was London, where he spent three months associating with the English savants there.[27] Terrall aptly describes the trip as an intellectual adventure, one consonant with his ongoing formation as an aristocrat and worldly savant. His letter of introduction to the Royal Society of London captured well the intellectual orientation he took with him. It described him as one with "a taste for natural history, [experimental] physics, and mathematics."[28] While in London, he also took advantage of the intellectual diversity that the city offered, meeting everyone from Desaguliers and Samuel Clarke to the journalist Pierre Desmaizeaux and the instrument maker George Graham. The Frenchman also met Henry Pemberton, whose widely read *A View of Sir Isaac Newton's Philosophy* had just appeared. Nicole's mathematical networks, however, appear to have been especially important for him since he met and formed close ties with many of the leading English mathematicians. Brook Taylor, Rémond de Montmort's close friend and philosophical sparring partner, became a particularly important contact in England, and through him Maupertuis made the acquaintance of James Stirling, another important Newtonian mathematician. The importance of these ties was revealed when Maupertuis was admitted to the Royal Society, an admission orchestrated by another of Rémond de Montmort's English mathematical colleagues, Abraham de Moivre.[29]

Maupertuis' initial understanding of Newtonianism was no doubt crystallized during his English visit, just as it was for Voltaire in the same years. The career paths that led the two men here could not have been more different, however, and they reveal much about the historical entanglements that joined the two savants together while also distancing them from each other during these years. Born François-Marie Arouet in 1694, Voltaire was raised within a family that personified the new administrative professionalism Louis XIV's monarchy had made possible. Voltaire's father, François, was a notary, and as such repre-

---

26. *HARS-Hist.* (1727): 57–61.

27. On the London trip, see Brunet, *Maupertuis,* 13–15; and Beeson, *Maupertuis,* 64–66. La Beaumelle claims that Maupertuis spent six months in London, but Beeson, using the evidence of his surviving correspondence, thinks the trip lasted no more than twelve weeks.

28. Terrall, *Man Who Flattened the Earth,* 42.

29. Ibid.

sentative of the new nontitled class of royal servants that staffed the administrative state. While not by any means a central figure either at court or in the royal administration, he did receive a passing reference in Saint-Simon's *Mémoires* and was also on familiar terms with many of the more important courtiers and state officials.[30] More important for the future development of his son, François was active in the literary culture of the period, both in Paris and at Versailles.

From early in his youth, Voltaire aspired to become a playwright like his idols Molière, Racine, and Corneille. Voltaire's father opposed the idea, wanting to install him instead in a position of public authority. First as a law student, then as a lawyer's apprentice, and finally as a secretary to a French diplomat, Voltaire attempted to fulfill his father's wishes. In each case, however, he ended up abandoning his posts, sometimes amidst scandal.[31] Escaping from the burdens of these public positions, Voltaire would retreat into the libertine sociability of Paris. His wit and congeniality were legendary, even as a youth, so he had few difficulties establishing himself quickly as a popular figure in elite society. He also learned how to play the patronage game so important to men of letters in this period, and thanks to some artfully composed writings, a couple of well-made contacts, more than a few bon mots, and a little successful investing, especially during John Law's Mississippi Bubble fiasco, he was able to establish himself by 1720 as an independent man of letters in Paris.

During these years, there was nothing in Voltaire's intellectual and cultural biography that distinguished him from other independent men of letters in the city—with the possible exception of his success. Moreover, throughout the 1720s, Voltaire conducted himself in a manner utterly typical of the literary culture of period. His theatrical debut occurred in 1718 with the performance and publication of his *Oedipe*, a reworking of the ancient tragedy that evoked the French classicism of Racine. Performed first at the home of the Duchesse du Maine at Sceaux, this work also confirmed Voltaire's reputation in the elite society that congregated in sites such as these.[32] His other activities also placed him in the mainstream of the elite, literary society of the period. His circle included the group of libertine poets who met each week at the former monastery of the Knights Templar in Paris, a gathering where aging worldlings and younger libertines met to eat, drink, and read poetry. The members of this circle included blue-blooded aristocrats such as the Chevalier d'Aydie, future

30. Pomeau, *D'Arouet à Voltaire*, 65.

31. In 1708 Voltaire's father obtained a *lettre de cachet* ordering the confinement of his son after he became involved in a scandalous love affair with a young Dutch girl while serving a French ambassador during a sensitive diplomatic negotiation. Ibid., 60–63.

32. On Voltaire at Sceaux, see ibid., 92–94, 102–106.

judges such as Charles-François Hénault (Réaumur's early entrée into Parisian society), and men of more common stock as well. In this way, the Temple epitomized the union of elegant sociability and literary *mondainité* so typical of early eighteenth-century Parisian society.[33]

Voltaire's intellectual work also captured perfectly the dual drives of this cultural world. He devoted his considerable literary talents to a mix of poems and plays that shifted between playful libertinism and serious classicism seemingly without pause. This aesthetic balanced perfectly the values of pleasure, *honnêteté*, and taste—the watchwords of elite Parisian society in the 1720s— and attached to these values Voltaire and his work prospered. The English aristocrat, freethinker, and Jacobite Lord Bolingbroke was also a key influence on the young Voltaire during this period.[34] Living in exile in France between 1715–1723, Bolingbroke encouraged Voltaire to become a frequent visitor to La Source, the Englishman's estate near Orléans. As its name suggests, La Source was an important center for the dissemination of ideas in many fields, and the chateau also served as a reunion point for a wide range of intellectuals.[35] Before his exile, Bolingbroke had helped to found the Brothers' Club, a literary society that was, like the Temple, emblematic of the literary sociability so important to the Republic of Letters at this time. Listing "wit and learning" as the membership criteria, it attracted writers such as Alexander Pope, Jonathan Swift, and John Gay to its program of "improving friendship" and "encouraging letters." La Source was a comparable environment. It was here as well that the crucial contributions that Bolingbroke would make in English politics after 1723 also took shape. Voltaire was at the center of these shifts from the beginning.

Bolingbroke was also a flagrant *libertine* and rake, and both the Brothers' Club and La Source exemplified these impulses as well. The English lord's professed models were Alcibiades and Petronius, and he willingly accepted the

33. On Voltaire and the Temple, see Pomeau, *D'Arouet à Voltaire*, 77–79; Wade, *The Intellectual Development of Voltaire* (Princeton, 1969), 120–129; and Gay, *Voltaire's Politics*, 38. On the egalitarian nature of elite sociability in Old Regime France, see Daniel Gordon, *Citizens without Sovereignty: Equality and Sociability in French Thought, 1670–1789* (Princeton, 1994).

34. On Bolingbroke, see H. T. Dickinson, *Bolingbroke* (London, 1970); Rex A. Barrell, *Bolingbroke and France* (Lanham, 1988); Bernard Cottret, *Bolingbroke: Exil et écriture au siècle des Lumières* (Paris, 1992); and Dennis J. Fletcher, "The Fortunes of Bolingbroke in France in the Eighteenth Century," *Studies on Voltaire and the Eighteenth Century* 47 (1966): 207–232.

35. On the importance of La Source in the dissemination of ideas in France, see Dennis J. Fletcher, "Bolingbroke and the Diffusion of Newtonianism in France," *Studies on Voltaire and the Eighteenth Century* 53 (1967): 29–46.

nickname "Man of Mercury," a double entendre given the common practice of using mercury to treat syphilis in the eighteenth century. Even when Bolingbroke abandoned the libertine lifestyle in practice, he never gave up his image as a man of pleasure. As he once wrote: "I have no very great stock of philosophy, and am far from being a stoic. Pain to me is pain, and pleasure pleasure."[36] Bolingbroke established the same *esprit* at his gatherings at La Source. For all his distractions, however, Bolingbroke was also a serious intellectual and an important political thinker.[37] After his return to England in 1723, he became very active in the more public and openly contested politics that characterized England in the years after the Hanoverian Succession. In particular, he helped to support a number of journalistic enterprises during the 1720s with explicitly Tory agendas. His move to the Scriblerus Club in this period further reflected his new agendas. Here, according to Dennis Fletcher, "friends affected a rugged common sense and geared discussion to the intellectual level of the plain man, to whom [Bolingbroke] continually refers as the ultimate arbiter in any discussion."[38] This new ethos supported Bolingbroke's move into journalism, and during these years he helped to pioneer a new kind of political writing geared toward the ordinary citizen.[39]

Voltaire was personally involved in these developments before 1723, and when the Frenchman visited England between 1726 and 1729 Bolingbroke was his principal contact in the country. Many also believe that Voltaire first considered modern natural philosophy generally, and the work of Locke and the Newtonians specifically, at Bolingbroke's estate. It was certainly true that these ideas, especially in their more deistic and *libertine* configurations, were at the heart of Bolingbroke's identity as well.[40]

Voltaire's associations with Bolingbroke were also crucial in fostering his important relationships with other leading figures in Parisian society at the same time. While in France, Bolingbroke was a co-founder, along with abbé Pierre-Joseph Alary, of perhaps the first English-style political club in France. The Club de l'Entresol, as it was called, met in Paris from 1722 until its sup-

36. Cited in Dickinson, *Bolingbroke*, 9.

37. Pomeau describes Voltaire's admiration for Bolingbroke this way: "[He] appeared to him to be the model of a cosmopolitan thinker, a man who knew how to reconcile pleasure, political action, and literary taste." *D'Arouet à Voltaire*, 164.

38. Fletcher, "Bolingbroke and Newtonianism," 32.

39. Ibid. See also Dickinson, *Bolingbroke*, 184–186.

40. On the importance of Bolingbroke in exposing Voltaire to English natural science and philosophy, see Fletcher, "Bolingbroke and Newtonianism," 31–33; and Wade, *Intellectual Development*, 129–135.

pression by the government in 1731.[41] At the Entresol, the all-male membership presented papers and discussed politics in a manner quite different from either the Parisian *salons* or the literary gatherings of the Temple. The Entresol was a forerunner of the predominantly male political clubs that would become important in France later in the century, but in the context of the 1720s its creation reveals the extent to which English-style politics and political sociability was increasingly infiltrating France.

There is no evidence that Voltaire himself participated in meetings at the Entresol, but he was on intimate terms with many of its members. Most important was the Marquis d'Argenson, the club's de facto recording secretary, whose record of the meetings at Alary's home in Paris, preserved in his *Mémoires,* provides the best account of the membership and character of the gatherings.[42] Born the same year as Voltaire, d'Argenson descended from a long line of public servants.[43] His father had been the lieutenant of police in Paris, and there were ministers, magistrates, ambassadors, and royal intendants on both sides of his family tracing back generations. D'Argenson once wrote that he had spent his whole life preparing to become controller-general of France, but while he never obtained this office, he did serve as a counselor to the Parlement of Paris, as a counselor of state, and as minister of foreign affairs. Voltaire and d'Argenson were already close friends by the time the Entresol was founded, and thus even if Voltaire was not in fact a member of the club, and even if he never enjoyed d'Argenson's access to political power, he certainly moved very close to the center of his circle.

D'Argenson's sociability beyond the club and the halls of state, however, was more important in cementing his relationship with Voltaire. Both men circulated in Parisian society, frequenting the salons of Mme. Lambert and Mme. Tencin and the courtly gatherings of the Duchesse du Maine at Sceaux. Both d'Argenson and the abbé Saint-Pierre also shared close ties with the Comte de

41. On the Club de l'Entresol, see Nick Childs, *A Political Academy in Paris, 1724–1731: The Entresol and Its Members* (Oxford, 2000); Nannerl Keohane, *Philosophy and the State in France* (Princeton, 1980), 361–363; Isaac Kramnick, *Bolingbroke and His Circle: The Politics of Nostalgia in the Age of Walpole* (Cambridge, 1968), 15–17; Robert Shackleton, *Montesquieu: A Critical Biography* (Oxford, 1961), 63–64; and E. R. Briggs, "The Political Academies of France in the Early Eighteenth Century with Special Reference to the Club de l'Entresol and Its Founder, the Abbé Pierre Joseph Alary," (Ph.D. diss., Cambridge University, 1931).

42. E. J. B. Rathéry, ed., *Journal et mémoires inédits du Marquis d'Argenson,* 9 vols. (Paris, 1859–1867).

43. On the political career and philosophy of d'Argenson, see Keohane, *Philosophy and the State in France,* 376–391.

Boulainvilliers, who was so important in the dissemination of clandestine manuscripts in France in this period. Voltaire no doubt moved in this circle as well, a fact attested by his famous libertine tract *Le dîner du Comte de Boulainvilliers.* D'Argenson's administrative service also connected him with the Phélypeaux clan, particularly Maurepas, and since the latter enjoyed the company of men of letters, Voltaire benefited from these ties as well. D'Argenson would eventually become Voltaire's political mentor, imparting to his more literary friend the political lessons that his years as a royal intendant had taught him. His *Considérations sur le gouvernement,* which circulated widely in manuscript after 1730, echoed Saint-Pierre's notion of rational, scientific administration and the ethos of administrative monarchy that informed it. The book quickly became Voltaire's political bible.[44]

By 1726, therefore, Voltaire was a well-connected and esteemed man of letters enjoying the pleasures of Parisian literary life. He involved himself in the literary quarrels of the day, including the "querelle d'Homère," a reprise of the ancients and moderns battle, which occurred around 1715. He also witnessed from the center the famous "battle of the couplets" that placed Saurin and his circle at the Café Gradot into the public eye in a dramatic way.[45] Voltaire made his share of friends and enemies as a result of these activities, and he spent some time in the Bastille for his indiscretions. He also helped to put others there while helping to save some of his friends from a similar fate. Yet before 1726, there was little in Voltaire's background that anticipated his radical transformation after 1734. The changes in Voltaire's intellectual and cultural life were nevertheless beginning to occur, and the first break involved an episode of disgrace. In 1726, the young writer challenged the honor of a very powerful aristocrat, the Duc de Rohan. The scandal that erupted forced Voltaire into exile. The details of this encounter, so often recounted, are less important than the result.[46] In order to save face and avoid more serious prosecution, he agreed to leave the country indefinitely. Thus, in the spring of 1726, Voltaire left Paris for England.[47]

44. D'Argenson, *Considérations sur le gouvernement ancien et présent de la France* (Amsterdam, 1765). Peter Gay asserts that this edition is virtually identical with the manuscript copy that Voltaire read in the 1730s. On the importance of d'Argenson for Voltaire's political thought, see Gay, *Voltaire's Politics,* pp. 103–108, 126.

45. See Pomeau, *D'Arouet à Voltaire,* 67–70.

46. Pomeau offers a full account, *D'Arouet à Voltaire,* chap. 13.

47. On Voltaire in England, see Archibald Ballantyne, *Voltaire's Visit to England, 1726–1729* (London, 1919); J. Churton Collins, *Voltaire, Montesquieu, and Rousseau in England* (London, 1908); and more recently, André Michel Rousseau, *L'Angleterre et Voltaire (1718–1789)* (Oxford, 1976), esp. chap. 3. Also Pomeau, *D'Arouet à Voltaire,* 212–257.

Voltaire arrived in London around June 1, and he proceeded immediately to the home of Bolingbroke, whose address he left in Paris as his own forwarding address. Once established, he found himself amidst Bolingbroke's circle of literary luminaries and fledgling political pundits at a key moment in the circle's history.[48] Between June and August of 1726, Jonathan Swift's *Gulliver's Travels*, Alexander Pope's *Dunciad*, and John Gay's *Fables* all appeared. Each of these works was a literary satire of the current political scene, and all of these authors were close to Bolingbroke. In November of the same year, Bolingbroke also brought out the first issue of the *Craftsman*, a political journal that served as the public platform for his circle's Tory opposition to the Whig oligarchy in England. With the publication of the *Craftsman*, Bolingbroke helped to create English political journalism in the grand style.[49] For the next three years, Voltaire moved in this circle, absorbing the culture and sharing in the public political contestation that was exploding all around him.

He did not, however, restrict himself to Bolingbroke's circle of Tory publicists. Virtually none of his correspondence from the English period remains, and thus accounting for his activities is notoriously difficult. Nevertheless enough evidence survives to assert the following generalizations. After Bolingbroke, Voltaire's primary contact in England was a merchant by the name of Everard Fawkener. Fawkener appears to have introduced the exiled Frenchman to a side of London life entirely different from that offered by Bolingbroke. Voltaire, for example, developed a deep appreciation for English commerce during his visit, and Fawkener was a central figure in this development. Fawkener also had ties to the Whig circles opposed to Bolingbroke, and it is clear that Voltaire developed ties with this group as well.[50] In particular, he met Samuel Clarke,

---

48. The most useful map of English political culture in this period that I have found is J. G. A. Pocock, "The Varieties of Whiggism from Exclusion to Reform: A History of Ideology and Discourse," in *Virtue, Commerce History: Essays on Political Thought Mostly in the Eighteenth Century* (Cambridge, 1981), 215–310. On Bolingbroke's work in this context, see Alexander Pettit, *Illusory Consensus: Bolingbroke and the Polemical Responses to Walpole, 1730–1737* (Newark, 1997); Kramnick, *Bolingbroke and His Circle*; and Dickinson, *Bolingbroke*, esp. chaps. 10–14.

49. Jürgen Habermas describes Bolingbroke's significance this way: "The press was for the first time established as a genuinely critical organ of a public engaged in critical political debate: as the fourth estate." *The Structural Transformation of the Public Sphere: An Inquiry into a Category of Bourgeois Society*, trans. Thomas Burger (Cambridge, 1989), 60. On Bolingbroke and the *Craftsman*, see Kramnick, *Bolingbroke and His Circle*, 17–24; Dickinson, *Bolingbroke*, 181.

50. On this relationship, see Norma Perry, *Sir Everard Fawkener, Friend and Correspondent of Voltaire* (Oxford, 1975).

and although he did not meet Newton himself before the latter's death in March 1727, he did meet his sister—learning from her the famous story about Newton's apple.[51] Voltaire also came to know the other Newtonians in Clarke's circle, and since he became proficient enough in English to write letters and even fiction in the language, it is very likely that he immersed himself in their writings as well. Voltaire also visited Holland, forming important publishing contacts with Dutch journalists and publishers and meeting 'sGravesande and the other Dutch Newtonians.[52] Given his other activities, it is also likely that Voltaire frequented the coffeehouses of London, and thus it would not be surprising to learn that he attended the Newtonian public lectures of Desaguliers or those of one of his rivals.

The exiled poet returned to France secretly in the early spring of 1729, and when he was granted permission by Maurepas to reenter Paris, he was devoid of pensions and banned from Versailles. Yet he was also a changed man. It is no doubt overly grandiose to say, with Lord Morley, that "Voltaire left France a poet and returned to it a sage."[53] It is also an exaggeration to say, with the traditional Voltaire scholarship, that he was transformed from a poet into a philosophe while in England. For one, these two sides of Voltaire's intellectual identity were forever intertwined, and he never experienced an absolute transformation from one into the other at any point in his life. More important, the emergence of Voltaire's new philosophe persona really began in 1734, not 1729, and while the English trip was a critical step in this development, his new identity was not produced by the English experience alone. In fact, viewed historically, Voltaire's life in Paris in the years immediately after his return from England has more in common with his life before 1726 than after 1734.

Once his legal residence was secured in Paris, Voltaire worked hard to restore his sources of financial and political support. The financial problems were the easiest to solve. In 1729, the French government, directed by Cardinal Fleury, staged a sort of lottery to help amortize some of the royal debt. Voltaire's companion Charles-Marie de La Condamine, whose ties to Maupertuis will be

51. On this relationship, see William H. Barber, "Voltaire and Samuel Clarke," *Studies on Voltaire and the Eighteenth Century* 179 (1979): 47–61; and Margaret C. Jacob, *The Radical Enlightenment: Pantheists, Freemasons and Republicans* (London, 1981), 105–106. On the connections between Newtonians such as Samuel Clarke and the Whig oligarchy opposed by Bolingbroke, see Margaret Jacob, *The Newtonians and the English Revolution, 1689–1720* (Ithaca, 1976).

52. J. Vercruysse, *Voltaire et la Hollande* (Oxford, 1966); Jacob, *Radical Enlightenment*, 102–104.

53. Cited in Wade, *Intellectual Development*, 180.

discussed shortly, perceived an opportunity for investors in the structure of the government's offering. At a dinner attended by Voltaire, he formed a society to purchase shares. Voltaire participated, and in the fall of that year when the returns were posted he and La Condamine had made a fortune. It is not clear exactly how much Voltaire made, but it was certainly a very large sum of money.[54] As he quipped at the time: "In order to make a fortune in this country, one only needs to read the royal decrees [*les arrêts du Conseil*]." Voltaire's inheritance from his father also became available to him at the same time, and from this date forward he never again struggled financially.[55]

This result was no insignificant development since Voltaire's financial independence effectively freed him from one dimension of the patronage system so necessary to other men of letters in the period. Other writers were required to appeal to powerful patrons with their writings in order to secure their material livelihood. Voltaire, by contrast, was largely free from these constraints after 1730. There was, however, more to the patronage structures of Old Regime France than mere economics, and restoring his honor and *crédit* within the elite society upon which the reputation of writers depended was far less simple.[56] Gradually, however, through a combination of artfully written plays, poems, and literary essays and careful self-presentation in Parisian society, he began to regain his public stature. By the fall of 1732, when he wrote his first letter to Maupertuis, Voltaire was residing at Versailles, a sign that by this date his reestablishment in French society was all but complete.

During his rehabilitation, Voltaire also formed a new relationship that was to prove profoundly influential in the subsequent decades. He became reacquainted with the daughter of one of his earliest patrons who had married in 1722 to become Émilie Le Tonnier de Breteuil, la Marquise du Châtelet.[57] Emilie du Châtelet was twenty-nine years old in the spring of 1733 when Voltaire began his relationship with her. She was also a very uniquely accomplished woman by that time. Du Châtelet's father, the Baron de Breteuil, had hosted a

54. Pomeau speculates that it was a million livres or more. *D'Arouet à Voltaire*, 261.
55. Ibid., 259–261.
56. On the complex structures of writing and patronage in early modern France, see Gregory S. Brown, *A Field of Honor: Writers, Court Culture, and Public Theater in French Literary Life from Racine to the Revolution* (New York, 2005).
57. The best account of Emilie du Châtelet's life is René Vaillot, *Madame du Châtelet* (Paris, 1978). See also Elizabeth Badinter, *Emilie, Emilie, l'ambition féminine au XVIIIe siècle* (Paris, 1983); Esther Ehrman, *Mme. du Châtelet: Scientist, Philosopher and Feminist of the Enlightenment* (Leamington Spa, 1986); and Wade, *Voltaire and Madame du Châtelet*.

gathering of men of letters during her youth that had included Voltaire. Emilie, ten years younger than him, benefited from these associations. Her father also ensured that she received an education that was exceptional for girls at the time. She studied Greek and Latin and trained in mathematics, and by 1733, she was a very knowledgeable thinker in her own right. Her own intellectual career had not yet begun when Voltaire met the mature Emilie in Parisian society, yet her mind combined with her vivacious personality drew Voltaire to her.[58]

The couple's vivacious intellectualism was the glue that held their relationship together, but important to Voltaire as well were the political connections that du Châtelet offered. Her husband was an important military officer who was conveniently away from home a great deal, and even when the Marquis was at home, he appears to have accepted his wife's relationship with Voltaire entirely. This meant that Mme. du Châtelet was free to use her many powerful connections on Voltaire's behalf, an asset that would prove tremendously valuable when the scandal triggered by the *Lettres philosophiques* unfolded in the summer of 1734.

Du Châtelet also served to link Voltaire with Maupertuis since the latter also had an intimate relationship with du Châtelet. Maupertuis' relationship with the Marquise was also grounded in the exchange of scientific ideas as well, yet the academician and aristocrat had far less need than Voltaire for the political protection du Châtelet could offer. Maupertuis first met du Châtelet in 1733, and the meeting was facilitated most likely by Voltaire. Voltaire and Maupertuis had already met several years earlier, and while they continued to travel in many of the same circles, the walls of the Royal Academy of Sciences continued to separate them. Maupertuis accordingly reconnected with his colleagues in the Royal Academy after his sojourn in England, and from this base he resumed his project of academic self-fashioning. This led to another trip, this time to Basel to meet and work with Johann Bernoulli.

Maupertuis' time in Basel, which began the fall of 1729, was perhaps even more transformative than his travels in England. When the young academician arrived, Bernoulli was in the twilight of his career, but he was still without question the greatest living practitioner of analytical mathematics in Europe. Maupertuis' prior mathematical training had prepared him to study with Bernoulli, and in May 1729, prompted perhaps by Nicole, or maybe Dortous de Mairan, who was one of Bernoulli's correspondents, he wrote to the master for the first time seeking advice on the algebraic treatment of curves.[59] Unbe-

58. The most complete modern biography of du Châtelet is Judith Zinsser, *La Dame d'Esprit: A Biography of the Marquise du Châtelet* (New York, 2006).

59. Beeson, *Maupertuis*, 68.

knownst to Maupertuis, Bernoulli recognized the name at the bottom of the letter. As Maupertuis' missive was making its way to Basel, Bernoulli received a letter from his Parisian friend Thiancourt describing Maupertuis as "one of the zealous partisans in favor of *vis viva*," a reference to the dispute about the correct measure of force in moving bodies that had been raging for decades in the Republic of Letters and was starting to erupt again within the Paris Academy. Learning also that Maupertuis had just visited England and that he had been made a fellow of the Royal Society during his stay, Bernoulli responded that "this cannot have pleased the English . . . since they are the mortal enemies and despisers of the doctrine of *vis viva*."[60] In any case, Bernoulli welcomed the entreaties of the young mathematician when they arrived, and a correspondence between the two men ensued.

By September 1730 Maupertuis was in Basel, enrolled at the university (an oddity for a French royal academician), and living at Bernoulli's home.[61] The two mathematicians quickly formed a lasting relationship. Immediately after Maupertuis' departure, Bernoulli wrote to Dortous de Mairan that he was "charmed by his conversation" and that the two men had "made new discoveries in mathematics which would never have arisen without Maupertuis' presence."[62] Maupertuis' reactions to the visit do not survive, but it is clear that his experience with Bernoulli was deeply transformative. He remained closely associated with Bernoulli throughout his life, visiting him in Basel several times before the Swiss savant's death in 1748. He also became closely associated with Bernoulli's son, Johann II, and it was at the younger Bernoulli's home that Maupertuis died in 1759.

When Maupertuis returned to Paris in July 1730, the impact of his study with Bernoulli also became manifest in his academic work. As Terrall explains in her thorough account of the partnership, "Maupertuis arrived at just the right time to take on the role of disciple."[63] Soon after his return to Paris, Maupertuis resumed his correspondence, writing that "the desire, or rather a kind of requirement, to read something to the academy from time to time has made me hurry too much."[64] Consequently, he pledged to refrain from presenting anything that had not first been vetted and approved by Bernoulli. By November he was ready to offer his first academic paper in almost two years, a new

60. Cited in Terrall, *Man Who Flattened the Earth,* 44, n. 32.
61. Beeson, *Maupertuis,* 68.
62. Cited in Greenberg, *Problem of the Earth's Shape,* 246.
63. Terrall, *Man Who Flattened the Earth,* 47.
64. Ibid.

treatment of the curves assumed by moving bodies in resisting media.[65] Maupertuis let Bernoulli edit the paper before delivering it, and his mentor praised the work, while contributing his own paper on the subject that Maupertuis read to the academy himself.[66] Both works were published. Since Bernoulli and Varignon had already published work on this topic, Maupertuis' paper positioned him as a direct heir to the analytical mechanics program in France. The next year, he produced three more papers in analytical mechanics—a paper on indeterminates in differential equations, a second on the mathematics of ballistics, and a final paper on the motion of celestial bodies. All of this work was thoroughly vetted by Bernoulli, and all of it was published.[67] Maupertuis did not abandon his other scientific identities entirely, delivering a natural history of the scorpions of Languedoc in May 1731.[68] Yet overall Maupertuis was becoming more focused than ever before on becoming an analytical mathematician and *mécanicien* in the model of Bernoulli.

Fontenelle's account of the work of this budding successor to the French program in analytical mechanics is revealing of the changes that had occurred in France over the previous two decades. Between 1726 and 1728, Fontenelle's outspoken support for analytical mechanics was transformed into his public Cartesian campaigns against Newtonianism. Nowhere are the effects of this shift more evident than in his responses to Maupertuis' science. According to Fontenelle, Maupertuis' analytical work was to be seen as a continuation of the important mathematical achievements of the French. Indeed, Varignon is invoked explicitly as Fontenelle explains that Maupertuis has "added even more universality to an area that has already been treated with Varignon's customary attention to universal solutions."[69] Nowhere, however, does the secretary connect these achievements with Cartesian physical principles as he once did.

65. Maupertuis, "La courbe descensus equabilis dans un milieu résistant comme une puisssance quelconque de la vitesse," PVARS, November 29, 1730 and *HARS-Mem.* (1730): 233–242.

66. Johann Bernoulli, "Méthode par traiver les tantochrones, dans des milieux résistants, comme les quarrés des vitesses," PVARS, December 16, 1730 and *HARS-Mem.* (1730): 78–101. On the exchanges between Maupertuis and Bernoulli, see Beeson, *Maupertuis,* 70–71.

67. Maupertuis, "Sur la séparation des indéterminées dans les équations differrentielles," *HARS-Mem.* (1731): 103–109, idem, "Ballistique arithmetique," *HARS-Mem.* (1731): 297–298, idem, "Problème astronomique," *HARS-Mem.* (1731): 464–465.

68. Maupertuis, "Expériences sur les scorpions," PVARS, May 30, 1731 and *HARS-Mem.* (1731): 197–221.

69. *HARS-Hist.* (1730): 94–95.

His descriptions merely praised the mathematical achievements of the papers while isolating them altogether from any physical assumptions that might be read into them.

In the same *histoire,* the other side of Fontenelle's now dichotomous approach to the mathematical sciences was revealed. Reporting on Du Fay's most recent work on the physics of magnets, Fontenelle stated that "this theory is not merely a pure theory that produces nothing. Using it, M. du Fay shows the best way of magnetizing a compass needle, and one can do it oneself so long as one has a clear idea of the magnetic vortex, its direction, and some iron filings. One works in a superior manner on such questions when one employs a system in this way." [70] Similarly, in the section of Fontenelle's *histoire* devoted to *mécanique,* the secretary presented the empirical work of Pitot on the movement of waters and the comparably inspired work of Couplet on the structures of vaults in buildings.[71] Maupertuis' work, by contrast, was presented as pure *géometrie.* Fontenelle also defended the empirical orientations of these *mécaniciens,* noting in particular that Couplet had first approached his topic in pure geometric terms before "returning to reality." As he summed up: "Geometry can only ally itself with mechanics in supposing something more absolute and more precise than the actual truth." [72]

In adopting this approach, Fontenelle was mirroring the culture of the academy as it had developed in the years leading up to 1730. However, a transformation in the relationship between mathematics and mechanics had occurred as a result of these shifts, and in the case of Maupertuis this outcome positioned the meaning of his work in mathematical mechanics in a very precise way. In the decade after 1699, Varignon's work often appeared under the rubrics *mécanique* and *astronomie,* and Fontenelle adopted a far different attitude toward the physical assumptions implied in this mathematical work. Indeed, he very often conflated Varignon's very similar analytical work with vortical mechanics, placing both at the center of the public academy. By 1727, however, these two programs had become distinct. The result was a very important alienation between the new work in analytical mechanics being developed by Bernoulli, Maupertuis, and Nicole in the 1720s and the authoritative public discourse of the academy controlled by Fontenelle at the same time.

This separation was crucial because it effectively isolated Maupertuis and Nicole from the discourse about vortical mechanics at the very moment when it was becoming more strident and polemical. Maupertuis' work, like Nicole's,

70. Ibid., 1–2.
71. Ibid., 107–115.
72. Ibid., 107.

was for all intents and purposes a continuation of the French program in analytical mechanics begun by Varignon three decades earlier, and given a different intellectual climate it is easy to imagine Fontenelle recognizing and celebrating Maupertuis and Nicole as the heirs to this program. Far more concerned, however, with the emerging Cartesian opposition to Newtonianism than with any analytical mechanics renaissance, Fontenelle let their work remain at the margins of French public science. This neglect created an institutional and intellectual chasm between Maupertuis and Fontenelle despite their shared devotion to the French tradition in analytical mechanics. The emergence of this alienation was one of the foundational ironies responsible for the emergence of the Newton wars in France after 1730.

Yet despite Fontenelle's neglect, Maupertuis' work nevertheless helped trigger a revival of the analytical sciences in the academy after 1730.[73] By the end of 1731, a number of shifts had occurred that placed Maupertuis at the center of a new and growing power center within the academy, one organized around analytical mechanics. Given his perfect combination of intellectual ability, public appeal, and ministerial connections, Maupertuis' ascent to the top of the academy hierarchy happened quickly. It was completed in July 1731 when Saurin resigned his *pensionnaire* position and asked to be made a veteran. The king agreed, and further accepted the academy's nomination of Maupertuis to fill his seat.[74] From this date forward Maupertuis was a major figure within the institution.

Making French Academic Newtonianism in the 1730s

Whether connected to Maupertuis' ascent or not, Nicole's fortunes, and those of analytical mathematicians more generally, also improved in this period. In 1729 and 1730 Nicole served as an officer of the academy, the first analytical mathematician to do so since Varignon had served in 1719. He served again in 1733 and 1734, and Maupertuis succeeded him in 1735 and 1736. Réaumur and

---

73. Greenberg confirms this assessment with the aid of the statistical evidence provided in James E. Maclellan, "The Académie Royale des Sciences, 1699–1793: A Statistical Portrait," *Isis* 72 (1981): 375–391. As Greenberg writes: "By the late 1730s what we now recognize, based on statistics, to be the highest level of interest in mathematics at the Paris Academy ever attained during the ancien regime was reached." John Greenberg, *The Problem of the Earth's Shape from Newton to Clairaut: The Rise of Mathematical Science in Eighteenth-Century Paris and the Fall of "Normal" Science* (Cambridge, 1995), 246.

74. PVARS, July 24, 1731.

Dortous de Mairan had virtually monopolized these offices in the 1720s, and thus the rise of Nicole and Maupertuis to positions of authority in the academy after 1730 marks an important institutional shift.[75] Also indicative of the new climate was the sudden reappearance of the abbé Bragelogne on the pages of the academy *mémoires* after two decades of complete silence. Bragelogne had entered the academy in the first years of the eighteenth century under the wing of Varignon, and before 1715 he showed promise as one of several young and up-and-coming analysts. As the fortunes of the mathematical science he practiced waned inside the academy, however, his notoriety was eclipsed, and when the 1716 restructuring of the academy assigned Bragelogne to the marginal class of *adjoint surnuméraires*, he all but disappeared from the academic record. In 1730, however, whether encouraged by Maupertuis or not, he made a stunning reappearance, publishing two *mémoires* in analytical mathematics. In 1731 he published a third, before again receding into the historical darkness. Whatever the precise reasons for Bragelogne's renaissance, it further confirmed the revival of analytical mathematics during the period of Maupertuis' academic ascension.[76]

Maupertuis also worked to strengthen his wider public position in the same period. In the summer of 1731, he decided to publish one of his mathematical works (in Latin) in the *Philosophical Transactions of the Royal Society*. This move prefigured his decision to publish his *Discours sur les différentes figures des asters* outside the publication channels of the Paris Academy. The paper itself also marks his transition to the subject matter of this treatise. The paper focused on the formation of planets, and as such it was included, in a French translation, as the last two chapters of his work on the figure of celestial bodies.[77] The paper also implicated Maupertuis in an emerging dispute about the precise and accurate shape of the earth. This dispute will be analyzed in detail later, but when Maupertuis set to work on his paper, the figure of the earth was still nothing like the *cause célèbre* that it would soon become. Indeed, it was largely thanks to Maupertuis that this of all issues became the great public measure of the validity of the Newtonian and Cartesian systems in the 1730s.

Yet it is clear that from the outset Maupertuis saw in these questions a key test of Newtonian physics. As he prepared his paper, Maupertuis tried to study Newton's demonstration of the shape of the earth in the *Principia*, and he dis-

75. Sturdy, *Science and Social Status*, 421–423.

76. For Bragelogne's papers, see "Examen des lignes du quatrième ordre out Courbes des troisième génre," *HARS-Mem.* (1730): 158–171, "Examen des lignes du quatrième ordre," 363–390; and "Examen des lignes des quatrième ordre," *HARS-Mem.* (1731): 10–23.

77. Maupertuis, *Discours sur les différentes figures des astres*, 1: 148–170.

cussed these efforts in his ongoing correspondence with Bernoulli. He claimed to be thoroughly baffled by Newton's reasoning.[78] He then tried, with the assistance of Bernoulli, to apply the Continental methods of mathematical analysis to the same problem. His Royal Society paper was the final product of this early work.[79]

Significantly, Maupertuis' analytical methods nevertheless reinforced Newton's physical conclusions. This led him to reconsider Newton's overall system, especially the principle of gravitational attraction at its center. Institutional factors also played a part in this development. For as he wrote to Jean Bernoulli at the time, "[French academicians] think that vortices explain everything without bothering about all the insults that one has to swallow in order to reconcile them with phenomena."[80] In fact, Maupertuis confessed that he sent the paper to England rather than publishing it in France precisely because he felt that the intellectual climate there was more receptive. In the same letter, dated July 1731, Maupertuis also revealed that he had "composed an apology for attraction that would no doubt shock people here [*bien revolté les esprits*]." This was the first mention of Maupertuis' *Discours*.[81] Responding in a subsequent letter to Bernoulli's shock at such a prospect, Maupertuis also added this elaboration: "I have not tried to defend the idea of attraction so much as to show that the reasons with which our French philosophers so haughtily reject it are not well founded. And as far as playing court to my compatriots [a reference to one of Bernoulli's sardonic quips], I do not believe that the love of country [*patrie*] must rule the world in matters of philosophical opinion."[82]

By the end of 1731, therefore, Maupertuis had both established himself in the Paris Academy and the wider Republic of Letters and begun the institutional repositioning that would eventually make him a provocative presence both inside and outside the institution. He had also found the topic and the specific issue—the Newtonian theory of attraction as it pertained to the shape of the earth—that would propel his public breakthrough. At precisely this moment, however, another topic reared its head—*vis viva*—and Maupertuis' response

78. Newton's argument is found in *Principia*, 424–428. Maupertuis' frustrations were expressed to Johann Bernoulli in a letter from March 1731. Greenberg, *Problem of the Earth's Shape*, 11.

79. The paper itself was Pedro Ludovico de Maupertuis, "De figuris quas fluida rotata induere possunt, problemat duo," and it appeared in *Philosophical Transactions* 37 (1731–1732): 240–256. For a detailed discussion of this work, see Beeson, *Maupertuis*, 76–82.

80. Cited in Greenberg, *Problem of the Earth's Shape*, 222.

81. Cited in Beeson, *Maupertuis*, 82–83.

82. Ibid., 83.

to it reveals much about his strategies of self-fashioning at this critical moment. Bernoulli provoked Maupertuis to return to *vis viva* when he read the new defenses of the Cartesian position in this debate published by Dortous de Mairan in 1728 and Louville in 1729. These papers only reached Bernoulli, who held the opposing view, in 1731, and when he read them he was outraged. Writing of Dortous de Mairan, he said that "it is impossible to misunderstand the true meaning we give to the term 'living force' [i.e., *vis viva*], nevertheless he takes it in the opposite meaning all the time, as if on purpose to smother the truth out of fear that it might triumph over popular error."[83] Louville's paper made him even more angry. "These gentlemen sing out their victory at full force," he wrote, "seeing that no one dares to respond to them in defense of *vis viva*."[84] Maupertuis was in a position to offer the opposition Bernoulli craved, and thus in October 1731 he began pushing Maupertuis to publicly take on the Parisian opponents of *vis viva* inside the Royal Academy.

Bernoulli's pressure put Maupertuis into an awkward position. He was indeed sympathetic to *vis viva* and to Bernoulli's position regarding the simplistic and stubborn reasoning of Dortous de Mairan and Louville. On the other hand, he saw no hope of winning this battle. As he wrote to Bernoulli: "to convert Mairan is not possible; I have tried to engage him several times on this matter and he did not want to hear of it."[85] He also saw no clear way of winning the argument, especially since sound empirical and mathematical reasoning supported each position. He further worried that any defense of *vis viva* was doomed to be misunderstood given the limited mathematical abilities of their opponents. He thus implored Bernoulli to just pen a few lines of critique himself, but his elder colleague was inconsolable. "A rebuttal, written by your pen, will have a thousand times more grace," he wrote. He further provoked Maupertuis by lamenting the fact that no one "dared to respond to them."[86] This challenge to Maupertuis' courage had the desired effect, for it provoked the academician to reluctantly pen a critique of Louville's position, which he sent to Bernoulli as a demonstration of his valor. Early 1732 saw the exchange of several letters between the two men designed to hone the language of their intended intervention. During this period Saurin was also enlisted in the cause. He expressed his support for Maupteruis' *mémoire;* however, just as the response was reaching completion, illness intervened making a direct assault on Louville inappropriate. In October, the academician died, prompt-

83. Cited in Terrall, *Man Who Flattened the Earth*, 52.
84. Ibid.
85. Ibid.
86. Ibid., 62.

ing Maupertuis to write that the Chevalier had "gone to learn dynamics in heaven."[87] Their proposed refutation was shelved, and both men soon became preoccupied with other matters.

Louville's death thus ended this skirmish in the ongoing *vis viva* controversy, but the episode reveals several things about Maupertuis at this juncture. One was his complex relationship to polemical struggles, and especially his strong desire to avoid such battles unless he could do it on his own terms. As Terrall states, "Maupertuis had no personal stake in this question," and consequently he did not see the value in provoking an academic battle on this point.[88] Later, as he came to better understand the scientific reasons behind Bernoulli's vehement defense of *vis viva,* his views would change, and this too reveals an important point about Maupertuis' thinking. In 1731, he was still a mathematician much more than a student of physics and mechanics, and this same orientation would also shape his early conception of Newtonianism. Finally, as his strategizing with Bernoulli throughout 1732 reveals, Maupertuis was acutely focused on how to ensure the proper public perception of his ideas. The style of any intellectual intervention was as important to him as its substance, and this reveals how Maupertuis' self-conception as an aspiring savant was as important to him as any particular scientific position. The same themes would be present in his initial Newtonian interventions as well.

The year 1731 further brought the death of Houdar de la Motte, and Maupertuis' succession as the leader of the Café Gradot circle.[89] La Beaumelle, the source for this story, further contends that Maupertuis began using sites such as these as institutional counter-weights to the Royal Academy.[90] After 1732, the gatherings at these extra-academic sites increasingly became Newtonian party meetings as Maupertuis' outspoken defense of Newtonianism in the *Discours* was translated into a new, public campaign on its behalf. However, even before 1732, the existence of these extra-academic, public networks centered around Maupertuis was significant. For one, they encouraged the academician's important decision to publish his *Discours sur les différentes figures des astres* outside of the usual academic publishing structures. They also permitted Maupertuis to establish his precise position as a public savant in France, one linked and oriented simultaneously to official circles and those outside of them.

After 1732, Maupertuis increasingly used his newly created authority both inside the academy and in the wider public sphere to his own advantage. The

---

87. Ibid., 63–64.
88. Ibid., 52.
89. La Beaumelle, *La vie de Maupertuis,* 12.
90. Ibid., 33–34.

nonacademic publication of his *Discours,* the first assertion of his new identity, helped him to solidify his position by producing a new Newtonian constituency in France that rallied around Maupertuis as its pole star. Equally important, however, was the emergence inside the academy of a group of allies sympathetic to Maupertuis' positions and willing to defend them professionally and publicly as well. Two of Maupertuis' most important allies— Alexis-Claude Clairaut and Charles-Marie de La Condamine—found their way into the academy as the result of a remarkable set of circumstances. At the center of their peculiar debut was a now almost forgotten Parisian organization known as the "Société des Arts." [91] Like Maupertuis' own ascent to intellectual prominence, this obscure society illustrated well the shifting relationship between the academy and the public, and between science, society and state in France in the 1720s.

The story begins sometime around 1730 when one of the Princes of the Blood, Louis de Bourbon-Condé, Comte de Clermont, began supporting an independent society in Paris devoted to the cultivation of the arts. Most reports confirm that this "Société des Arts" contained one hundred members and that it met twice weekly at the count's Parisian *hôtel.* Very little documentary evidence about the society itself or its work survives, however. [92] Whatever its precise nature, it is clear that by 1730 a large organization devoted to the arts was

91. Given its obscurity, the Société des arts has not been the subject of much scholarly work. The best introduction is Roger Hahn, "The Application of Science to Society: The Societies of Arts," *Studies in Voltaire and the Eighteenth Century* 25 (1963): 829–836. See also idem, "Science and the Arts in France: The Limitation of an Encyclopedic Ideology," *Studies in Eighteenth Century* 10 (1981): 7–93. Also useful is the discussion in Marianne Roland Michel, *Lajoüe et l'art rocaille* (Neuilly-sur-seine, 1984), 30–33. My own understanding of the Société des Arts was most helped by a seminar on this topic held at the Archives of the Academy of Sciences in Paris in the spring of 1996. I am grateful to Irène Passeron, who presented her work on the society at this seminar, and to the other participants, especially Eric Brian, Alice Stroup, Jeanne Pfeiffer, Françoise Bléchet, and head archivist Christiane Demeulenaere-Douyere, who invited me.

92. The most important source is the official regulations of the society which were published as *Réglement de la Société des Arts* (Paris, 1730). Manuscript copies of these regulations are also found at Bibliothèque de l'Arsenal, MSS 12,727, ff. 981–91; and in the papers of the abbé Bignon, BN, MSS f. fr., 22225. Hereafter I will refer to these regulations as "RSA." Bignon's papers also contain a list of the members of the society. Another list of members dated much later is found in the Stockholm Nationalmuseum, C.C. 354. The Bibliothèque de l'Arsenal also contains several papers that were allegedly presented to the society, "Observations diverses, Mémoires et Pièces Detachés conçernant les Arts et Sciences," MSS 6130.

meeting regularly at the home of the Comte de Clermont in Paris. Both Clairaut and La Condamine were members of this organization, and the precise orientation of the assembly was significant. In 1730, the Society published a pamphlet containing its regulations and goals, and the dominant theme running throughout was utility. The official title of the organization, for example, was the "Société des Arts et des Sciences," and the precise ordering of the terms said everything. As the regulations indicated, "the unique object of the society will be the perfection of the arts." [93] The more theoretical sciences, by contrast, were to be included only insofar as they were useful in the achievement of this goal. Article 44 of the regulations even went so far as to declare that "since the arts are the unique focus of this society, *mémoires* and other works which treat pure science alone will be rejected." [94] Overall, then, the society announced itself as an institution devoted to applied knowledge with an overt bias against more theoretical pursuits.

The composition of the organization fulfilled this mandate. The regulations stipulated that the hundred members would be chosen according to a precisely defined set of disciplinary classifications. Roughly a third of the members were to be drawn from disciplines comparable to those at the academy of sciences— geometry, *mécanique,* astronomy, *physique,* botany, etc. The rest, however, were taken from disciplines excluded from the Royal Academy such as engineering, architecture, ship building, navigation, and the technical artisanal disciplines (the famous clockmakers of the Sully and Le Roy families were members). The regulations also explained how the more theoretically inclined should be put to work in the service of applied, utilitarian work. Thus, astronomers were explicitly assigned to work with lens makers, and geometers were assigned to work with clockmakers to give but two examples. All the disciplines were explicitly linked in this way. [95]

Voltaire called the society "une académie," and it is clear that in conception and membership the Société des Arts offered itself as a rival institution to the *compagnie royale des sciences.* Yet while the simple establishment of the society itself challenged the academy, the precise intellectual orientation of the organization was even more provocative. From as early as the Colbert ministry, the state had been attempting to create institutions devoted to the applied sciences. These efforts had been increasingly concentrated within the Royal Academy in the period 1690–1730, yet what the appearance of the Société des Arts revealed was the inadequacy of the academy in this respect. Whole bodies of knowl-

93. RSA, art. II.
94. Ibid., art. XLIV.
95. Ibid., passim.

edge crucial to the advancement of French society—agriculture, economics, animal husbandry, the manufacturing sciences, precision instrumentation—were included in the Sociéte des Arts while they remained excluded from the century-old structure of the Royal Academy. Thus, when these disciplines were institutionalized publicly in the rival Société des Arts, a bright light was cast on the narrowness of the academy's definition of useful scientific knowledge.

Of course, the academy had another mandate as well: the pursuit of pure science for its own sake and the greater good of truth. The institution could certainly fall back on this orientation when necessary and find many sympathetic to it. However, since the utilitarian dimension of the academy was fundamental as well, especially given the increased pressures in this direction applied by the state during the Pontchartrain ministries, the Société des Arts nevertheless constituted a provocation. The inadequacies of the academy's own program in the mechanical arts, begun in the 1660s under Colbert, revived by the Pontchartrains in the 1690s, and then renewed again under Bignon and Réaumur after 1715, were particularly exposed. The project to produce authoritative descriptions of the arts and industry in France, first announced in 1675, had still not produced a single volume despite two decades of renewed efforts under Réaumur. What the creation of a new and independent Société des Arts suggested to many, then, was that perhaps the academy was not the right institution to meet these crucial needs.[96]

All of this was made even more intriguing by a set of events that occurred in 1731. During that year, a number of the members of the Société des Arts, including Clairaut and La Condamine, were admitted to the Academy of Sciences within months of each other and under very unique circumstances. Furthermore, according to the academy's first modern historian, this transfer of membership was no accident. Joseph Bertrand reports in his 1869 history, without revealing his source, that Réaumur worried about the success and influence of this new company. He therefore proposed to the Prince de Clermont that the academy would choose, as much as was possible, its new members from among the theoreticians of this society. They would do so, however, only under the condition that the academy would possess these members exclusively. This arrangement was not found acceptable and was rejected. According to Bertrand, Réaumur then decided that he would simply eliminate [*faire tomber*] the society. His method for doing so was very simple. The academy successively admitted La Condamine, Clairaut, [Grandjean de] Fouchy, Nollet, and de Gua to the academy but obliged each to make an exclusive commitment to their new institution. The intended result occurred quickly. Deprived of its most active members, the Société des Arts soon weakened and then completely dissolved without producing any work that would perpetuate its memory.[97] Despite the absence of specific documentation, there are good reasons to believe that something like what Bertrand describes did happen in 1731. Several facts can be offered in support of a modified version of the story. First, the academy by 1730 had a long history of absorbing rivals to its authority in France. Perhaps the earliest academician to enter the academy in this way was the popular public physics demonstrator Pierre-Sylvain Régis. Born in 1632, Régis was already an important public savant when the Academy was established in 1666. However, because of the stridency of his then controversial Cartesian views, along with the perceived amateurism of his public scientific identity, he was excluded from the academy until 1699. By this date Régis was a sick old man, and he would die six years later without producing any academic work. Nevertheless, by admitting him into the institution in the year of the academy's own turn toward the public, the institution indicated its new willingness to embrace the kind of public science that he pioneered.[98] The popular salon mathematician Jacques Ozanam was another example, joining the academy in 1707 after he had first established himself as public figure through his works of "mathematical recreations."[99] Thus, if the academy did attempt to absorb its rivals in 1731, it would not have been an unprecedented move.

Each of these cases illustrates the same pattern, and there is no reason to think that these developments did not occur in 1731 as well. The fact that Réaumur was at the center of the story is particularly telling, for he had the most to lose by the success of the Société des Arts. A new energy in the academy's project to publish volumes describing the mechanical arts is also noticeable after 1731, and this activity can plausibly be connected to the challenge offered by

97. Joseph Bertrand, *L'Académie des sciences et les académiciens de 1666 à 1793* (Paris, 1869), 96–97.

98. On Régis, see Paul Mouy, *Le développement de la physique Cartésienne, 1646–1712* (Paris, 1934), 145–167; "Éloge de M. Régis," *Oeuvres de Fontenelle*, 6: 143–152. Said Fontenelle of his entrance to the academy in 1699: "Because of his illness he was unable to perform any academic function. But his name added luster to a list that the public was surprised had not included it before" (6: 150).

99. Jacques Ozanam, *Récreations mathématiques et physiques*, 2 vols. (Paris, 1694). For a discussion of Ozanam's career, see Barbara Maria Stafford, *Artful Science: Enlightenment Entertainment and the Eclipse of Visual Education* (Cambridge, 1994), 45–47, 56; "Éloge de M. Ozanam," *Oeuvres de Fontenelle*, 6: 419–426.

the rival society. With respect to Maupertuis' Newtonian initiatives after 1732, however, the precise details of this history are less important than the truths that no one doubts. A scientific society containing many of the most talented nonacademic savants from a variety of intellectual disciplines did form between 1728 and 1730. This fact illustrates how the Royal Academy was increasingly being pushed and pulled by the wider expansion of public science. Furthermore, it is a documented fact that between December 1730 and July 1731, for whatever reason, three members of the Société des Arts were admitted to the academy.[100] Two of these new members quickly established themselves as close allies of Maupertuis, and for this reason alone the demise of this society was significant.

The first to arrive was La Condamine, who was made an *adjoint* chemist in December 1730. La Condamine shared with Maupertuis a wider interest in Parisian society, and he soon became one of Maupertuis' most loyal public supporters. During Maupertuis' many public battles, especially those relating to the figure of the earth in the 1730s, La Condamine served as a public advocate on his behalf. He did the same when Voltaire publicly satirized Maupertuis in the 1750s. Likewise, when Maupertuis died in 1759, La Condamine inherited the deceased academician's papers, a testimony to his trust in him. Using these documents, La Condamine worked hard to shape the legacy of his deceased friend and colleague. He published posthumous works directed against Voltaire, and he arranged for the publication of Maupertuis' letters and collected works. He also supported La Beaumelle, helping to protect him from the French authorities when necessary, and supporting him when his failure in this regard forced the writer into the Bastille.[101] La Beaumelle's sympathetic biography of Maupertuis was very much a product of La Condamine's efforts, and it reflects the intimate knowledge of the man that only he possessed. In all these ways, then, Maupertuis found perhaps his most intimate ally when La Condamine joined the Academy of Sciences in 1731.[102]

100. The record of these admissions is found for La Condamine at PVARS, December 9, 1730, for Grandjean de Fouchy at PVARS, April 28, 1731, and for Clairaut at PVARS, July 14, 1731).

101. See La Beaumelle's dossiers in the Archives de la Bastille, Bibliothèque de l'Arsenal, 11,830, 12,493, 12,495, 12,498, and 12,499.

102. This relationship is chronicled in detail in Brunet, *Maupertuis*. A very rich record of La Condamine's correspondence with the Bernoulli family is in BN, MSS N.A. f. fr. 21015. This is an important source for tracing his efforts to support Maupertuis after the latter's death. A substantial collection of the papers that Maupertuis left to La Condamine

Although appointed as a chemist, La Condamine was also a competent mathematician. In his first academic paper, delivered in April 1731, he offered a new way of treating conic sections that employed the differential calculus. He delivered the paper to the academy on the same day that Maupertuis, just back from Basel, read his new work on the indeterminates in differential equations.[103] Both papers were published, along with other mathematical papers by Nicole and Clairaut and the work by Bragelogne noted above.[104] Fontenelle, however, offered no accounting of this new work in his *histoire*. Instead, he devoted his attentions to Cassini II's new theory regarding "the real motion of comets," stating that the work was directed toward confirming "the Cartesian theory of the vortices, which is so worthy of being preserved."[105] Yet while La Condamine's debut appeared to place him squarely alongside Maupertuis and Nicole in the camp of the new analysts, he was neither by training nor interest a serious practitioner of analytical mathematics. He soon became an ardent defender of Newtonian attractionist physics, however, and throughout the 1730s he defended Maupertuis' Newtonian positions here. Yet as a mathematician, La Condamine was a minor figure at best, and more important, therefore, in supporting Maupertuis' mathematical approach to mechanics and physics was another figure who moved from the Société des arts to the academy in 1731: Alexis-Claude Clairaut.[106]

---

have recently been acquired by the Archives of the Academy of Sciences. For an inventory and other papers pertaining to these two academicians, see Ac. Sci., dossier Maupertuis and dossier La Condamine.

103. La Condamine, "Sur une nouvelle manière de considerer les sections coniques," PVARS, April 28, 1731, and *HARS-Mem.* (1731): 240-249.

104. Nicole, "Sur les sections coniques," *HARS-Mem.* (1731): 130-143; Clairaut, "Nouvelle manière de trouver les formules des centres de gravités," *HARS-Mem.* (1731): 159-162, idem, "Sur les courbes que l'on forme courbe quelquonque par un plan donné de position," *HARS-Mem.* (1731):483-494. On Bragelogne, see note 75 above.

105. *HARS-Hist.* (1731): 54-55.

106. Given the deeply technical nature of all of Clairaut's work, very little general scholarship about him exists. The only real biography of Clairaut that I am aware of is Pierre Brunet, *La vie et l'oeuvre de Clairaut (1713-1765)* (Paris, 1952). But this is a short and rather cursory work. Clairaut's rivalry with d'Alembert after 1740 is featured in Thomas Hankins, *Jean d'Alembert: Science and the Enlightenment* (Oxford, 1970), esp. 30-42. My understanding of Clairaut has also benefited from reading Irène Passeron, "Clairaut et la figure de la terre au XVIIIe siècle: Cristallisation d'un nouveau style autour d'une pratique physico-mathématique," (doctoral thesis, Université de Paris VII–Denis Diderot, 1994). I am grateful to Dr. Passeron for permitting me to acquire this work. More technical in his

Clairaut's admission to the Academy occurred in July 1731, two weeks before Maupertuis assumed Saurin's *pensionnaire* position.[107] His was the most exceptional of the appointments made in 1731. Clairaut was only eighteen years old at the time, and since the academy regulations stipulated a minimum age of twenty, an exemption had to be obtained in his case. The exceptional nature of Clairuat's appointment may suggest support for Bertrand's back-room machinations; however, there was little reason to doubt that Clairaut was intellectually ready to become an academician in 1731 despite his age. Like Maupertuis, he had been introduced to the new analytical mathematics by Guisnée's *L'application de l'algebre à la géometrie*.[108] Yet while the Breton aristocrat was taught this mathematics directly by the author at the age of twenty, Clairaut learned it from the text alone beginning at the age of nine. Clairaut's mathematically inclined father, a Société des Arts member as well, no doubt helped him in this work. But even before he was a teenager Clairaut was so adept at the new analysis that he was producing original mathematical work. In 1726, Clairaut even read an original paper before the Royal Academy of Sciences even though he was only eleven years old.[109] This debut precedes even the earliest estimates for the foundation of the Société des Arts, and thus even if it is clear that Clairaut was a member of this organization, it is not necessarily the case that these connections were the most decisive in his admission to the academy.

Once admitted, Clairaut and Maupertuis became close colleagues and collaborators. Clairaut produced two papers in 1731 that applied analytical methods to the mathematics of curves. This work was published alongside the mathematical work of Nicole, Maupertuis, Bragelogne, and La Condamine.[110] It is also clear that Maupertuis, Clairaut, and Nicole became closely associated with one another almost immediately. In early 1732, for example, all three mathematicians presented academic papers offering solutions to the same problem posed to Maupertuis by Johann Bernoulli. All three papers were then

---

treatment of Clairaut is Craig B. Waff, "Universal Gravitation and the Motion of the Moon's Apogee: The Establishment and Reception of Newton's Inverse-Square Law, 1687-1749" (Ph.D. diss., Johns Hopkins University, 1976); and Greenberg, *Problem of the Earth's Shape*, esp. 132–224.

107. PVARS, July 14, 1731.

108. N. Guisnée, *Application de l'algèbre à la géometrie, ou méthode de determiner, par l'algèbre, les théoremes de géometrie* (Paris, 1705)

109. PVARS, April 13, 1726.

110. See note 103 above. Clairaut likewise shared in the inattention that each received in Fontenelle's *histoire* (see note 105 above). Clairaut's appearance in 1726, by contrast, received a full accounting. See *HARS-Hist.* (1726): 45–46.

published in the academy's *mémoires*.[111] As the 1730s progressed, Maupertuis and Clairaut also became close friends as well. In 1734–1735 they traveled together to Basel to visit with Bernoulli and his son. In the next year, they also journeyed together to Lapland as part of the royal expedition to determine the shape of the earth.[112] Maupertuis further introduced Clairaut to the Marquise du Châtelet, bringing the young academician into the network that tied each together with Voltaire. Clairaut also became du Châtelet's mathematical colleague and editor, and when she died in 1748, leaving her French translation of Newton's *Principia* unpublished, it was Clairaut who performed the final revisions that brought the work into print.[113]

In the shorter term, the arrival of Clairaut in the academy in 1731 added one more young, talented analyst to the new community of analytical mathematicians increasingly centered around Maupertuis. And what an addition it was! More single-minded in his devotion to analytical mathematics than anyone else in France, and truly one of the most gifted mathematicians in Europe in a century famous for its gifted mathematicians, Clairaut in many respects made Maupertuis' revival of analytical mechanics in the academy possible. He certainly made Nicole and Maupertuis better mathematicians, and his genius helped to elevate the fortunes of the entire circle. Furthermore, the single-mindedness of his intellectual work insured that French Newtonianism, once established, would always be strongly attached to the mathematical program in analytical mechanics in a way unique in Europe. Maupertuis also helped to give French Newtonianism this distinctive hue, but since his more diverse approach to science offered a number of different trajectories for French Newtonianism to take, it was Clairaut who really cemented the deeply mathematical character of French Newtonian science. For this reason his admission to the Royal Academy in 1731 was a crucial turning point.

But it was also an important moment in the ascension of Maupertuis. Through the odd history of the Société des Arts, Maupertuis' circle in the academy doubled in number in 1731. Given the talents of his new allies, moreover, the power and prestige of his circle increased even more than these numbers indicate. Maupertuis' position was also strengthened by another appointment made to the academy in 1731, one with a very different history. The academi-

---

111. Clairaut, "Solution d'un problème de géométrie," *HARS-Mem.* (1732): 435–436; Nicole, "Solution du même problème," *HARS-Mem.* (1732): 437–441; Maupertuis, "Solution de deux problèmes de géométrie," *HARS-Mem.* (1732): 442–445.

112. Brunet, *Clairaut,* 10–12.

113. Ibid., 14–16. The manuscript copy of du Châtelet's translation, with Clairaut's editorial notes, is in the BN, MSS f. fr.12266–68.

cian appointed was Pierre Bouguer, who was appointed an *associé mécanicien* in September 1731. Bouguer was the only person after Dortous de Mairan to have entered the academy above its lowest rank of *adjoint*.[114] He was appointed to the seat vacated after Maupertuis' promotion to *pensionnaire*, and Bouguer's admission was further exceptional in that he was not the first choice of the academy.[115] Usually the Crown honored the academicians by approving the winner of the academy election, but in the case of Bouguer, the king exercised his privilege to appoint the nominee that he believed most worthy. Dortous de Mairan admitted that his own appointment had resulted from favor, but in the case of Bouguer he insisted that talent alone had justified the exemption.[116] There is good reason to believe that Dortous de Mairan had it right.

By 1731, Bouguer was already an established savant with a number of important works to his credit. Thus, in certain objective respects, he was more qualified than Camus, the winner of the academy's election, to obtain the seat, even if Camus was the more senior academician. Born, like Maupertuis, in Brittany in 1698, Bouguer had become a royal hydrographer at Brest in the early 1720s. From this post, he began contributing actively to the scientific work of the kingdom. He was involved in a number of state supported projects having to do with navigation, ship building and other maritime sciences, and in 1727 he won the academy's prize on the masting of ships.[117] He also had wider interests and talents as well, and in 1728, while on a visit to Paris, he participated in an academic debate even though he was not yet an academician.[118] In 1729, he also published *Essai d'optique sur la gradation de la lumière,* a theoretical work on the nature and behavior of light. This work was reviewed in the *Journal des savants* in 1730, and it reveals the extent to which Bouguer was already an established scientific practitioner before his admission to the Royal Academy.[119]

The obstacle keeping Bouguer out of the institution was the Parisian residency requirement. Yet by 1731 this hurdle had been overcome. Bouguer never gave up his affiliation with the institutions at Brest, but to judge by his activities

114. PVARS, September 1, 1731.

115. This Camus was in no way related to the Camus expelled from the academy on the eve of Maupertuis' admission.

116. He expressed this belief in a letter to Bouillet in March, 1734. Cited in Abby Rose Kleinbaum, "Jean Jacques Dortous de Mairan (1678–1771): A Study of an Enlightenment Scientist" (Ph.D. diss., Columbia University, 1970), 11.

117. PVARS, April 27, 1727.

118. PVARS, January 31 and February 4, 1728.

119. Pierre Bouguer, *Essai d'optique sur la gradation de la lumière* (Paris, 1729); reivew of Bouguer, *Essai d'optique,* in *Journal des savants* (1730): 67–73.

in the academy he started spending a large amount of his time in Paris. From the perspective of Maupertuis' career, this change was significant for a number of reasons. Most important, Bouguer was a talented practitioner of the new analytical mathematics, and his entrance into the academy thus increased the size and prestige of the analyst community in the institution. In his first official academic paper, Bouguer presented a new analytical treatment of the movement of bodies in resisting media that used the same mathematical approach that Varignon had pioneered after 1699.[120] This paper was published in the academy's *mémoires* for that year accompanied by a lengthy account in Fontenelle's *histoire*.[121] A few weeks, later both Bouguer and Maupertuis spoke on the same day, delivering mathematical papers on the curves traced by moving bodies.[122] In this case, only Maupertuis' paper was published, but the affinities between their scientific orientations were nevertheless undeniable.

Bouguer's academic work throughout the next three decades would continue along this same trajectory, and for this reason, he deserves to be considered alongside Nicole, Maupertuis, and Clairaut as one of the key figures in the revival of analytical mechanics that occurred in France after 1730. In other respects, however, Bouguer's scientific orientation made him less clearly attached to Maupertuis than Clairaut, Nicole, and La Condamine. The sources of these differences are revealed in Bouguer's work on the movement of bodies in resisting media. Varignon's work on this topic was very much an expansion upon Newton's similar, if nonanalytical work, in book 2 of the *Principia*. Working, however, at a time when the imperative to defend or refute Newtonian physics did not exist, Varignon developed his science without any concern for the critique of Cartesian vortical astronomy implicit in Newton's argument. He simply pursued the mathematical approach to these questions pioneered by Newton and ignored the rest. His ultimate goal was to produce a more universal mathematical account of these movements, one applicable to any physical account one might give of them. Fontenelle, not yet driven by the Cartesian/Newtonian opposition of the 1730s, framed Varignon's work the same way. He celebrated the superiority of Varignon's analytical approach while in no way making his work either a refutation of Newtonian attraction or a defense of the Cartesian vortices.[123]

By 1730, however, the intellectual climate had changed. When Bouguer approached the topic of bodies moving in resisting media, the clear context for this

120. Bouguer, "Sur le mouvement curviligne des corps dans les milieux qui se menvent," PVARS, December 30, 1731 and *HARS-Mem.* (1731): 390–416.

121. *HARS-Hist.* (1731): 76–81.

122. PVARS, January 30, 1732.

123. See, for example, *HARS-Hist.* (1700): 97–100.

work was now vortical celestial mechanics. As he stated himself in the opening of his *mémoire:* "No one has yet considered geometrically the movement of
bodies in resisting media that are themselves moving, and for this reason this
topic is of interest. But also, if we look at the skies, we are forced to recognize
that [the planets] move in a fluid that circulates with extreme rapidity. This
geometry is applicable to these motions"[124] Not surprisingly, Fontenelle seized
on this angle in his presentation of Bouguer's work. He used his 1731 *histoire*
to build an anti-Newtonian defense of Cartesian vortical astronomy out of this
mathematical work. "Ever since astronomy ceased being simple astronomy and
became astronomical physics [*astronomie physique*]," he wrote, "the greatest
mathematicians have devoted themselves to studying the mechanics of celestial
motions and to finding the algebraic formulas which represent them. Out of
this work was born the sublime and elegant theory of central forces. For since
the planets move in orbits that are related [*se rapport*] to central points, one
must consider them as bodies pulled toward this point by some force. This
force is called [a planet's] *pesanteur.*"[125]

This genealogy makes Bouguer's analytical approach to central force mechanics a response to the physical problem of explaining planetary motion.
Bouguer suggested that this was one motive for his work, but he also contended
that there were pure mathematical reasons for exploring these questions as well.
Given the increasing intensity of the rhetorical environment in 1731, however,
the academy secretary made sure to present things in their most provocative
light. Bouguer's work suggested that fluid vortices could produce planetary
motions that agreed with both Kepler's laws and Newton's inverse-square relation. This, for Fontenelle, was further proof that the Cartesian system could
still be sustained while the monster of gravitational attraction across empty
space could be shunned. As he stated the point: "If they move in a void, as
M. Newton believes, then they obey Kepler's laws as we know they do. . . . But
there is nothing less certain than the idea that they move in a void. This idea
provokes mathematicians as well, and it leads them to consider whether the
same laws hold in the plenum when bodies are subject to resisting media."[126]
This presentation makes Bouguer's work an explicit defense of Cartesianism
even if the paper itself adopted a more complex stance. Fontenelle also acknowledged this complexity warning his readers "not to pretend to pull the
true system of nature from M. Bouguer's theory." "Mathematics is always more
rich than physics," he explained, so let us remember that "it is only a method

124. Bouguer, "Sur le mouvement curviligne des corps," *HARS-Mem.* (1731): 390.
125. *HARS-Hist.* (1731): 76.
126. Ibid., 76–77.

for proving quickly and surely what will follow given a set of observed facts or a set of reasonable suppositions that one might imagine. One has here only a touchstone on the path toward the treasure." [127]

Although not intended as such, this final note about the disparity between geometry and physics was actually deeply significant. It pointed to the fact that Bouguer's analytical mechanics, the same mechanics pioneered by Varignon and then revived by Maupertuis, Clairaut, Bouguer and others after 1730, was not in and of itself either Newtonian or Cartesian. It was, rather, a purely mathematical science applicable to a number of different physical systems interchangeably. Thus, Fontenelle was right to locate the ultimate battlefield for Newtonians and Cartesians in the physical debates about the nature of causality, and not in the mathematics used to model these behaviors. At the level of the mathematical practice itself, neither Maupertuis, nor Clairaut, Bouguer, Nicole, nor anyone else devoted to analytical mechanics was acting in an especially Newtonian or Cartesian manner in their work. Indeed, the analytical mechanics that they practiced was an amalgam of Cartesian, Newtonian, Leibnizian, and Malebranchian ideas, and its relationship to physics was never defined absolutely. Meanwhile, the Cartesian versus Newtonian opposition had arisen over debates about the physics of universal gravitation. Analytical mechanics, therefore, as a purely mathematical science, was neither Newtonian nor Cartesian as such according to the terms that defined this debate. It could rather be claimed under either label depending upon which physical picture one adopted to conceive of the pictureless mathematical categories (such as force) that it deployed.

The result was a twofold outcome for Bouguer with respect to Maupertuis. As a practitioner of analytical mechanics, Bouguer became a key member of Maupertuis' circle, one who played an important role in the renaissance of analytical mechanics inside the academy after 1731. However, since Bouger was less comfortable than Maupertuis with attractionist physics, he could find himself positioned as an opponent to Maupertuis and Newtonianism, as he was by Fontenelle in the presentation of his work in 1731. Bouguer was not the only savant to suffer an identity crisis in this respect as the Newtonian/Cartesian dichotomy took hold. Johann Bernoulli also returned to the center of the scientific community in France after 1730 thanks in large part to Maupertuis and his young coterie of link minded analysts. [128] In this position, he continued to

127. Ibid., 81.
128. Says Greenberg: "In the late 1720s, Bernoulli's fortune in Paris began to undergo a change for the better, . . . largely as a result of the interest, zeal, and personal ambitions of Maupertuis." Greenberg, *Problem of the Earth's Shape*, 246.

do important mathematical work that advanced the French program in central force mechanics in important ways. When called upon to speak physically about these forces, however, Bernoulli remained a convinced vorticist throughout his life.

In 1730, in fact, the academy awarded its prize for the best explanation of why the planets move in elliptical orbits to Bernoulli's *Nouveaux pensées sur le système de M. Descartes.*[129] Bernoulli grounded the entire argument of this essay in vortical mechanics, yet he did so while applying the analytical central force mechanics that he and Maupertuis studied together throughout much of 1729. Moreover, Bernoulli also claimed in his private correspondence that his physical commitments in this paper were anything but deeply held. Bernoulli had struggled throughout the 1720s to win one of the academy's annual prizes, and by 1730 he had become convinced that his failure to do so was a result of a cabal inside the academy, one that slavishly adhered to certain ideas while prejudicially excluding others. Thus, to win the prize in 1730, he claimed to have produced a paper that deployed a physical theory very different from his own because "the attractions and void of M. Newton would only generate horror among Messieurs the French. I therefore gave them reheated meat with a new sauce, in a word a *chapon en capilotade* [a dish of leftover chicken], which is to say that I adroitly reproduced the vortices of Descartes in a slightly modified form and added the appearance of a new luster to them." He adopted this approach, he said, because this is what he believed the academicians in Paris wanted.[130]

His victory in the prize contest confirmed his conspiracy theory, and if Bernoulli's actual position on the physics of planetary motion was actually much more complex, this case showed how the independence of mathematics and physics in his work, as with Bouguer's, could allow for a variety of settlements between them. The same was true for Maupertuis, for so long as he and Bouguer were talking about mathematics they could agree while debates about physical theories often made them antagonists. To understand the absence of a necessary link tying mathematics to physics in the work of Bernoulli, Maupertuis, and Bouguer, then, is to appreciate the problems that arise whenever one reads the history of French science in this period too neatly in terms of the binary Newtonian/Cartesian labels of the period. Bouguer maintained a com-

129. PVARS, April 19, 1730. This work is discussed in Brunet, *Newton en France,* 186–199.

130. Bernoulli to Gabriel Cramer, 15 April 1730, Bernoulli Archive, Öffentliche Bibliothek der Universität Basel, L I a 655, nr.9. Bernoulli told different versions of the same story to many of his correspondents in this period. My quotation is taken from Otto Spiess's transcription of the Bernoulli correspondence available at the office of the Bernoulli Edition.

plex stance with respect to the Cartesian/Newtonianism battles of 1730s, and in 1731, before joining the academy, he submitted an essay for the academy's prize contest that reveals much about his intellectual outlook in this period. Parisian academicians could not compete for the annual prizes, and Bouguer's submission was therefore disqualified as soon as he was admitted to the company. It was nevertheless given an honorable mention, and published in 1752 with the other prize winning papers.[131] Viewed as a barometer of his thinking, the text reveals a great deal about Bouguer's Newtonian and Cartesian sentiments as he entered the academy.

Perhaps the most revealing aspect of the work is that Bouguer chose to present his views as a dialogue. He defined three characters in his text: Ariste, a zealous defender of Cartesianism; Théodore, an equally zealous defender of Newtonianism; and Eugène, a moderate who attempts to balance the views of the other two. In the end, it is Eugène's judicious compromise that prevails. Overall, this complex, perspectival approach to the problem reflected Bouguer's own complex stance on these issues. Théodore, for example, articulated a defense of attraction that was very radical in 1731. He first offered the usual attack on those that confused gravity with the occult qualities of the scholastics, but instead of claiming attraction as an empirical fact or a mathematical principle, he asserted that "the attracting or moving force of which we speak is nothing other than the will of the Author of nature."[132] This established Théodore as the most radical kind of pantheist Newtonian, and not surprisingly both Eugène and Ariste stood opposed to his views.

In a similar fashion, Ariste represented the excesses of Cartesianism. He tried to account for the precise obliquity of the ecliptic by invoking a complex chain of vortical mechanisms, but Eugène, the moderate, challenged him, arguing that the system cannot account for the facts. He further explicated the empirical inadequacies of Ariste's system, and this provoked Theodore to remind his interlocutors that Newton's theory of universal gravitation easily accounted for all of these detailed observations. Eugène, however, challenged Theodore's claim as well, citing Newton's *Principia* as an authority in arguing that attraction had nothing to do with this aspect of planetary motion. Eugène also went on to criticize Théodore's erroneous assertion that the mechanics of impulsion could not explain the facts. This served as the launching pad for

131. Pierre Bouguer, "Entretiens sur la cause de l'inclinaison des planètes," in *Recueil des pièces qui ont remporté les prix de l'Académie Royale des Sciences,* 4 vols. (Paris, 1752), 1: 4–41. This work is discussed in Aiton, *Vortex Theory of Planetary Motions,* 219–221, 224–228.

132. Bouguer, "Entretiens sur la cause de l'inclinaison des planètes," 30.

Eugène to assert his own views. The dialogue ended with Bouguer's moderate interlocutor offering a modified vortical account of this phenomena that ultimately won the Cartesian Ariste's approval. Théodore's Newtonian objections to it, however, were left largely unanswered.[133]

This text, especially in its dialogic character, captured well Bouguer's thinking about Newtonianism and Cartesianism in the early 1730s. It also articulated the complex stance maintained by many others in France in the same period. Cognizant of the insufficiencies of the prevailing vortical account of celestial mechanics, but unable for philosophical, theological, and even moral reasons to accept the Newtonian theory of universal gravitation across empty space as a valid philosophical explanation, many French "Eugène's" attempted to develop moderate, compromise solutions that reconciled the two systems. The key challenge was to save the hypothesis of the vortices and the clear and distinct mechanics of impulsion that anchored it, while nevertheless accounting for the observed facts of planetary motion as completely as Newton's theory did. Bouguer's work on the movement of bodies in resisting media was one example of this approach; Johann Bernoulli's work, which won the academy prize pursued by Bouguer in 1732, was another.[134] Equally representative of these attempts at synthesis, however, was the work of Privat de Molières, and here an important distinction must be drawn.

Privat de Molières' work was self-consciously and polemically Cartesian in a way that both Bouguer's and Bernoulli's was not. His vortical astronomy also eschewed the complex analytical mathematics that these other mathematicians placed at the center of their work. Indeed, in his magnum opus, *Les Lécons de physique,* which began to appear in 1733, Privat de Molières presented his system as a deductive, physical system demonstrated from "self-evident" first principles in the manner of Euclid.[135] Very little mathematics was included at all. Privat de Molières' work, therefore, was intended as an explicit defense of French Cartesianism in both its physical and methodological/epistemological aspects. He also made sure that the public conceived of it this way through the stridency of his own anti-Newtonian language.

133. Ibid.

134. Johann Bernoulli, "Essai d'une nouvelle physique céleste," in *Recueil des pièces qui ont remporté les prix de l'Académie Royale des Sciences,* 3: 5–88. On this work, see Aiton, *Vortex Theory of Planetary Motions,* 228–235; and Brunet, *Newton en France,* 272–293.

135. Joseph Privat de Molières, *Les leçons de physique,* 4 vols. (Paris, 1733–1739). To give an example, one self-evident first principle of mechanics for Privat de Molières was the proposition that motion cannot be transferred without the physical contact between bodies.

Bouguer and Bernoulli did not share Privat de Molières' Cartesian zeal. It was thus possible, given this distinction, to conceive of a merger between Bouguer's and Bernoulli's analytical mechanics and the emerging Newtonianism of Maupertuis while Privat de Molières' work remained impossibly alienated from any such compromise. Maupertuis' work could be joined because it first reinforced the analytical approach to these problems that he and Clairaut would establish at the heart of academic French Newtonianism after 1732. It could also be merged because as mathematical analysis itself, all that was required was a new physical picture of the mathematical category called central forces for a gestalt shift to be effected that transformed the ostensibly Cartesian and mechanical nature of this science into a framework supportive of a Newtonian and attractionist outlook instead. Such a conversion was simply not possible with Privat de Molières' more essentially Cartesian approach to the same questions. For example, while forces for Privat de Molières were always physical entities understood in terms of the necessary mechanisms of impact, Bouguer and Bernoulli treated central forces mathematically, attributing no essential physical meaning to them whatsoever. For this reason, Privat de Molières' work was irreducibly Cartesian while Bouguer's and Bernoulli's could be coopted to a Newtonian world-picture by simply changing the physical meaning attached to the mathematical concepts deployed.

Furthermore, the essential agnosticism about physical causality at the heart of their analytical mechanics gave both Bernoulli and Bouguer a justification for accepting a limited belief in attraction that was simply not available to Privat de Molières. His entire system was founded upon the premise that gravitational attraction at a distance across empty space was an irrational proposition. Nature simply could not behave this way, he believed. His *évident* synthetic demonstrations proved it. The mathematical approach of Bouguer and Bernoulli, however, was not at all constrained by these rationalist necessities. Defining their terms mathematically, and letting mathematics alone dictate the epistemology of their work, they could remain credibly agnostic about the physics of centripetal forces while at the same time producing important new work in mechanics. Privat de Molières, by contrast, was a slave to his synthetic physical demonstrations.

Given their primary identity as analytical mathematicians, therefore, Bernoulli and Bouguer were in a position to shift allegiances easily on the physical question of impulsion or attraction. This ease supported Bouguer's own gradual conversion to attractionism by the 1740s, while Privat de Molières went to his grave in 1743 a strident vorticist. Furthermore, viewed from this perspective, one also sees why Maupertuis' *Discours sur les différentes figures des astres* was so important in France. Before analytical, central force mechanics could become Newtonian mechanics in France, a credible philosophical and scientific picture of gravitational attraction was needed, one that supported picturing analytical mechanics in Newtonian as opposed to Cartesian terms. Fontenelle had spent three decades constructing a Cartesian picture of this mathematical science, one rooted in vortical mechanisms, and by 1730 this discourse was being intensified and polemicized. To challenge it, an alternative Newtonian picture of central force mechanics was needed, one that was equally powerful. This is what Maupertuis offered his colleagues and the French public in 1732.

From this perspective, the most important section of Maupertuis' *Discours* was chapter 2. Here the author offered what he called a "metaphysical discussion of attraction." One of the clear goals of this discussion was to offer a defense of Newtonian attraction compatible with the methods and assumptions of French analytical mechanics.

Maupertuis opened his discussion by framing the issue in opposed Newtonian and Cartesian terms. The Cartesians, he wrote, recognize correctly that the impact of one body provides a force capable of moving another. Yet they "try to explain everything by this principle and to show that even *pesanteur* follows from it." This gives their system "an advantageous simplicity," but "great difficulties are found with it at the detailed level of phenomena."[136] Newton, fully aware of these difficulties, and dissatisfied with the Cartesian neglect of them, "established another principle of action in nature." He asserted that "all the parts of matter gravitate [*pesent*] toward one another." With this one principle, all the phenomena were marvelously explained." Indeed, Maupertuis continued, "the more one goes into detail, the more profound his system becomes, and the more it seems confirmed." Certainly, the notion of attraction itself has "unsettled people [*effarouché les esprits*]." Many "fear seeing the rebirth in philosophy of the doctrine of occult qualities."[137] However, these fears were unjustified. In the rest of the chapter, Maupertuis offered a systematic explanation of why this was so.

He first established a precise set of boundaries for understanding Newton's theory of universal gravitation. "We owe it M. Newton," he wrote, "to acknowledge that he never claimed that attraction was an explanation for the *pesanteur* between bodies. He asserted continually that he employed this term only to designate a fact and not a cause. Indeed, he employed the term precisely to avoid systems and explanations. He even suggested that this tendency might be caused by a subtle matter that flows through bodies and thus be the effect of a veritable impulsion. But however one conceives it, [gravity] remains a primary fact which

---

136. Maupertuis, *Discours sur les différentes figures des astres,* 90–91.
137. Ibid., 91.

one must accept before explaining the other effects that follow from it."[138] This description set up the rest of the argument. Rather than defend Newton's conception of attraction directly, Maupertuis instead worked to expose the erroneous reasoning of those who attacked it philosophically. More precisely, Maupertuis refuted those who "reject[ed] attraction as a metaphysical monster." He argued instead that the reality of attraction is as plausible as its cause is unknown.

Overall, the key to Maupertuis' argument was his use of Malebranche's general skepticism about human understanding in the construction of his brief. "People are not surprised when they see a body in motion communicate that motion to another," he explained. "The habit of seeing such things all the time removes the marvelous quality from it. But philosophers should guard themselves from thinking that an impulsive force is easier to conceive than an attractive force. What is an impulsive force? How does it reside in bodies? How can one be sure where it resides before seeing two bodies collide? The site of other material properties is no more clear than this."[139] This was a clear echo of Malebranchian skepticism, and in Maupertuis' hands it became a vehicle for accepting Newtonian attraction. For if, philosophically speaking, the principle of action at a distance was neither more nor less comprehensible than that of mechanical impulsion, then one could not refute it on philosophical grounds alone. This was Maupertuis' point. He thus used philosophical skepticism of a Malebranchian sort to level the epistemological playing field for Newtonianism and Cartesianism alike.

What properties of bodies can we be certain about, Maupertuis asked? Certainly, bodies contain extension, for how can a body be a body without this attribute. Similarly, bodies possess impenetrability because otherwise bodies would not present themselves to us as distinct and separate entities. However, while these two attributes were necessary to bodies as such, they were not necessarily related. More important, bodies possessed countless other attributes that were not necessary to them in this way. We say bodies are hot and cold, for example, but do bodies need to be hot or cold to be bodies? Clearly not, Maupertuis argued. Thus, given our limited human understanding, we are always confronted with a hierarchy of bodily attributes, some necessary and others not. Science functions, therefore, not through necessary demonstrations, but through the study of those attributes and relationships that appear regularly to our senses in the same way.

Regarding gravitational attraction and action at a distance, then, it was only justifiable to deny these possibilities when a clear philosophical contradiction

138. Ibid., 91–92.
139. Ibid., 98.

presented itself. For example, since we see bodies in motion everywhere, it is not justifiable to claim bodies as immobile entities. Likewise, since a body never completely penetrates another body upon impact, we cannot assert that bodies do not possess impenetrability. But since no such contradiction applies in the case of attraction, we cannot deny its existence on philosophical grounds alone. As Maupertuis summed up: "Everything that has just been said does not prove that there is attraction in nature. Indeed, I have not set out to prove such a thing. I have only undertaken to examine whether attraction, even considered as an inherent property of matter, is metaphysically impossible. If it is, then the most readily apparent phenomena of nature will fight against it. But if neither its impossibility nor its susceptibility to contradiction is clear, then one must freely inquire into whether the phenomena prove it or not. In other words, attraction is now only a question of fact. It is in the system of the universe that one must look in order to determine whether it is a principle which holds in nature, and if so how far it goes in explaining the phenomena, or whether one introduces it uselessly to explain facts that can better be explained without it."[140]

This was a brilliantly artful argument given the state of public scientific discourse in France in 1732. Most important was the fact that Maupertuis did not offer a defense of Newtonianism per se, but a powerful critique of the Cartesian attacks increasingly being made against it. Figures such as Fontenelle, Dortous de Mairan, Privat de Molières, Cassini II, and others railed publicly against attraction because they believed the principle was philosophically irrational. Yet it was precisely this claim that Maupertuis refuted. Furthermore, by approaching this philosophical critique with an equally rational philosophical alternative, Maupertuis countered these charges on their own terms. The very obvious associations with Malebranchianism in Maupertuis' discourse only reinforced the strength of his position. Dortous de Mairan had severed his ties with Malebranche early in his life, but both Fontenelle and Privat de Molières, albeit in very different ways, remained strongly wedded to the work of the Oratorian Father. Fontenelle in particular had used a version of this same phenomenalist skepticism in his own public academic discourse on behalf of analytical mechanics earlier in the century. For Maupertuis to turn this philosophy back upon Fontenelle, therefore, was a powerful move.

Maupertuis' skill as a stylist also revealed itself on every page of his *Discours,* serving him in equally potent ways. Fontenelle had disseminated Cartesianism in the French public sphere by developing a clear, accessible, and stylistically elegant discourse on its behalf. Maupertuis did the same with Newtonianism. He further did so while also maintaining a thoroughly *honnête* decorum

140. Ibid., 103–104.

throughout, and this despite the polemics swirling around the idea of gravitational attraction in France. In all these ways, then, Maupertuis' *Discours* turned the Cartesian discourse back against itself. Maupertuis' argument was also very different from the defense of Newtonianism common among the English and Dutch Newtonians, and this distinctiveness further made it compatible with French sensibilities. In many respects, it had much in common with Brook Taylor's rejoinder to Rémond de Montmort in 1719, especially in its combination of mathematical phenomenalism and philosophical skepticism. Yet Maupertuis also Cartesianized (or perhaps more precisely Malebranchianized) Newton in certain key respects, making his science much more compatible with prevailing French assumptions. This move at once made his Newtonianism compatible with French analytical mechanics while also distancing both from the rationalist, mechanistic form of Cartesianism championed by Fontenelle and Dortous de Mairan. Maupertuis' Malebranchian approach towards Newtonian gravity also created a space between French Newtonianism à la Maupertuis and the various strands of English and Dutch Newtonianism that still conjured fears of active matter materialism and religious heterodoxy for many.

In this respect, Maupertuis constructed something like a "French orthodox" version of Newtonianism that paralleled the "English orthodox" Newtonianism of figures such as Samuel Clarke and Willem 'sGravesande. Each was suspicious of philosophical rationalism in science, and each spoke a language of empiricism that made the truth or falsity of Newton's theories a matter to de determined by careful observation of natural phenomena alone. Each also distanced the "Newton" they invoked from the more physical and metaphysical conception of attractive force that radicals like Toland advocated. However, by drawing more strongly upon Malebranche, Maupertuis nevertheless developed an alternative French version of this empirical, phenomenalist Newtonianism that was distinctive in important ways. Bouguer's character Theodore revealed how Clarke's views about the liberty and omnipresence of God could be construed as a heretical and pantheistic defense of active matter materialism. Maupertuis, however, avoided the Socinian and Spinozistic dangers that Leibniz had exposed in Clarke's position by adopting Malebranche's skepticism instead. We have no idea how God effects motion in nature, Maupertuis argued, and thus it is rash to conclude that impulsion is more worthy of God's creation than attraction. Yet it is equally rash to consider rationalism to be the royal road to atheism and fatalism since God clearly works according to the laws of mathematics. By invoking Malebranchian phenomenalism as a third alternative distinct from either position in the Leibniz-Clarke debate, therefore, Maupertuis distanced his Newtonianism from the theological quagmire at the center of their dispute.

He also altered the meaning of Newtonianism in this way as well. Clarke's Newton believed that God's free will was involved in every natural occurrence. Maupertuis' Newton, by contrast, believed that God was a mathematically minded creator, free to establish whatever natural order he saw fit, but one who was then constrained ever after by the mathematical laws that guided his decision. As a result, Maupertuis never mentioned, let alone defended, the idea, so central to the Newtonianism of Samuel Clarke, that God intervened on occasion to correct the workings of an essentially unstable universe. Instead, he built his defense of Newtonianism upon the regularities of the heavens that only Newton's simple mathematical principles could explain. There was certainly more than a hint of Cartesian, Malebranchian, Leibnizian, and even Spinozist rationalism in this formulation; however, attached to mathematical clarity rather than physical or metaphysical necessity, and tempered as well by a deep commitment to empiricist phenomenalism, Maupertuis' Newtonian vision of nature was actually an original formulation. It was also tailor-made for French intellectual sensibilities of the 1730s, and especially for those who practiced analytical mathematics inside the Paris Academy.

Maupertuis' stance with respect to the experimental empiricism of Desaguliers, 'sGravesande, and the other Newtonian experimental philosophers was equally nuanced. 'sGravesande argued in his *Elements of Physics* that we cannot know the underlying causes of things, thus we must restrict our science to finding the regularities in those things that we can observe. Maupertuis argued much the same thing in his *Discours.* However, he parted company with the discourse of Newtonian experimental philosophy in other ways, and these differences were important. Most significantly, he did not echo the conviction, common to the English Newtonians, that experimental demonstration was a crucial tool in the establishment of scientific truths. He in fact never mentioned experiment at all in his *Discours,* and he avoided altogether 'sGravesande application of experimental philosophy to the principle of Newtonian attraction. Maupertuis instead defined the "fact" of gravitational attraction in nonexperimental terms. His guarantee for its truth was neither empirical regularity nor experimental consistency, but mathematics. As he wrote: "Every regulated effect, even if its cause remains unknown, can be the object of study for a Mathematician because he treats everything susceptible to increases and decreases in magnitude no matter what its nature is. Furthermore, the usage that he makes of it will be no more certain if its nature is absolutely known. Thus, if he is not permitted to treat such things, the limits of philosophy will be strangely constrained."[141]

141. Ibid., 92.

Here the real distinctiveness of Maupertuis' position emerges. He in effect argued that attraction is a valid scientific hypothesis primarily because it leads to a set of quantifiable mathematical relations that sustain a powerful mathematical description of nature. This formulation privileges mathematics as the epistemological guarantor of empiricism in a manner very different from English and Dutch "experimental philosophy." It also reveals Maupertuis' pedigree in the Malebranchian world of French analytical mechanics. Early in the century, figures in the circle of Father Malebranche developed a deeply mathematicized and phenomenalist interpretation of Newton's *Principia* as a centerpiece to their origination of analytical mechanics.[142] In a very real sense, Maupertuis was the great heir to their work. His defense of Newtonian attraction in the *Discours* articulates the hidden assumptions that led Varignon, l'Hôpital and others to pioneer the analytical approach to central force mechanics in the decades around 1700. Read retrospectively, moreover, their understanding of Newton's *Principia* also makes more sense when interpreted through this lens.

Maupertuis' primary audience, however, was the new generation of analysts that began to congregate around him inside the academy after 1731. Thus, one of the key achievements of the *Discours* was its articulation of a Newtonian explanation of analytical mechanics that reinforced the general renaissance of this science then occurring inside this institution. Maupertuis' Newtonianism, rooted as it was in the reduction of nature to simple, mathematical relationships, was ideally suited for a French mathematical community still operating within a scientific field defined by Malebranche's mathematical phenomenalism. Not surprisingly, therefore, the *Discours* quickly triggered the conversion of large portions of this mathematical community to the Newtonian cause. Maupertuis' text, through its unprecedented public offering of a French defense of Newtonianism, also created a unique place for this constituency within the public sphere. The other chapters of the *Discours* helped to fuel the polemics that these developments triggered.

In chapters 3 and 4, for example, Maupertuis scrutinized the Newtonian and Cartesian systems side by side. He maintained an *honnête* objectivity throughout, yet as Fontenelle noted "the presentation clearly favor[ed] the views of the English philosophe."[143] Perhaps most devastating was Maupertuis' use of the Cartesians' own methods to systematically critique the prevailing vortical mechanics. He used the two works awarded the academy's prizes in 1728 and 1730 (by Georg Bulfinger and Johann Bernoulli, respectively) to expose a host of difficulties with vortical celestial mechanics. He did the same with the Cartesian

142. See Shank, *Before Voltaire,* chap. 3.
143. *HARS-Hist.* (1732): 93.

theory of the comets and of *pesanteur*. These theories, Maupertuis asserted, could only save the phenomena by assuming multilayered vortices where the layers themselves each moved at different speeds. In some cases, moreover, the motions of these fluids were clearly at odds with one another. Thus, while Newton's single principle of universal gravitational attraction explained all of these phenomena with ease, the Cartesians, Maupertuis suggested, were compounding complexity upon complexity in their effort to avoid Newton's simple, quantitative law.

Framed this way, Newton became the real avatar of Cartesian clear and distinct reasoning while the Cartesians were shown to be lost in a haze of mechanist confusion. Indeed, Maupertuis' indictment of the confused and obfuscating vortical mechanics of the Cartesians resonated strongly with the standard critique of medieval scholasticism even if he never, unlike the English Newtonians, drew this connection explicitly. Maupertuis' defense of Newtonian celestial mechanics in chapter 4 was equally artful. He did not defend Newton's theory of universal gravitation directly as a physical principle since his rhetorical strategy throughout was to avoid polemical defenses of this sort. Instead, he simply enumerated all of the ways that the observed behaviors of the stars and planets could be accounted for using this one principle alone. When combined with his metaphysical defense of the possibility of attraction in chapter 2, however, the point was clear. As Fontenelle reported "one finds here attractionism without a veil."[144] However, one also found a devastating presentation of the empirical power of Newton's system and the inadequacies of the Cartesian alternative.

Each of these moves allowed Maupertuis to step into a discursive position prepared by at least two decades of previous discussion of Newtonianism in France. Furthermore, by assuming this position in such a calculated, public way, he channeled all of the public attention directed toward these ideas into his own intellectual position. The result was a solidification of the public image of Newtonianism in France around the public discourse and image of Maupertuis. The same occurred within the academy of sciences. This solidification made Maupertuis the first authentic French Newtonian, and it also established French Newtonianism as a coherent, viable entity. The French Newton wars had begun.

Maupertuis' initiative also linked him in a new way with the other central protagonist in this battle. In October 1732, just months before the appearance of Maupertuis' *Discours*, Voltaire wrote to Maupertuis seeking counsel. "Being now at court while not playing the courtier, and reading philosophical books

144. Ibid.

without being a philosopher, I have been thinking about you during my moments of doubt. I am very angry that I cannot talk to you personally. It concerns M. Newton's principle of attraction."[145] Voltaire included with his letter a brief *mémoire* that he had composed, perhaps a preliminary draft of his letter on attraction that would appear in the *Lettres philosophiques* the next year. He also asked for Maupertuis' advice regarding its contents. "I beg your pardon for my impetuousness," he wrote, "but I beg of you only a moment of your time to clarify things for me. I await your response before I decide whether I believe in attraction or not. My faith will depend on you. If I am as persuaded of the truth of this system as I am of your merit, I will assuredly become the most convinced Newtonian in the world."[146]

Maupertuis' response does not survive, but apparently it was both rapidly sent and persuasive. For a few days later Voltaire wrote again to Maupertuis thanking him for giving two demonstrations when only one was requested. "I thank you from the bottom of my heart for your generosity, for it is easy to see that there are prodigious riches here. You have clarified my doubts with a most luminous precision. Consider me a Newtonian like you. I am your proselyte, and my profession of faith is in your hands! Indeed, given the way you write, I have no doubt that your book will produce other disciples as well. You are so intelligible that without doubt *unusquisque audiet linguam suam*. I will merely have the honor of being instructed before the others and of being the first neophyte. One can no longer stop oneself from becoming Newtonian, and it is necessary now to eradicate the chimera of the vortices."[147] Voltaire ended his letter by asking a few more questions. He also raised one or two final doubts. Apparently Maupertuis again responded quickly, for on November 8 Voltaire wrote his third letter to Maupertuis in little more than a week, further declaring his Newtonian convictions enthusiastically." Pardon me sir, my temptations have returned to the devil from which they came. Your first letter baptized me in the Newtonian religion, and your second letter has served as my confirmation. Here's thanking you for your sacraments. Burn, I beg you, my ridiculous objections. They were the proclamations of an infidel. I will guard forever your letters. They are from a great apostle of Newton: *lumen ad revelationemen gentium*. I am with great admiration, gratitude, and shame, your very humble and unworthy disciple."[148]

145. Voltaire to Maupertuis, 30 October 1732, in Voltaire, *Corr.,* 86: D533.

146. Ibid.

147. Voltaire to Maupertuis, 3 November 1732, in Voltaire, *Corr.,* 86: D534. The Latin quotation is a biblical reference meaning roughly "his language will be heard by all."

148. Voltaire to Maupertuis, 12 November 1732, in Voltaire, *Corr.,* 86: D535. The Latin quotation is from the Biblical Vulgate (Luke 2.32): "Light for Revelation to the Gentiles."

This letter terminated the correspondence for the time being, but the exchange reveals much about the public effect of Maupertuis' defense of Newtonianism in the *Discours sur les différentes figures des astres.* Clearly, Voltaire's embrace of Newtonianism was of a very different sort than the conversions that had begun to occur inside the academy in 1732. Even if many of the excesses of Voltaire's discourse in these letters can, and should, be attributed to the poet's exuberance with a playful rhetorical trope, he clearly conceived of Newtonianism as a creed or an intellectual identity in these declarations, more than as a scientific or philosophical position. His desire to attach himself to Maupertuis was also motivated by a vastly different set of priorities than those that brought figures like Clairaut, Bouguer, La Condamine, and others into the Newtonian constituency inside the academy. Maupertuis had done little to explicitly cultivate relationships with individuals such as Voltaire, even if his esteem within the literate public sphere was important to him. He instead devoted his energies almost exclusively toward building his Newtonian constituency within the academy. Yet by positioning himself and his Newtonianism in the precise public way that he did, Maupertuis invited these associations to form.

He also did little to check the wide dissemination of his public Newtonian persona once it was launched. He had become a major figure in Parisian society by 1732, and he enjoyed the company of *gens des lettres* such as Voltaire. Consequently, he did little to hinder his increasing public acclaim, and in fact began to actively encourage it. The apparent convergence of Maupertuis and Voltaire in late 1732 around the discourses of Newtonianism was anything but surprising, therefore. It also provoked little confrontation, or even attention, since the two men routinely crossed paths in many locales, both virtual and material, even if they also continued to follow two seemingly different intellectual trajectories. Yet driven by his different agendas, and seeing in Maupertuis' newly crafted Newtonian persona a powerful set of cultural opportunities, Voltaire began to take the Newtonianism first defended in France by the senior academician and to transform it into something very different. The eventual result was the philosophe movement that came to define the French Enlightenment.

5

Making the "Philosophe"
Voltaire's Newtonianism and
the Scandal of Lettres
philosophiques

The text that triggered Voltaire's transformative initiatives was his *Lettres philosophiques*, which appeared in the spring of 1734, about eighteen months after Maupertuis' *Discours*. "This work initiated among us a period of revolution," wrote the Marquis de Condorcet two years before the fall of the Bastille.[1] Modern historiography has agreed, canonizing the work as the first bomb thrown against the Ancien Régime. The text did trigger a scandal that established Voltaire as the first Enlightenment philosophe. It also launched an unprecedented philosophical campaign in which Voltaire and his followers, employing the ideas introduced in this "manifesto of Enlightenment" (another label often attached to the *Lettres philosophiques*), attempted to transform France socially, culturally, and intellectually. But how exactly did this text trigger such an explosion? Voltaire opened the French Enlightenment in 1734, I argue, by defining and asserting a new intellectual persona: the philosophe. Furthermore, I contend that his strategic use of Newtonianism in the *Lettres philosophiques*, when situated within the context of the Cartesian/Newtonian debate already underway in France, played a decisive role in initiating this outcome.[2]

Voltaire entered the crucial year of 1734 occupying an intellectual and social position that was thoroughly commonplace. The scandal caused by the *Lettres philosophiques* in this year cannot, therefore, be attributed to any reputation that preceded the work. Yet like all lasting generalizations, the idea that the *Lettres philosophiques* signaled Voltaire's prior passage from a poet into a phi-

---

1. Condorcet, *Vie de Voltaire,* ed. Elizabeth Badinter (Paris, 1994), 46. The French is: "Cet ouvrage fut parmi nous l'époque d'une révolution."

2. Other works about Voltaire that have been useful to me besides Pomeau, *Voltaire en son temps,* include Jean Ehrard, *L'idée de la nature en France dans le premier moitié du dix-huitième siècle* (Paris, 1964); Gay, *Voltaire's Politics*; René Pomeau, *La réligion de Voltaire* (Paris, 1969); Ira O. Wade, *Voltaire and Madame du Châtelet: An Essay on the Intellectual Activity at Cirey* (Princeton, 1941), idem, *Studies on Voltaire with Some Unpublished Papers of Mme. de Châtelet* (Princeton, 1947), and idem, *The Intellectual Development of Voltaire.*

FIGURE 19. *Maurice Quentin de La Tour,* Portrait de Voltaire.
Louvre, Paris, France. Courtesy of Art Resource, NY.

losophe contains a germ of truth. Voltaire did, for example, premeditate at least
some aspects of the intellectual provocation of the work. Ideas were one component of the challenge, for Voltaire infused the *Lettres philosophiques* with a
set of claims that he knew to be disturbing given the state of public discourse
in France. At the center of this effort was his particular fluency in the discourse
of Anglo-Dutch Newtonianism, including its radical materialist variant, a lan-

guage that was as yet publicly unspoken by any Frenchman. Equally important, however, was the new ideal of the man of letters that he defined through the articulation of this precise language. Voltaire had watched as Bolingbroke's circle of literati had become full-fledged critical intellectuals, and he imported a similar tone and stance into the *Lettres philosophiques.* The result was the emergence of a new species of intellectual in France—the philosophe—inspired by Voltaire's new critical, public stance.[3]

The change, however, did not occur all at once, and to trace its development the history of Voltaire's relationship with this text must be followed in some detail.

The Making of an Enlightenment Manifesto

The idea of publishing a brief collection of reflections on English culture occurred to Voltaire while in exile across the Channel.[4] In the spring of 1728, less than a year before his return to France, he began writing such a work, calling it "an account of my journey to England." He composed fourteen letters in English that were published in London in 1733 as *Letters concerning the English Nation.* These letters were in Voltaire's possession when he reentered

3. For an interesting study of the same theme, see Paul M. Cohen, *Freedom's Moment: An Essay on the French Idea of Liberty from Rousseau to Foucault* (Chicago, 1997). Part of Voltaire's significance resides in becoming arguably the first examplar in France of what Cohen calls "the consecrated heretic."

4. Pomeau, *D'Arouet à Voltaire,* chap. 19, and Vaillot, *Avec Madame du Châtelet,* chap. 1, treat the early history of the *Lettres philosophiques.* More detailed accounts can be found in Harcourt Brown, "The Composition of the *Letters Concerning the English Nation,*" in *The Age of Enlightenment: Studies Presented to Theodore Besterman* (London, 1967), 15–34; André-Michel Rousseau, "Naissance d'un livre et d'un texte: les *Letters concerning the English nation,*" *Studies on Voltaire and the Eighteenth Century* 179 (1979): 25–46; and William Hanley, "The abbé Rothelin and the *Lettres philosophiques,*" *Romance Notes* 23 (1983): 1–16. A modern, critical edition of the *Lettres philosophiques* has not yet appeared in the Besterman edition of Voltaire's *Oeuvres complètes.* Until it does, the most thorough critical edition of the text remains Gustave Lanson, ed., *Lettres philosophiques. Édition critique avec une introduction et un commentaire,* 2 vols. (Paris, 1909). Lanson's introduction summarizes the history of the text, but it has been superseded by the more recent work cited above. In citing from the *Lettres philosophiques,* I will make use of the translation offered by Ernest Dilworth, trans. and ed., *Philosophical Letters* (Indianapolis, 1961), unless otherwise noted. But my textual citations will be to the Lanson edition, and will be noted by the letter number and marginal notation that he employs.

Paris in April 1729.[5] The English letters would form the foundation of Voltaire's French *Lettres philosophiques,* which appeared in two different editions in 1733 and 1734. But as the change in titles indicates, a major reconceptualization preceded the publication of the French work. The differences between the two are deeply revealing.

Letters concerning the English Nation includes all the letters on the sects of England that were published in the French edition, as well as the letters on the English parliament, on commerce, on Chancellor Bacon, on tragedy and comedy, and on the different English literary figures. Missing, however, are the important letters on Locke, Newton, Descartes and natural philosophy more generally, not to mention the two concluding letters treating academies and the consideration owed to men of letters. The omission of the letters on Descartes and the Paris academies is anything but surprising given the English focus of the original conceptualization; however, the absence of Voltaire's letter on inoculation for smallpox, an important English practice, is less so. Indeed, reducing the *Lettres philosophiques* to the contents of the earlier English work alone eliminates almost everything that was controversial and transformative about the text. What the disparities between the two editions reveal, therefore, is the extent to which Voltaire's real transformation into a new kind of French intellectual only occurred after he was back in Paris after 1729.

One impetus for this transformation was the emerging polemic about Newtonianism that crystallized in France at just around the time of his return. Voltaire writes in the *Lettres philosophiques* about the reception of Fontenelle's "Éloge de M. Newton" in England.[6] This discourse was delivered in Paris in November 1727, and it is possible that Voltaire participated in debates triggered by this work in London before returning to France in early 1729. When he arrived in Paris, the *éloge* was already available in print, and the academy volumes containing Fontenelle's newly polemicized defense of Privat de Molières' Cartesian celestial mechanics would appear soon after. Voltaire appears to have followed these scientific debates closely, for in the fall of 1730 his publisher, Formont, noted that Voltaire had conceived the odd idea of adding a supplement to his epic poem *La Henriade,* which explained Newtonian philosophy. The explanation never appeared, but Formont claimed that the work was "very good [*fort bien*]."[7] Voltaire appears to have continued working sporadically

5. See Brown, "Composition of the *Letters*"; and Pomeau, *D'Arouet à Voltaire,* 254.

6. Voltaire, *Lettres philosophiques,* 2: 5–8.

7. Jean Baptiste Nicolas Formont to Pierre Robert Le Cornier de Cideville, November 1730, in Voltaire, *Corr.,* 86: D380.

on this piece throughout 1731–1732 while beginning work on the new letters that would appear in the newly titled and conceived *Lettres philosophiques*. His time was also occupied with the production of an important play, his tragedy *Adélaïde du Guesclin,* and with the publication of a major work of poetry, *Le temple du goût*. Newtonian natural philosophy, therefore, was by no means his exclusive or even his primary occupation during this time.

The publication of Maupertuis' *Discours sur les différentes figures des astres* in the summer of 1732 further sparked Voltaire, and it was in the context of his own, as yet unpublished, efforts to defend Newtonianism that Voltaire first wrote to Maupertuis and sought his counsel. The contacts with Maupertuis intensified in the winter of 1732–1733 as Voltaire prepared to submit the final manuscript of his *Lettres philosophiques* to his publishers. In late November 1732 he wrote to Maupertuis that he had read his *Discours* that morning with the "pleasure of a girl who reads a novel, and with the faith of a *dévot* who reads Scripture."[8] Soon after, he sent Maupertuis an essay, whose contents are not known, asking that the academician review it while remaining his "master in physics and his disciple in affection."[9] Later in December, Voltaire passed Maupertuis actual drafts of the letters on Newtonianism that would soon be published in the *Lettres philosophiques*.[10] Indicating, however, the different audience for and conceptualization of his work, Voltaire also sent copies of the letters to Bolingbroke as well.[11]

Introducing the letters to Maupertuis, Voltaire offered a revealing précis of the nature of their relationship at this time, "You are accustomed to giving me lessons," he wrote. "Let me, therefore, impose upon you to offer your judgment regarding several letters that I wrote while in England and which are about to be published in London. I have since corrected them somewhat, but they strike me as much in need of your examination. . . . I have selected only the letters that pertain to the areas that you honor by studying. Not that you do not enjoy your own reign over more than one region of Parnassus. It is just that I did not want to bore you by giving you too much. I want to try your patience incrementally. The next time you would like to dine together at M. du Fay's home with that *honnête* Muslim who understands French so well [i.e., La Condamine], I will come obediently. I will read my *Temple de goût* to you. This is a country well known to you, but unknown to the vast majority of mathematicians.

8. Voltaire to Maupertuis, 15 November 1732, in Voltaire, *Corr.,* 86: D538.

9. Voltaire to Maupertuis, 1 December 1732, in Voltaire, *Corr.,* 86: D541.

10. Voltaire to Maupertuis, 15 December 1732, in Voltaire, *Corr.,* 86: D546.

11. Lanson, introduction to *Lettres philosophiques,* 1: viii.

M. Newton does not know it at all, and M. Leibniz only knows how to travel there in German. *Adieu, Monsieur*—you do not have a more ignorant, docile, and tenderly attached disciple than me." [12]

In this letter's appeal to both Maupertuis' authority in technical science and his prestige in *mondain* literary society, the text captures perfectly the nature of Voltaire's relationship with the senior academician on the eve of the publication of the *Lettres philosophiques*. Within a few weeks, Voltaire had received Maupertuis' comments. He thanked him for his suggestions, and further asked for a confirmation regarding Newton's theory of light. He told Maupertuis that his copy of Pemberton's *View of Sir Isaac Newton's Philosophy*, which was "right by his side," reported findings different from his. But he deferred to his friend in asking that Maupertuis clarify the truth that must be alluding him. Voltaire then offered a precise description of his conception of the *Lettres philosophiques* as a whole. "I cannot even understand myself in these letters," he wrote, "whether on this point or on so many others. I would have to write an entire work of philosophy to do so, yet I must struggle even to understand yours. I only felt myself obliged in speaking about all the fine arts to make M. Newton a little bit better known among the ignorant such as myself." [13] In closing, he thanked Maupertuis for his assistance and expressed again his devotion to him. He also expressed an interesting reservation about the labor of the philosopher. "I love and admire you," he explained, "but I fear I must abandon all of this philosophy. It is a practice [*métier*] that requires a great deal of health and leisure, and I have neither." [14]

Voltaire did not correspond with Maupertuis again until after the appearance of the *Lettres philosophiques*, yet these pronouncements say much about Voltaire's thinking as he prepared to publish his text. Most important, they articulate the precise bonds uniting and distinguishing these two public figures at the moment when they first made French Newtonianism a public phenomenon. For example, by situating Maupertuis within "Parnassus" and by praising his ability, unique among mathematicians, to enter the "world of the poets," Voltaire was signaling the importance of Maupertuis' wider cultural position in French society. By further defining himself as a practitioner of the beaux arts whose only philosophical ambition was to popularize Newton among amateurs like himself, Voltaire was also clarifying the boundary that separated the official scientific academician from the independent man of letters. He did the same when he implicated himself with academicians like Du Fay and La Con-

12. Voltaire to Maupertuis, 15 December 1732, in Voltaire, *Corr.*, 86: D546.
13. Voltaire to Maupertuis, 20 December 1732, in Voltaire, *Corr.*, 86: D550.
14. Ibid.

damine through dinner engagements and poetry readings rather than through joint scientific labor. He did so also when he lamented the difficulty and obscurity of philosophy, suggesting that he might leave its pursuit to those, such as these academicians, with the health and leisure to succeed in it. At issue in these very cordial, yet nevertheless highly strategic, definitions was a crucial negotiation among savants about the authority of independent men of letters such as Voltaire to treat the serious natural philosophy usually reserved for expert academicians. The redefinition of this authority would be one of the central achievements of the *Lettres philosophiques.*

The events triggered by the appearance of the text itself were the major catalysts in this reconfiguration.[15] In the spring of 1733, Voltaire's letters were ready to be published; however, a set of publication difficulties intervened at this point, which made Voltaire virtually powerless over the text and its reception. By this time, he was officially negotiating with two different publishers, while a third in London possessed copies of Voltaire's fourteen original English letters. By the summer of 1733, all three editors were eager to publish their respective manuscripts. In August, an unauthorized version of the *Letters concerning the English Nation* appeared in London. Annoyed by this development, his other London publisher, Thieriot, brought out a rival edition despite Voltaire's pleas that he wait. Thieriot's edition, published in September as *Lettres écrites de Londres sur les Anglais et autres sujets,* included all the additional letters written after 1729, especially those on Locke, Newton, and Descartes. But it did not include the provocative final letter on Pascal. As works published without royal approval, these editions were prohibited from circulating in France, but they did generate some press coverage. For example, before the end of the year, the *Lettres anglaises,* as Thieriot's edition came to be called, had earned a review in the *Bibliothèque britannique* and another in the inaugural volume of abbé Prévost's *Le pour et contre.*[16]

Voltaire's French publisher, Jore, was fully aware of the threat that these two London publishers posed. He, therefore, moved his copies of the *Lettres philosophiques* into print in July 1733. But he also agreed to put the volumes in a Rouen warehouse while Voltaire worked behind the scenes to win approval for the work with the royal censors. These negotiations with the royal authorities reveal a great deal about the controversial nature of the book. Voltaire read selected passages to both Cardinal Fleury and Maurepas during the winter of

---

15. The most economical account of this history is offered in Pomeau, *D'Arouet à Voltaire,* 321–331.

16. *Bibliothèque britannique* (1733), 2: 16–35, 3: 104–137; *Le pour et contre, ouvrage périodique d'un goût nouveau* 1 (1733): 241–264, 261–288.

1732–1733. Each found the selections "amusing" but harmless. More important, Voltaire also consulted the abbé Rothelin, who, although not an official royal censor, was a doctor of theology, a protégé of Cardinal Polignac, and a member of the Académie française. His approval would offer tacit sanction for the work, yet Rothelin responded that royal permission was unlikely unless he "softened" certain of the book's arguments. In particular, the letter on Locke appeared to him dangerous and excessive. In the end, Rothelin's partial acceptance was as close as Voltaire ever got to an official approval for his work.[17]

While these efforts continued into early 1734, Jore became increasingly frustrated. His editions continued to languish in the Rouen warehouse while rival editions were being discussed in the French press. Voltaire also appears to have begun accepting the inevitable soon after Thieriot's London edition appeared. "I remain constantly in my retreat at St. Gervais where I lead a philosophical life troubled by fits of coughing and the holy inquisition which reigns at present over literature," he wrote to Cideville in September 1733. "It pains me to suffer, but it is even harder to be prevented from thinking with an honest liberty [*une honnête liberté*], the most beautiful privilege of humanity [*le plus beau privilege de l'humanité*]. . . . The life of a man of letters is liberty. Why is it necessary to be subjected to the rigors of slavery in the most agreeable [*aimable*] country in the universe? . . . Thieriot enjoys the fruit of my labors in London while living in peace; yet I am trapped in Paris. . . . I often envy Descartes his solitude at Egmont, even if I have no envy at all for his vortices and his metaphysics. In the end, I will no doubt finish by either renouncing my country or my passion to think at a high level. This is the wisest course of action. . . . Happiness and suffering are real, while reputation is only a dream. [*Le bonheur ou le mal est réel, et la reputation n'est qu'un un songe.*]"[18]

True to Voltaire's fears, matters took a turn for the worse in the spring of 1734. Jore discovered that counterfeit copies of Thieriot's *Lettres anglaises* were being smuggled into France. He soon decided that he could not wait any longer to release his edition. In April, while Voltaire was away from Paris attending the wedding of a leading patron, the Duc de Richelieu (a marriage that had been arranged by the Marquise du Châtelet), copies of the *Lettres philosophiques* began to be sold in Paris. At this point, matters spiraled out of control. The response of the French authorities to the challenge of a scandalous work published without royal approval was rapid and decisive. Angered by Voltaire's audacity, the keeper of the royal seals, Chauvelin, arranged for the issue of a *lettre de cachet* on May 9 that would have sent Voltaire to the Bastille had it

---

17. On this exchange, see Hanley, "The abbé Rothelin and the *Lettres philosophiques*."
18. Voltaire to Cideville, 15 September 1733, in Voltaire, *Corr.*, 86: D555.

been executed. Helped by Richelieu, however, Voltaire secretly escaped to du Châtelet's manor home in Champagne (Cirey), and the letter never reached its destination.

The prosecution, however, did not stop here. On June 9, the head of the Paris police, Hérault, already immersed in a bitter war with the clandestine publishers of Jansenist texts, sent Jore to the Bastille. He also ordered the confiscation of all copies of the *Lettres philosophiques* in Paris. These searches turned up a number of copies of Voltaire's work, and while one suspects that Hérault was more pleased with the copies of the Jansenist *Nouvelles ecclésiastiques* that were also confiscated during these raids, the persecution ultimately reinforced Voltaire's status as an outlaw. The next day, on June 10, the Parlement of Paris enacted the ultimate punishment, ordering a copy of the *Lettres philosophiques* publicly burned by the hangman on the steps of its judicial chambers. Voltaire wrote letters to all of his most important contacts during this period, reassuring them that he intended nothing improper by his work. His entreaties, however, produced no concessions. For the remainder of the decade, Voltaire would live in exile, a fugitive from the royal authorities.

Viewed retrospectively, then, a central component of Voltaire's new philosophe persona—namely, his status as a rebel and outsider challenging the established authorities in France—emerged through a set of accidents. Indeed, Voltaire in no way aspired to become an intellectual outlaw in 1734. He no doubt believed deeply that he possessed an inherent liberty to express himself freely, a conviction nurtured in England and expressed with passion in the *Lettres philosophiques* itself. But as he expressed to Cideville, he was not desirous of combat on this point. Had someone been willing to offer him royal approval for his work, he no doubt would have performed the editorial dances necessary to obtain it and avoid scandal altogether. Furthermore, as Rothelin's comfort with the majority of the book's contents revealed, the *Lettres philosophiques* did not really contain the sort of claims that warranted such persecution. The authorities treated many more scandalous works more leniently, and the excessiveness of the official reaction to Voltaire's work remains puzzling even today. Certainly Voltaire's challenge to the royal control of the book trade goes farther in explaining the official reaction than either the ideas or even the intent of the text itself. Accordingly, a different publishing history for the *Lettres philosophiques* might very well have generated an entirely different outcome for the text and its author in France.

Intellectual issues were not entirely beside the point, however. For one, Voltaire did realize the provocative nature of many of his claims, and when pushed to the brink he ultimately sided with his right to assert his ideas freely. Here the convictions that he forged during his immersion in English public life revealed themselves most clearly. Voltaire also infused the *Lettres philosophiques* with a new critical tone that was likewise premeditated. It too reflected his experience in the more open, contestatory, and libertarian intellectual culture of England. The content of the *Lettres philosophiques* also contributed to the work's shock value even if its ideas alone cannot explain the scandal. Rothelin had emphasized the dangerous character of the letter on Locke, and what shocked people about it was its celebration of Locke's metaphysics, especially the idea that matter might be sufficient to explain thought. This defense suggested strong affinities between Voltaire and the most radical materialists of the day, and the author did little to quell these suspicions. Leibniz, for example, had drawn a strong connection between Lockean materialism and Newtonian gravitation in his letters to Samuel Clarke. The widely read French translation of Locke's *Essay concerning Human Understanding* published in 1723 also cemented this connection since Coste's text accentuated the association already being made in the public sphere between Locke's metaphysics and active matter materialism.[19] This brand of Lockeanism was strongly tied to the materialist reading of Newtonian physics, and in this way Coste's translation of Locke reinforced Leibniz's claim that Locke and Newton constituted the twin sources of the English materialist contagion. Voltaire seized upon this equation as well, while attempting to turn it toward very different ends. The resulting Newtonianism that Voltaire articulated thus resonated strongly with the materialist discourses hovering around Newton in the 1730s.

Stated simply, what Voltaire did in the *Lettres philosophiques* was to offer himself as the first French defender of the more controversial English and Dutch style of Newtonianism. He did so while also identifying his intellectual positions with the new and more stridently critical, intellectual tone that had increasingly become integral to Newtonianism outside France. Maupertuis' *Discours sur les différentes figures des astres* had already offered French readers one defense of Newtonianism, but his was rooted in a deeply Malebranchian mathematical-phenomenalist orientation toward natural philosophy. It was also geared toward his institutional ambitions inside the academy, and therefore dependent upon the codes of *honnêteté* that were still dominant there. Voltaire's agendas were very different. His Newtonianism was derived from the critical polemics of the English and Dutch Newtonians, and he adopted the critical style of the coffeehouse and the periodical press. He also directed

19. John Locke, *Essai philosophique conçernant l'entendement humain, où l'on montre quelle est l'étendue de nos connoissances certaines, et la manière dont nous y parvenons,* trans. by Pierre Coste (Amsterdam, 1723). On this connection, see John Yolton, *Thinking Matter: Materialism in Eighteenth-century Britain* (Minneapolis, 1983).

his arguments toward the public at large, especially those engaging in critical, philosophical debates within the wider Republic of Letters. The analytical mathematicians and other savants inside the Royal Academy, by contrast, concerned him little. From the outset, therefore, Voltaire was a different kind of Newtonian with a different set of motives and agendas. This, in turn, allowed him to turn his Newtonian advocacy toward the creation of a new kind of intellectual identity in France, that of the critical, independent philosophe.

His different approach appeared very early in the *Lettres philosophiques*, even before the crucial letters on natural philosophy.[20] From the very first page, Voltaire employed a strong, first person voice, adopting the stance of a traveler reporting from a foreign land. This narrator also acted throughout as a critical judge, spicing the narrative with critical remarks and reflections on many occasions. The second of his four opening letters on the Quakers illustrates well this dimension of Voltaire's rhetorical style. Here he has a Quaker preacher explain to him the nature of the Quaker faith. "When thou movest one of thy limbs, does thine own power move it?" he has his Quaker preacher ask. "Certainly not," the narrator replies, "for the same member often acts involuntarily." "It is He who created thy body, then, who gives motion to this clay," the Quaker continued. "And what of the ideas the soul receives? Is it thou who forms them? Even less so, for they come in spite of thee. So it is the creator of thy soul who gives thee thine ideas."[21] "Ah, it's the purest Father Malebranche!" Voltaire's narrator declares, to which the Quaker retorts, "I know thy Father Malebranche. He was a bit of a Quaker, but not enough of one."[22]

Overall, this exchange is typical of the early letters in the *Lettres philosophiques*. They evoke throughout the *mondain* philosophical repartee common in Voltaire's other writings. These were not the letters where Voltaire defined

---

20. It is useful at the outset to outline the precise contents of the *Lettres philosophiques*. It contains twenty-five separate letters that each could stand alone. But I will consider the work as functioning according to a set of implicit sections. The final letter on Pascal was added at the very last minute and it really should be seen as a text apart. But before this letter, the book falls comfortably into four main groupings. The early letters (1–7) treat the religious sects of England. The next group (8–11) treats English social and political life and includes discussions of English commerce and government. Voltaire's letter on the inoculation for smallpox (12) is included here. The third group (13–17) is the most intellectually challenging and provocative, treating English philosophy. Voltaire's important letters on Bacon, Descartes, Locke, and Newton are found here. The final group (18–22) considers English literature and letters and ends with two important programmatic statements on academies and the "consideration owed to men of letters" (23–24).

21. *Lettres philosophiques*, 1: 24–25.

22. Ibid., 1: 25.

his new and more assertively critical philosophe persona. Indeed, we know that both Maurepas and Cardinal Fleury read the letters on the Quakers and found them both amusing and harmless. Clearly Voltaire was not making himself an outlaw here. If anything, he was confirming his status as a master of the bon mot. Witty satire would remain crucial to Voltaire's intellectual style throughout his life, but it was not the same thing as his Enlightenment critical voice. The latter contained serious, authoritative philosophical criticism, a language that was compatible with ironic wit but not the same thing as it. The early letters in the *Lettres philosophiques* stressed the former, while also planting the seeds for his philosophical discourse to come.

Newton and the English Newtonians also appeared in these early letters, acquiring a provocative identity that further colored their reappearance later in the work. In letter 7, Voltaire cited Newton, Samuel Clarke, John Locke, and Jean Le Clerc as defenders of the Arian, or anti-Trinitarian, religious position that this letter sought to describe. Writing about Newton, Voltaire explained that "the party of Arius is beginning to revive in England as well as in Holland and Poland. The great Mr. Newton did its doctrine the honor of favoring it [because] that great philosopher believed the Unitarians reasoned more geometrically than we."[23] Voltaire then presented Clarke as "the most vigorous defender of Arianism in England." "This man of rigid virtue and gentle disposition prefers to play the intellectual amateur rather than create proselytes." He was also single-minded in his devotion to scientific calculation and demonstration, "a veritable reasoning machine," according to Voltaire. Clarke's theological views were revealed in "a little understood but highly esteemed book on the existence of God" and in another, "better understood but more criticized," on the truth of the Christian religion.[24]

These elaborations softened somewhat the intimation that Clarke and Newton were loyal to the most unorthodox Christian doctrine in all of Christendom. Nevertheless, the charge itself was too powerful to be qualified out of existence. Furthermore, Voltaire's manner of presentation did more to heighten the theological danger than to defuse it. The Jesuits in particular had long located the roots of heresy in an excess of rationalism, particularly geometrical rationalism, and French theological discourse more generally agreed with them on this point. Voltaire implicitly aligned himself with these enemies by connecting the mathematical sensibilities of Newton and Clarke and their embrace of radically heretical theology. He did the same by also including the label "Socinian" in the title of the letter even if the text itself never drew out the implications of this

23. Ibid., 1: 79.
24. Ibid.

term in relation to Newton or Clarke. Both Leibniz and Castel had connected Newtonianism and Socinianism in a damning way, and by including this label in the title, Voltaire accentuated the same association.

The overall tone of the letter certainly pointed more to *mondain* play than religious subversion, yet Voltaire's final, ironic conclusion nevertheless allowed for a more provocative understanding. "Isn't it amusing," he wrote, "that Luther, Calvin, Zwingli, all of them writers that no one can read, nevertheless founded sects that today divide Europe; . . . [Yet] Messrs. Newton, Clarke, Locke, Le Clerc, etc., the greatest philosophers and the best writers of their day, have only managed to acquire a little flock of followers that, small as it is, dwindles every day. This is what it means to come into the world at the right time. If Cardinal de Retz reappeared today, he would not be able to rouse up ten women in all of Paris. If Cromwell were reborn, the man that once had the head of his king cut off and made himself sovereign would be a simple London tradesmen."[25] Playful manipulation of the prevailing theological discourses in this manner was a staple of French literary *mondainité* by 1734; however, in the context of the work as a whole, these particular presentations only intensified the provocation of Voltaire's defense of Newtonianism later on.

In these early letters, then, written in London before 1729, Voltaire was merely playing the typical intellectual game of the eighteenth-century man of letters, turning serious arguments in contemporary philosophy and theology into witty *jeux d'esprit* appealing to *mondain* readers. These letters thus reflected Voltaire's pedigree in the literary culture of Old Regime France, but they also prepared the way for the new critical program that his later letters would offer. The move to Enlightenment criticism began in Letter 7, "On the Parliament," and a noticeable shift in tone also accompanied this shift into topics of politics, philosophy, and science. The early letters on the Quakers, for example, read almost like dialogues, with characters engaging in a give and take about different ideas. By letter 7, however, this approach has been entirely replaced by a more unified, explanatory voice and by a language of direct, objective presentation. The witticisms and word play of the earlier letters also give way to a more direct, authoritative narration. Voltaire's irony, of course, never disappeared. But the irony was now placed more overtly in the service of critical, constructive positions. Furthermore, Voltaire's narrator assumed a more direct, authoritative presence as the writer marshaled his rhetorical tools to craft a unified intellectual persona that argued for particular positions rather than ironizing with available possibilities.[26]

25. Ibid., 1: 80–81.
26. Ibid., 1: 90.

The new tone and the new content merge in the letter on the English Parliament, which also opens up a crucial theme in the work as a whole—the nature of political liberty. "[The English] people are not only jealous of their liberty, they are concerned about the liberty of others," Voltaire wrote. English liberty would appear again and again in the work, becoming a key rhetorical weapon in the author's general defense of intellectual liberty as a whole. Ultimately, Voltaire's own assertion of his intellectual liberty would be a major feature of the book as a whole, and his letter on Bacon, which opened the sequence of philosophical letters, initiated this turn. In this text, the tone of libertine criticism was preserved, but it was now attached to a new assertive and positive, if no less provocative, intellectual agenda. The focus of this new agenda was contemporary natural philosophy.

The letter opened by praising Isaac Newton. "Not long ago, in a company of well-known persons, a worn-out and frivolous question was asked," Voltaire explained: "who was the greatest man, Caesar, Alexander, Tamerlane, Cromwell, etc.? Someone replied that without question it was Isaac Newton. This man was right; for if greatness consists in having received a powerful genius from Heaven and in having used it to enlighten oneself and others, then a man such as Mr. Newton was truly the greatest for such a man has not been seen in ten centuries."[27] This passage began the long and complex defense of Newton to come, and in view of its historical context it was a striking statement. John Keill's 1718 *Introduction to the True Astronomy*, which was reviewed in both the *Journal de Trévoux* and the *Journal des savants*, opened with a similarly glowing celebration of Newton.[28] Fontenelle's "Éloge de M. Newton" of 1727 was likewise celebratory. But neither of these texts, not even Keill's, matched Voltaire's in its zeal on behalf of Newton. In fact, Voltaire's celebration of Newton, although so apparently ordinary to modern eyes, was largely unprecedented in France. Coming from a Frenchman, it was doubly unique.

The subtext of this pronouncement was just as important as the praise itself. Voltaire used Newton to make a much more general point about the value of intellectuals in general. In his letter on the consideration owed to men of letters, Voltaire reinforced the idea, noting via Newton the immense respect that the English have for their men of learning. "Mr. Newton was honored during his lifetime," he wrote, "and after his death as he should have been. The leaders of the nation disputed the honor of being his pallbearers. Go into Westminster

27. Ibid., 1: 152.

28. Review of John Keill, *Introductio ad veram astronomiam*, in *Journal des savants* (1719): 437-440; and in *Journal de Trévoux* (January 1720): 123-144, (March 1721): 643-665.

Abbey. It is not the tombs of kings that one admires there, but the monuments erected with the gratitude of the nation to the greatest of the men who have contributed to its glory. You will see their statues, as one saw at Athens those of such men as Sophocles and Plato. I am convinced that the mere sight of these glorious monuments has roused more than one mind and been the making of more than one great man." [29]

Taken together, these arguments offer Newton as a new kind of hero: the philosophical hero, who will soon serve as the prototype for the Enlightenment hero celebrated by the philosophes. Unlike the honor and martial prowess of the warrior or the self-sacrificing virtues of the Christian saint, the philosophical hero finds his virtue through excellence in philosophy and science. In particular, his heroism derives from the use of knowledge to improve and serve the public good. The philosophical hero might be a man of wit and pleasure as well, but his heroic status derives from his service on behalf of knowledge. Voltaire makes Newton the icon for this new identity, initiating as a result an important Enlightenment trope. Casting Newton of all people in this precise role, however, was no self-evident choice. Voltaire in fact engaged in a sleight of hand worth noting when choosing Newton to be the hero of publicly useful science. The Englishman's intellectual gifts were certainly known to all, yet his scientific work was not as obviously useful to the broader public as the invention of the smallpox inoculation discussed only pages earlier. Newton's science was in fact unabashedly abstruse, and it offered achievements that few even understood let alone found meaningful for improving their daily lives. Voltaire nevertheless ignored these tensions, making Newton the paragon of the publicly useful scientist. In this way, he laid the foundation for a powerful Enlightenment ideology and gave this ideology its first icon.

Voltaire's letter on Francis Bacon, where these initial encomia about Newton were offered, further focused the meaning of Voltaire's interpretation. The discussion began by calling Bacon's *Novum organum* "the best of his works," if "the least read." It then offered the text as "the scaffolding by which the new philosophy has been built." Voltaire further noted that this scaffolding has since become redundant, but he gave credit to Bacon for "pointing out all the paths that lead to [nature]." [30] Key to this pursuit was experimental philosophy, and Bacon was its father. "Nobody before Chancellor Bacon had understood experimental philosophy," Voltaire claimed. [31] Many wondrous inventions had been invented, but these were the result of accident. The universities,

29. Voltaire, *Lettres philosophiques*, 1: 158–159.
30. Ibid., 1: 154–155.
31. Ibid., 1: 156.

moreover, founded for the perfection of human reason, "continued to ruin it with their *quiddities,* their *abhorrence of a vacuum,* their *substantial forms,* and all the other impertinent words that ignorance made respectable and religion, through a ridiculous mixture, made sacred."[32]

In fact, Voltaire continued, academic philosophy was not as important to human progress as was often thought. "It is to a mechanical bent, natural to most men, that we owe all the arts; we do not owe them to philosophy. What prodigious use the Greeks and Romans made of mechanics! And yet in their day it was believed that the skies were crystal, and the stars were little lamps that sometimes fall into the sea."[33] At one level, this presentation offered a typical tradesman's critique of effete theory. Yet having undermined philosophy in this way, Voltaire then rescued it by making Bacon not a heroic artisan but a new, and more useful, kind of philosopher. As he concluded: "Of all of the physical experiments that have been made since his time, hardly one was not suggested in his book. Several of them he made himself, . . . [and] shortly afterward experimental physics suddenly began to be cultivated in almost all parts of Europe at once. It was a hidden treasure of which Bacon had some expectations, and which all the philosophers, encouraged by his promise, labored to unearth."[34]

With experimental philosophy, therefore, Bacon brought into the world a new system of learning that allowed science to achieve a new truthfulness while serving the public good. This formulation gives mechanical thinking and experimental practice a centrality in the origins of modern science in a manner very different from previous French approaches to the same thing. Fontenelle, for example, always celebrated the utility of science, but he did so while maintaining the clear superiority of philosophical reasoning over the mere labor of mechanics. Not surprisingly as well, Descartes rather than Bacon was the hero of Fontenelle's stories.[35] Voltaire agreed with Fontenelle in presenting the work of simple "mechanics" as ignorant and haphazard, yet he nevertheless treated Bacon as the father of modern science, locating his importance in the philosophical systematization of empirical, mechanical methods. This conceptualization, while quite typical of experimental Newtonians such as Desaguliers,

32. Ibid., 1: 155.

33. Ibid., 1: 156.

34. Ibid., 1: 157.

35. See, for example, Fontenelle's *Entretiens sur la pluralité des mondes* (*Oeuvres de Fontenelle,* 2: 7–130), which offers an account of the rise of modern science without mentioning Bacon once.

was virtually unknown in France in 1734. It was also deeply influential since it created an origin story for modern science that treated explicitly utilitarian, experimentalist, and mechanical work as superior to "vain philosophy" in the progress of knowledge.

Importantly, Voltaire also drew an explicit connection to Newton's work in this context. He claimed to find a defense of Newton's theory of universal gravitation in Bacon's writings even though the chancellor died many years before the development of the theory in Newton's *Principia*. Citing two passages in which Bacon appeared to defend the Newtonian idea that bodies attract one another in the void, Voltaire suggested that Bacon had anticipated Newton's achievement. Even more important was Voltaire's method of accounting for this convergence. He took considerable liberty with his citations, but in one he has Bacon assert that "we ought to find by experiment whether the same clock, which is moved by weights, will go faster on the top of a mountain or at the bottom of a mine. If the force of the weights diminishes on the mountain and increases in the mine, then it is likely that the earth has a real power of attraction." [36] The French astronomer Richer had discovered variations in identical pendulum clocks placed at the equator and more northern latitudes, and these experiments were used in several important demonstrations in Newton's *Principia*. Bacon's ideal experiment certainly would have resonated with those familiar with this history, and these references therefore helped to present Newtonian attraction as an empirical fact proven by experimental philosophy. The precise presentation of the theory of universal gravitation in letter 15 of the *Lettres philosophiques* would reassert this idea even more emphatically.

Having thus framed the development of modern and publicly useful philosophy in a very particular way, Voltaire then turned to the controversial philosophy of John Locke. The precise arrangement of these letters was no accident. Following Bacon's experimentalism and preceding Newton's physics, Locke's metaphysics was presented as the implicit linchpin in the new philosophical trinity that Voltaire was constructing. He began with a provocation. Having already positioned himself as an advocate for experimental reasoning in a manner clearly provocative to the French Cartesians, Voltaire opened his letter on Locke by firing another anti-Cartesian volley. "Perhaps there has never been a wiser, more orderly mind or a logician more exact than Mr. Locke," he asserted, "and yet he was no great mathematician. He never could submit to the drudgery of calculations, nor to the dryness of mathematical truths, which in and of themselves offer nothing concrete to the understanding. Yet no one has

36. Voltaire, *Lettres philosophiques,* 1: 157.

better demonstrated than he that one may have a geometrical intellect without the help of geometry." [37]

This was a provocative formulation given the state of French scientific culture in 1734. It at once challenged Maupertuis' phenomenalist, analytical Newtonianism and Fontenelle's mechanistic, synthetic Cartesianism. Indeed, this characterization placed Voltaire most comfortably alongside Father Castel, who often used the same relationship between natural, nonmathematical thinking and an easy, ordered mind to criticize both analytical mathematics and mechanistic rationalism in France. The origins (not to mention intent) of Voltaire's presentation, however, are not to be found in Castel's Jesuit critique. Instead, it was the discourse of experimental Newtonianism common in England and Holland that gave Voltaire his language. Desaguliers commonly used the term "geometrical reasoning" to describe the deductive, yet thoroughly unmathematical, demonstration of truths using experimental facts. Voltaire made Locke the metaphysician who embodied this philosophical approach. For as he wrote in a formulation destined to define an entire paradigm of Enlightenment discourse: "After so many deep thinkers had fashioned the romance [*roman*] of the soul, there came a wise man who modestly recounted its true history [*histoire*]: Locke has unfolded to man the nature of human reason as a fine anatomist explains the powers of the body. Throughout his work he makes use of the torch of science. He dares sometimes to affirm, but he also dares to doubt. Instead of defining at once what we know nothing about, he examines bit by bit what we want to understand." [38]

This characterization echoed Côtes' presentation of experimental philosophy in the "Preface" to the second edition of the *Principia*. It also resonated with the epistemological proclamations of Clarke and 'sGravesande, discourses that had become well known to French readers by 1734. In fact, Voltaire's conflation of Lockean metaphysics and experimental Newtonianism was in no way original. Jean Le Clerc used similar tropes frequently in his journalistic writings throughout the 1710s and 1720s, and other self-proclaimed "Lockeans" and "Newtonians" did the same. Voltaire did not offer, therefore, a new examination of Locke's metaphysics cast in the language of Newtonian experimental philosophy in the *Lettres philosophiques*. Instead, he offered himself as a new and native-born French speaker of the still exotic and often controversial discourse of Anglo-Dutch Newtonianism.

Voltaire's treatment of Descartes and Malebranche in the letter on Locke further sealed this persona. "Our Descartes," he wrote, implicitly mocking the

37. Ibid., 1: 166.
38. Ibid., 1: 168–169.

parochial prejudices of his countrymen, was "born to bring to light the errors of antiquity and to put his own in their place. Being led astray by that spirit of system that blinds the greatest of men, he imagined he had demonstrated that the soul is the same thing as thought, just as matter, according to him, is the same as extension. He maintained that we are perpetually thinking, and that the soul makes its arrival in the body already provided with every possible metaphysical notion, knowing God, space, and infinity, as well as the whole range of abstract ideas. It is filled, in other words, with splendid knowledge, all of which it unfortunately forgets as it leaves its mother's womb." [39] This satirical formulation was pure Voltaire, yet in style and substance it also echoed clearly the anti-Cartesian discourse that the English and Dutch Newtonians had made commonplace after 1715. Voltaire's discussion of Malebranche in this context was even more English and "Newtonian." "M. Malebranche of the Oratory in his sublime hallucinations not only allowed the existence of innate ideas, but was certain that all we perceive is in God, and that God, so to speak, was our soul." [40] This made Malebranche a dreamer and *romancier* rather than a rigorous philosopher. It also connected his thought explicitly with Spinozist pantheism.

Both charges were well-worn tropes in the Anglo-Dutch Newtonian discourse of the time, and by invoking them here Voltaire was fashioning a new intellectual identity for himself, that of the radical French Newtonian. Equally typical was the author's description of Locke as one who "consults only his own experience, the consciousness of his own thought." [41] Voltaire cited a long passage in which Locke admitted his own inability to fathom nature's complexities. Voltaire then strongly aligned himself with Locke's "Newtonian" empirical skepticism. "As for me," he stated, "I am proud of the honor of being in this respect as stupid as M. Locke, for no one will ever convince me that I am always thinking." [42] Empirical skepticism also framed Voltaire's treatment of the controversial materialist tendencies in Locke's philosophy. "Having done away with innate ideas," he wrote, "and having altogether renounced the vanity of believing that we are always thinking, Locke proves that all our ideas come to use through our senses." Locke also asserts, however, that the precise nature of human consciousness is beyond our comprehension. "Possibly we shall never be able to know whether any material being thinks or not," Voltaire explained, and yet "these sage words struck more than one theologian as a scandalous

39. Ibid., 1: 168.
40. Ibid.
41. Ibid., 1: 169.
42. Ibid.

declaration that the soul is material and mortal."[43] "All people had to do was consider without bitterness whether there is any contradiction in saying 'Matter can think,' and whether or not God is able to fuse thought into matter. But theologians have a bad habit of saying that God is outraged when someone simply fails to be of their opinion."[44]

This presentation of Lockean materialism, not as a dogmatic materialism per se but as an epistemological defense of its possibility, echoed the skeptical Newtonianism of Clarke, 'sGravesande and many other English and Dutch Newtonians. It also resonated with Maupertuis' *Discours sur les différentes figures des astres,* but in this case the differences were more important than the similarities. First, unlike Maupertuis, Voltaire directly confronted the specter of materialism that was then hovering around both the philosophies of Locke and Newton. Furthermore, in the manner of the Anglo-Dutch experimental philosophers, he deflected these charges by invoking empirical skepticism rather than the image of a free and active God capable of constructing matter however he saw fit. He also directly challenged those who attempted to use rationalist philosophy to explain away the contingencies of nature. Maupertuis had invoked God's freedom as an argument in defense of the possibility of attractive matter, and this placed him alongside Clarke and the skeptical Malebranchian position that Rémond de Montmort used in his debate with Brook Taylor. Voltaire took a different tact entirely. He mocked the abstraction and worldy detachment of mathematical phenomenalism while also calling Malebranche's philosophy little more than a metaphysical fairy tale. In its place Voltaire also vigorously defended the experimentally based empiricism and antirationalist skepticism of the English and Dutch Newtonians. In doing so, he defined a conception of Newtonianism that was very different from Maupertuis'.

Thus, from the very outset, French Newtonianism was a fractured and contested entity. Positioned as he was in the *Lettres philosophiques,* Locke became for Voltaire the metaphysical anchor of his Newtonianism, one conceived according to the canons of Anglo-Dutch experimental philosophy. Maupertuis instead put Malebranchian phenomenalism at the center of his conception of the same thing, making his Newtonianism a mathematical philosophy consonant with the goals of French analytical mechanics. Newton remained the common denominator in each formulation, yet Voltaire articulated a very different conception of Newtonianism and then let the language of the *Lettres philosophiques* accentuate it so as to further his wider intellectual ambitions. The three letters that followed the letter on Locke treated Newtonian philosophy directly,

43. Ibid., 1: 169–170.
44. Ibid., 1: 170.

and these presentations completed the program. Having introduced both Ba-con and Locke via the discourse of Anglo-Dutch experimental philosophy, and having further made each thinker a founding father of modern Newtonian thought (a genealogical innovation in 1734, it must be remembered), Voltaire then proceeded to introduce Newton's philosophy itself in letters 14–17. These letters constitute the conceptual as well as the literal center of the work.

Alert readers, familiar with the writings of the English and Dutch Newto-nians, would not have read anything in these presentations that they had not heard before. They would have been treated, however, to a new and always salty voice speaking on behalf of these ideas. By using his gifts as a witty po-lemicist to heighten the particular French valence of these charged discourses, Voltaire defined a new discursive position in France by making Newtonianism the preferred philosophy of the free philosophical thinker.

The famous opening passage of letter 14 on Descartes and Newton illustrates all the dimensions of Voltaire's achievement. "A Frenchman arriving in London finds quite a change, in philosophy as in all else," he wrote. "Behind him he left the world full; here he finds it empty. In Paris one sees the universe composed of vortices of subtle matter; in London one sees nothing of the sort. With us it is the pressure of the moon that causes the rising of the tide; with the English it is the sea gravitating toward the moon, so that when you think it ought to give us high tide, these gentlemen think it ought to be low; none of which unfortunately can be verified, for in order to know the truth of it we should have to examine the moon and the tides at the first moment of creation. You will notice that the sun, which in France has nothing to do with this business, over here contributes its twenty-five percent or so. According to your Cartesians, everything is done by means of an impulse that is practically incomprehensible; according to M. New-ton it is by a kind of attraction, the reason for which is no better known. In Paris you picture the earth as shaped like a melon; in London it is flattened on both sides. Light, for the Cartesian, exists in the air; for a Newtonian it comes here from the sun in six-and-a-half minutes. All the operations of your chemistry are owing to acids, alkalis, and subtle matter; in England the concept of attraction dominates even in this. The very essence of things is totally different. You agree neither on the definition of the soul nor on that of matter. Descartes assures us that the soul is the same thing as thought, and Locke pretty well demonstrates the contrary. Descartes declares, again, that matter is nothing but extension; to that Newton adds solidity. Here are some tremendous contradictions. *Non nostrum inter vos tanatas componere lites.*" [45]

45. "It is not for me to settle such high debates," ibid., 2: 1–2. The Latin quotation comes from Virgil, *Eclogues*, III, 108.

The first thing to note about this justly memorable passage is the degree to which it sums up and thus clarifies the Newtonian/Cartesian debate that was already at least a decade old in 1734. Attraction in both astronomy and chemistry; the figure of the earth debate that was just beginning to erupt; the tides; the vortices; the plenum and the void; the nature of matter; the theology of the soul; the proper epistemology of science; England versus France—here in one economical and eminently quotable passage are all the key dichotomies that had formed over two decades of public argument about Newtonian philosophy. However, while the playful reinscription of these increasingly notorious polarities was one reason for the book's influence, the *Lettres philosophiques* also manipulated this debate in accordance with Voltaire's wider ambitions. Most important, he used this letter and the others on Newton to assert more categorically than before the critical and authoritative role of the independent philosopher in these disputes.

Voltaire's use of pronouns in the opening to letter 14 offers one window into his maneuverings. He opened the account by positioning himself above the entire scene, describing another person, a French traveler, who moved between Paris and London. This placed the narrator as a neutral, cosmopolitan observer. The narrator's use of the third-person "he" when speaking of this traveler further fit comfortably with the overall "travel narrative" structure of the work as a whole. In his discussion of the tides, however, Voltaire slipped out of this neutral voice, saying that "with us" the pressure of the moon causes the tides. This located the narrator more strongly as a Frenchman, a stance that resonated with his voice in the earlier letters on the sects. In this case, the narrator was a Frenchman commenting from a parochial vantage point upon English mores. Later in the letter, however, Voltaire offered still another perspective, giving his narrator a non-French identity when speaking authoritatively about Newtonianism. He spoke, for example, about "your Cartesians" on two occasions, and contrasting French chemistry with attractive chemistry, he also called the French approach "your chemistry." In these passages, Voltaire's narrator resided outside France in a space populated by loyal Englishmen.

The overall result was to locate the narrator neither in France nor in England per se, but in the space that included both—that is, the public space of the Republic of Letters. Yet even here mobility and independence are central values. The narrator ultimately acquires his authority through his ability to exchange identities and cross boundaries of intellectual and national prejudice. He moves freely in this letter from Paris to London, and as he does so he evinces no consistent loyalty to either location. He can invoke "we the French" on one occasion while speaking of "you the Cartesians" on another because unlike other intellectual figures, he is beholden neither to nation nor intellectual party. He is a cosmopolitan philosophe who sees things as they are and describes them from the universal viewpoint that subsumes all others. This transcendent perspective also locates the authority of the philosophe precisely in the cosmopolitan freedom and independence of the Republic of Letters. Only with this freedom can one cross all boundaries and reject all claims to authority except those grounded in universal truth. By ending his comparison with the citation from Virgil, Voltaire confirms this autonomous yet authoritative stance. The philosophe is not a creature of nation, party, or institution. His responsibility rests exclusively with the public and its demand of universal reason.

The final Latin invocation thus grounds ideal neutrality as the perspective of the philosophe; however, in many respects the statement itself is patently disingenuous. Voltaire's goal in the text is not simply the isolation and articulation of a new independent status for the man of letters, but the deployment of that status in the service of a cause—the defense of Newtonianism in France. It is for this reason that the Newtonian/Cartesian debates were so crucial in preparing for Voltaire's emerging philosophe campaign. In the opening letter on Newton, Voltaire positioned the philosophe narrator above all nations and parties as a means of establishing the universality of his authority. Once established, however, this universal intellectual authority was then used to defend a particular position in a newly contentious debate: Newtonianism. Constructed this way, Voltaire's Newtonianism does not become one philosophy among many but rather the one and only true philosophy recognized as such by those who think as freely and independently as the philosophe. Likewise, his French Newtonianism does not become simply a national crusade. It transcends this local importance to become a new, universal ideology. In short, it becomes a programmatic part of cosmopolitan Enlightenment.

The character of the preexisting French debate about Newton was crucial in the achievement of this outcome. Because the English and others in the Republic of Letters had so vigorously equated France with Cartesianism, Voltaire was able to exploit this provincialism in the defense of his new philosophical stance. He did so on the one hand by taking on the dangerous identity of the Newtonian "other." Equally important, however, was the way that he also made this identity the clear choice for those who viewed the world from the universal, philosophe perspective. In this way, English- and Dutch-style Newtonianism became both the marker of the philosophe and the cause around which cosmopolitan intellectuals needed to rally. The debate prior to 1734 had created this opportunity for Voltaire, but the Enlightenment that he triggered—an

Enlightenment rooted in the new conception of the philosophe as the critical voice of truth—resulted from the precise way that he defended Newtonianism in this context.

One last feature of his self-presentation was also important: scientific competence. To have remained merely a wit who poked fun at the philosophical follies of the day would have been to stop short of the new intellectual authority that Voltaire in fact asserted. To be perceived as an authoritative philosophe, Voltaire needed to show himself as a knowledgeable and technically competent philosophical thinker. His letters on Newtonian science were devoted particularly toward achieving this precise goal. In letter 15, entitled "On the System of Attraction," Voltaire's philosophe narrator revealed his mastery over the reigning philosophical and scientific discourse by offering a thorough account of the Cartesian and Newtonian positions on this topic. Here Voltaire's manner of presentation echoed Fontenelle's and Maupertuis' in its precision, clarity, and accessibility. Nevertheless, while Voltaire adopted a more neutral and objective tone in this letter, he nevertheless tailored the presentation to his wider agendas. Most important, he offered an experimental and empirical explanation of the origins of Newton's theory of gravitational attraction that reinforced the Anglo-Dutch character of his Newtonian identity.

Voltaire employed a developmental narrative in explaining the theory, recounting the famous story of Newton's meditation upon a falling apple, and then locating the genesis of the theory of universal gravitation in this epiphanic empirical insight. Voltaire's letter also offered a very accurate account of Newton's crucially important demonstration of the fall of the moon to earth in book 3 of the *Principia*, a discussion that accentuated throughout the empirical logic of Newton's reasoning.[46] He noted, for example, that Newton attempted to demonstrate the law of gravitation in 1666 after his encounter with the falling apple but that he could not yet confirm the theory given the insufficiency of his data at the time. To his credit, he reported, Newton abandoned his idea until Picard's geodesic data became available to him. At this point, he returned to his empirical investigations, and this time he was able to demonstrate the quantitative law of gravity that he previously had only intuited. Narrated this way, the law of universal gravitation became not so much a mathematical law as an empirical fact discovered primarily through highly disciplined experimental reasoning.

This was exactly how Voltaire understood it, for as he concluded, in an obvious reorientation of Maupertuis' more mathematical account of the same thing, "It has always seemed extraordinary to me that such sublime verities should

46. *Principia*, book III, proposition 4 (trans. Cohen and Whitman, 407–409).

have been discovered with the aid of a quadrant and a little arithmetic."[47] Voltaire here made Newton primarily an experimentalist, a discoverer of physical facts about nature achieved through inductions from the rigorous study of observed experience. "The force of gravitation . . . is a truth M. Newton has demonstrated by experiments [*une verité . . . demontrée par des experiences*]," he stated categorically.[48] This distinguished his understanding of Newtonian philosophy from Maupertuis', and having parted company with him on the question of method, Voltaire also increased the separation by drawing very different conclusions about the meaning of universal gravitation as well.

For Maupertuis, Newtonian attraction was a fact demonstrated mathematically. We cannot know the nature of bodies as they actually exist, he contended, and thus we have no conception at all about forces. As a result, Newton's theory of universal gravitation says nothing at all about the physical nature of matter in the world. Not surprisingly, Voltaire's views were very different. After illustrating with noteworthy clarity the explanatory power of the theory of universal gravitation, and after further outlining with comparable knowledge all the problems raised by the Cartesian alternative to it, Voltaire framed his ultimate conclusion this way. "If this force of gravitation, of attraction, is at work in all the celestial globes it doubtless acts on all parts of these globes; for if bodies attract one another in proportion to their masses, it can only be in proportion to the quantity of their parts; and if the force resides in the whole, it certainly resides in the half, in the quarter, in the eighth part, and so on to infinity. In addition, if this power were not distributed equally in every part, there would always be some part of the globe that would gravitate more than others—which doesn't happen. Therefore this force really exists in all matter, and in the smallest particles of it. So there is *attraction*, the great means by which all nature is moved." Here Voltaire asserts categorically that which only the most radical and physicalist Newtonians were willing to assert in 1734; namely, that the force of gravity is an inherent property of matter and gravitational attraction a physical fact of nature. Maupertuis had avoided this physicalist and essentialist reading of attraction altogether, but Voltaire's defense of it was clear and unequivocal.

The rest of the letter on attraction, which offered a thorough critique of the Cartesian arguments against this essentialist view, pursued a similar line. According to Voltaire, Newton knew that his "demonstrations of the reality of this principle" would generate controversy, and to avoid it he explicitly distinguished his theory of gravity from the occult qualities of the ancients. As a

47. *Lettres philosophiques*, 2: 21.
48. Ibid., 2: 23.

result, the protestations of Fontenelle and Saurin, whose 1709 academic public assembly discourse Voltaire cited directly, were misguided.[49] Furthermore, the Cartesians were mistaken in asking, as they often did (Fontenelle's 1727 "Éloge de M. Newton" being a case in point), why Newton did not speak of impulsion in explaining centripetal forces. This idea is much clearer, they contended, and to claim attraction as a causal principle is to avoid rigorous explanation altogether.[50] In response, Voltaire again invoked the philosophical stance of the English and Dutch Newtonians. First, he insisted, the idea of impulsion is no more understandable than the idea of attraction since no one understands philosophically how motion is transferred. This was the same argument that Maupertuis had made. Yet Voltaire drew very different conclusions from it. Philosophical debates, he contended, were beside the point; what mattered were empirical facts. Impulsion was impossible because the demonstrations of Newton's *Principia* proved as much. For the same reason, the only responsible conclusion was to accept attraction as a fact of nature confirmed by the regularity of the effects it produces.

"I use the word attraction only to express an effect that I have discovered in nature," Voltaire has Newton proclaim. "[It is] the certain and indisputable effect of an unknown principle, a quality inherent in matter, of which cleverer men than I will find the cause, if they can. . . . I have shown you that the mechanics of central forces makes bodies gravitate [*peser*] toward one another in proportion to the quantities of their matter, and that these forces alone move the planets and the comets in their regular orbits. I have also demonstrated that it is impossible that there should be another cause of gravity and of movement of all the heavenly bodies. . . . Not one of these bodies ever shows a single degree of motion, of speed, of determination, which has not been demonstrated to be the effect of central forces. Thus, it is impossible that there should be another principle."[51] This presentation reinforced Voltaire's position as a defender of the essentialist reading of Newtonian attraction, the most provocative at the time. It also confirmed Voltaire's self-fashioning as a very radical and English Newtonian, a stance that placed him at odds with the prevailing intellectual climate of 1730s France.

Other features of the work further accentuated Voltaire's "dangerous outsider" image. At the end of his letter on English commerce, for example, his narrator asked which was more useful to the state, "a well-powdered lord who knows precisely what time the king gets up in the morning and what time he

49. Ibid., 2: 27.
50. Ibid., 2: 27–28.
51. Ibid., 2: 28.

goes to bed, and who gives himself airs of grandeur while playing the role of the slave in a minister's antechamber, or a great merchant who enriches his country, sends others from his office to Surat and Cairo, and contributes to the well-being of the world." [52] This passage used English commercialism to challenge the aristocratic fabric of France. Voltaire also used English liberty in the arts to challenge French authoritarianism. Letting his narrator speak as a Frenchman at the end of his letter on the consideration owed to men of letters, Voltaire wrote that "we dishonor plays in which Louis XIV and Louis XV have been actors, [and] we give Satan the authorship of works that have been reviewed by the strictest magistrates and acted before a virtuous queen; when foreigners hear about this insolence, this want of respect for royal authority, this Gothic barbarity that people dare call Christian severity, what do you expect them to think of our country?" [53]

Against this barbarous image of France, Voltaire juxtaposed an idealized image of England, the land of liberty, prosperity, and truth. His eagerness to celebrate England in this way and to use it as a foil for criticizing France was central to the critical agenda of the *Lettres philosophiques* as a whole. It appeared in countless places in the text and was used in the service of numerous other agendas that have not been discussed. However, at the center of this English-inspired assault on French culture was the very English character of Voltaire's defense of Newton. This defense became, after 1734, the first real, public crusade of the Enlightenment. Indeed, it was the campaign that launched the movement.

Fighting for Newton in France after 1734

The *Lettres philosophiques* thus triggered the Enlightenment in France because it took advantage of the ripening Cartesian/Newtonian contestation in France to assert a new, critical ideology supportive of a new role for the man of letters. Voltaire's Newtonianism was the centerpiece of this new ideology, and for this reason its articulation in the *Lettres philosophiques* really constitutes the work's polemical heart.

That the work also appeared illegally, and made its author, no doubt unwittingly, an outlaw as well, only intensified this result. Also supportive of it was the increasing rancor of the Cartesian/Newtonian debate in France. Without this chasm, Voltaire's outspoken defense of Newtonianism would not have made him such a clearly oppositionalist thinker, and after his work appeared it only worked to add fuel to this already burning fire. Maupertuis' prior defense of Newtonianism, different though it was, further prepared the way for these innovations. He crafted a Newtonianism tailored primarily to his ambitions as a mathematician at the Royal Academy, and the differences between it and Voltaire's understanding allowed the Academy to avoid entanglements with Voltaire at first. Maupertuis also used his Newtonian advocacy to start a debate that placed the public and the official scientific establishment into a relationship that served his very different intellectual agendas. Voltaire exploited this dynamic as well, while Maupertuis did what he could to turn Voltaire's initiatives to his own advantage. Yet crucially these two Newtonian campaigns remained isolated at first.

Voltaire's most important decision, made while living with the Marquise du Châtelet in exile at Cirey, was to stand firm behind his work and fight the persecution. As he planned this resistance, he appealed to Maupertuis for support. In the spring of 1734, just after the scandal broke, he wrote to the academician lamenting his fate. "These English letters are going to keep me exiled," he explained. "In truth, I believe that someday people will be very ashamed that they persecuted me over a book that you corrected. I am beginning to think that it is the partisans of the vortices and the innate ideas who are driving my persecution. Cartesians, Malebranchians, Jansenists—everybody is aligning against me. But I hope I still have your support. If you are so inclined, I think you should become the leader of a party [*chef de secte*]. You are an apostle of Locke and Newton, and an apostle of your skills with a disciple such as Madame du Châtelet could return sight to the blind." [54]

The mention of the Marquise du Châtelet in this passage was revealing, for it reflected the acquaintance, perhaps intimate, of she and Maupertuis during this period. [55] This development implicated the academician at the center of the newly illicit community at Cirey; nevertheless, Maupertuis appears to have avoided any direct alliance with Voltaire during this period. Instead, he focused his energies inside the Royal Academy and through it to the wider world of the French intellectual establishment. He further chose the figure of the earth, not Voltaire's liberty, as the focus for his initial campaign, and the royal astronomers, not the enemies of the *Lettres philosophiques*, as his central antagonists.

54. Voltaire to Maupertuis, 29 April 1734, in Voltaire, *Corr.*, 86: D728.

55. Du Châtelet's role in all these developments is chronicled most fully in Judith Zinsser, *La Dame d'Esprit: A Biography of the Marquise du Châtelet* (New York, 2006). See also Elisabeth Badinter, *Emilie, Emilie: L'ambition féminine au XVIIIème siècle* (Paris, 1983).

Voltaire's scandals, in fact, arrived in the middle of Maupertuis' own efforts to manage the results of his own Newtonian initiatives in 1732. Although published outside of the academy and discussed publicly in the periodical press, the impact of Maupertuis' *Discours sur les différentes figures des astres* was felt most immediately inside the Royal Academy. Very soon after its publication, Maupertuis also presented comparable arguments to his colleagues inside the institution, and quickly a veritable Newtonian academic constituency formed with Maupertuis at its center.[56] The group was overwhelmingly drawn from the young analytical mathematicians but if La Beaumelle is to be believed, the circle was not restricted to mathematicians, or even to mathematical philosophizing alone. "On the days of the academy assemblies," La Beaumelle reminisced, "Maupertuis collected several young Newtonians at his home for dinner and then led them to the Louvre [where academic assemblies were held] full of gaiety, presumption, and good arguments. He then turned them loose on the old academicians, who could not even open their mouths without being attacked by these lost children, ardent defenders of attraction. One of them composed epigrams against the Cartesians, and another demonstrations.[57] The former, quick to seize on the opportunity for ridicule, aped the gestures, comportment, and language of his adversaries and thus responded to them through mimicry. The latter, opposing only the mockery of their ancient system, insisted that the base of the true system had been found and that it was ridiculous to oppose it. This little band was animated with the occasionally caustic good cheer of their leader, and in this way Maupertuis established Newtonianism in the academy."[58]

Whether anything like this really happened is not clear. Certainly a group of young academicians grouped around Maupertuis did start defending Newtonianism inside and outside the academy after 1732. Yet it was more likely the precise arguments of Maupertuis' *Discours* than any carnivalesque mockery of the Cartesians that carried the day. Father Castel recognized the precise significance of Maupertuis' intervention as well, and he seized upon the academician's deeply mathematical approach to Newtonian attraction in his April 1733 review of the *Discours* in the *Journal de Trévoux*. "[Maupertuis] ends . . . by insinuating that he believes attraction well established by the facts," the Rever-

56. Maupertuis, "Un eclaircissement sur les principes de M. Newton," PVARS, December 23, 1732, idem, "Sur la loi générale de l'attraction Newtonienne," PVARS, January 21, 1733. The latter was published in *HARS- Mem.* (1733): 343–362.

57. Brunet thinks that the former is La Condamine and the latter Clairaut. Brunet, *Clairaut,* 10.

58. La Beaumelle, *La vie de Maupertuis,* 33–34.

end Father wrote. "But what he calls facts are for the most part only mathematical hypotheses. . . . The facts upon which the Cartesians have always based their hypotheses resemble much better facts than those that M. Maupertuis calls by that name. For purely mathematical facts are better called abstract possibilities than facts." [59]

Others, more sympathetic to Maupertuis' analytical approach, however, reacted differently. The *Discours* gave them a clearly defined argument for framing their work in Newtonian terms, while also providing a new and very proximate foil for the increasingly strident anti-Newtonianism of Fontenelle, Dortous de Mairan, and the other Cartesians inside the academy. Thus, as Maupertuis himself described the result, "my *Discours* only made me enemies in France." [60] Other aspects of Maupertuis' *Discours* fueled these polemical tensions. The title of Maupertuis' text announced it as a treatise on the shape of celestial bodies, and in the first chapter he reinforced this general framework by summarizing the current debate on the figure of the earth. In the final chapters, after presenting his philosophical defense of Newtonianism, he also returned to this question, presenting a defense of the claim, made Newtonian in recent years, that planetary bodies possessed an oblate, or grapefruit-like, shape. To argue this point was to enter into what had become by 1732 a very hot theater in the emerging Newton wars.

That the shape of the earth would become by 1732 the great measure of the validity of the Newtonian and Cartesian systems of natural philosophy was in no way evident a decade earlier. Instead, the question acquired its significance as a cosmological litmus test through a set of precise historical alignments that ended up focusing a number of intrinsically unrelated issues around this one question. [61] Most plausibly connected to the actual physics of Descartes and Newton was the hydrodynamics of planetary formation. In the *Principia,* Newton had used his conception of universal gravitation to build a hydrodynamical account of the earth's formation, one that suggested that it was shaped like a grapefruit, with flattened poles and a wider mass at the equator. [62] Cartesians such as Huygens, however, arrived at similar shape using Descartes' very different theory of gravity, so the hydrodynamic approach alone did not set up

59. Castel, review of Maupertais' *Discours,* in *Journal de Trévoux* (April 1733): 711–712.

60. Maupertuis, *Lettres,* in Maupertuis, *Oeuvres de M. de Maupertuis,* 2: 132.

61. In addition to Greenberg, *Problem of the Earth's Shape;* and Terrall, *Man Who Flattened the Earth;* see I. Todhunter, *A History of the Mathematical Theories of Attraction and the Figure of the Earth from the Time of Newton to That of Laplace* (London, 1873).

62. The key demonstration is found in book 3, propositions 18–20 of the *Principia,* 821–832.

the shape of the earth as a test case between Newton and Descartes.[63] Instead, it was the introduction of a set of empirical arguments, particularly those made by French royal astronomers, that changed the discussion.

Returned to their mapmaking projects by the Regency government in 1717, the French astronomers of the Royal Observatory resumed their geodesic surveying operations along the meridian the passed through Paris.[64] In 1718, the head of this project, Cassini II, announced a new description of the earth drawn from these measurements. He suggested that the earth was oblong, not oblate, or that it was shaped like a lemon rather than a grapefruit. Cassini II's argument was nothing more than an empirical hypothesis justified through inductions from the observed measurements, and in this respect his theory embodied the scientific assumptions that still guided work at the French Royal Observatory during these years. The academy as a whole also treated the theory as a major breakthrough, inviting Cassini II to present his findings at the fall public assembly for 1718. They then published a full elaboration of his *mémoire* as a separate volume appended to its annual publication for that year (Fontenelle's *Éléments* would receive similar official publication a decade later).[65] The volume did not appear until 1720, and while it was going to press Dortous de Mairan added a mechanical justification for Cassini II's theory by publishing a *mémoire* of his own that showed how Cassini II's conception of the shape of earth could be derived hydrodynamically using the Cartesian conception of *pesanteur*.[66] Cassini II made his own Cartesian mechanical sympathies public in the same period through his published, though equally empirical and inductive, theories regarding the tides and comets, and in this way the coincidental agreement between he and Dortous de Mairan on the shape of the earth further supported a growing perception, fueled by the Cassini II/Fontenelle discourse on the tides as well, that the Paris Academy was a Cartesian monolith.

In 1725, Desaguliers articulated, and thus solidified, this very perception when he published in the *Philosophical Transactions of the Royal Society of London* an explicit refutation of Dortous de Mairan's work, one that used

63. Christiaan Huygens, *Discours de la cause de la pesanteur* (Leiden, 1690).

64. See Josef W. Konvitz, *Cartography in France, 1660–1848: Science, Engineering and Statecraft* (Chicago, 1987); and Monique Pelletier, *Les cartes des Cassini: La science au service de l'état et des régions* (Paris, 2002).

65. PVARS, November 12 1718. Jacques Cassini, *De la grandeur et de la figure de la terre*, in *HARS* (1718).

66. Dortous de Mairan, "Récherches géometriques sur la dimunition des degres terrestres en allant de l'équateur vers les poles," in *HARS-Mem* (1720): 231–277. See Greenberg, *Problem of the Earth's Shape*, who discusses this *mémoire* at length (15–51).

Newtonian mechanical principles to defend Newton's rival conception of an oblate earth.[67] Desagulier's challenge to Dortous de Mairan set the stage for the great Newtonian/Cartesian battle on the shape of the earth that would erupt a decade later; however in 1728 there were few if any indications that such a climactic struggle was yet brewing. Maupertuis might have learned about Desagulier's work during his trip to England, and his own *Discours* was certainly targeted toward this emerging debate, even if it was targeted in many other directions as well. He was also no more polemical on this point than on any other in the *Discours*. Yet by framing the text the way he did, Maupertuis nevertheless made it clear that among the things he intended by his work was an intervention into this emerging battle.

Father Castel accentuated this polemicism in his review of Maupertuis' work. He at first sided with Maupertuis, asserting that it was "more natural to assume that the earth had a spherical character than to pretend, as M. Mairan had done in his work admirably refuted in the *Philosophical Transactions* of 1725, that the ellipsoid was anterior to the spheroid."[68] Castel was referring to Desagulier's Newtonian critique of Dortous de Mairan's paper on the hydrostatics of the earth, and in this instance he was expressing his familiar Anglophilia by agreeing with Maupertuis' Newtonian position. Elsewhere, however, he criticized Maupertuis' argument. In general, Castel had little patience for Maupertuis' mathematical approach to these questions. He noted that "the author had a mania for analysis [*manie fort bien l'Analyse*]," and he also cautioned that "a little less calculus and a little more explanation would make things better understood."[69] The Jesuit further mocked Maupertuis for his habit of conjecturing beyond what the empirical record justified. As he opined in response to Maupertuis' claims about the universality of his views: "We have absolutely no observations about the shape of the fixed stars. Yet M. Maupertuis conjectures nevertheless that they are completely flattened [at the poles]." In the end, Castel blamed Newton himself for triggering these errors. "[He] likewise seemed incapable of presenting an idea in any other way than enveloped in the most profound mathematics."[70]

Castel's eagerness to provoke controversy on this point was anything but surprising since by 1732 polemics against mathematical philosophy had be-

67. "A Dissertation concerning the Figure of the Earth," *Phil. Trans.*, 33: 201–222, 239–255, 277–304. See also Greenberg, *Problem of the Earth's Shape,* 51–78.

68. Review of Maupertuis, *Discours sur les différentes figures des astres,* in *Journal de Trévoux* (April 1733): 705–706.

69. Ibid., 712.

70. Ibid., 716–717.

come his trademark. What was significant in this case, however, was how Castel was not so much provoking a new controversy as mirroring one that was already underway. By 1732, in fact, the figure of the earth had become ripe for contestation. As noted above, the question itself was in no way rooted in any intrinsic disparities between Cartesianism and Newtonianism, but for a host of different reasons the question crystallized more than any other topic the central polarities of this dispute. Attraction in the void versus impulsion by vortical fluids; mechanism/empiricism versus mathematics; England versus France: these quickly became the points of contention for the figure of the earth debate, and Maupertuis' work moved these issues further into the public eye by seizing upon this of all topics as the central vehicle for his unprecedented Newtonian proclamations.

He also furthered these developments through his own activities inside the academy. During the roughly twenty-five-month period between December 1732 and February 1735 (the period more or less when Voltaire was producing the scandal of the *Lettres philosophiques*), Maupertuis presented more than a dozen academic *mémoires* that explored different technical aspects of the shape of the earth. He was joined in these efforts by many of his Newtonian colleagues in the academy, and the overall result was the establishment of the shape of the earth question as the central theater for the first stage of the Newton wars in France.[71] Maupertuis' approach to this problem was complex. On the one hand, he pursued the analytical, mathematical approach to this problem pioneered by Huygens and Newton, and he helped to transform it through the new application of the differential calculus to these problems. He was ably supported in these efforts by Clairaut, who also began treating the hydrostatics of planetary bodies using the methods of differential analysis. This work would culminate in his 1743 *Théorie de la figure de la terre, tirée des principes de l'hydrostatique*, a book that was in many respects the first great masterpiece of the analytical, mathematical strand of Newtonianism in France.[72] Another young mathematician, Louis Godin, was admitted to the academy in 1726, and he also contributed to these analytical efforts, along with Bouguer. Thanks to Maupertuis, therefore, the analytical community in the academy became increasingly focused on the shape of the earth after 1732.

At the same time, Maupertuis also attempted to pursue and refine the geodesic approach to the problem pioneered by the Cassinis. On more than one occasion, he declared that the problem was only "a question of fact that ought

71. For a discussion of this work, see Terral, "Representing the Earth's Shape," 224.

72. Alexis-Claude Clairaut, *Théorie de la figure de la terre, tirée des principes de l'hydrostatique* (Paris, 1743).

to be decided by actual measurements."[73] Proclamations such as these, which were also directed at nonacademic critics such as Castel, located the ultimate epistemological authority not among analytical mathematicians but in the empirical practice of the positional astronomers. Further reinforcing this position, Maupertuis also worked hard after 1733 to position himself as an authoritative practitioner of astronomical geodesy. In fact, as John Greenberg has shown, Maupertuis helped to revive the program in geodesy in France during these years and to bring it more under his control. New methods pioneered by the Italian Giovanni Poleni suggested a way of improving upon the existing methods of geodesic measurement. Maupertuis seized upon Poleni's work and used it as an opportunity to implicate himself at the center of the geodesy program inside the academy.[74]

Between 1733 and 1736, Maupertuis contributed a host of papers on geodesy to the Royal Academy in addition to his ongoing production in analytical mechanics as well. This work very often maintained an agnosticism about the actual shape of the earth, arguing instead about the correct methods for determining the truth accurately; however, since the geodesy program remained the exclusive preserve of the Observatory-based astronomers, Maupertuis' efforts, as Mary Terrall has argued, nevertheless helped him and his allies to further challenge the institutional power structures inside the institution.[75]

To actually challenge the Cassini Cartesian theory of the earth on geodetic grounds, Maupertuis needed to produce better data. Operations such as these were costly, however, and accurate data were difficult to obtain. The measurements themselves involved calculating the curvature of a terrestrial meridian through comparisons between precisely measured baselines and equally precise stellar observations. The work was both time-consuming and tedious, but it fit perfectly with the empirical orientation of the observatory community. Accordingly, they pioneered surveying operations such as these in France, extending their work as far north as Dunkirk and as far south as the Pyrenees. The geodesic operations themselves were driven by state pressures to improve maps, but the data collected also served as the foundation for French theories about the shape of the earth. In particular, an empirical measurement of the arc

73. This quote is taken from Maupertuis' public assembly address of April 1736 on the shape of the earth as reported in the *Mercure de France*. Cited in Terrall, "Representing the Earth's Shape," 223.

74. John Greenberg, "Geodesy in Paris in the 1730s and the Paduan Connection," *Historical Studies in the Physical Sciences* 13 (1983): 239–260.

75. Terrall, "Representing the Earth's Shape."

of the meridian that showed a greater curvature near the equator than at more northern latitudes offered empirical proof for Newton's oblate picture of the earth, while the opposite result lent support to the opposite theory. Cassini II's own theory had been induced in precisely this way, and thus as Poleni's innovations gave reason to question the accuracy of Cassini II's measurements, geodesy was revived in France and focused in a new way on the question of the shape of the earth.

The central problem in resolving the matter this way was the complexity of the measurement process itself. Simply establishing the base measurement was an arduous task, requiring the isolation of a sufficiently flat and expansive area of land and then the pursuit of a series of highly precise measurements. The possibilities for error here were numerous. The astronomical measurements were equally complex, requiring precise calibration of instruments and the meticulous attention of the observer. Here as well, the potential for error was great. The final calculations also involved some intricate mathematical manipulations, and while the mathematics itself was routine, computers in this period were fallible human beings, not electronic machines. Errors of calculation were thus commonplace, with errors further exaggerated by their repetition throughout the remaining calculation process. Maupertuis' more mathematical approach to this problem was one solution to these technical difficulties, but even he hesitated to claim that geodesic work could be overturned through pure mathematics alone. What was needed, then, was a means of acquiring rival data, and in this way the two great international scientific expeditions of the 1730s were born.

In June 1733, in a paper that he read before the academy, Maupertuis raised the idea of an overseas expedition to conduct geodesic measurements as a means of resolving once and for all the question of the earth's shape.[76] Once articulated, this idea quickly became a rallying point for Maupertuis' colleagues. Given his ties to Maurepas, Maupertuis' proposal was quickly turned into a well-funded reality. Two trips were authorized by the king in 1735. One, under the direction of Godin, included La Condamine and Bouguer. They would travel to the equator and make measurements of an arc of the meridian there. A second trip, headed by Maupertuis, would include Clairaut. This group would go to Lapland north of the arctic circle in order to do the same. Once back in Paris, the data collected from these remote sites would be compared with the data collected in France. Through these means, it was suggested, the figure of the earth would be determined once and for all. Both expeditions

76. PVARS, June 8, 1733.

left in 1736, and before departing Maupertuis offered the public a full explanation of the voyages and their justification at the spring public assembly of the academy.[77]

This public address and the intense public attention given to these expeditions and the figure of the earth dispute they were designed to resolve marks one climax of the public campaign Maupertuis began with his *Discours sur les différentes figures des astres*.[78] By this date, the debate about the figure of the earth had become the centerpiece of French discussions about Newtonianism, and Maupertuis had shaped events to his own advantage by building a powerful and multidimensional constituency within the academy devoted to a number of his positions in this debate. Indeed, many were astonished at Maupertuis' ability to shape these developments toward his desired ends. Neither Maupertuis nor Clairaut had any experience with detailed astronomical observations, yet the state had entrusted them with an expensive and highly sophisticated international scientific expedition. Johann Bernoulli was particularly stunned by this outcome since he knew Clairaut to be nearsighted, a handicap not conducive to accurate astronomical work.[79] Maupertuis, however, made sure that the project served his institutional agendas. As Terrall writes: "[He] made it clear from the start that his expedition to Lapland would not be modeled on the French surveying efforts. Rather than calling on individuals trained apprentice-style at the observatory, . . . mathematical experience and interest in Newtonian physics took precedence."[80] In this way, Maupertuis turned the argument opened up by his *Discours* into a vehicle for solidifying the power of his Newtonian constituency in France.

Voltaire did the same in the years following the publication of his *Lettres philosophiques,* yet his site of activity was not the Royal Academy and its public but the wider public of the Republic of Letters. He also acted as philosopher in exile at Cirey, and not, like Maupertuis, as a royal academician with the full backing of the French establishment. Voltaire's campaign nevertheless mirrored Maupertuis' in many respects. Both, for example, worked hard after 1734 to establish their respective Newtonian identities on a firm, authoritative footing, and both campaigned to construct the supportive constituency necessary

77. Maupertuis, "Sur la figure de la terre," PVARS, April 11, 1736 and *HARS Mem.* (1736): 302–312

78. A full account of the expeditions and their aftermath is offered in Terrall, *Man Who Flattened the Earth,* chaps. 4–5.

79. Terrall, *Man Who Flattened the Earth,* 101.

80. Terrall, "Representing the Earth's Shape," 224.

to secure their broader ambitions. They pursued these parallel campaigns, however, while having virtually no contact with one another.

Voltaire's life during his exile at Cirey was full of intellectual activity, and while he remained outside the literal eye of the Parisian public, managing his reputation in this space was a top priority. A new Francophone periodical that appeared in June 1733 can serve to illustrate the challenges that Voltaire faced. Published in The Hague and devoted to "works by the savants of Great Britain," *La Bibliothèque gritannique* was at one level just one more learned Dutch Francophone periodical.[81] Yet in the journal's opening "Avertissement," it justified its focus on England in a way indicative of the changing climate of the times. "More than any other country," the editors wrote, "England is fertile in works remarkable for their novelty, singularity, and strength of conviction [*hardiesse des sentiments*]." The "unique liberty" that is found in England, the editors continued, a "liberty to examine everything and to bring all matters before the tribunal of reason alone," explained this singularity. "This liberty is advantageous," the "Avertissement" continued, "for it allows one to dig deeply into subjects, to make known the strengths and weaknesses of any opinion, and to judge well the importance of a topic on its own terms and independent of any external consideration."[82]

For all these reasons, the *BilbiIothèque britannique* further aligned its own journalistic practices with this same freedom, promising to show readers how English liberty has produced "great progress in the most sublime metaphysics and profound theology," even if this progress is not yet known to those outside England.[83] These pronouncements equated the journal with the agendas of Voltaire's *Lettres philosophiques;* however, tempering this libertarianism, the journal also vowed to honor the journalistic codes of neutrality and empirical objectivity as well. If topics of religious or philosophical discord ever emerged (and there was no doubt that they would in a journal of this nature), the editors promised to give an exact accounting of the issues involved without declaring either for or against any position. They also promised to banish from their reviews dishonorable epithets or malign and malicious insinuations, preferring to adopt the stance of the "faithful reporter and disinterested historian" rather than the authority of "the prosecutor or the judge." In the same spirit, the journal promised to ban satirical works and libels from their pages to "avoid at-

---

81. *Bibliothèque britannique, ou Histoire des ouvrages des savans de la Grand-Bretagne* 25 vols. (La Haye, 1733–1747).
82. Avertissement, *Bibliothèque britannique* (1733): 1: 3–4.
83. Ibid.

tracting the disgust and indignation of those *honnête gens* who express nothing but disapproval for such odious practices."[84] This orientation fought against the witty, satirical tone of Voltaire's *Lettres philosophiques.*

In practice, the editors adhered to their mission statement, reviewing a number of important English texts from the years 1733–1740 (including many that were scandalous) in a style that was decorous, factual, and nonpartisan. When it chose to devote two articles in late 1733 to the London edition of Voltaire's *Letters on the English Nation,* therefore, their agenda was neither to rebut its arguments nor to enflame the controversies contained within it. Accordingly their review was cordial and matter of fact and did little to either retard or provoke the controversy that surrounded the work. Other journals and journalists in this period, however, adopted a different stance with respect to the increasingly prevalent discourse of English intellectual liberty, and in their works Voltaire was treated very differently.

Among those sympathetic to Voltaire was the abbé Prévost, who created his own journal in 1733 similarly oriented toward English notions of intellectual liberty.[85] Like Voltaire, Prévost had spent a significant period of time in England as a French exile, acquiring in the process a proficiency with the English language and a love of English culture. Also like Voltaire, he began to see opportunities for playing the role of cultural bridge and arbiter between England and France. His focus was on English literature rather than philosophy, and his greatest single work was an English-style novel, *Manon Lescaut,* inspired in no small part by the novels of Richardson. Journalism also captured his attention, however, and returning to England in January 1733 with a host of financial difficulties to resolve, he conceived (with the help of a Parisian publisher) a project for a weekly periodical devoted to discussions in the Republic of Letters. His avowed models were English periodical works like the *Bee,* or to use its full and most descriptive title *Universal Weekly Pamphlet. Containing an Abridgement of every Thing Material, and all the Essays worth Reading in the Weekly Papers; The strongest Arguments on each Side of the Question, in all Disputes of a Publick Nature, placed in a fair and impartial Light.* This and a host of similar works were flooding the English public sphere in the early 1730s, and Prévost

84. Ibid., 1: 4–5.

85. On Prévost, see Steve Larkin, introduction to *Le pour et contre (nos. 1–60),* ed. Steve Larkin (Oxford, 1993). My citations from this journal will be from this edition. See also Marie-Rose de Labriolle-Rutherford, *Le pour et contre en son temps* (Oxford, 1965); Jean Sgard, *Le Pour et contre de Prévost* (Paris, 1969); and Henry Harisse, *L'abbé Prévost: Histoire de sa vie et de ses oeuvres d'après des documents nouveaux* (Paris, 1896).

imagined a Francophone work, published in Paris, that would bring the immediacy and avowedly critical tone of this journalism to the French market.

Nothing like the *Bee* existed in France or even among the Francophone periodicals that were falling off of Dutch printing presses in ever greater numbers after 1730. Market opportunities thus existed for similar works addressed to French readers. At the same time, Prévost was also a man of letters who admired traditional learned journalism and the savants, such as Bayle and Le Clerc, who had made journalistic writing and editing an esteemed pathway to honor and glory within the Republic of Letters. Accordingly, his journal, which he called *Le pour et contre*, was a compromise between the free and bawdy journalism of the English weeklies and the more sober and judicious commentary of the established learned press.

Liberty was nevertheless central to Prévost's journalistic conception, and his particular relationship to it made his periodical innovative. His first issue, which appeared in June 1733, included a long but highly indicative subtitle: *Periodical work in a new style [d'un nouveau goût] in which everything that might interest the curiosity of the public regarding science, the arts, books, and authors is freely [librement] explained without taking any position or offending anyone.* It also opened with an articulation of the journal's mission, a statement that similarly used liberty as its guiding theme. "Although the French are a free nation [*un nation libre*]," Prévost explained, "and although they enjoy many advantages under the present administration, which is full of gentleness and moderation, the fate of the N. du P. has shown me that this liberty has his limits." [86] The "N. du P." was the *Nouvelliste du Parnasse*, abbé Desfontaines' first periodical that was suppressed by royal authorities in March 1732. Prévost invoked the journal's history throughout his first issue in order to position his own journalistic endeavors alongside it. However, since the fate of this journal and its editor were also relevant to the larger history of this period, it is important to digress briefly to consider it.

Desfontaines was about ten years older than Prévost, Voltaire, and Maupertuis and like the first two he established himself as a man of letters after 1715 through a combination of literary writings, translations, and essays. He made a particular name for himself through his French translations of English literary works such as Swift's *Gullliver's Travels* and Pope's *Rape of the Lock*. Like Prévost as well, Desfontaines also moved from English literature into journalism, serving as one of Bignon's team of reviewers at the *Journal des savants* from 1724 to 1727. A sexual scandal involving a young boy sent him to the Bicêtre prison in 1724, but despite this spot on his reputation, he remained

86. Larkin, *Le pour et contre,* 54.

a largely successful, if always controversial, figure in the Parisian literary scene throughout his life.[87]

Desfontaines' first, independent journal *La nouvelliste du Parnasse* was born in the fall of 1730 as a result of a combination of literary ambition and financial need.[88] Earlier in the year, Desfontaines had written to his patrons expressing financial hardship and by the end of the year he had earned a privilege to begin publishing a new weekly. Adopting the genre of the letter because it allowed for "an easy and free style" that encouraged "vivacity and sparkle [*éclat*] in the reflections," Desfontaines offered a weekly assessment of the "most recent books, plays, and pamphlets" that were then drawing attention in "*le monde.*" He preferred "the liberty of reflection" to the "regularity of reviews" and he thus self-consciously avoided effecting what he called "the air of the journalist," which is to say the sober tone of factual objectivity characteristic of the established learned press.[89] His letters were indeed boisterous, personal, and critical in ways that no other officially sanctioned French periodical of the time was, and since he quickly made clear his classical and anti-Modern sensibilities, singling out Houdar de la Motte and the writers who congregated around him at the Café Gradot for particular criticism, his periodical soon attracted opponents. In January 1732, one of his letters was censured by the royal authorities for its libelous content, and three months later, when Desfontaines offered his most sustained and direct attack yet on the moderns camp yet—the context was Trublet's eulogy for Houdar de la Motte, who had died four months earlier—the Crown shut down the journal permanently.[90]

Later sources, including Trublet who participated directly in the events themselves, claimed that Fontenelle used his political influence to silence Desfontaines.[91] Whatever the actual events, the demise of this popular and successful journal sent an important shock wave throughout the French literary scene. Given that the first number of *Le pour et contre* appeared little more than a year after this episode, Prévost's invocation of this history in his own self-presentation was anything but surprising. Like Desfontaines, he wanted to produce a work that emphasized critical judgment and the free exercise of the mind and the pen; however, given the contemporary intellectual climate in

---

87. Waddicor, *Voltairomanie,* vii–xxiii.

88. Ibid., xxiv; Morris, *L'abbé Desfontaines et son role dans la literature de son temps,* 102.

89. *Le nouveliste du Parnasse,* 3 vols. (Paris, 1731–1732), 1: 278–9.

90. Morris, *L'abbé Desfontaines,* 47–48.

91. Ibid.

France, one had to tread carefully. Prévost's encounter with the French censor charged with approving his work in March 1733 was revealing in this respect. In addition to a work that discussed "sciences, the arts, books, and authors without taking sides or offending anyone," Prévost had also proposed the inclusion of discussions of "civil, political, and ecclesiastical affairs" as well. The censor excised these terms from the title. He also encouraged Prévost's promise, delivered to the public in his first letter, that he would never reason about political affairs in a way that "disrespected the powers that one must respect" or "indiscreetly revealed that which they judge inappropriate to reveal." "Voilà, the limits [of French liberty]," he concluded.[92]

These abjurations help to explain why Voltaire's far less restrained discourse in the *Lettres philosophiques* was treated so harshly by the authorities in 1734. Prévost further made it clear that provoking scandals of this sort was not his goal. He in fact offered a defense of the mixed nature of French liberty in his first issue, writing that "even among those nations that glory in their liberty without limits, there is not one sensible person who does concede that reason, justice, morality [*honnêteté des moeurs*], religion, and the public interest all support the way things are done in France." Thus, he continued, if the liberty of the French is constrained, it is not because of rigorous government restraint. Rather, "it is because of the just and delicate idea of liberty held in France, one that does not consist, like other nations imagine, in the power to think however one wants and to openly say and write whatever one thinks, but in the discerning and moderate exercise of the talents one has received for the good of one's society."[93]

There is no reason to think that this proclamation was completely sincere. It captures well the thinking of many at the time who were actively seeking a compromise between unrestrained, and thus reckless and destructive, English-style liberty on the one hand and the complete unfreedom of absolutist coercion on the other. Voltaire attempted to negotiate a similar compromise in the years after the publication of his *Lettres philosophiques,* and if his letters from this period are consequently full of appeals to the principle of liberty, they are also full of sincere confessions of respect for self-imposed standards of moderation, respect, civility, and *honnêteté*. Prévost's review of the *Lettres philosophiques,* which appeared in his eleventh, twelfth, and thirteenth letters of *Le pour et contre* on September 21, 28, and October 5, 1733, respectively, reveals

92. Larkin, *Le pour et contre,* 2–4.
93. Ibid., 54.

well the cultural field that joined these two thinkers around these attitudes.[94] They also reveal the fissures that cut through this agreement as well. The result is a window onto Voltaire's complex struggle to fashion himself as a new kind of philosophe after 1734.

Prévost offered three lengthy reviews of Voltaire's text based upon the English edition of 1732. He claimed to possess the French originals, however, and this revelation allowed his review to constitute the first public discussion of the French edition even though the actual text appeared only six months later. Overall, Prévost was neither celebratory nor damning in his assessment of Voltaire's work. He ended his review by noting that there were "a thousand beauties that he had not touched upon" and that, therefore, he had "exhausted neither his praise, nor his criticism."[95] Yet he did offer a thourough treatment of several dimensions of the work. He saw the letters falling naturally into two parts bridged by the letter on smallpox inoculation (11). He described the first set as focused on questions of religion and the second on philosophy and politics. Introducing his approach to the work as whole, he noted that the book was pleasant to read and full of "*esprit*" and "*agrément.*" However, he also stressed that a full judgment would require "knowing whether exactitude is found in the facts, justice in the criticism, etc." In the end, Prévost insisted, the key was determining whether the proper balance had been struck between "*le bel esprit* and exactitude."[96]

The review in its entirety surveyed and evaluated all the letters with this criteria in mind, but a precise set of biases also guided Prévost as well. Having opened up a discussion of Voltaire's writing about the religious sects of England in his first review, he abruptly closed it by saying that "the necessity that I have imposed upon myself to guard a respectful silence regarding matters of religion and politics does not permit me to extend my discussion as much as I would like in making this work known." Voltaire's other letters, which Prévost promised to discuss in his next issue, "will open a freer field for me," he declared, but here too his presentation was constrained.[97] He discussed at length Voltaire's letter on Bacon, echoing in large part Voltaire's own celebration of the lord chancellor and his claim that Newton's "système de l'attraction" was first articulated in Bacon's works. Prévost also indicated, however, his particular opinion regarding this matter by repeating Voltaire's contention that Bacon's moral philosophy is less read than it should be because of the appeal of

94. Review of Voltaire, *Lettres philosophiques*, in ibid., 154–157, 167–173, 177–181.

95. Ibid., 181.

96. Ibid., 155.

97. Ibid., 157.

La Rochefoucauld's more satirical and Montaigne's more skeptical approach to moral discourse.[98]

To celebrate Bacon's plainspoken approach to morality over and against that of these famous Frenchmen was to make an argument for the superiority of English values over prevailing French attitudes. Voltaire certainly intended such an argument, and Prévost reinforced it in his review. But elsewhere he cut against the grain of Voltaire's agendas. Speaking of Voltaire's letters on philosophy, for example, a critical discourse emerged that was representative of the wider struggles that Voltaire faced. "To be blunt, this twelfth letter [on Bacon] is constructed so agreeably, despite the serious philosophy that is mixed within it, that it shows the failures of the next five." These were the letters on Locke and Newtonian philosophy, and Prévost had little affection for Voltaire's work here. "In speaking of M. Locke and Newton," the reviewer wrote, "M . . . wanted to be nothing more than a *philosophe*. He exposed their systems, showed their reasoning, and reasoned with them. He showed them saying very good things and he also said very good things about them himself." Prévost, however, was not happy with this approach. "*Le beau sexe* who make up half the world, and at least three-quarters of those who make up the other half, would willingly have dispensed with this apparatus of philosophical science." This audience would have preferred, he argued, that Voltaire imitate "a certain singer who makes the dead speak and carries his reader so agreeably to the moon."[99]

Prévost was alluding here to Fontenelle's *Conversations on the Plurality of Worlds*, a work that used serious philosophical questions to engage in playful and entertaining literary sport. Voltaire was attempting to construct a different kind of philosophical discourse in his *Lettres philosophiques,* one that avoided the *mondain* play of traditional libertine discourse so as to assert and defend authoritative philosophical positions. Accordingly, he bristled at Prévost's suggestion that he "temper the dryness of his [philosophical letters] with some agreeable fiction . . . since it costs so little to a beautiful imagination."[100] "[This journalist] lacks style and learning," Voltaire retorted, "and he especially needs to acquire judgment, intellect, and talent."[101] In Prévost's estimation, Voltaire should have focused his writing less on utility and more on *agréments.* Had he done so, "the seven-eighths of the world of which I spoke would not have found the need to skip through so many pages." Prévost described the other one-eighth as "deep and well-instructed minds that are difficult and a bit ar-

98. Ibid., 168.
99. Ibid.
100. Ibid.
101. Voltaire to Cideville, 1 July, 1733, in Voltaire, *Corr.,* 86L D626.

rogant [*orgeuilleux*]." [102] He certainly believed that they would find the letters accessible, yet even here he did not think Voltaire's work was successful. "They were not at all happy with the attempt to explain all of ancient and modern philosophy in five short letters," he opined. They further found it "disrespectful to figures such as Newton, Descartes, Locke, etc., to attempt to give 'a light idea' of their profound speculations." Accordingly Prévost believed that Voltaire ultimately gave "too much to the first and not enough to the second." [103]

The review did not end here, however. Prévost devoted the rest of his second installment to a long and largely praiseworthy discussion of Voltaire's letters on English tragedy and comedy. In his third and final installment, he also discussed the final letters of the *Lettres philosophiques,* noting both the "inappropriate dryness" of Voltaire's letter on the consideration due to men of letters but also the "éclat" of the letter on the scientific academies of England and France. Overall, the review sustained no single critical agenda, but Prévost nevertheless revealed an important critical slant in how he chose to praise and criticize Voltaire's effort. Most important was the journalist's assumption that a strong divide existed between philosophy and *le beau monde,* a divide that marked out distinct audiences, discursive conventions, and epistemologies for the writer. Of further significance was his claim that *le beau monde* was more important for a writer like Voltaire. This assumption led to his most important criticism: that Voltaire's attempt to navigate between serious philosophy and worldly society, and even to attempt a merger of them, was a failure. Central to the new critical, intellectual identity inscribed in the *Lettres philosophiques* was the claim that writers could channel the public authority of *mondain* society into intellectual authority with respect to philosophy. Prévost, however, found such a project schizophrenic. He further undermined it by holding Voltaire's attempts at philosophical discourse up to the intellectual values that defined *mondain* society, namely, *esprit,* wit, *agrément,* and stylistic panache. By this standard, Prévost insisted, the *Lettres philosophiques* had failed.

Others in the period approached the *Lettres philosophique* in the same way, and accordingly Prévost's review illustrates well Voltaire's struggles in the 1730s to win respect as a serious philosophical thinker and an authoritative speaker about science. He was fully aware of this difficulty and set out almost immediately to remedy it. But other aspects of his identity and work also created obstacles in this effort. Most difficult was the struggle to sustain the critical discourses that were central to his identity as a leading writer and *bel esprit* while also establishing the *honnête* authority necessary to be taken seriously as

102. Larkin, *Le pour et contre,* 169.
103. Ibid.

a credible philosophe. Institutional power made this project especially difficult for Voltaire since he possessed neither the clerical authority necessary to authorize his statements about theologically laden philosophy, nor the academic or university seat that would have sanctioned his discourse about science. Without such support, Voltaire sought to ground his authority in the wider public of the Republic of Letters, but this created as many problems as solutions.

To the extent that it gave him a position from which to speak, it allowed him to realize his ambitions. Furthermore, to the extent that established institutions such as the Royal Academy had made this wider public an authority to which it submitted itself, it also held out the promise of a new kind of independent authority rooted in the public at large. However, to the extent that this institutional positioning also forced Voltaire to navigate the divides between literature and philosophy, truth and pleasure, and liberty and authority in a particularly ungrounded way, it also created new challenges for him. Voltaire's precise struggles in the years immediately following his retreat to Cirey illustrate well his dilemmas.

Almost immediately, he set to work writing what he called "a little metaphysical tract," a work that revealed his commitment to English philosophy and deism.[104] Locke's sensationalism and materialism, as well as Clarke's voluntarist natural philosophy, were key components in Voltaire's work, for overall the text echoed in countless ways the English thought that he had celebrated in the *Lettres philosophiques*. Du Châtelet was similarly at work in this period on a book of rational biblical criticism, and together they joined forces in thinking and writing their way toward a fully developed scientific deism.[105] They also kept these writings resolutely private, even if Voltaire did write a long letter to his old Jesuit philosophy professor, Father Tournemine, seeking his blessing for his increasingly Lockean and Newtonian views on physics, metaphysics, and natural theology.[106] The gesture reveals Voltaire's longing throughout this period for official sanction of his views. But it also reveals how, in the face of his growing reputation as a radical philosophical thinker, he increasingly embraced rather than fled from this identity.

The couple also attempted to construct a public foundation for articulating their philosophical views in other ways as well. Beginning in late 1734,

104. Voltaire, *Traité de metaphysique*, in Beuchot ed., *Œuvres complètes de Voltaire*, new edition, 52 vols. (Paris, 1877–1885), 22: 189–230.

105. On these collaborations, see especially Wade, *Voltaire and Madame de Châtelet*; Vaillot, *Avec Madame du Châtelet*, 33–43.

106. Voltaire to René Joseph Tournemine, August 1735, in Voltaire, *Corr.*, 87: D901. See also the reply, Tournemine to Voltaire, September 1735, in Voltaire, *Corr.*, 87: D913

they began ordering scientific instruments from the leading vendors in Paris to construct an experimental laboratory at Cirey for serious scientific investigation. Part of the impetus behind these purchases was the Royal Academy of Science's biannual prize contest for 1734–36, which focused on the nature and propagation of fire. Both Voltaire and du Châtelet decided to compete for the prize, and in April 1736, when the winners were announced, the couple learned that they were each awarded an honorable mention.[107] Voltaire made particular use of experiments performed at Cirey in defending his theories, and since the results of the prize were widely publicized, this worked to elevate his status as a serious scientific thinker.[108] In this context, he also conceived of his next great intervention into the Newtonian debates of the period. He began to compose a popular exposition of Newtonian natural philosophy in the manner of the great English expositor Henry Pemberton. Voltaire first mentioned this work in his correspondence in January 1736, and by the end of the year he had departed for Holland to attend 'sGravesande's public lecture courses in Newtonian experimental philosophy and thus perfect the work.[109]

Yet even while Voltaire was laying the groundwork for his claim to speak as an authoritative, Newtonian philosophe, other, equally powerful aspects of his identity were undermining his effort. In January 1737, just days after his arrival in Holland, the *Gazette d'Utrecht* noted his presence by declaring on the "authority of several people" that "M de Voltaire will never return to France again since he would prefer to live in foreign lands where it is possible to write with a full liberty of thought and feeling."[110] Voltaire had expressed similar sentiments to his correspondents, noting to one his persecution in France as an atheist and to another his love for the "liberty of Amsterdam."[111] These comments notwithstanding, however, Voltaire was not fleeing French persecution in going to Holland. Quite the contrary, he was trying to establish a base from which to become a legitimate philosophic presence within the country. Yet this did not stop his enemies and rivals from insinuating something very different.

107. PVARS, April 11, 1736.

108. Review of the papers that won the Paris Academy of Sciences prize for 1736 on the nature and propagation of fire. Desfontaines offered a full account. See *OEM* 18 (n.d.): 337–353, 19 (n.d.): 97–112, 20 (n.d.): 169–186.

109. Vaillot, *Avec Madame du Châtelet*, 63–71.

110. This episode is recounted in Jeroom Vercruysse, *Voltaire et la Holland* (Oxford, 1966), 35.

111. Voltaire to Frederick, 1 January 1737, in Voltaire, *Corr.*, 88: D1249; Voltaire to Prévost, 16 March 1737, in Voltaire, *Corr.*, 88: D1298.

The announcement in the Utrecht press was in fact a sign of a brewing scandal, and at the center of it was a set of bawdy, libertine verses called *La pucelle d'Orleans*. Voltaire certainly wrote the verses, but he had no intention of circulating them. Nevertheless, they fell into the wrong hands and triggered a scandal in Paris and elsewhere. In particular, since the verses indulged in a host of sexually explicit references and innuendos, Voltaire was forced to defend himself against charges of immorality.[112] "Rousseau and Desfontaines are behind this," he fumed, a reference to the journalist Desfontaines, Voltaire's one-time ally, and Jean-Baptiste Rousseau, who was living in Brussels after his exile from France as a result of the "affair of the couplets" involving Saurin and the Café Gradot.[113] Whatever their involvement in the scandal of *La Pucelle,* Rousseau and Desfontaines certainly made life difficult for Voltaire during these years. In early 1736, for example, Rousseau published a libel in the *Bibliothèque française* that purported to reveal a scandalous relationship between the poet and a fictious "ami" who was alleged to be the author of a recently published scandalous tract.[114] Desfontaines' critique was more subtle, but he also used an early issue of his new journal, *Observations sur les écrits modernes* launched in 1735, to discuss a work by an Englishman critical of Voltaire's *Letter philosophiques.* He also published, without Voltaire's approval, an *Epistle to M. Algarotti* that pointed embarrassingly to the unorthodox relationship between Voltaire and du Châtelet at Cirey.[115]

These insinuations only furthered Voltaire's reputation as a reckless freethinker and immoral libertine. Such perceptions were further catalyzed by the circulation of not only *La pucelle,* but another poem, *Le mondain,* that openly celebrated hedonistic living and luxury as the engines of moral and social progress.[116] Taken as whole, these scandals fueled the sense that an unholy trinity existed between libertinism, irreligion, and free literary expression and that Voltaire was an aspiring patron saint of this anti-Church. Newtonian natural philosophy was also routinely associated with the same scandalous trinity (especially by critics, like Leibniz, who wanted to make it the philosophical highway to religious and moral corruption). Since Voltaire's *Lettres philosophiques* struck matches near this very gunpowder, Voltaire's Newtonianism exacerbated rather

112. Vaillot, *Avec Madame du Châtelet,* 67–68
113. Cited in ibid., 67.
114. Ibid., 47.
115. *OEM* 2 (n.d.): 299–305, 3 (n.d.): 142–144.
116. A copy of the poem and an analysis of it is found in André Morize, *L'apologie du luxe au XVIIIe siècle: "Le Mondain" et ses sources* (Paris, 1909).

than alleviated these perceptions. Rousseau and Desfontaines exploited these attachments in their budding rivalry with Voltaire, and in early 1737 each began to publicize a story that charged the author of the *Lettres philosophiques* with further scandalous conduct. The report claimed that 'sGravesande had been forced to expel Voltaire from his classes after the French visitor had instigated improper discussions about the existence of God and the immortality of the soul. Formont reported that the story was circulating in Paris in January 1737, and even though 'sGravesande agreed to issue a disclaimer dismissing the validity of the entire story, the damage was already done. These scandals made it hard for Voltaire to assert and defend a philosophical position with respect to Newtonian philosophy without the taint of libertinism, irreligion, and immorality being joined to it.[117]

Given this situation, one understands why Maupertuis maintained such a careful distance from Voltaire before 1738, despite the intellectual and social interests they had in common. One also understands why academicians like him worked so hard to keep the debate inside the Paris Academy about the figure of the earth detached from the wider public discourse that made this a test case over Newtonian philosophy. By 1738, Newtonianism was becoming increasingly associated in France with reckless libertinism, immorality, and irreligion, not least because its core elements, the theory of universal gravitation and the physics of corporeal attraction across empty space, were perceived to be at best philosophically suspect and at worst veiled forms of immanent, pantheist materialism. Voltaire's perceived attachment to all of these suspect positions, therefore, accentuated rather than tempered these connections, as did his controversial circulation within the often viciously critical world of lettered sociability. Maupertuis, by contrast, let his aggressive Newtonianism subside in this period while working successfully to achieve intellectual respectability within the establishment circles that looked condescendingly upon Voltaire's pursuit of *honnête* intellectual liberty. After 1738, however, things began to change.

117. Vercruysse, *Voltaire et la Holland*, 41–43.

6

A French Culture War
The Battle over Newtonianism

By 1738, all the key theaters of the Newton wars had opened and were hot with conflict. Maupertuis' *Discours* and the discussions of Newtonianism that it triggered inside and outside the Paris Academy; the emergence of a coherent and publicly discussed Cartesianism that asserted itself against the Newtonian position; the isolation of a set of clear and concise "test cases"—the figure of the earth most prominently—that could serve to focus the war; a new critical, libertarian spirit manifest in English thought and Francophone journalism, as well as its increasing circulation within France; Voltaire's *Lettres philosophiques* and its scandalous adoption of this liberty in the name of English philosophy, science, and culture; the rise of the figure of the earth question as a central focus of the Paris Academy; Maupertuis' activities in this debate, and his perceived connection to the wider academic public that included Voltaire and du Châtelet; the controversies that continued to follow the two as they navigated the terrain of French public science: all of these factors led to the beginning of the public battles over Newton that erupted in full force after 1738, yet no one of them overdetermined the rest. Instead, they all converged to launch France into a bitter and consequential culture war over the nature and limits of true natural philosophy, one with transformative consequences.

The Newton Wars in France: The First Battles

Maupertuis' public assembly address in November 1737, just weeks after his return from Lapland, marks one milestone in this transformation. In the years leading up to this event, Maupertuis worked vigorously to sustain his views on the shape of the earth in a way that made his academic honor and scientific prestige rest upon this question alone.[1] The victory of his Newtonian position after

---

1. Maupertuis' conduct during the figure of the earth debate is analyzed in detail in Terrall, *Man Who Flattened the Earth*, chaps. 4–5. I will follow her account closely here.

1737 thus solidified his reputation and those of the other Newtonians in France who supported him. Yet because he and his closest allies had devoted themselves most intensively toward winning this battle inside the Royal Academy, the victory did not necessarily translate automatically into a victory for Voltaire as well. In fact, throughout these years, Maupertuis maintained a guarded distance from the rebel Voltaire and worked hard to restrain any equation of their public Newtonian identities. Maupertuis and the other academic Newtonians were also away on their various expeditions during the initial years when Voltaire was working hard to establish his Newtonian identity. This further separated them. It also allowed the newly self-conscious Cartesians aligned against each to enjoy a brief monopoly over academic discussions of terrestrial and celestial mechanics in France.

For Privat de Molières, this situation launched his career. After assuming Varignon's chair in mathematics at the Collège royal in 1723, he began delivering a set of annual lectures devoted to physics and mechanics.[2] Drawing upon his Oratorian training, he adopted Malebranche's vortical system as the mechanical basis for his explanations. There was nothing controversial or even noteworthy about this choice in 1723 since vortical physics remained the conventional framework for these topics in France. However, as the more strident polemicism about Newtonianism and Cartesianism began to emerge over the course of the 1720s, Privat de Molières found his position transformed. Fontenelle acknowledged the new situation in 1730 and 1731 when his public descriptions of Privat de Molières' avowedly "Cartesian" approach to mechanics were published in the academy's *histories* for 1728 and 1729. These presentations celebrated Privat de Molières' Cartesianism, while also framing it in terms of the anti-Newtonianism that was now assumed to be implicit in it.

Privat de Molières took this framework and ran with it. In his *Leçons de physique*, which appeared in four volumes between 1733 and 1738, the former Oratorian offered a complete account of terrestrial and celestial mechanics rooted throughout in the causal mechanisms of the Cartesian vortices.[3] He also offered a methodological defense of this physics by presenting his *leçons* in a deductive arrangement, one that alleged synthetic, Euclidean certainty as the epistemological glue binding his demonstrations. The *Journal des savants* noted this orientation in its review of volume 1, writing that "one is accustomed to consider physical systems as conjectures that explain the natural effects." "Descartes himself spoke of his system as a romance (*roman*) of nature," and

2. Bonnardet, "Joseph Privat de Molières," in Ac. Sci., dossier Molières.

3. Privat de Molières, *Leçons de physique, contenant les elements de la physique déterminées par les seuls lois de mécanique, expliqués au Collège Royale*, 4 vols. (Paris, 1734–1738).

even "M. Newton, who proved that the system of Descartes is subject to contradictions, imagined his own built upon the principal qualities of matter without ever revealing the causes of them." Privat de Molières hoped to advance the "progress of this science" by "substituting certain principles for the conjectures that have thus far been proposed." Accordingly, the journal explained, he presented his lessons "in the same way that Euclid presented geometry" so as to "fix forever the number and the quality of the physical principles."[4]

The reviewer praised the utility of Privat de Molières' efforts while also noting the ambitiousness of his undertaking. In doing so, he also pointed to the way that philosophical method had become a key weapon in the battle against Newtonian physics. Leibniz had accused Clarke and the Newtonians of offering no explanation whatsoever for the theory of gravitational attraction, and in France the absence of such a causal, mechanical account became by the 1730s a key reason to reject universal gravitation altogether. Maupertuis had defended the contrary position in his *Discours,* arguing that philosophical considerations were irrelevant to the truth or falsity of attraction as a physical principle. Yet Privat de Molières spoke for many French savants when he rejected such a view, claiming that the only true and tenable physics was a philosophically rational and mechanical physics. He also buttressed this view by adopting the "Cartesian method" literally and polemically in the presentation of his own, anti-Newtonian mechanics.

The result was the construction of a rallying point for the newly energized French Cartesians. Desfontaines also spoke for many when he used his biweekly newssheet, *Observations sur les écrits modernes,* to praise Privat de Molières and his work. Stressing the confidence that the book's "added rigor" provided, he further noted how Privat de Molières had not so much disproved Newton as isolated the true from the untrue in his system. Like many post-*Principia* vortical theorists, Privat de Molières was convinced that Newton's quantitative inverse square relationship was indeed a law of nature, and he thus integrated this law into his rival vortical account of celestial motions.[5] For Desfontaines, this was evidence that the Euclidean deductive method could reveal valid truths in an unprejudiced and nonpartisan way. For a man of letters like Desfontaines, such neutrality was particularly important, for it showed how proper philosophical method, rigorously applied, could remove the taint of party warfare that was already hovering over the Newtonian/Cartesian debate.[6]

4. Review of Privat de Molières, *Leçons de physique,* in *Journal des savants* (1734): 455.

5. On these further developments of the vortical theory of celestial mechanics in France, see Aiton, *Vortex Theory;* and Brunet, *L'introduction des théories de Newton.*

6. Review of Privat de Molières, *Leçons de physique,* in *OEM* 13 (n.d.): 305–312.

Desfontaines' review had a wide circulation, and his coverage of Privat de Molières on this and other issues attests to the celebrity that the former Oratorian and academician acquired after 1733 as a result of his Cartesian views. Privat de Molières' lectures at the Collège royal were also public events, and while no hard evidence exists about the size of his audience, one suspects that attendance surged during these years. His own assessment that he was, by 1739, a European scientific celebrity, was not wholly exaggerated. In fact, so long as one takes into consideration Privat de Molières' typically Paris-centered view of European opinion as a whole, he was indeed a well-known figure within it. In France, his work was much talked about, and this translated into a general elevation of his status inside the academy and in the wider public sphere. His celebrity perhaps reached its high point in 1741 when a man of letters named Prémontval began to host a set of free public courses devoted to mathematics and mechanics, courses that used Privat de Molières' *Leçons* as their core text. Another *mondain* periodical, *Nouveaux amusemens du coeur et de l'esprit*, ran an advertisement for the courses, noting the precise day, time, and location of the sessions.[7] Prémontval also published his own accompanying texts, and while nothing more is known about the nature or success of these courses, they illustrate Privat de Molières' notoriety within Parisian society of the time.[8]

Inside the academy, Privat de Molières' position was also improved even if he never obtained the *pensionnaire* seat he so coveted. Supported intellectually and professionally by Dortous de Mairan, Fontenelle, and Cassini II, to name only three of the most powerful, he came to embody the Cartesianism that these academicians increasingly defended for themselves and for the academy as a whole. In 1732, Privat de Molières was joined in this camp by another outspoken Cartesian: Étienne-Simon de Gamaches. Gamaches' biography is rather obscure, but he first appeared in the academy registers in the spring of 1732 when he was already sixty years of age. Prior to that, he had served as the *chanoine régulier* at Sainte-Croix de la Bretonnerie and had authored three books, the first a work of philosophy (1704), the second a work on rhetoric and

---

7. "Sundays from 10–11:30 AM and 3–5 PM 'sharp' in the second apartment from the last on the rue Sainte Genvieve, next to the Collège de la Marche." *Nouveaux amusemens du coeur et de l'esprit* (1734–1745), 9: 471–475. Prémontval published two works based on these courses, *Discours sur l'utilité des Mathématiques. Pronocée par Monsieur de Prémontval, à l'ouverture de ses conferences* (Paris, 1742) and *Discours sur la nature des quantités que les mathémathiques ont pour objet* (Paris, 1742). But the advertisement in the literary journal indicated that the course was based on Privat de Molières' *Leçons de physique*.

8. On Prémontval, see his obituary in *Le necrologes des hommes célèbres de France, par une société de gens de lettres*, 18 vols. (Paris, 1766–1784), 5: 95–118.

grammar (1718), and the third a treatise on mechanics (1721).[9] The last linked Gamaches to the ongoing debate about *vis viva* and the laws of impact, but it does not appear to have exerted much influence on these discussions.[10] What role he saw himself playing in these debates, or in the scientific life of the period more generally, is similarly unknown.

In March 1732, however, he presented to the academy a work entitled *Physical Astronomy*, earning a favorable review from the academicians—Nicole and Réaumur—assigned to review the text.[11] The timing appears to have been perfect, for eight years later Gamaches published a similarly titled work with the added subtitle *General Principles of Nature Applied to the Mechanisms of Astronomy and Compared to the Philosophical Principles of M. Newton.*[12] If one assumes that this text in draft form was the manuscript that Nicole and Réaumur approved in 1732, then it was this treatise that won Gamaches a seat in the Royal Academy. Two months after the academy approved his text, he was admitted as an *associé libre,* a rare designation that gave the chair holder full academic privileges without assigning them to a particular class or creating a permanent seat for them.[13] From this position, Gamaches became a very active presence at the academy sessions over the next decade, constantly presenting work in celestial mechanics.

In his first academic paper, delivered in June 1732 (one month before Maupertuis' *Discours* was approved for publication by the academy), Gamaches spoke on "the mechanism of the heavens."[14] Similarly in August 1734, just two weeks after Maupertuis delivered his first paper on geodesy and the figure of the earth, Gamaches spoke on "physical astronomy."[15] After the departure of Godin, Bouguer, and La Condamine for Peru in May 1735 and Maupertuis, Clairaut, and Le Monnier for Lapland in May 1736, Gamaches was even less constrained to use the academy to air his views. He read a variety of different

9. Étienne-Simon de Gamaches, *Système du coeur, ou Conjectures sur la manière dont naissent les différentes affections de l'âme* (Paris, 1704), idem, *Les agrémens du langage réduits à leurs principes* (Paris, 1718), idem, *Sisteme du movement* (Paris, 1721).

10. De Gamaches' *Sisteme du mouvement* appeared as an appendix in Jean-Pierre Crousaz, *Essay sur le mouvement, ou l'on traitte de sa nature, de son origine, de sa communication en général* (Groninge, 1726; 2nd ed., 1728).

11. PVARS, March 12, 1732.

12. Étienne-Simon de Gamaches, *Astronomie physique, ou Principes généraux de la nature, appliqués au mécanisme astronomique, et comparés aux principes de la philosophie de M. Newton* (Paris, 1740).

13. PVARS, May 17, 1732.

14. PVARS, June 14, 1732.

15. PVARS, August 18, 1734.

papers on issues relating to celestial mechanics, and during their absence he focused particular attention on the Newtonian theory of gravitational attraction as it related to *pesanteur*.[16] On one interesting occasion in the winter of 1737, a series of academic sessions devoted to Gamaches' anti-Newtonian work on "the mutual attraction of the planets" was interrupted by the arrival and presentation of letters from Peru and Lapland. These included an important report by Bouguer and La Condamine that confirmed Picard's pendulum experiments showing that the mass of mountains exerted an influence on the force of moving bodies.[17]

Such interruptions, however, appear to have done little to hinder the increasing aggressiveness of Gamaches' Cartesian program. He also joined with Privat de Molières in the review of a Cartesian work in mechanics by a savant named Deidier, a work that was soon after reviewed favorably in the *Journal des savants* as well.[18] None of this work was published in the academy's *mémoires*, but in March 1740 Gamaches asked the Academy to approve his *Astronomie physique* for publication.[19] It appeared soon after, accompanied by a preliminary discourse that celebrated the Royal Academy of Sciences for its refusal to embrace any physical system. In the same text, however, Gamaches also alleged "to demonstrate that the principles of Cartesian philosophy are the only ones to be adopted in mechanical astronomy." Only Cartesianism, he contended, offered the appropriately clear and distinct foundations for explaining the natural complexity of nature.[20]

The appearance of *Astronomie physique* coincided with the apotheosis of Privat de Molières in the French public sphere, and Gamaches' reception was equally favorable. The *Journal des savants* used its "Nouvelles de la republique des lettres" section in March 1740 to announce the book's appearance, promising that "we will speak incessantly of this work in the journal."[21] In September and October it published its promised reviews, continuing a trend of favorable

---

16. See PVARS, November 1, 1734; July 21, 1736, August 4, 1736, January 30, 1737; February 13, 1737; July 24, 1737. Interestingly, none of this work was published in the academy volumes for 1734, 1736, or 737.

17. PVARS, February 13, 1737.

18. L'abbé Deidier, *La méchanique générale* (Paris, 1741). The report of Gamaches and Privat de Molières is noted in PVARS, November 18, 1739. See also review of Deidier, *La méchanique générale,* in *Journal des savants* (1742): 490–493. Desfontaines also reviewed this work; see review of Deidier, *La méchanique générale,* in *OEM* 26 (n.d.): 37–46.

19. PVARS, March 9, 1740.

20. De Gamaches, *Astronomie physique,* i–ii.

21. "Nouvelles de la republique des lettres," *Journal des savants* (1740): 191.

press for Cartesian mechanics in France.[22] Three months earlier, the journal had also published a two-part review of volume 4 of Privat de Molières' *Leçons de physique,* and treated as a group these reviews echoed a number of common themes.[23] Newton, for example, was presented in each as a brilliant mathematician who erroneously let mathematical calculation stand in for rigorous, causal physics. As the journal declared through a citation from *Astronomie physique:* "[For Newton], a phenomena analyzed mathematically becomes a phenomenon explained. In this way, moreover, this famous rival of M. Descartes took great satisfaction in being a great philosophe, when in fact he was only a great mathematician." [24] This distinction between mathematical analysis and causal, mechanistic physics also supported another common claim: that both Privat de Molières and Gamaches accounted for the real innovations of the Newtonian system, namely, the inverse square law, while also providing the mechanical account of it that Newton's system lacked. As the journal expressed with respect to Gamaches, "the rules of Kepler are the laws of astronomy," and accordingly Gamaches' treatise "acquires its force only from its complete conformity with these laws." [25]

The reviewers at the *Journal des savants* continually praised this mechanistic approach, and Desfontaines did as well in his account of Gamaches' work. "The immense riches that this famous rival of Descartes pulled from the most sublime geometry are prodigious beyond measure," the journalist wrote. Newton, by contrast, was a "dangerous philosopher" because he offered "a chain of principles that appear to subsume all of the phenomena of nature" but are in fact only "a seemingly harmonious chain" of empirical associations. "The geometrical analysis of Kepler's law confirms that the planets gravitate [*pese*] toward the sun, but this in no way proves that the sun must gravitate toward to the planets. To suppose it, as Newton has done, is simply to divine it." [26] Both Desfontaines and the *Journal des savants* also reinforced the nationalist categories that were increasingly central to this debate. The official French journal called Gamaches an "honor to the nation." It also quoted him in alleging that Newton, impatient with having to submit himself to the instruction of a foreign nation, had developed his philosophy in an effort to "liberate

22. Review of de Gamaches, *Astronomie physique,* in *Journal des savants* (1740): 547–563, 620–634.

23. Review of Privat de Molières, *Leçons de physique, vol. 4,* in *Journal des savants* (1740): 263–269, 387–393.

24. Review of de Gamaches, *Astronomie physique,* in *Journal des savants* (1740): 550.

25. Ibid., 551–552.

26. Review of de Gamaches, *Astronomie physique,* in *OEM* 26 [1740]: 65–66.

his homeland (*patrie*) from the necessity of having to borrow from us the art of clarifying the steps of nature." [27] For his part, Desfontaines called Gamaches' *Astronomie physique* an honor to "the nation, the Academy, the religious house of the author, and the genius which led to its birth." [28] He further expressed a debt of patriotic gratitude since "the honor of our nation and of philosophy is at stake in his demonstration that the founding principles of Cartesian philosophy are the only ones suitable to mechanical astronomy." [29] "The systematic spirit, which since the time of Descartes has so advantageously characterized the genius of our nation, is rapidly making progress among those that emulate the illustrious savants of France," Desfontaines enthused. Newtonianism, however, represented a threat to this French philosophical spirit, and Gamaches was to be praised for resisting its "degradation among those who allow isolated principles to stand in for the explanation of difficult phenomena." [30]

These same themes were echoed in other venues as well, and taken as a whole they allowed Privat de Molières and Gamaches to become prominent spokesmen for the newly clarified French Cartesianism that opposed itself to the French Newtonianism of Maupertuis and Voltaire. When Maupertuis and his Lapland team returned to Paris in September 1737, this Cartesianism had begun to assert itself more loudly and aggressively in the French public sphere; ironically, however, Maupertuis and his allies chose not to contest it directly but instead to focus on other battles instead. By 1737, the question of the figure of the earth had been posed in many people's minds as a test between the two great natural philosophic systems. Nevertheless, this is not how Maupertuis pursued the matter, at least before 1740. Cassini II and the royal astronomers saw it as a simple question of empirical fact, and Maupertuis accepted this definition, while trying to establish his own credibility in the empirical sciences that mattered in this arena, namely, geodesy. He had worked hard between 1734 and 1737 to establish an authoritative reputation in this area, and when he returned from his arduous work in Lapland with a set of empirical findings that confirmed the thesis of a flattened, grapefruit-shaped earth, he expected to be rewarded with new prestige and glory within the academy. Instead, Cassini II and his allies questioned the credibility of Maupertuis' work, drawing attention to the errors that may have been produced by Maupertuis' new and untested instruments, not to mention the inexperience of the observers. [31]

27. Review of de Gamaches, *Astronomie physique*, in *Journal des savants* (1740): 547, 550.
28. Review of de Gamaches, *Astronomie physique*, *OEM*, 26 [1740]: 67.
29. Ibid., 26: 63.
30. Ibid., 26: 64.
31. See Terrall, *Man Who Flattened the Earth*, 130–142.

What ensued, therefore, was a bitter and increasingly personal debate within the academy about the true shape of the earth, but one that focused in no way on the hydrodynamical and other physical theories that had framed the question as a rivalry between Newtonian and Cartesian mechanics. Furthermore, while all indicators suggest that the battle itself was bitter and intense—Madame de Graffigny described it as "a civil war worse than one can say"—it was largely contained within the walls and gentlemanly decorum of the academy itself.[32] Maupertuis reported his findings to Maurepas and his correspondents in September on the eve of the academy's long fall break, and the *Mercure de France* reported them immediately to the wider public, saying that the team had determined "that the earth is a spheroid flattened at the poles as Mssrs. Huygens, Newton, and several other great astronomers had thought, based on theory."[33] Other than that, however, discussion of the measurements or the conclusions that Maupertuis drew from them was absent from the public sphere until November, when Maupertuis used the academy's public assembly to deliver an eloquent presentation of his team's work.

The *Mercure* reported on the "marked interest of the most numerous assembly that there has ever been in any academic meeting," and the journal further praised "the finesse and exactitude of the operations, and the clear and elegant manner in which M. de Maupertuis made everyone capable of judging it."[34] Réaumur agreed. Writing to Bignon, who had only a year earlier retired from his position as the ministerial manager of the French learned establishment, he noted that "M. de Maupertuis' report lasted more than an hour and a half, yet everyone in the audience found it too brief." He further described the size of the crowd, writing that "the gallery was filled with those who were not able to enter into the hall." Describing the combination of scientific sobriety and *bel esprit* that characterized Maupertuis' discourse as well, he further observed that "had this account been printed just after these gentlemen arrived, many of the nasty comments made in the cafés during this vacation would not have been spoken."[35]

This reference to the wider public buzz regarding the trip and its conclusions is revealing of the broader notoriety that the figure of the earth debate had acquired by this time. However, in the more formal organs of public discourse such as the learned journals, discussion was far more subdued. Overall, an *honnête* tone characterized what little public discourse about the debate escaped from the academy. In his public address, Maupertuis treated the Cassinis and

32. Cited in ibid., 147.
33. *Mercure de France* (1737): 2032.
34. Ibid., 2462.
35. Cited in Terrall, *Man Who Flattened the Earth*, 134–135.

their work respectfully, noting simply that his team's findings revised theirs. Inside the academy's walls, by contrast, the arguments were no doubt vitriolic as Cassini II refused to accept Maupertuis' results and called instead for a new survey using more trustworthy instruments as the only means for resolving the question reliably. These rebuttals infuriated Maupertuis, who saw in them little more than an authoritarian use of academic power to thwart what he knew to be the truth. As he wrote privately at the time: "Cassini's paper [delivered inside the academy] dishonors him and the institution; it also does not make the earth any less flattened at the poles." [36]

Cassini II was certainly defending a pet theory in these rebuttals, but in other ways he was also offering a reasonable scientific challenge to Maupertuis' alleged revisions to his work. As an empirical matter of fact, the precise shape of the earth could only be determined through a consensus about the empirical results in question. Cassini II had real reservations about the reliability of Maupertuis' work, and in this respect his caution about leaping to overly hasty conclusions was justified. Also legitimate was the assumption that Maupertuis contain his desire for immediate scientific glory until all the facts were in and confirmed. Maupertuis found this imperative difficult to live with, even if he largely respected it. As a result, while he privately made clear his utter frustration with Cassini II and his position, publicly he worked hard to overturn the arguments of his opponents according to the terms set by Cassini II and his epistemological assumptions.

Maupertuis continued to rebut Cassini II's critique of his measurements inside the academy, and in this effort he was joined by Clairaut among others. He also encouraged the acquisition of new geodesic measurements, working in particular to win royal financial support for new expeditions that would vindicate his earlier work. However, when called upon to speak publicly about the matter, either orally or in print, he maintained an *honnête* decorum with respect to Cassini II and his work. He did persuade the Swedish astronomer Anders Celsius, an assistant during the Lapland expedition, to publish a pamphlet critical of Cassini II and his work.[37] The French astronomer J. N. Delisle, who was struggling to stay abreast of the controversy from his position at the St. Petersburg Academy of Sciences, read Celsius's tract and found it to be "very rude." He hoped that Celsius would soften it before publishing it. Celsius replied that Cassini II "brought the attack on himself," and he published the text without emendations.[38] Yet since the pamphlet was written in Latin,

36. Cited in ibid., 135.
37. Anders Celsius, *De observationibus pro figura telluris* (Uppsala, 1738).
38. Cited in Terrall, *Man Who Flattened the Earth*, 137–138.

it exerted little apparent influence on the debate outside the academy. None of the more accessible Francophone journals reported on its contents, and even though Cassini II devoted several academic sessions to a fierce rebuttal, publishing the result, in French, as a pamphlet of his own, it too appears to have generated no wider comment.[39]

Meanwhile, at the public assemblies of the Royal Academy held in the spring of 1738, 1739, and 1740, respectively, the antagonists refrained from any open sparring even if they addressed the question of the figure of the earth in each of these sessions. Eventually new measurements conducted by Cassini II's son, Cassini de Thury, vindicated Maupertuis' position, and these results coincided with the retirement of the elder Cassini, a move that allowed the mantle of the elongated earth to pass to a new generation. Cassini de Thury proved more artful than his father in resolving the controversy, and in the spring of 1740 he conceded that the earth was indeed an oblate sphere, while also emphasizing that it was the accumulation of reliable, empirical evidence, especially through surveys that he and the other royal astronomers had conducted themselves, that ultimately resolved the question.[40] This allowed the Cassini family, and the dynasty that they still managed at the Royal Observatory, to lose the battle while winning the larger epistemological war. It also allowed the academy to escape from a bitter scientific battle with its *honnête* codes of nonpartisan independence and gentlemanly decorum intact.

Maupertuis benefited personally from this resolution and from his conduct overall throughout the controversy. After his arrival from Lapland but before his public assembly address of November 1737, he was offered a pension from the Crown in recognition of his service. He refused it. Du Châtelet called the award "mediocre" and thus supported Maupertuis' decision to decline this "demeaning" reflection of his real honor and status. She also insinuated that it was "the persecution" of "Cassini and his Jesuit allies" that had triggered the Crown to act in such a niggardly fashion.[41] Whatever the real story behind the pension, the decision placed Maupertuis in a precarious position with respect to Maurepas, who had otherwise been a generous patron and supporter. As the controversy over the figure of the earth raged, Maupertuis also had reason to worry that his contentiousness on this matter would further alienate the minister.

In 1739, while Mauptertuis was campaigning for royal support for his new geodesic measurements, Maurepas approved the Cassinis' request while ini-

39. PVARS, April 30, 1738, and May 3, 1738; *Réponse à la dissertation de M. Celsius* (Paris, 1738).

40. See Terrall, *Man Who Flattened the Earth*, 151–154.

41. Cited in ibid., 134.

tially refusing Maupertuis' similarly conceived initiative. By August, however, Maupertuis had won over his ministerial patron, for during the academy's fall vacation of 1739, as the academician prepared a new paper defending the theory of the oblate earth, he was rewarded for his diligence. While visiting the royal court at Fontainebleau, Maupertuis was notified that he had been awarded a lucrative pension and an honorific office as the royal savant in charge of the perfection of navigation. "It is a position created expressly for me by M. de Maurepas," Maupertuis gushed to Johann Bernoulli. "Everything was done in the most gracious way. It makes me very happy and quite comfortable."[42] This recognition preceded Cassini de Thury's concession with respect to the shape of the earth by over six months, and the reward thus acknowledged Maupertuis' honorable and office-worthy conduct in the dispute as much as any particular scientific theory or accomplishment he had made.

Outside the official institutions of French science, however, matters were perceived very differently. "You [M. Maupertuis] are extremely *à la mode*," announced the duchesse de Saint-Pierre from her residence with the royal court at Fontainebleau.[43] Such opinions were not entirely arrived at accidentally, for Maupertuis worked vigorously to create this favor, even if he also worked hard to craft an identity as a serious academician as well. One crucial moment in this wider campaign occurred in December 1738 when his frustrations with the Cassinis led him to flee the academy for a period. "Since I do not have the approval of the academy," he wrote to the younger Johann Bernoulli, "I am going to take advantage of this time to wander around and divert myself."[44] What precisely he did is not at all clear, but he certainly pursued a great many activities in *le beau monde,* and very few that involved Graham quadrants or detailed trigonometry, his weaponry in his academic battles. Whatever his recreations, his wanderings immersed him in the public discussion of the figure of the earth that was then raging, a discussion that made it a battle between Newtonian and Cartesian physics rather than over the instrumental prowess of French academicians. Within the terms of this public understanding, he was an avowed Newtonian challenging the narrow, Cartesian prejudices of Cassini II and the Royal Academy. This was, of course, a caricature of the actual debate inside the company, which never really operated as a battle between Newtonians and Cartesians or as a test case between these two rival physical systems. Nevertheless, since the debate was read in these terms in the wider public sphere, and since Maupertuis gained cultural capital by positioning himself as a defender

42. Cited in ibid., 150.
43. Cited in ibid.
44. Cited in ibid., 149.

of Newtonianism amidst an academy increasingly perceived as the slavish servant of Descartes, it is likely that he donned these costumes with little or no resistance.

Voltaire, though still living in exile in Cirey, had contributed enormously to creating this wider public understanding. He had also acquired notoriety as a Newtonian advocate as a result of his *Lettres philosophiques*. Voltaire captured perfectly the public perception of Maupertuis that went with his Newtonian identity when, with characteristic wit, he congratulated the academician for flattening the earth and the Cassinis with one blow. In his own *Eléments de la philosophie de Newton*, which appeared in the summer of 1738 and which will be discussed in detail below, Voltaire also devoted a chapter to the question of the figure of the earth, showing how Maupertuis' oblate earth followed naturally from Newton's theory of universal gravitation.[45] Within the context of Voltaire's own, increasingly aggressive Newtonianism, therefore, Maupertuis' triumph over the Cassinis would mean a victory for Newtonianism as a whole and a defeat for his and others' Cartesian enemies, this despite the very different framework for the debate inside the Royal Academy. Maupertuis, however, refrained from framing the debate in Voltaire's terms, at least in his public statements and his writings sanctioned by the Royal Academy. His book on the topic, which finally appeared in the summer of 1738 alongside Voltaire's differently oriented exposition of Newtonianism, illustrates well these differences.[46]

Maupertuis' *The Figure of the Earth Determined by Observations* offered a detailed account of the empirical issues at stake and a defense of his own geodesic conclusions (and eventually those of the Cassinis as well). But it contained no wider Newtonian polemic. Bignon appreciated Maupertuis' efforts, and he praised both the text and the savant's overall conduct throughout the dispute. Writing to Réaumur, he expressed satisfaction that his "dear Académie" had enjoyed such great success with respect to this difficult and contentious matter. He especially praised the judiciousness of Maupertuis' public discourse and his tendency to "substitute facts that are accessible to everyone for overly profound discussions." At the same time, he also admired Maupertuis' detailed accounts of the operations, calculations, and observations since without them his work and that of the academy would earn neither the praise of "the true savants" nor justify "to the eyes of the entire universe the utility of such great expense on the part of the king." "Please send him a thousand compliments in

45. Voltaire, *Eléments de la philosophie de Newton*, ed. Robert L. Walters and W. H. Barber in *Œuvres complètes de Voltaire* (ed. Besterman), vol. 15 (Oxford, 1992).

46. Maupertuis, *La figure de la terre déterminée par les observations de MM. de Maupertuis, Clairaut, Le Monnier, Outhier, Celsius au cercle polaire* (Paris, 1738).

my name," Bignon concluded.[47] In clebrating Maupertuis in this way he was also celebrating the wider values of *honnête* academicism to which Maupertuis, and most French academicians, still adhered throughout the 1730s.

In private, however, and especially during his circulation in *le beau monde*, Maupertuis conducted himself very differently. His decision to include a visit to Cirey among his wanderings during his absence from the academy was, therefore, deeply significant. During these years Voltaire lived in exile, but he did not remain isolated as a result. His life was in fact full of intellectual activity, but he pursued his work, including a number of different scientific projects, away from Paris and without any association with Maupertuis or any other French academician. For this reason, Maupertuis' visit to Cirey in September 1738 was an important turning point.

"I thank you for the lessons in Newtonian philosophy," Voltaire wrote to Maupertuis after his departure, his first substantive missive to the academician since the exchanges four years earlier in the wake of the scandal over the *Lettres philosophiques*.[48] These discussions occurred within the context of Voltaire's own Newton war in response to the critics of his *Lettres philosophiques,* and the interesting thing about the letter is the way that it reconnects Maupertuis to the latter's battles rather than summarizing some joint struggle. Maupertuis had in fact distanced himself from Voltaire after the academician had declined the offer to lead a Newtonian sect in 1734. Therefore, in 1738, when Maupertuis returned to Cirey, he was doing so as an outsider reconnecting with old friends. For his part, Voltaire had also been consumed since 1734 with the struggle to manage the scandal provoked by his *Lettres philosophiques*. His Newtonian pronouncements were not read as serious scientific interventions but rather as libertine scandal-mongering. This interpretation was further reinforced by the severity of the book's persecution and by Voltaire's departure into exile. In going to Cirey, therefore, Maupertuis was attempting to reconnect with a very different public, and in hosting him Voltaire was likewise doing the same.

The learned journals that arbitrated philosophical esteem within the Republic of Letters paid close attention to Maupertuis' work throughout the 1730s while deeming Voltaire's *Lettres philosophiques* unworthy for discussion. Other periodicals with different orientations certainly confirmed that the public sphere for science in France was not restricted to the serious organs of the learned establishment; yet the attention that Voltaire received in these works often hindered rather than helped his ambitions as an aspiring philosophe. Accordingly, Maupertuis represented for Voltaire a much sought-after intellectual authority. To acquire

47. Cited in Terrall, *Man Who Flattened the Earth,* 142.
48. Voltaire to Maupertuis, 1 October 1738, in Voltaire, *Corr.*, 88: D1622.

this authority for himself, Voltaire needed to secure his new identity in France—that of the independent yet authoritative philosophe. Yet in 1738, there was more work yet to be done on this project, even if the Newton wars that Voltaire entered into with new vigor gave him a perfect sphere within which to work.

Inventing the Enlightenment Philosophe

To stabilize his philosophe persona, Voltaire needed to reconcile two competing perceptions. One was articulated in 1727 by the playwright Marivaux in a prose work *L'indigent philosophe*. This text painted a portrait of the philosophe familiar to many eighteenth-century elites, and it was this conception that Voltaire set out to change through his own Newtonian campaigns.[49] Marivaux's indigent philosopher is a pauper who lives at the margins of society as well as a pariah who combines obsessive narcissism with nihilistic skepticism. As Jack Udank describes the character, "he seems to primp and compose himself in every utterance, to take his unsteady, equivocal bearing and identity from circumstantial and imaginary projections, from illusion itself—or from allusions to other texts." In short, the philosophe is a creature of fashion with "no place of grace, no true independent existence, except perhaps in the parodic rehearsal of what has found favor in the eyes his readers."[50] This negative and lowly image of the philosophe was widespread in early eighteenth-century France, as the *Dictionnaire de l'Académie française* attested in 1718. It characterized as "philosophical" any opinion or conduct that was "natural, shocking, irreligious, or self-concerned." The 1734 edition of the Jesuit *Dictionnaire de Trévoux* repeated the same, defining the philosophe as any man who is "surly, dirty, uncivil, and unconcerned with the duties and properties of social life."[51]

It became easy to associate Voltaire with this conceptualization of the philosophe in the wake of his own, self-titled *Lettres philosophiques*. The scandals over the alleged improbity and impiety of the work further triggered this perception, as did the eventual retreat of the author into a suspiciously adulterous (if in no way impoverished) exile away from the Parisian center of *le monde*. Not all of these characterizations were unfair, moreover, since the ideal of intellectual liberty that increasingly inspired Voltaire after 1730 was a key element of the

49. Pierre Carlet Marivaux, *L'indigent philosophe*, in *Marivaux. Journaux et oeuvres diverses*, ed. Frédéric Deloffre and Michel Gilot (Paris, 1969).

50. Jack Udank, "Portrait of the Philosopher as Tramp," in *A New History of French Literature*, ed. Dennis Hollier (Cambridge, 1989), 422.

51. Cited in ibid., 423.

indigent philosophe persona as well. Indeed, what gave Marivaux's philosophe humanity, at least for those capable of sympathizing with him, was his ability to see through the hypocrisies, prejudices, and oppressions of the world. Also estimable was his refusal to submit to error under any circumstances. A critical spirit, therefore, and the willingness to use it, was often highly valued, as was the inclination to follow nature even when it seemed to contradict recognized authority. When carefully balanced with the Stoic values of self-control, sacrifice, and service, the critical, intellectual liberty of the philosophe could in fact become a central feature of the *honnête homme*.[52] This meant that Marivaux's philosophe was less the "other" against which the *honnête homme* was defined, and more an extreme case that illustrated what happened when philosophical liberty reigned unchecked by the tempering virtues of honor and civility.

Voltaire struggled throughout his life to reconcile these conflicting principles of liberty and honnêteté. Two essays published around 1740 illustrated well the other pole toward which he strove after 1734. The first was published in two parts in 1739 in the *Bibliothèque britannique*. It used an English work entitled *The Moral Philosopher*, or *Le philosophe honnête-homme* as the editors translated it, to explore the nature of this combination.[53] The English work that triggered the essay was a dialogue that discussed natural religion through a conversation between a "Christian Deist" and a "Christian Jew." In its first installment, published in late 1737, the journal reviewed the book by citing the many criticisms that had been launched against it. "If our philosophe had limited himself to arguing modestly, then I would have found no reason to pick up the pen," wrote one pamphleteer. "But since he violates every law of truth, propriety, and *honnêteté*, he must suffer by being told so and by being proven guilty."[54] Another critic chastised the author for taking "great liberty with reli-

---

52. This theme is discussed at greater length in Marc Fumaroli, *L'âge de l'éloquence: Rhétorique et "res literaria" de la Renaissance au seuil de l'époque classique* (1980; reprint, Paris, 1994); Anne Goldgar, *Impolite Learning: Conduct and Community in the Republic of Letters, 1680–1750* (New Haven, 1995); Daniel Gordon, *Citizens without Sovereignty: Equality and Sociability in French Thought, 1670–1789* (Princeton, 1994); Lawrence Klein, *Shaftesbury and the Culture of Politeness: Moral Discourse and Cultural Politics in Early Eighteenth-Century England* (Cambridge, 1994); Peter Miller, *Peiresc's Europe: Learning and Virtue in the Seventeenth Century* (New Haven, 2000), idem, "Friendship and Conversation in Seventeenth-Century Venice," *Journal of Modern History* 73 (2001): 1–31; and Rene Pintard, *Le libertinage érudit dans la première moitié du XVIIe siècle* (Paris, 1943).

53. "Le philosophe honnête-homme," pt. 1, *Bibliothèque britannique* 10 (1739): 1–19. The English title was *The Moral Philosopher. In a dialogue between Philalethes, a Christian deist, and Theophanes, a Christian Jew* (London, 1738).

54. "Le philosophe honnête-homme," 5.

gious subjects" and for pleading "every day for the liberty to think and dispute on religious questions." [55] Further critics were cited, and since their indictments throughout were vitriolic, the journal decided to appeal to those "whose curiosity about this topic has not been satisfied" by publishing its own reflections on the proper character of the *honnête philosophe.*[56]

"By a moral philosopher," the journal wrote, "one means a Sage, or one who studies wisdom; a philosopher for whom philosophy embraces the practical and the theoretical, and for whom theory informs his practice." He is a philosophe who "has morals, and who recognizes the necessity of moral duties, not just toward men but also toward God." Lastly he is one who "wants neither to be confused with those who move from philosophy into libertinage, or those for whom libertinage produces philosophy." "In short," the journal concluded, we are talking about "a *philosophe honnête homme,* or to say the same thing another way, an *honnête homme philosophe.*"[57] "Our *philosophe honnête homme* is not to be classed among the atheists," the journal continued, because "his love of truth leads him directly to religion." He knows that God created the world and that he governs it with his providential hand. As a result, "he combats, often heatedly, those who deny this fundamental truth of all religions." [58] Yet the true philosophe also takes pride in being counted among the freethinkers [*Libres-Penseurs*], or even to be called a Skeptic or a Pyrrhonnien." He in fact willingly adopts these identities "so long as these terms signify a man who seeks the truth, believes in nothing without reason, and, ceding nothing to prejudice, dares to reject the most authorized opinions until such time as a free and impartial examination of them has led him to conclude that they are reasonable, or at least probable." [59] Building on this judicious balance between free thought and probity, the essay ultimately argued for reasonableness and moderation as the values most representative of the true philosophe.

Even more influential in defending the positive value of the philosophe was an article, written perhaps by Dumarsais, which was first published in 1742 in the anonymously edited collection *Nouvelles libertés de la pensée.*[60] This article

55. Ibid., 10–11.
56. "Le philosophe honnête-homme," pt. 2, *Bibliothèque britannique* 12 (1739): 331–40, quotation on 331.
57. Ibid., 332–333.
58. Ibid., 336–337.
59. Ibid., 333–334.
60. This essay had a long and storied history, and the details of it are meticulously collected and analyzed in Herbert Dieckmann, *Le Philosophe: Texts and Interpretations* (St. Louis, 1943). Dieckmann also reprints the four most famous published versions of the

would eventually define the term "philosophe" in Diderot and d'Alembert's *Encyclopédie*, and appearing in 1742 in this widely noticed collection it further supported the reconstruction of the philosophe persona that Voltaire was attempting. "[The philosophe] is an *honnête homme* who wishes to please others and render himself useful," the brief essay declared in a representative passage.[61] Overall, the work was full of quotable lines such as these, and the goal of the essay as a whole was to join such claims to an eloquent defense of how intellectual liberty could produce a socially beneficial human being rather than a destructive and antisocial skeptic.

Dumarsais' philosophe was not a reckless libertine hell-bent on destroying revealed religion and its moral authority. Instead, he was a careful thinker who sought truth at all times, but with a full understanding of the limitations that must always restrain one in this quest. "A philosopher acts only after reflection, even in moments of passion; he walks through the night, but he is preceded by a torch."[62] The philosopher in particular forms his principles from "an infinity of particular observations. He esteems the science of facts," and from this basis he "applies himself to understanding both the universe and himself."[63] Likewise, the real sage has no interest in "vain speculations," "the vain disputes of the schools," or in "books that only explore vain questions." He is shocked at "the wars and disorders caused by those who chase chimeras," and he wishes that "subtle points of theory" were less central to intellectual discourse than "practical matters of utilty."[64]

The sober, empirical search for knowledge in fact defines the essence of the philosophe in Dumarsais' essay, but the philosopher also knows how to separate himself from those "very intelligent persons who are always judging." "When we judge without having sound grounds for judgment, we simply

---

text in a convenient side-by-side presentation, making possible an analysis of the changes that the text underwent during its lifetime. My analysis is built on the earliest of the published versions, which appeared in Nicolas Fréret's anonymously published *Nouvelles libertés de la penser* (Paris, 1742), 173–204. This version contains discussions excised in the later versions, and since it is chronologically most appropriate to this discussion, I have focused on this version in my analysis. Many of the translations are my own, but when passages from the 1742 version are replicated in the version of the article that appeared in Diderot and d'Alembert's *Encyclopédie*, I have drawn upon the translation offered in John Lough, ed., *The "Encyclopédie" of Diderot and d'Alembert. Selected Articles* (Indianapolis, 1969), 284–289. My citations, however, will be to the French text found in Dieckmann, *Le Philosophe*, 30–65.

61. Dieckmann, *Le Philosophe*, 44.

62. Ibid., 32.

63. Ibid., 32, 36.

64. Ibid., 42–44.

guess," the essay stated. And since judges of this sort "do not know the reach of the human mind and think it can know everything," the philosopher pursues a different agenda. "He accepts as true what is true, and false what is false, as doubtful what is doubtful, and as probable what is merely probable."[65] He also contents himself with judging infrequently but judiciously, knowing that it is better "to judge and to speak less," but to do the first "more accurately" and the second "well." "When he does not have any proper basis for an opinion," the text declared, "[the philosophe] knows when to suspend judgment, and this is his most perfect trait."[66]

Reasoned skepticism and judicious restraint were the hallmarks of the *honnête homme,* and here the article made disciplined philosophical inquiry compatible with this ethic. Equally important to this persona, however, were social norms regarding community, sociability, and service. "Reason demands that [men] know and study the qualities of sociability and endeavor to acquire them."[67] Similarly the philosophe properly defined is not a "monster" who lives "in exile in this world." "He wishes to find pleasure in the company of others," and "he seeks to adapt himself to those with whom he lives."[68] The real philosophe, in fact, wants nothing more than to serve his fellowman, for in the same way that grace is the principle that determines the actions of Christians, "civil society is for him like a divinity on earth." An urge toward probity, therefore, is a natural feature of the true philosophe, for since "civil society is his unique God," "he is concerned, far more than other men, with directing all his efforts toward achieving the ideal of the *honnête homme.*"[69] The debauched, the superstitious, even the devout: each of these men is a creature of passion. The philosophe, by contrast, finds constancy in his life since he knows himself and how to live with others. "Do not fear that [the philosopher] will engage in acts contrary to probity," the text declares. "No! Such an action is not in accord with his makeup." The philosopher is rather "filled with concern for the good of civil society," and "his heart is nourished by religion, to which he has been brought by the natural light of reason."[70] The philosophe, in short, is a model citizen and subject. He is "an *honnête homme* who follows reason in all his actions and who combines a reflective and precise mind with the manners and qualities of a sociable man."[71]

65. Ibid., 38.
66. Ibid., 38.
67. Ibid., 42.
68. Ibid., 44.
69. Ibid., 52.
70. Ibid., 52–54.
71. Ibid., 58.

Dumarsais' essay (if in fact he wrote it) ultimately defined an ideal type, one that contrasted strongly with the negative image of philosophers prevalent in eighteenth-century France. This positive image, though widely influential, did not completely replace the negative image popularized by Marivaux and others; rather, it defined with it two rival positions in a complex field of cultural contestation. Voltaire entered this field after the publication of his *Lettres philosophiques*, and it was through his Newtonianism that he began to assert his understanding of philosophy and the philosopher. In Voltaire's mind, Dumarsais' ideal was his ideal as well, but to his critics he was either a failed philosophe, a disingenuous subversive, or perhaps a little bit of both. Also at issue in these struggles were the incompatibilities between Voltaire's dual identity as a poet and a would-be philosophe, the same tensions that Prévost pointed to in his discussion of Voltaire's letters on Locke and Newton. In the years immediately following Voltaire's exile at Cirey, this confusion plagued him, and it crystallized around his Newtonianism during his encounters with the Venetian man of letters, Francesco Algarotti.

Voltaire first met Algarotti when he visited Cirey during October and November 1735.[72] Still in his early twenties, the Italian had made a name for himself in 1729 when he repeated Newton's optical experiments before a skeptical audience of Cartesian savants in Bologna. In 1733 he began to travel, meeting among others Martin Folkes, vice-president of the Royal Society during Newton's tenure as president. He also came to know Anders Celsius, the Swedish astronomer. In 1734, he accompanied Celsius to Paris, where he spent some time working with Maupertuis and Clairaut before accepting their invitation to join them on their trip to Lapland.[73] At the last minute, Algarotti chose not to make the journey, and it was in this context, perhaps through the recommendation of Maupertuis, that he found himself at Cirey instead.

"We have with us here the Marquis Argalotti [*sic*]," Voltaire wrote incorrectly to Thieriot on November 3, 1735. "He is a young man who knows the languages and mores of every country, who writes verse like Ariosto, and who

72. On this visit, see Vaillot, *Avec Madame du Châtelet*, 26–27. On Algarotti, see Paolo Casini, "Les débuts du Newtonianisme en Italie, 1700–1740," *Dix-huitième siècle* 10 (1978): 85–100; Vincenzo Ferrone, *The Intellectual Roots of the Italian Enlightenment: Natural Science, Religion, and Politics in the Early Eighteenth Century*, trans. by Sue Brotherton. (Atlantic Highlands, 1995); Massimo Mazzoti, "Newton for Ladies: Gentility, Gender and Radical Culture," *British Journal for the History of Science* 27, no. 2 (June 2004): 119–146; and Robert L. Walters and W. H. Barber, introduction, in *Voltaire. Eléments de la philosophie de Newton*, in *Œuvres complètes de Voltaire* (ed. Besterman), 15: 41–47.

73. Terrall mentions this invitation in *Man Who Flattened the Earth*, 104.

knows his Locke and his Newton. He is reading us some dialogues that he wrote concerning the interesting parts of philosophy, . . . and I am giving him my little course on metaphysics. . . . We have also been reading several stanzas from *Jeanne la pucelle,* a tragedy of my making, and a chapter from the *Siècle de Louis XIV.* From this we return to Newton and Locke, but not without a little champagne and our dearly beloved [*excellente chère*], since we are very voluptuous philosophes. . . . There in a nutshell is a fairly exact account of my life."[74] The mix of sociability, poetry, and philosophy revealed in this account captures well the general framework of Voltaire's intellectual orientation during these years. Especially revealing was the mention of Algarotti's dialogues since these would become a trigger for Voltaire's next major philosophical undertaking.

The dialogues in question were no doubt drafts of Algarotti's widely read and discussed *Newtonianism for Ladies* that was first published in Milan, with a Naples imprint, in early 1737.[75] As early as spring 1736 Voltaire and du Châtelet were asking Algarotti for copies of the manuscript, for each admired Algarotti's work. "It is full of intelligence, grace, imagination, and science," du Châtelet wrote, and Voltaire honored the text by praising the "surety" of Algarotti's capacity to "instruct and to please."[76] Privately, Algarotti's example also stirred Voltaire's own ambition to publish a French work explaining Newtonian philosophy. "There are not even twenty Frenchmen who understand Newton," Voltaire complained to his former teacher the abbé d'Olivet during Algarotti's visit. "We incorrectly dispute against him without ever taking the time to read his geometrical demonstrations." But referring to Algarotti, he also noted that "we currently have here a noble Venetian who understands Newton like the elements of Euclid. Would it be shameful for the French to do so as well?"[77]

For many, Algarotti's dialogue answered this need since it was learned, accurate, witty, and accessible to a broad public. Du Châtelet said in 1736 that she was working on her Italian with the goal of perhaps translating the work

74. Voltaire to Thieriot, 3 November 1735, in Voltaire, *Corr.,* 87: D935.

75. Francesco Algarotti, *Il newtonianismo per le dame ovvero dialoghi sopra la luce e i colori* (Naples, 1737). An excellent facsimile digital edition, edited and with an introduction by Massimo Mazzotti, is found at http://www.cis.unibo.it/cis13b/bsco3/intro_opera .asp?id_opera=32.

76. Du Châtelet to Algarotti, 20 April 1736, in Voltaire, *Corr.,* 87: D1065. Voltaire made his remarks in his 1736 dedicatory epistle to *Alzire, ou Les américains,* in *Œuvres complètes de Voltaire* (ed. Besterman), 14: 111–112.

77. Voltaire to Joseph Thoulier d'Olivet, 30 November 1735, in Voltaire, *Corr.,* 87: D950.

into French one day.[78] Voltaire too was clearly influenced by Algarotti in the creation of his own Newtonian exposition. The Italian had adopted the as yet unprecedented approach of using optics—particularly the Newtonian theory of refraction as a consequence of the attractive forces operative in dense media—as a vehicle for explaining Newton's theory of universal gravitation more generally. Voltaire adopted the same focus in his exposition of Newtonianism, and in this way too he showed his debt to Algarotti. In other respects, however, Voltaire broke crucially with the example set by the Italian, defining as a result his very different understanding of what it meant to be a philosophe. Most significant was his decision to produce a prose work modeled on the example of Henry Pemberton rather than either a *mondain* dialogue or an expository poem. Each of these genre choices was available to Voltaire, and to a certain degree his choice was consciously deliberated. Gender played a key role in his deliberations. Algarotti's dialogue mirrored Fontenelle's *Conversations on the Plurality of Worlds,* with a learned man leading an aristocratic lady through a pleasurable discussion of the details of Newtonian philosophy. By adopting this approach, Algarotti positioned himself as a bridge between serious male learning and the feminized world of *le beau monde,* a stance that had become commonplace as a result of the success of Fontenelle's text over the previous four decades.[79] Algarotti was exceedingly comfortable in this role, and accordingly his work earned wide esteem to the extent that it confirmed these preexisting expectations.

Desfontaines was in a position to recognize Algarotti's achievement, and he showered praise on the work and its author in his biweekly newssheet. "Here is an Italian *bel esprit* who, wanting to credit the dogmas of [Newton] in his homeland, chose to impart them first to women since their vote is so important to the success of any novelty."[80] The journalist then alluded to Fontenelle's example, noting that Algarotti's work possessed "the same taste and the same dialogue form, not to mention all the same gallantry, salt, and pretty details, as well as all the learning, science, and clarity of expression."[81] Arid science and

---

78. Du Châtelet to Algarotti, 20 April 1736, in Voltaire, *Corr.,* 87: D1065.

79. For an analysis of this relationship, see Mazzotti, "Newton for Ladies." For a different interpretation of Fontenelle's *Mondes,* see J. B. Shank, "Neither Natural Philosophy, nor Science, nor Literature: Gender and Natural Knowledge in Fontenelle's *Entretiens sur la pluralité des mondes,*" in *Men, Women, and the Birthing of Modern Science,* ed. Judith Zinsser (DeKalb, 2005).

80. Review of Algarotti, *Il newtonianismo per le dame,* in *Observations sur les écrits modernes* 14 (n.d.): 217.

81. Ibid., 14: 218.

belles lettres are rarely compatible, he concluded, which is why one rarely sees someone who possesses both "serious scientific learning and a delicate mind." Nevertheless, "M. Fontenelle, and now M. Algarotti, show that the two can be allied," and for Desfontaines this marriage was the achievement of Algarotti's work.[82] The Italian, Desfontaines continued, had the harder assignment, in fact, for whereas Fontenelle was charged with clarifying the Cartesian philosophy "recently perfected by M. Privat de Molières," Algarotti had to explain the "abyss of the void," the "chaos" of action at a distance, and the mysterious and unconscionable revival of the "long lost occult qualities of the Peripaticians."[83] "The book offers a complete course in Newtonian philosophy," the review declared, and even if it was surprising to Desfontaines to see "such an *homme d'esprit*" sympathetic to these philosophical abominations, the dialogue itself was an example of "philosophical gallantry par excellence." The book, he recommended, should quickly be translated into French by a writer who is up to the task.[84]

Voltaire also found his name mentioned among those who combine scientific learning with *bel esprit,* yet as much as the poet cum philosophe wanted to bring these two principles together, he wanted no part of the feminized, *mondain* approach to philosophy that Fontenelle and Algarotti personified. "Let Algarotti instruct the ladies," he wrote. "Maupertuis can instruct the men, and I will instruct the children."[85] In other letters, he was more assertive about his desire to speak directly as an authoritative, masculine philosopher. "I am going to work on a corrected edition of the *Eléments* of Newton," he wrote to Thieriot in May 1738 after a faulty edition of the work had been published in Holland. "It will be neither for the ladies, nor for everyone, but in it one will find the truth expressed with method."[86] In this first, unauthorized Dutch edition of Voltaire's *Eléments de la philosophie de Newton,* an editor had added the subtitle "Accessible to Everyone." Voltaire found this to be a gross misrepresentation of his text. "It is a book that one must study," he explained to his friend Berger, not a book for everyone.[87] "When M. Algarotti read to me his dialogues on light, I gave them the praise that they merited since they spread an infinite quantity of *esprit* and *clarté* over this beautiful part of physics." However, while "dialogues such as these are charming works," as works of physics

82. Ibid., 14: 219–220.
83. Ibid., 14: 220, 227.
84. Ibid., 14: 228–229.
85. Voltaire to Thieriot, 5 May 1738, in Voltaire, *Corr.,* 89: D1492.
86. Ibid.
87. Voltaire to Berger, 14 May 1738, in Voltaire, *Corr.,* 89: D1502.

they are "lightly researched and superficially developed." "I chose to speak instead to a real philosopher," Voltaire asserted, and from this perspective neither Algarotti's imaginary marquise nor his address "to the ladies" was at all appropriate.[88]

Voltaire's *Eléments de la philosophie de Newton* was indeed a very different kind of work than Algarotti's, and a key reason for the differences rested in Voltaire's desire to be a true philosopher, as he understood the term, and not just a gallant popularizer who made austere philosophy palatable to the *mondain* public. "There are more truths in ten pages of my text than in all of [Algarotti's] work," Voltaire explained, and he constructed his text in such a way as to drive home these serious scientific ambitions.[89] His text opened, for example, with a pictorial frontispiece and two dedicatory pieces to the Marquise du Châtelet, one in verse and one in prose. Yet even these "artistic" flourishes announced a work very different than Algarotti's. The frontispiece depicted an image of enlightenment radiating through the head of a cosmos-measuring Newton before bouncing off a mirror held by a female figure onto a table where a seated figure is writing. Visually, the work introduced the focus on light and its laws that formed one theme of Voltaire's treatise. Symbolically, however, the image also set up the gender and genre reconfigurations that Voltaire was attempting as well. In the dedicatory poem and prose "Avant propos" that came after this image, Emilie du Châtelet was named as the female who reflected enlightenment onto Voltaire the writer. "You call me to you, vast and powerful genius/ Minerva of France, immortal Emilie/ Disciple of Newton, and of truth/You penetrate my senses with the fire of your clarity."[90]

In these presentations, du Châtelet was given a certain authority over Voltaire's science, since she was the source of his knowledge and he merely the recorder of her enlightenment. Algarotti had also dedicated his work to du Châtelet, yet given the nature of his text, his dedication positioned her very differently. In the Italian's work, du Châtelet merged with the marquise who receives instruction in the text. Yet as Voltaire retorted at the time, "she knows at least as much as him, and could well correct the presentations of his book."[91] Voltaire accordingly reconfigured this gendered relationship, presenting his "beloved and immortal Emilie" as a philosopher in her own right. Thus, writing to du Châtelet in the opening preface, he promised her "neither an imaginary marquise, nor an imaginary philosophy here. The solid study that you

88. Ibid.

89. Voltaire to Thieriot, 18 May 1738, in Voltaire, *Corr.*, 89: D1505.

90. Voltaire, *Eléments,* in *Œuvres complètes de Voltaire* (ed. Besterman), 15: 186.

91. Voltaire to Thieriot, 5 May 1738, in Voltaire, *Corr.*, 89: D1492.

FIGURE 20. *Frontispiece from Voltaire,* Éléments de la philosophie de Newton *(Amsterdam, 1738). Courtesy of the James Ford Bell Library, University of Minnesota Libraries.*

have made of several new truths and the fruit you have taken from your respectable work is what I offer to the public for your glory, and that of your sex." He also cautioned her against looking for *agréments* in the text. "Others know how to cover the nettles of science with flowers; I will limit myself to trying to fully understand the truths and to explaining them with order and clarity." [92]

Voltaire further emphasized the strong ties that bound his stylistic and genre choices with his veracity and credibility as a philosophe. His goal, as Voltaire stated it, was to follow Newton in as clear and precise a manner as possible. His *Eléments,* therefore, was "not a complete course of physics," but a more limited work focused only on offering "a precise idea of [Nature's] exceedingly delicate and powerful springs, and their fundamental laws," to those who "only know Newton and his philosophy by name." [93] Precision, accuracy, and clarity were thus his watchwords, and while Voltaire admitted that digressions might be required to understand certain particularities, his ultimate goal was to offer a clear and direct exposition free of any rhetorical flourishes or diverting digressions whatsoever. Just in case anyone still held hopes that Voltaire's poetic side would be deployed, he ended his "Avant propos" by referring readers to "the excellent works in physics by 'sGravesande, Keill, Musschenbroek, and Pemberton" should further instruction be desired. [94]

The text itself conformed to this program in both style and content. In managing its reception, moreover, both Voltaire and du Châtelet worked hard to insure that the book was read and received properly. The first appearance of the treatise in an unauthorized, error-filled edition in May 1738 created the first occasion for these efforts. [95] Less than two weeks earlier, the scientific authority of Voltaire and du Châtelet had been elevated when each was awarded honorable mention by the Paris Academy as a result of the essays on the theory of fire that each had submitted to the academy's prize competition. The essays, though quite different, were the result of the scientific collaborations that the couple had pursued at Cirey since 1735. The announcement of their success, made at the academy's spring public assembly, therefore created quite a stir. The academy took advantage of the occasion to emphasize its own admiration for the work of talented amateurs, including the work of learned ladies. The announcement also emphasized the institution's openness to the worthy

92. Voltaire, "Avant propos," in *Œuvres complètes de Voltaire* (ed. Besterman), 15: 547.

93. Ibid.

94. Ibid., 549.

95. The critical edition of Voltaire's *Eléments,* ed. Walters and Barber, in *Œuvres complètes de Voltaire* (ed. Besterman), vol. 15, chronicles the history of the text. I will follow the Walters and Barber edition closely, therefore, in my account.

philosophical aspirations of the wider public, while also reasserting its ongoing commitment to free philosophical inquiry and its stark refusal to embrace any single philosophical system.[96] Voltaire also presented himself as a sage and neutral arbiter of truth in his *Eléments,* and when, therefore, within days of his recognition by the Royal Academy, an uncorrected and unauthorized version of his text began to circulate, the author felt a strong urge to intervene.

He moved on several different fronts simultaneously.[97] Behind the scenes he worked to secure a *privilège* for a corrected French edition, a move that, if successful, would have allowed Voltaire's corrected explanations to carry the imprimatur of the French intellectual establishment. The text had originally been sent to the censor in June 1737, but in January 1738 the request was denied after the notoriously cautious and narrow-minded Chancellor d'Aguesseau personally read the text (a rare occurrence) and then vetoed the recommendation of two other censors (another rarity) who had both recommended publication.[98] Among the reasons for the refusal, beyond the chancellor's fussiness, were the theological views attributed to Newton in the final chapter; the great age that Voltaire assigned to the earth, in contradiction to orthodox Christian dating; and the overall anti-Cartesian tone of the work. The last in particular was seen as enflaming public debate in a climate that had already become far too contentious. Du Châtelet speculated that d'Aguesseau "feared a union of Voltaire and Maupertuis that would subjugate the world," and she wondered whether France would soon see "an *arrêt de Parlement* against Newtonian philosophy."[99] She also told Algarotti that Voltaire's real crime was "treating Descartes with insufficient respect."[100] Yet since d'Aguesseau granted a *privilège* to Algarotti's Newtonian dialogue while rejecting Voltaire's text, it is likely that Voltaire's tone and general reputation were the operative criterion in his censure far more than any conviction about Newtonian or Cartesian philosophy.

After the unauthorized Dutch edition of the *Eléments* appeared in May, Voltaire renewed these efforts using other contacts. In August, he succeeded in

96. The declaration is transcribed in PVARS, April 16, 1738.

97. In addition to Walters and Barber, introduction, see Vaillot, *Avec Madame du Châtelet,* 84–92.

98. Of d'Aguesseau, Condorcet wrote, "His superstition, his timidity, his respect for ancient practices, his indecision, these narrowed his views with respect to reforming the law and thwarted his activity." *Vie de Voltaire,* 213. Cited in Walters and Barber, introduction, 69, n. 21.

99. Du Châtelet to Algarotti, 10 January 1738, in Voltaire, *Corr.,* 88: D1421.

100. Du Châtelet to Algarotti, 2 February 1738, in Voltaire, *Corr.,* 89: D1441. On this history more generally, see Walters and Barber, introduction, 65–72.

370 } Chapter Six

getting an authorized French edition into print. In the interim, however, the Dutch edition began to sell wildly and stir up discussion in France and the wider Republic of Letters. Voltaire was, therefore, forced to respond to this edition despite his repudiation of it. In June, the *Journal des savants* noted the publication of the work under "Holland" in its "Nouvelles de la république des lettres" section.[101] The journal also followed its announcement with "a note submitted to us by the author." Overall, the missive emphasized the "very defective" character of the Dutch edition, stressing that "the number of errors was particularly problematic given the nature of the work." Voltaire also distanced himself from his Dutch editor's decision to add "*mise à la porté de tout le monde*" (put at the level of everyone) to his chosen title. These were not his words, he asserted, and echoing his many statements in his contemporary correspondence, he described his work as something much more substantial than a popularization.[102]

Voltaire also offered a long defense, which the journal printed verbatim, of his use of the line about "imaginary marquises and philosophies" in his opening dedication. "I had no intention of criticizing the author of *La pluralité des mondes*," he wrote, "and I declare here publicly that I regard [Fontenelle's] book as one of the best that has ever been written. Its author is also one of the most estimable men that has ever lived." Instead, he continued, it was Algarotti that he intended to target with this description. "When I had the honor," he explained, "of hearing at Cirey the Italian dialogues of M. Algarotti, in which the foundations of Newtonian philosophy are established with great *esprit*, while those of Descartes are destroyed with force, I decided to engage myself in the same cause in France. . . . Since his work is a dialogue with an imaginary marquise written in the style of *La pluralité des mondes*, he addressed his work to M. Fontenelle. However, as M. Algarotti and the illustrious woman to which the book is dedicated can confirm, I was very annoyed [*tres faché*] to see a phantom Marquise [*en air*] in his work, and I told him that he should not put an imaginary being at the center of solid truths. This is why I began my *Eléments* the way I did."[103]

In forcing Voltaire to focus less on the philosophical details of his position and more on his intellectual and literary style, these criticisms were harbingers of the battles to come. Nevertheless, Voltaire's primary focus during the summer of 1738 was on the philosophical substance of his text. He thus worked

101. "Nouvelles de la république des lettres. "Letter from M. de Voltaire, author of the *Eléments de la philosophie de Newton," Journal des savants* (1738): 381–382.

102. Ibid.

103. Ibid., 382.

hard to ensure readers access to an accurate edition of his ideas. One strategy involved the circulation to various journals of a carefully prepared set of "necessary clarifications" in the hope that his correct views would reach the wider public. In July, the *Journal de Trévoux* complied, publishing the revisions along with Voltaire's request that readers include them with any copy of the text.[104] In October, the *Journal des savants* again used its "Nouvelles de la république des lettres" section to further publish "a letter from M. de Voltaire" that the journal believed "could not be refused."[105] Much of the letter (labeled "De Cyrey en Champagne") was a response to the many criticisms of his text that Voltaire had received, and these will be discussed in detail below. But his *apologia* also began with a set of further corrections regarding his text. Concluding a long list of detailed corrections, he asked only that "the public be served with exactitude." "My only concern is for the perfection of the arts to which we all aspire," he stressed.[106]

Voltaire especially emphasized the linkage between textual inaccuracy, misinterpretation, misunderstanding, and destructive disputation. Yet while Voltaire's text certainly triggered a great deal of controversy, and while the author's struggle to stabilize an accurate text was one theater for these battles, another involved stabilizing the book's interpretation as well. This was a far more difficult proposition, and in his October letter to the *Journal des savants* he evaded this challenge, writing that "I will not respond at all here to the objections made in France about the truths contained in the *Eléments de Newton*."[107] One reason he could dodge the debate here was because a response had already been made by one of his closest allies. In September 1738, the Marquise du Châtelet, undoubtedly supported by Voltaire, succeeded in publishing an anonymous letter in the same journal that presented Voltaire's text in very sympathetic terms.[108]

The letter was framed as an unsolicited and anonymous book review, and the editors published it "with pleasure" while explaining in a footnote the anomalous character of the submission.[109] To maintain the conceit of impar-

104. "Eclaircissements nécessaries donnez par M. de Voltaire," *Journal de Trévoux* (July 1738): 1448–1470. They are also reprinted in *Œuvres complètes de Voltaire* (ed. Besterman), 15: 655–672.

105. "Nouvelles de la république des lettres," *Journal des savants* (1738): 636.

106. Ibid., 637–638.

107. Ibid., 638.

108. [Marquise du Châtelet], "Lettre sur les Eléments de la philosophie de Newton," *Journal des savants* (1738): 534–541.

109. Ibid., 534.

tiality, du Châtelet did not shy away from criticizing Voltaire's text. "It seems to me that there would have been more order in the book had M. de Voltaire divided it into two parts," she wrote. Reiterating differences that were found in their respective essays on fire as well, she also accused Voltaire of going beyond the facts in attributing weight to light. Du Châtelet further sided with those critics who accused Voltaire of "speaking a bit too harshly" about "great men" such as Descartes and Malebranche. She attributed this fault to "an excessive zealousness for the truth" rather than any malignity of spirit, and "while the philosophes who have defended Descartes, not to mention Descartes himself, certainly deserve our respect," she wrote, "their errors deserve no courtesy. For as M. de V says, *the primary thing that must be respected is the truth.*"[110] Du Châtelet likewise corrected Voltaire on several technical points regarding the reflection of light in mirrors and the nature of interparticle attractive forces. Overall these challenges established her as an honest reviewer.

Yet du Châtelet's review, despite its criticisms, was ultimately designed to praise Voltaire, not to bury him. Her arguments were most powerful in supporting her partner's claim to be explaining and defending the true philosophy of nature in a country that was both ignorant of and prejudicially biased against it. "The only glory that yet remains for Newton is to be more widely known," she wrote, for "while the largest part of the learned world rendered its praise onto him long ago, his philosophy, bristling with calculations and algebra, has nevertheless remained something of a mystery known only to the initiated. M. Pemberton, who tried to make Newton known to readers in England, is often harder to understand than Newton himself. But in a kind of miracle, it has been reserved to M. de Voltaire to change this. He has made Newtonian philosophy—the only philosophy worthy to be studied since it is the only one that is proven—accessible, not to everyone as the booksellers in Holland announced, but to any reasonable and attentive reader."[111]

Du Châtelet also described the stylistic and genre choices that made Voltaire's miracle possible. In doing so, she reiterated publicly his intentions to write and argue like a true philosophe, and not like a poet who produced pretty philosophical romances. "Some have accused M. de Voltaire of a crime in writing that there was neither an imaginary marquise nor imaginary philosophy in his work," she wrote. "No matter whether he intended to refer to the author of *Des Mondes* with these words, all one has to do is to read his *Eléments* to know that M. de Voltaire has too much merit not to appreciate his as well." Du Châtelet also found it "important to remark" that "the style of *des mondes,*

110. Ibid., 541.
111. Ibid., 534.

as graceful as it is in the mouth of the Marquise, becomes strained when it is forced to become too long-winded. The judicious author of this charming book would, therefore, have adopted instead a clear and precise style had he wanted to descend into the profound details of his subject." [112]

This formulation made Fontenelle's style appropriate to the light and entertaining book he intended while legitimating Voltaire's very different style as equally appropriate to his very different philosophical ambitions. Du Châtelet added a further sting to this formulation by drawing another distinction between the different philosophical contents of the two books. "I daresay also that the vortices, these offshoots of Descartes' sublime and active imagination, seem to lend themselves to graceful style in ways that more severe truths grounded in mathematical calculation absolutely resist. I doubt, for instance, that anyone could make a good witticism out of the phrase *in inverse proportion to the square of the distance.*" [113] This formulation distanced Voltaire from Algarotti and Fontenelle by making his clear, precise philosophical prose the only style worthy of Newton's thoroughly veracious science. It also implicitly echoed the prevailing discourse of Newtonianism by implying a connection between the Cartesian vortices and the songs of *romanciers.*

In the subsequent discussions, the marquise built upon this position, making Voltaire's *Eléments* not only the most authoritative account of Newtonian philosophy available but also the scientific book that France most needed. "In the countries where the philosophy of Descartes has already been abandoned, one might find it surprising that M. de Voltaire found it necessary to spend as much time as he did refuting it. But those who might make this critique should make a trip to France, for there they will see that the French, even some of the most respectable, are still very attached to this philosophy, and that M. de Voltaire cannot take too much care in refuting it." [114] Indeed, she continued, "it is the very goal of his work to demonstrate to the French—for it is to them alone that it is addressed—the impossibility of the plenum and the vortices and to introduce them to Newton, who is too little known." [115] She also made it clear that the stakes involved could not have been any higher. "We are currently experiencing a veritable revolution in physics, and consequently one must avow that it has become indispensably necessary for all Frenchmen who are interested in the glory of their nation and who wish the continuation of the gentle and happy government under which we currently live, to concern themselves with

112. Ibid., 534–535.
113. Ibid., 535.
114. Ibid.
115. Ibid.

it."[116] In a gesture of patriotic appeal, she also noted that France would benefit from such a turn. "We have long worked for the glory of Newton, but the time has come now for us to participate in it. . . . We should blush at being the last to render homage to this great man, which is rather little more than rendering homage to the truth."[117]

Voltaire's project, therefore, "was perhaps the greatest service that one could render to our nation in matters of philosophy." Du Châtelet further noted the irony that it was "to one of our greatest poets that we owe the obligation of becoming philosophes." "Yet is there any other way of being [a philosophe] than to abandon error for the truth?"[118] This rhetorical question launched her into her own explication of the truths and errors that Voltaire had illuminated. She still had criticisms to offer about Voltaire's presentation of optics, but the heart of her polemic began when she turned to "the second and most important part of the book." Here Voltaire "attacked Cartesianism down to its very roots," and du Châtelet found "nothing more clear, methodical, or forcefully reasoned than what he had to say about the plenum and the vortices." She further offered her own technical yet abbreviated exposition of why mechanical impulsion could not account for the phenomena of terrestrial *pesanteur*, and explained why Newtonian attraction was the only reasonable theory that could be held.

Continuing her defense, she also emphasized that Newton had judiciously refrained from offering an explanation for gravity since no such explanation was available. She then staked out a more radical position than most when she added that no such explanation would ever be found since the goal of explaining gravity was as quixotic as trying to find a perpetual motion machine.[119] She wished that Voltaire had drawn a clearer distinction between gravitation, or the force that exists between all bodies, and attraction, the effect that this force manifests to our senses. In drawing this distinction, she was both demonstrating her own scientific acumen, while improving upon Voltaire's exposition. Nevertheless, the ultimate thesis of her discussion was clear: Newtonianism was the one and only true philosophy of nature, and those who denied its validity were either ignorant, prejudicial or both. Similarly, Voltaire's *Eléments*, despite its faults (she likened them to "small stains on a Raphael painting"), offered an accurate and authoritative account of this philosophy.[120] To criticize Voltaire or his arguments, therefore, was to criticize truth itself.

116. Ibid., 535–536.
117. Ibid., 536.
118. Ibid., 535.
119. Ibid., 540.
120. Ibid., 538.

Published prominently in the official journal of record in France, du Châtelet's letter was a powerful piece of advocacy on Voltaire's behalf. However, as if this was not challenge enough, she also concluded her letter with a direct and public (not to mention subtly protofeminist) challenge to the Royal Academy. "It is surprising given all the savant men in France that not one of them thought of giving to his nation the same service to theirs that Mssrs. 'sGravesande, Musschenbroek, Wolff, Keill, and others have given. It is true that the *mémoires* of our Academy of Sciences provide excellent materials, yet none of the learned men who compose this illustrious body has yet undertaken to construct a building from them." As if to explain this absence, but in no way making this connection explicitly, she then concluded her letter with the following observation: "Few of them have yet to shake off, at least openly, the yoke of Cartesianism, and the respect that they have for the opinions of Descartes works like a remora halting this great vessel in its course."[121]

Framed in terms of the nationalist discourse that permeated du Châtelet's letter throughout, this was a provocative charge. It attributed a parochial Cartesian bias to the French Royal Academy at a time when, du Châtelet claimed, a revolution in physics was occurring, one that was revealing the true system of the universe to savants throughout Europe. It also articulated public incredulity about the academy's stated ideology of philosophical neutrality, while drawing a clear institutional borderline between "official and prejudicial Cartesianism" on the one hand and "independent, outsider, and veracious Newtonianism" on the other. As we have seen, the academy was anything but a Cartesian monolith in 1738, but the effect of Voltaire's self-fashioning as a non-academic and nonauthorized defender of Newton in a country that, he and du Châtelet claimed, was prejudicially and institutionally attached to Cartesianism, worked to make this framework the operative one for discussions of his *Eléments*. As the Newton wars unfolded, it was this framework, invented as it may have been, that proved decisive.

Forging the Philosophe Persona:
Voltaire's Newtonianism in the Public Sphere

Voltaire and du Châtelet were hosting Maupertuis at Cirey at the time that du Châtelet's letter appeared. The academician's reports about the "Cartesian" enemies of his Newtonian theory regarding the shape of the earth no doubt colored their thinking as they drafted their text. Maupertuis would soon ex-

121. Ibid., 541.

ploit this same public perception regarding French academic Cartesianism in ways that will be discussed shortly, but for Voltaire and du Châtelet his sympathy meant that they had the support, at least privately, of a senior French academician as they constructed their self-conception as heroically embattled Newtonian outsiders. As du Châtelet described the psychology in a letter to Maupertuis from early 1738: "We are philosophical heretics. I admire the temerity with which I say 'we' here, but the underlings of the army are saying that *we must fight the enemy.*"[122]

Not that Voltaire or du Châtelet wanted to actually launch a rebellion against the Royal Academy or its science; quite the contrary, the image of the Newtonian army fighting the "enslaved Cartesian Academy" was simply a self-congratulatory trope that united them with those sympathetic to their view. When Réaumur, in the summer of 1738, began arranging for the official academic publication of the essays on fire produced by Voltaire and du Châtelet, the correspondence of the couple became filled with glowing and respectful commentary about the academician despite his well-known Cartesian leanings. The "Cartesian Academy," therefore, was always a gross, polemical exaggeration, and the image never mirrored either the actual intellectual constitution of the institution or Voltaire and du Châtelet's real perception of it. It did, however, serve the polemical agendas of the "Newtonian heretics" at Cirey, while also putting academicians into a difficult bind. For the latter, the trick was defending the academy and Cartesianism against the rival claims of the Newtonians in ways that evinced cosmopolitan disinterestedness with respect to the Newton war that was brewing. Voltaire espoused his own commitment to the same values in an important exchange of letters with Dortous de Mairan in August–September 1738.[123]

Dortous de Mairan initiated the correspondence, offering a set of criticisms of the *Eléments* that stressed the philosophical and scientific inadequacies of the Newtonian system. Voltaire responded in kind and at much greater length, but his defense nevertheless stressed his deep respect for Dortous de Mairan, not to mention his gratitude at the respect shown to his work by the senior academician. Voltaire's letter was also steeped in the language of gentlemanly *politesse,* even if his goal throughout was nothing less than the conversion of Dortous de Mairan to Newton's attractionist physics. The academician was unmoved by Voltaire's arguments, but the exchange reveals the shared bonds of *honnêteté* and sociability that still bound Voltaire to the Cartesians inside the

---

122. Du Châtelet to Maupertuis, 10 January 1738, in Voltaire, *Corr.,* 88: D1422.

123. Dortous de Mairan to Voltaire, 11 August 1738; and Voltaire to Dortous de Mairan, 11 September 1738, in Voltaire, *Corr.,* 89: D1518, D1611.

academy even as he, du Châtelet, and Maupertuis were also beginning to define themselves publicly as the strident opponents of this very group.

The public discussion of the *Eléments de la philosophie de Newton* that erupted soon after its publication further complicated these negotiations. Most striking was how much more crowded, contentious, and attuned to questions of natural philosophy the French public sphere had become, a trend that had noticeably accelerated since the "bomb" of the *Lettres philosophiques* just four years earlier. At least seven different periodicals, including both the *Journal des savants* and the *Journal de Trévoux*, published discussions of Voltaire's work, and to these were joined three widely noticed critical pamphlets that triggered responses from Voltaire as well.[124] It would have been hard for a literate French reader to have missed the intense public debate about Newtonianism that was raging by the fall of 1738, and while the impact of this discussion was anything but straightforward, it did work to cement Voltaire's identity as a Newtonian with a powerful philosophical voice.

But even if the controversy ultimately established Voltaire as the credible Newtonian philosophe he aspired to be, it did so in ways that colored that identity as well. Most important was Voltaire's inability to ever eradicate the dangers of irreligion, moral subversion, and intellectual dishonor from his Newtonian position despite his intense efforts to achieve this very end. Maupertuis had avoided this pitfall by adopting (at least publicly) a brand of Newtonianism that was safer and more chaste. He also took care not to overly provoke the *honnête* structures of the Royal Academy, or the public that supported this authority. Voltaire had no such reliable foundation, even if he did work hard to construct one for himself. Even when critics indulged in broader and more personal attacks on his character as a Newtonian, Voltaire often responded by restricting himself to the technical, scientific challenges while ignoring the larger polemics embedded in them. He also attempted on more than one occasion to frame the debate as a simple matter of philosophical truth, one to be determined by

---

124. In addition to the *Journal des savants* and the *Journal de Trévoux*, the other periodicals that devoted reviews to Voltaire's work were Prévost's *Le pour et contre*, Desfontaines' *Observations sur les écrits modernes*; *Bibliothèque française*; *Refléxions sur les ouvrages de littérature* (a Parisian weekly edited by a friend of Desfontaines, l'abbé Granet); and d'Argens' *Mémoires secrets de la république des lettres*. The pamphlets launched against Voltaire are listed in Walters and Barber, introduction, 81–82, n. 39. Voltaire's published responses to this critical material, several of which will be discussed shortly, are conveniently collected and reprinted in the Walters and Barber critical edition of the *Eléments* in a section entitled "Rejoinders to Critics," *Œuvres complètes de Voltaire* (ed. Besterman), 15: 633–762.

calm, unprejudiced reflection by *honnête gens*. In this way, he conveyed a sincere desire to become a famous Newtonian *philosophe* without becoming an infamous one.

The nature of the French public sphere and Voltaire's place in it, however, fought against such hopes. Consequently, the public Newton wars that he and others fought after 1738 established neither a dispassionate understanding of Newtonianism in France nor a calm consensus regarding Voltaire's relationship to it. The philosophical positions themselves were still treated most often with mockery and scorn in 1738, and thus the benefit of the doubt remained squarely aligned against Voltaire. "What is certain is that the doctrine of Newton is not having much success in Paris," the l'abbé Granet wrote in late 1738 in his *Réflexions sur les ouvrages de littérature*. "All the professors in the university, and among the Jesuits, and everywhere else, reject it out of hand [*hautement*], for they see no distinction between it and ancient Greek philosophy."[125] Voltaire was disparaged in this context as well, and even if the abbé Le Blanc reported a different reception—claiming that a prior at the Sorbonne had spoken before an audience of bishops and other respectable people in praise of Newton, his new philosophy, and even Voltaire and du Châtelet—Voltaire's was anything but an easy struggle.[126] Exacerbating such tensions was the questionable nature of Voltaire's authority to even address such questions. His attempt to defend Newtonian attraction, space, and empirical epistemology was very often rebuked not with philosophical rebuttals but through attacks on his character and integrity.

Most benign in this respect was the abbé Prévost, who devoted a section of *Le pour et contre* to a discussion of Voltaire's *Eléments* soon after it appeared. Prévost had earlier criticized Voltaire's attempts at philosophical discourse in the *Lettres philosophiques,* preferring instead the more overtly literary dimensions of the text. In the case of the *Eléments,* however, he was confronted with a text that made no concessions to the literarily inclined. Prévost nevertheless chose not to provoke further antagonism on this point. He instead called the book "estimable" and devoted most of his discussion to the controversies surrounding the two "London" and "Amsterdam" editions of the text, a discussion that supported Voltaire's claim to be a victim of malicious and greedy publishers. At the same time, he avoided altogether the substance of Voltaire's philosophical arguments, insinuating strongly that he was neither equipped nor inclined to comment on such matters. In the end, however, his review sided

125. L'abbé Granet, *Réflexions sur les ouvrages de littérature* (1738): 6: 200–201.

126. Cited in Hélène Monod-Cassidy, *Un voyageur-philosophe au XVIIIe'siècle: L 'abbé Jean-Bernard Le Blanc* (Cambridge, 1941), 317; and in Walters and Barber, introduction, 92, n. 59.

with Voltaire. He described Voltaire's treatment of Descartes and Malebranche as "measured and well-supported," and he left his readers with a positive impression of the author and his work.[127]

Others did the same, including Father Castel, who devoted two long reviews to the work in the *Journal de Trévoux* in August and September 1738.[128] "Here is a work," Castel trumpeted in his opening line, "that well justifies our earlier declaration that this is the century of profound science, reliable arts, and useful discoveries, as opposed to the last century that was the century of the fine arts and belles lettres."[129] Eventually Castel settled into a thorough and celebratory account of the book, but even more important from Voltaire's perspective was how he framed his presentation. Picking up on one of his favorite themes— the exceedingly arcane nature of Newtonian mathematical philosophy—Castel used Voltaire to offer a few jabs at Cartesianism and Newtonianism alike. He noted first how Voltaire's conversion from a poet into a philosophe would have been less stunning had he been able to draw on the "taste for romance" inherent in Cartesian philosophy. Bashing Newtonianism as well, he then noted the difficulty of the Frenchman's project given the "abyss of difficulty" into which Newton forced his students to descend. "Newton measured, calculated, and weighed; he did not use words." This was to his credit, since there is nothing more dangerous and prone to misuse in philosophy than words. Nevertheless, by speaking only through the language of mathematics, Newton unduly restricted access to his truths. Moreover, even though mathematicians all over Europe have "descended into the abyss, pierced its shadows, and deciphered, commented upon, and knowledgably explained Newton," this knowledge was restricted to those who "speak using the signs a, x, +, -, etc." "Finally, however, M. de Voltaire spoke, and soon Newton was understood, or at least was ready to be."[130]

Castel also offered a positive assessment of Voltaire's reception, writing that "all of Paris is trumpeting Newton, repeating Newton, studying and learning Newton." Accordingly, "there is nothing that is not praiseworthy about M. de Voltaire's effort to make himself into a philosophe, or to make, as much as is possible, the entire universe into Newtonians."[131] Indeed, Castel continued,

127. Review of Voltaire, *Eléments de la philosophie de Newton*, in *Le pour et contre*, 15: 231–240.

128. Review of Voltaire, *Eléments de la philosophie de Newton*, in *Journal de Trévoux* (1738): 1669–1709, 1846–1867.

129. Ibid., 1669.

130. Ibid., 1673–1674.

131. Ibid., 1674.

while many in the public were treating Voltaire's hybrid combination of talents inequitably, such "professional criticisms are always unjust, and always place things in the worst possible light." Such indictments, he suggested, were mostly made by those secretly nervous about their own intellectual inferiority. Castel also placed himself with Voltaire atop an intellectual pedestal above the small-minded savants who saw in the "Voltaire phenomenon" (his phrase) something dangerous and suspect. "To attack these new *Eléments*, one must attack Newton himself," Castel declared, and he further added that such critics "must also be skilled mathematicians and understand Newton themselves in order to be in a position to know what to critique." Those who attacked Voltaire did not possess this understanding, Castel implied, and accordingly they distracted attention from their own intellectual weaknesses by calling Voltaire "a poet who does not understand this great mathematician." "What a miserable spirit of chicanery infects our century and our nation," Castel lamented.[132]

Castel's motivation in taking this precise, public stance was no doubt complex, but from Voltaire's perspective his statements, disseminated through a widely read organ of public science and sanctioned by the Jesuit order, marked the appearance of a powerful public voice on his behalf. Not surprisingly, Voltaire often made reference to the review in his correspondence, repeating with particular pleasure the Jesuit's claim that criticizing Voltaire meant criticizing Newton, an undertaking that was foolhardy for most. Yet Voltaire also found Castel to be predictably unpredictable. In June 1738, before Castel's review appeared, Voltaire complained to Maupertuis that he had sent Castel some of his optical writings, hoping to start an exchange with what he thought was a sympathetic Newtonian, only to see Castel publish "cruel and insulting" remarks about him in his journal. "I have known ever since that this crazy mathematician is your declared enemy," he wrote.[133] He had a different view three months later once Castel's review of the *Eléments* had been published, and overall one should not generalize about Castel's position or that of the Society of Jesus as a whole, since each continued to pursue a complex stance with respect to the unfolding Newton wars of the period.

Other Jesuits, for example, adopted different stances with resperct to Voltaire, and one who was specially critical was Father Noël Regnault, the author of the *mondain* dialogues that did so much to fuse Jesuit empirical naturalism with literary *bel esprit* in France. Regnault published a pamphlet in the fall of 1738 entitled *Letter of a Physicist on the Philosophy of Newton* that addressed

---

132. Ibid., 1675.
133. Voltaire to Maupertuis, June 15, 1738, in Voltaire, *Corr.* 89: D1519.

Voltaire's *Eléments* directly.[134] In it, he found Voltaire's philosophy and his manner of philosophical discourse equally suspect. He also exposed what he saw as the irreligious implications of the Newtonian theory of attraction and space. These same arguments were also echoed in another pamphlet published at roughly the same time by the abbé Machi.[135] Each of these critiques reinforced the then commonplace associations between Newtonian physics and either Spinozist, deist, or Epicurean materialism (the precise label did not really matter since all exerted the same charge). Moreover, since they did so while also rebuking Voltaire's "arrogance" in assuming to speak as philosopher, along with his blithe avoidance of these "obvious" philosophical problems, they also worked to impugn his character by associating his Newtonian convictions with his general proclivity for libertinism, immorality, and irreligion. Others with a less obvious religious or political agenda reinforced these associations in other ways.

Most difficult, perhaps, for Voltaire to deal with were those members of the public who shared Voltaire's intellectual agendas, including his commitment to liberty, but who nevertheless created problems for him by emphasizing these associations in ways that hindered his equally strong desire to become an authoritative and respected scientific thinker. Voltaire's decision to produce a prose work in the plain style of English experimental philosophy was intended to encourage *honnête* agreement about a complex philosophical problem, one grounded in the clear evidence of nature and the disinterested logic of empirical reasoning. Many accepted this overture and attempted to engage with it on its own terms, be it critically or not. Others, however, playing upon Voltaire's other recognized identity as a libertine poet, freethinker, and *esprit fort*, either rejected his prolix empiricism or found in it the fuel for further criticism. For some, Voltaire had simply taken a wrong turn in abandoning poetry for philosophy. "Leave it to Clairaut to trace the lines of the light rays that hit our eyes," wrote a certain M. de Clément in a poem that he addressed to Voltaire and published in *Nouveaux amusments du coeur et de l'esprit*. "You, of a more amiable delirium, listen to the lessons of another Muse. . . . Let go of the compass and pick up the lyre! I would give up every Pemberton and all the calculations of Newton for one sentiment of *Zaire*."[136]

134. Regnault, *Lettre d'un physicien sur la philosophie de Newton* (n.p., 1738).

135. L'abbé Machi, *Réflexions sur la philosophie de Newton, mise à la portée de tout le monde* (n.p., 1738).

136. M. de Clément, poem, in *Nouveaux amusements du coeur et de l'esprit* 5 (1737–1745): 192–193.

Others found in Voltaire's Newtonianism a different articulation of the libertinism that defined his character. Most illustrative in this respect was the appearance of a set of poetic works that connected Newtonian attraction, libertinism, and materialism in ways that echoed Voltaire's non-Newtonian libertine poetry. One such work was entitled "Letter from a Lady Philosopher to One of Her Friends." It appeared in both *Nouveaux amusments du coeur et de l'esprit* and in Desfontaines' *Observations sur les écrits modernes* during the intense, post-1738 period of the Newton wars.[137] Through its conceit of a learned lady (du Châtelet was of course implied) responding to the philosophical inquiries of a lady friend, it articulated the libertine interpretation of Newtonianism while playing with the erotic and irreligious implications of this materialist conception of nature. Opening with a prose discourse that framed the letter as a response to those who attacked Newtonian attraction and space, the work then shifted into verse, offering a poetic celebration of Newtonian philosophy. "This attraction inherent in nature, to which the universe owes its structure," forms bodies through its "force" and then "unites them by making them friends." The whole is thus formed through a "joyful embrace of all of its parts." Beings find their wholeness in the same way, for this "contributes to the well-being [*bonheur*] of the assorted souls." "But from where does this complexity arise," the verses continued? "One need only recognize the universe for what it is: an animal animated by this general attraction." This was the pure language of Spinozist materialism, and further echoing it the poet ended by describing "this interior charm, this gentle instinct" as "the soul that constitutes its principle attribute."[138]

Other poetic works of the period published in similarly *mondain* venues echoed this same understanding. In Robbe de Beauvest's ode "La Newtonique," published in *Nouveaux amusments du coeur et de l'esprit,* one read that Newton had "unveiled the archetype by which the world was formed." "I see by your fecund principle [universal gravitation] the entire universe animated. You force error into silence, . . . and I see repaired the disasters of the plenum that from error was adopted. The living light of the planets now passes to us in liberty, . . . and from the *Eléments* of Descartes you discover the fiction."[139] In

137. "Lettre d'une dame philosophe à une des ses amies," *Nouveaux amusements du coeur et de l'esprit* 6 (n.d.): 261–263 and in *Observations sur les écrits modernes* 19 (n.d.): 46–48. My citations will be to the latter version.

138. "Lettre d'une dame philosophe à une des ses amies," 47–48.

139. Robbe de Beauvest, "La Newtonique," *Nouveaux amusements du coeur et de l'esprit* 6 (n.d.): 284–286.

his letter to Dortous de Mairan, Voltaire had also likened the world to a great watch, suggesting that attraction, not impulsion, was the spring that animated the entire universe. "It's the great agent of nature," he wrote, "and an agent absolutely unknown until Newton came along. He discovered the existence of this agent, calculated its effects, and used it to illuminate elasticity, electricity, etc." [140] These were strong assertions about the validity and metaphysical universality of the principle of universal gravitation, and they were repeated publicly in the *Eléments* and Voltaire's other public pronouncements. However, articulated as they were in the idioms of judicious philosophy, and couched always within a discourse of skeptical empiricism and philosophical modesty, Voltaire's own statements about gravity carefully guarded against the explicit discourses of Newtonian materialism. This poetry, by contrast, spoke that language directly, and to the extent that Voltaire's own poetry, especially his libertine poems such as *La Pucelle* and *Le Mondain*, resonated with it as well, it fought against the philosophical redefinition that he was attempting.

The suspect nature of Voltaire's liaison with Emilie du Châtelet, already a source of titillating gossip thanks to Desfontaines' unauthorized publication of Voltaire's *Épître à M. Algarotti,* with its insinuations about du Châtelet's sexual indiscretions, only heightened this aura of scandal. Sexual libertinism and gender inversion were also strongly allied with the perceived immorality of materialist philosophy, and this made the erotic insinuations of the "Letter from a Lady Philosopher" all the more shocking. Overall, the poem emphasized rather than softened the erotic dimension of the materialism it articulated, even going so far as to suggest a connection between Newtonian attraction and erotic desire. "Oh yes, everything cedes to attraction, that charming passion!" the poet exclaimed. "It is constant and mutual," and "love, although always desperate and unfaithful, also bears a relationship to it." Consider the magnet, the poet wrote, using a pun that linked the French word for magnets [*aimants*] with the word for lovers [*amans*]. "Two *amans* at twice the distance desire each other four times less," the poet sang, and this law held for both magnets and love since "proximity gives [love] its greatest strength while separation weakens its power." [141] The verses also emphasized the gender politics that was just as crucial to this libertine sexuality by issuing a call to women [*les belles*] to take up the defense of Newtonianism. "Newton, who the sages admire, is being attacked out of envy." It was up to women, therefore, to "avenge this injury! He has nature, calculation, and reason on his side, but rather than

---

140. Voltaire to Dortous de Mairan, 11 September 1738, in Voltaire, *Corr.,* 89: D1611.
141. "Lettre d'une dame philosophe à une des ses amies," 48.

searching for other weapons, he only needs your charms for attraction to be proven."[142]

Voltaire was certainly not averse to indulging in erotic play such as this in his correspondence and poetry, but when speaking about Newtonianism, either publicly or privately, he avoided it altogether. His desire to be taken seriously as an *honnête* philosophical thinker rather than as a libertine poet and *esprit fort* was, therefore, sincere. Yet the associations, be they accidental or intended, between his philosophical convictions and his other, more scandalous libertine writings nevertheless fought against his ambitions. Also problematic was the bias of those who found literary *bel esprit* of this sort far more estimable and worthy of Voltaire's attention than knotty and contentious questions of physics and metaphysics. French elites in this period very often esteemed style, artistic grace, and *bon goût* more than scientific acumen or philosophical learning. Writing to Voltaire's loyal patron, the Duc de Richelieu, du Châtelet found herself confronting this prejudice as she tried to explain why Voltaire's *Eléments* was a superior work to Algarotti's *Newtonianism for Ladies.* "Algarotti's dialogues are full of wit [*esprit*] and learning," she wrote, "but I must avow that I do not like this style in matters of philosophy. For I find the love of a lover decreasing in inverse relation to the square of the time and the cube of the distance a little hard to digest."[143] Voltaire would express a similar sentiment when he called Algarotti's text "a book perfectly suited to Cirey." Yet statements such as these did not change the minds of those, like Président Bouhier, who belittled the public fame that Voltaire had attracted with his *Eléments* while claiming that "connoisseurs" like himself found Algarotti's text "infinitely more worthy."[144]

Aesthetic preferences such as these fought directly against Voltaire's transformation into a new kind of philosophe, and further thwarting these ambitions were those who used many of these same arguments to challenge Voltaire personally, either for reasons of sincere intellectual opposition or, more destructively, because of personal rivalry and animosity. Most influential in this latter respect was Desfontaines, whose long-brewing rivalry with Voltaire reached a public crescendo during the debates over Newtonianism. Relations were not

142. Ibid. In this spirit as well is *Épître Newtonienne sur le genre de al philosophie propre à render heureux. A Madame ****, in *Nouveaux amusemens du coeur et de l'esprit*, 9: 415–425.

143. Du Châtelet to duc de Richelieu, 17 February (or August?) 1738, in Voltaire, *Corr.*, 89: D1591.

144. Jean Bouhier, *Correspondance littéraire*, ed. Henri Duranton (Saint-Etienne, 1974–1988), 2: 185–186, cited in Walters and Barber, introduction, 94.

always bitter between the two men of letters.[145] In 1725, when Desfontaines was imprisoned for alleged sexual improprieties, Voltaire used his connections to help secure his release. The two men also exchanged cordial letters at times, trading manuscripts and discussing literary news. However, as Desfontaines increasingly devoted himself to journalism, and as he began, therefore, to develop what was fast becoming the journalistic norm of the day—a sharp, critical tone in his published writing—a bitterness began to grow between the two men. Voltaire also expressed a sense of betrayal when Desfontaines, driven by his journalistic need to find content, published works without his consent that Voltaire deemed inappropriate for public circulation.

At one level, their sparring was an unsurprising illustration of the critical jousting that all men of letters engaged in within the Parisian wing of the Republic of Letters. However, since critical, literary squabbles of this sort were anathema to anyone who aspired to be an *honnête* philosophe, Voltaire's ambitions with respect to the latter were perpetually challenged by his susceptibility to fall into the former. Privately, Voltaire also made clear his disdain for Desfontaines' intellect and character. He accused him of dealing in "invective," and called his journalism "miserable and malicious."[146] Personally, he found Desfontaines "an ignorant pedant who was incapable of writing, thinking, or understanding me."[147] "Monster" was an adjective that was further employed, as was the label "enemy."[148] Desfontaines was described on another occasion as "a malcontent who wants to violate all the little boys and outrage all the reasonable people," a direct reference to the sexual scandals that had placed Desfontaines in prison.[149] On still other occasions, he assumed bestial form in Voltaire's mind as he was called "a rabid cur that bites his master"; "a dog chased by the public who only returns to lick or bite it;"[150] and a bestial loner who "lives alone like a lizard."[151] On one particularly bilious occasion, Voltaire "repented" his

145. M. H. Waddicor offers a thorough accounting of the long history of the Voltaire-Desfontaines relationship in the introduction to W. H. Waddicor, ed., *La Voltairomanie. Édition critique* (Exeter, 1983), vi–lviii.

146. Voltaire to Thieriot, 30 June 1731, in Voltaire, *Corr.,* 86: D417.

147. Voltaire to Thieriot, 24 September 1735, in Voltaire, *Corr.,* 87: 918.

148. Voltaire to Cideville, 25 March 1736, in Voltaire, *Corr.,* 87: D1044; Voltaire to Berger, 27 November 1736, in Voltaire, *Corr.,* 88: D1208. See also Voltaire to d'Argens, 20 December 1736, in Voltaire, *Corr.,* 87: D1228, where he pairs up Desfontaines with Rousseau as his "enemies."

149. Voltaire to l'abbé d'Olivet, 4 October 1735, in Voltaire, *Corr.,* 87: D923.

150. Voltaire to Berger, 2 February 1736, in Voltaire, *Corr.,* 87: D1000; Voltaire to Gilles Thomas Asselin, 3 March 1736, in Voltaire, *Corr.,* 87: D1028.

151. Voltaire to Thieriot, 18 November 1736, in Voltaire, *Corr.,* 88: D1202.

decision to rescue Desfontaines from the Bastille, writing that "it is better, all things being equal, to burn a priest than to bore the public, and if I had left him cooking I would have saved the public much foolishness [*sottises*]."[152]

Overall, Voltaire showed little restraint when impugning Desfontaines' character, at least in his private correspondence. Much of this vitriol can be attributed to the deep, personal disaffection that erupted between the two men after 1735, but more substantively it was also a product of the different relationship that each man held to the new critical journalism of the eighteenth century. Desfontaines had transformed himself after 1730 into a major public voice in French letters through his pursuit of critical journalism. Voltaire, by contrast, was attempting to assert an identity that made journalists potential allies to the extent that they supported him, or bitter enemies if they did not. Accordingly, as Desfontaines increasingly made Voltaire the subject of his otherwise commonplace critical, journalistic commentary, he found himself increasingly labeled a "malcontent," a "monster," and an "enemy" by Voltaire.

The fledgling philosophe was not so consumed in these personal struggles not to see and appreciate the wider forces that were driving them. In a letter to Cideville from September 1735, Voltaire labeled as "the autumn of good taste" the general trend that had "inundated the land of literature with pamphlets and newssheets. "*Le pour et contre* is more insipid than ever," he complained, "and the *Observations* of abbé Desfontaines are a set of outrages that he issues once a week against reason, equity, erudition, and taste. It is impossible to adopt a more fulsome tone and to understand things less." "In one, this poor student of English goes so far as to call an English work in support of religion an atheistic tract," he wrote. "There is nothing in these sheets but a mess of errors."[153] These last descriptions were directed at Desfontaines rather than Prévost, and even if Voltaire tended to generically treat all critical journalists as enemies of *honnête* discourse, he also used these two examples to construct a framework for judging good from bad journalism. Invoking Prévost, who increasingly became for him the figure of the "good journalist," Voltaire wrote that "I am as grateful to [him] for his criticisms as for his praise. . . . However, to criticize with finesse and without inflicting wounds, one must have an exceedingly delicate and polite spirit. I may not agree with him on many things, but my esteem for him has been redoubled by the same things that render ordinary authors irreconcilable."[154]

152. Voltaire to Cideville, 20 September 1735, in Voltaire, *Corr.*, 87: D915.
153. Ibid.
154. Voltaire to Thieriot, 20 March 1736, in Voltaire, *Corr.*, 87: D1040.

Voltaire made it clear that Desfontaines did not measure up to Prévost in this respect: he had neither talent nor *esprit*, and his morals were anything but *honnête*. Consequently, he became the personification of all that was destructive and distasteful in contemporary letters. Since Prévost's journalism was typically far less critical of Voltaire than Desfontaines', it is easy to dismiss all this criticism as petty sour grapes. Yet in his letters, Voltaire did offer a cogent framework that defined good and bad journalism in ways that reinforced his larger conception of himself as an *honnête philosophe*. "When one attacks my work, I have nothing to say. It is up to them to defend themselves either well or poorly. However, when my person, my honor, and my morals are publicly attacked in twenty libels that have inundated France and abroad, one must oppose these calumnies with facts, and impose a silence on lies." [155] If Prévost was a superior journalist to Desfontaines, therefore, it was because he had both the learning and intellect to judge things properly, and the right manners and character to know the difference between *honnête* criticism and petty, destructive slander. Voltaire echoed this preoccupation with *honnêteté* in the same letter, writing of Prévost that "I do not know if he has said of me, or felt compelled to say, that I am an *honnête homme,* but I know that I owe it to say it about him." [156] Given this shared bond in morals and manners, Voltaire also found *Le pour et contre* to be a reputable public organ, while those of Desfontaines, Rousseau, and his other enemies were described as little more than dishonorable rags.

There was certainly an instability in this dichotomy since there was a tendency toward criticism and controversy inherent in the literary journalism that each of these men practiced. Accordingly, so long as Voltaire was dependent on the public sphere to authorize his status as an *honnête homme,* he could never fully trust journalism or journalists. As he wrote to Thieriot in February 1736, "If I was certain that *Le pour et contre* would speak as strongly as is necessary, I would stay silent, for my cause would be better handled by it than by me. But how can this certainty be found?" [157] Voltaire had few other resources to turn to, and thus with no small amount of frustration he began to engage publicly with the critics and journalists that were shaping perceptions of his Newtonianism in the French public sphere.

Desfontaines was at first quiet on the topic, but the appearance of the Jesuit Regnault's pamphlet and the equally religious work of the abbé Machi triggered Voltaire's response. In August he published a letter (perhaps not sur-

155. Voltaire to Thieriot, 26 February 1736, in Voltaire, *Corr.*, 87: D1023.
156. Ibid.
157. Ibid.

prisingly) in Prévost's *Le pour et contre* that responded to these criticisms.[158] Offering himself as a reader of Prévost's journal who wanted to clarify certain aspects of its previously published review of the *Eléments,* Voltaire began by issuing the usual instructions about which edition was correct, where to find the necessary *eclaircissements,* and how to read the text appropriately. He then used an example about the misunderstandings that arise from typographical errors such as these to launch into his real project, an attack on Regnault's pamphlet and a defense of his own work. "I would very much like to know," Voltaire wrote, "how a man who calls himself a *physicien,* and who is writing, he says, about the philosophy of Newton, can begin by saying that I apologized for the murder of Charles I? What relationship, if I may ask, exists between the tragic and unjust demise of this monarch and refraction and the square of the distances?"[159] The answer, Voltaire explained, was found in the erroneous Amsterdam edition of the *Eléments* that contained a work on this execution as an unauthorized addendum. Voltaire believed that this text, which he did not write, had no relationship whatsoever to Newtonian philosophy, and he further mocked the reading skills of his Jesuit critic by noting that the text in question condemned rather than defended this execution. "It is with the same sense of justice that this author critiques me instead of my work throughout," he concluded.[160]

Especially frustrating to Voltaire was Regnault's tone in deriding Voltaire's views and integrity as a thinker while avoiding altogether the substance of his arguments. He quoted Regnault's retort that "it would be useless to offer a commentary about such a major mistake" since "everyone can see it and it would be too humiliating for M. Voltaire." "It would be curious to see what considerable mistake I should find so humiliating," Voltaire replied, going on to condemn in general "the injustice of such precipitous criticism." "An insult is not an argument," Voltaire reminded his readers, and his letter honored this principle by defending his own positions in detailed terms.[161] Regnault had used his pamphlet to defend the rival Cartesian understanding of light, but to these arguments Voltaire responded "if one were to make more reasonable objections against me, then I would respond to them, either in correcting myself or in demanding greater clarification, since I have no other goal than the truth."

---

158. Voltaire, "Letter concerning the *Eléments de la philosophie de Newton,* in *Le pour et contre,* 15: 337–349; reprinted in *Œuvres complètes de Voltaire* (ed. Besterman), 15: 677–686. My citations will draw from the latter edition.

159. Voltaire, "Letter concerning the *Eléments de la philosophie de Newton,*" 15: 682.

160. Ibid.

161. Ibid., 15: 683.

Instead, since Regnault's counter-arguments were specious at best, he instead constructed a further defense of his own position.[162]

Central to Voltaire's presentation was a brief in support of the concept of gravitational attraction. He believed rightly that this idea was "causing the greatest upset among his compatriots," and draping himself first in robes of academic legitimacy, he invoked Maupertuis' authority to ease the anxieties of his public. "I repeat again that one need only read the dissertation on the figures of planets by [this] illustrious [academician] to see if one has a better idea of the impulsion we are supposed to accept or the attraction that we must combat." He then demonstrated his own scientific mettle by inviting "mathematical physicists [*physiciens géomètres*]" to consider the quantitative measurements of falling bodies and pendulum motions according to the vortical system. "All the mathematical contradictions that emerge seem to annihilate the vortices while agreeing with the other, more doubtful hypothesis." "Where is the great difficulty, then, in saying that God gave gravitation to matter in the same way that he gave it inertia, mobility, and impenetrability?" This opened up the theological dimension of Newtonian physics that both Regnault and Machi found heterodox and dangerous. Voltaire responded by defending his own conception of Newtonian theology openly. "I believe that the more one reflects on the matter, the more one is led to believe that *pesanteur* is, like movement, an attribute solely given to matter by God." Voltaire noted rightly that Newton never argued this forcefully for gravity as an inherent property of matter. He stressed, therefore, that this was his own extrapolation from the physics of his master. Radical English deists like John Toland had further turned this position into a foundation for radical materialism, yet here Voltaire drew a line between his views and those of the radical Newtonians. He did so by positioning himself and his scientific views not within the discourses of radical libertinism and free thought but instead within the sober and *honnête* discourses of judicious, skeptical empiricism.[163]

Prévost reinforced Voltaire's stance by praising his text in an addendum to the letter in *Le pour et contre*. "This letter merits being published someday as a supplement to the *Eléments of Newton*." For "not only does it justify the author against the accusations of an anonymous *physicien*, it also shows his *politesse* and moderation."[164] The combination of scientific correctness, philosophical authority, and sociable manners described by Prévost was precisely what Voltaire was attempting to assert with his Newtonian campaigns, and other

162. Ibid.
163. Ibid., 15: 685–686.
164. Ibid., 15: 678.

journalists further reinforced Voltaire's efforts in this respect. Castel, for example, disagreed with Voltaire that the Cartesian plenum was more atheistic and Spinoizistic than Newtonian space, but he nevertheless praised him for speaking "with decency about God and his attributes."[165] Yet while journalists such as Prévost and Castel helped Voltaire to achieve his ambitions as an *honnête philosophe,* others worked to thwart them. Desfontaines was a case in point, and in the wake of the public debate over Voltaire's *Eléments,* the long-brewing rivalry between the two men erupted into a full-blown public war.

If one is to believe Voltaire's correspondence, his disgust with Desfontaines combined with his distaste for critical journalism overall led him to stop reading *Observations sur les écrits modernes* soon after the journal began appearing in March 1735. Desfontaines' treatment of Voltaire's tragedy *The Death of Caesar* was the precise excuse, but overall, Voltaire suggested, these "biweekly impertinences" no longer interested him. He would cease to read them, therefore, until the author showed more "probity toward the public" and stopped filling his pages with "satire and calumny."[166] Whether he in fact ignored the journal is not clear, but in March 1738, on the eve of the publication of his *Eléments,* he wrote to Thieriot from Cirey asking if "that monster Desfontaines continues to deliver his weekly rags [*malsemaines*]?"[167] In May, he wrote to Maupertuis expressing "neither surprise nor anger" that Desfontaines was trying to ridicule attraction, noting that "a man who is so little a physicist and who is so corrupted by the sin of antiphysics will always sin against nature."[168] This was an indication that Voltaire had been alerted to Desfontaines' views about Newtonianism, yet his exposure might very well have come through word of mouth or letters. Evidence that he had not yet read the journal himself was offered in June when he wrote to Prault, his bookseller in Paris, asking that he be sent a number of books at Cirey, including the "sheets by that miserable Desfontaines."[169]

Assuming that Prault complied, what did Voltaire find when he read the *Observations* in July 1738? It is hard to know which issues Voltaire received exactly, but in the preceding months Desfontaines had discussed a number of works that addressed Newtonianism either directly or indirectly. Among those that treated it directly was a work by two Minim friars from Rome, Fathers Thomas le Seur and François Jacquier. Desfontaines reviewed their jointly au-

165. Review of Voltaire, *Eléments de la philosophie de Newton,* in *Journal de Trévoux* (1738): 1677.

166. Voltaire to Thieriot, 24 September 1735, in Voltaire, *Corr.,* 87: D918.

167. Voltaire to Thieriot, 8 March 1738, in Voltaire, *Corr.,* 89: D1469.

168. Voltaire to Maupertuis, 22 May 1738, in Voltaire, *Corr.,* 89: D1508.

169. Voltaire to Prault, 28 June 1738, in Voltaire, *Corr.,* 89: D1535.

thored commentary on Newton's *Principia* in March.[170] Describing the plan of the book, he wrote that "many commentaries on Newton exist, but no one has yet tried to follow and clarify Newton's work in the *Principia* proposition by proposition." This is what these friars attempted, and Desfontaines praised the initiative, calling the result "the definitive edition of Newton's text."[171] Neither the authors nor their reviewer offered any judgment of the arguments themselves short of noting that much of the confusion surrounding Newton's thought derived from the mathematical difficulty of his text. Accordingly, this book was welcome, Desfontaines suggested, as corrective to many misunderstandings even if it offered nothing by way of argument either for or against Newton.

Desfontaines, however, did launch an indirect bomb by appealing for another philosophical commentary that would bring similar clarity to the work of a different philosopher. "The zeal that I have for the glory of the nation makes me wish on this occasion that some able philosopher would also work to illustrate, extend, and perfect the ideas of our great Descartes. We have become passionate about a foreign philosophe while we are today indifferent about our own." The work of Privat de Molières, he added, had begun to alleviate this shortcoming, and he ended his review by praising it glowingly. "*Les leçons de physiques* is a work of admirable profundity and judiciousness, and it does not allow us to put Descartes below Newton."[172]

Overall, Desfontaines was an ardent Cartesian of this sort, one who was quick to launch dismissive attacks against Newtonianism. This most likely made a clash with Voltaire inevitable, but matters came to a head when Voltaire's name was attached directly to the anonymously published *Epistles on Happiness,* a work that included an epistle entitled "On Liberty." In it, the author called the Cartesian vortices "learned chimeras" that "hardly anyone still accepts." He also defended human liberty in ways that echoed a debt to the discourse of the English deists.[173] This prompted Desfontaines to take up the patriotic mantle of Cartesianism and challenge Voltaire's reckless defense of the errors of Newtonianism directly.

"Is it appropriate to show such little respect for the greatest man that philosophy has ever had?" Desfontaines asked in opening. Conceding to at least

---

170. Thomas Le Seur and François Jacquier, *Philosophiae naturalis principia mathematica. Auctore Issaco Newtono,* 3 vols. (Geneva, 1739–1742).

171. Review of le Seur and Jacquier, *Philosophiae naturalis,* in *Observations sur les écrits modernes* 12 (n.d.): 229.

172. Ibid., 234.

173. [Voltaire], *Épîtres sur le bonheur* (Paris, 1738). For a review, which names Voltaire as the author, see *Nouveaux amusemens du coeur et de l'esprit,* 2: 227–262.

one part of Voltaire's position, he admitted that Newton offered a better explanation of light than his French rival. "But his philosophy in general offers nothing comparable to that of the French philosopher." "[Newton's] opinions on the essence of matter, space, the void, and especially attraction are the real chimeras," he continued later, "borrowed from the ancient and erroneous philosophy of the Greeks." He further promised to show this in full when "he had the occasion to review the *Eléments de la philosophie de Newton, mis a la portee de tout le monde* by M. de Voltaire."[174] This review actually appeared several months later, but in the issue of June 5 he offered a précis of what that discussion would contain. In this review, "we will make it known that Descartes was right to submit all of nature's effects to mechanism while Newton, following another method, wanted to return us to the gibberish [*galimathias*] of the occult qualities." We will also see "that the physics of Descartes is unique and that all the others are irrational [*insensée*]." Desfontaines also asserted that "Newton's arguments, however much he was the greatest experimental physicist, serve only to prove our ignorance." "One can never reach conclusions that contradict the great demonstrated system of the mechanism of nature. It must never be swept away, as we will soon put *à la porté de tout le monde* (at the level of everyone)."[175]

The promised review did not appear right way, however. Instead, Desfontaines devoted several long reviews to a discussion of Privat de Molières' *Les leçons de physiques,* framing it as the reasonable, Cartesian alternative to Voltaire's "absurd" *Eléments de Newton.*[176] Also indicative of his precise positioning was his review of two other Newtonian texts of the period. In late July he discussed Maupertuis' book on the shape of the earth, a work that defended the Newtonian position of the flattened, grapefruit-shaped earth through evidence drawn from empirical, geodesic surveys. Since the book also narrated the heroic adventures of the great Lapland expedition, Desfontaines focused his attentions here, noting how a "strong desire to be useful to the country" motivated the expedition and the courage of the participants. "What perils! What bravery! What determination! Nothing is more worthy of admiration," he enthused.[177]

174. Review of Voltaire, *Épîtres sur le bonheur,* in *Observations sur les écrits modernes* 13 (n.d.): 232–233.

175. Ibid., 233–234.

176. Review of Privat de Molières, *Les leçons de physiques,* in *Observations sur les écrits modernes* 13 (n.d.): 305.

177. Review of Maupteruis, *La figure de la terre déterminée par les observations,* in *Observations sur les écrits modernes* 14 (n.d.): 149.

Desfontaines also offered a matter-of-fact account of the pendulum experiments and empirical demonstrations that gave credence to Newton's theory of universal attraction. In doing so, he in no way inflamed Maupertuis' highly judicious and empirical account of this work or the conclusions he drew from it. Nor did he attempt to drag Maupertuis into his emerging critique of Newtonians such as Voltaire.[178] He did the same with his review of Algarotti that appeared in late August. He celebrated his *Newtonianism for Ladies* as a wonderfully accessible introduction to Newton's philosophy while distancing the author from any defense of the chimeras of Newtonian attraction. As discussed above, Desfontaines did use the review to continue his increasingly vociferous diatribes against the Newtonian theory of gravity, space, and physical action, but he did so in such a way as to divorce Algarotti the *honnête* writer and *bel esprit* from any direct connection to these views.[179]

With Voltaire, by contrast, his Newtonianism and his person were consubstantial. Moreover, since Voltaire likewise equated Desfontaines' malicious intellectual views with his deficiencies of character, responding to the first inevitably involved impugning the second. Frustrated, therefore, by what he perceived to be a sustained and orchestrated assault on his character launched via a journalistic attack on his Newtonian philosophical convictions, Voltaire arranged for the printing and circulation of a libelous pamphlet entitled *Le préservatif* sometime around the end of October 1738.[180] The anonymous work took the form of a set of numbered critiques that dissected with sarcastic wit Desfontaines' erroneous and vindictive journalism. The critical language by itself was not that libelous, for very often Voltaire simply exposed Desfontaines' errors while unfavorably comparing his work overall with the "objective, accurate, scientific, tasteful, and judicious" journalists of such respected organs as the *Journal des savants*.[181] Nevertheless, the work as a whole amounted to a scathing attack on Desfontaines' character, and it triggered a controversy with important consequences.

Voltaire's prime mode of attack was to demonstrate his enemy's mistakes while chiding him for the arrogance of criticizing that which he did not understand. Natural philosophy, therefore, figured centrally in the pamphlet, and

178. Ibid., 149–163.

179. Review of Algarotti, *Il newtonianismo per le dame,* in *Observations sur les écrits modernes* 14 (n.d.): 217–229.

180. Voltaire, *Le préservatif, ou Critique des Observations sur les écrits modernes* (La Haye, 1738). A copy of the text is found in *Oeuvres complètes de Voltaire* (ed. Beuchot), 22: 371–389. My citations will be to this edition.

181. Voltaire, *Le préservatif,* 22: 371–372.

while Voltaire's tone was sarcastic throughout, the content of his arguments was substantive. A representative illustration was Voltaire's citation of Desfontaines' claim that "the English flatter the void" and "attribute marvelous properties to this Nothing." "Where does Newton 'flatter' the void?" he asked. Newton never attributed marvelous properties to the void, but he rather demonstrated that bodies act at great distances in a nonresisting medium. "One should at least learn about the topic before insulting the great men whose books one has never read, nor never could have read," he complained.[182] Many of the other arguments Voltaire offered were equally substantive, and consequently the pamphlet worked at one level as a defense of his own views and a challenge to those of his critics. Desfontaines, for example, had championed Privat de Molières' *Leçons de physiques*, as the sane, Cartesian alternative to Voltaire's *Eléments*. Accordingly Voltaire made a point of showing the erroneous nature of the science it defended. He cited Privat de Molières' account of falling bodies on earth and the alleged agreement between it and Newton's account. He then demonstrated its quantitative errors and its disagreement with the real theory of Newton.[183] Similarly, Voltaire mocked Desfontaines' claim that Dortous de Mairan had "imitated the system of Newton with respect to light." "One must teach him that Newton never made any system regarding light," he chastised, "but rather gave a direct, step-by-step account of experiments and mathematical demonstrations. To speak of these discoveries as a system is like calling Euclid's geometry a system."[184]

In each of these criticisms, Voltaire was drawing upon and exploiting the philosophical and scientific differences that separated Newtonians and Cartesians in France. He was also doing so in a substantive way that positioned his own views with those held by authoritative Newtonian philosophers. However, since the pamphlet as a whole framed these challenges within the idioms of libel, they acquired a critical coloring that had nothing to do with the philosophical pronouncements themselves. The title of Voltaire's pamphlet, *Le préservatif*, set the tone, since this word carried two meanings in French, "preservative" and "condom." Throughout the text, Voltaire also emphasized the sexual scandals that impugned Desfontaines' character in the public mind, while connecting his alleged depravity to his intellectual and critical character. Defending the progress of the arts and sciences against critics like Desfontaines who saw them in decline, Voltaire called this perspective typical of a social and intellectual misfit. "The sciences are better than ever," he wrote. "Just consider the

182. Ibid., 22: 375–376.
183. Ibid., 22: 374.
184. Ibid., 22: 383.

work of the polar expedition" and the accolades awarded to "our best authors," "men such as Réaumur and Voltaire" (the pamphlet, it should be remembered, was published anonymously). If Desfontaines could not see this, it was because he had the "merits of a satyr" and was "despised by the public."[185]

Especially provocative was the "malicious engraving" (Madame de Graffigny's label) that Voltaire commissioned as the frontispiece of his work.[186] The libelous image offered an allegorical depiction of Desfontaines' flagellation at the Bicêtre prison. In it, a prostrate Desfontaines, dressed in clerical robes, exposes his naked buttocks to a man with a whip, while a naked female figure, perhaps personifying wisdom, knowledge, or reason, hides her eyes while hovering on a cloud above. A set of venomous rhyming couplets underneath the image explained the action. "Formerly a curate, formerly a Jesuit, and everywhere known and hunted," Desfontaines, the verses explained, has become a "parasitical author, and the public has grown weary of him. To repair the past, he became a sodomite," but "at Bicêtre he was well spanked." For in this act "God offered recompense for his merit."[187]

Whatever serious philosophical disputation Voltaire had hoped to achieve with the pamphlet was subverted by the licentious invective offered in the image and the text. For his part, Desfontaines only exacerbated this outcome by responding in kind to Voltaire's attack. In December, he began to circulate his own libelous pamphlet called *La Voltairomanie* that sold thousands of copies within two months.[188] Desfontaines' work was even more venomous than Voltaire's, using phrases like "angry dog" and "proud fool" and adjectives such as "crazy," "impious," "reckless," "brutal," "impetuous," "libelous," "enraged," and "shameful baseness" to describe Voltaire and his discourse. Moreover, rather than trying to defend himself against any of Voltaire's accusations, he simply expanded the list of Voltaire's errors, while heightening the ad hominem rhetoric attached to his critique. Readers of these pamphlets, therefore, would have been treated to a raucous and sometimes witty brawl, but they would have been hard pressed to find any clear and consistent philosophical positions.

The wider dispute triggered by these pamphlets similarly muddied the philosophical differences at issue. Desfontaines used the pages of his *Observations*

185. Ibid., 22: 378.
186. This is reprinted, along with Graffigny's commentary, in Waddicor, *La Voltairomanie*, xliii.
187. Ibid.
188. Desfontaines, *La Voltairomanie, ou Lettre d'un jeune avocat, en form de mémoire en reponse au libelle du sieur de Voltaire, intitulé* Le Preservatif (n.d). The text is reprinted with critical notes in Waddicor, *La Voltairomanie*, 1–70.

Jadis Curé, Jadis Jesuitte,
  Partout conu, partout chassé.
Il devint auteur parasitte,
  Et le Public en fut lassé.

Pour réparer le temps passé
  Il se déclara Sodomite,
A Bissetre il fut bien fessé.
  Dieu récompense le méritte.

FIGURE 21. *Frontispiece from* Le préservatif, ou Critique des observations sur les écrits modernes *(n.p., 1738). Courtesy of the James Ford Bell Library, University of Minnesota Libraries.*

to offer the promised review of Voltaire's *Eléments* in September, and while most scholars think that Voltaire had already penned his pamphlet before this review appeared, it provided the immediate context for its reception.[189] While critical, Desfontaines refrained from the ad hominem attacks that he would soon launch in *La Voltairomanie*. Picking up the lament that many expressed regarding Voltaire's abandonment of poetry for philosophy, the journalist expressed concern that such a luminous writer would embrace Newtonianism, "a philosophy damned by all the good philosophes of Europe."[190] "One can rest assured that there are not two other philosophes in France who agree with him."[191] He then offered what was by then the standard Cartesian critique, emphasizing in a long and very detailed analysis the absurdity and impiety of the Newtonian theories of space and attractive matter; the superiority of directly causal, mechanical accounts of physical change; the danger of groundless, speculative philosophical reasoning; and the arrogance of presuming to disparage great thinkers such as Descartes and Malebranche.[192]

These criticisms were fairly moderate, but after the appearance of *Le preservatif* a few weeks later, whatever moderation was to be found in Desfontaines' relations with Voltaire was lost. As a journalist who published under his own name, French law made it impossible for Desfontaines to criticize Voltaire in his journal as freely as he did in his anonymous pamphlet. Nevertheless, a noticeable intensification of his polemical tone did occur with respect to his discussion of Newtonianism after the appearance of Voltaire's attack. One occasion was the appearance in 1739 of a serious and detailed refutation of Voltaire's Newtonian theories of light by a Cartesian savant named Jean Banières.[193] Only a year before, Banières had published his own, rival theory of light in a treatise of his own, and building on that work Banières focused attention on what he perceived to be Voltaire's errors and theoretical inconsistencies in the *Eléments*. His discussion, therefore, was as much a critique of Newton as a critique of Voltaire.

In his review, however, Desfontaines refocused this criticism, offering on the one hand a celebratory platform for Banières' rival Cartesian views, and on the other a new set of criticisms of Voltaire's reckless and misguided phi-

---

189. See Waddicor, *La Voltairomanie*, xxxix; Vaillot, *Avec Madame du Châtelet*, 97.

190. Review of Voltaire, *Eléments de la philosophie de Newton*, in *Observations sur les écrits modernes* 15 (n.d.): 74.

191. Ibid., 75.

192. For the full discussion of Voltaire's *Eléments*, see ibid., 49–67, 73–89.

193. Jean Banières, *Examen et réfutation des Eléments de la philosophie de Neuton de M. Voltaire, avec une dissertation sur la réflexion & la réfraction de la lumière* (Paris, 1739).

losophy.[194] The appearance soon after of an anonymous anti-Newtonian work entitled *Examination of the Void, or Newtonian Space Relative to the Idea of God* gave Desfontaines further occasion to attack Voltaire.[195] In two reviews, he reiterated the work's thesis that Newtonian space was philosophically absurd and religiously suspect. He also implicated the defenders of this view with immorality and impiety. He also published a response by Davy de Fautrière, the anonymous author thanking Desfontaines for his defense of sane philosophy.[196] The "Letter from a Lady Philosopher" discussed above was also published by Desfontaines in this context, and overall this letter joined in the chorus of anti-Newtonian discourse that Desfontaines released from the pages of the *Observations* in the wake of his public battle with Voltaire.

Other pamphleteers also joined the fray, including the anonymous author of *Le portefeuille nouveau,* an anti-Voltaire libel that Graffigny called "terrible [*effroyable*]."[197] About the same time, the publisher of Voltaire's *Lettres philosophiques,* Jore, released something like the critical edition of the Voltaire-Desfontaines dispute, publishing *Le préservatif* and *La Voltairomanie* side by side in the same volume followed by a *"factum"* that evaluated the merits of each position.[198] Voltaire's position was supported by a critical discussion of Desfontaines' journalism published in the *Nouveaux amusements du coeur et de l'esprit.* The piece assumed the form of a salon conversation where Desfontaines was mocked as a "critical buffoon" [*haranguer boufon*], while his style was criticized as hard and clumsy (Voltaire had called it "base" and "dimwitted [*marotique*] in *Le préservatif*). "He writes like a German," the text repeated more than once.[199] Also supportive of Voltaire was the work of the probably fictional Josephe de Neufville de Montodor. He published the pro-Voltaire *La nouvelle*

194. Review of Banières, *Examen et réfutation des Eléments de la philosophie de Neuton,* in *Observations sur les écrits modernes* 18 (n.d.): 337–353; 19 (n.d.): 97–11; 20 (n.d.): 169–186.

195. *Examen du vide, ou Espace Newtonienne r'lativement à l 'idée de Dieu* (Paris, 1739).

196. *Observations sur les écrits modernes* 28 (n.d.): 18–23, 42–45; letter from the author of *Examen du vide, ou Espace Newtonienne,* in *Observations sur les écrits modernes* 28 (n.d.): 355–357.

197. *Le portfeuille nouveau, ou Mélanges choisis en vers et en prose* (n.p., n.d.). See also Vaillot, *Avec Madame du Châtelet,* 123.

198. *La Voltairomanie avec le Préservatif et Le factum de Sr. Claude Francois Jore* (London, 1739).

199. *Dialogue on Writing and Writers,* in *Nouveaux amusemens du coeur et de l'esprit* 1 (n.d.): 433–459.

astronomie du Parnasse français in 1740.[200] This text opened with Apollo seeking a successor and then choosing Voltaire because of his "luminous and determined brilliance." The "hideous venom" of *La Voltairomanie* was mentioned as a way of imploring the new Sun King to "use his will in a sovereign manner to govern all the untamed and jealous people who, like pernicious basilisks, seek only to destroy him." A new zodiac was then offered that described the *caelestis* that Voltaire would govern. In it, the *Journal des savants* was labeled Libra, or the scales, and the scorpion was the *Journal de Trévoux*. Sagittarius, or the wise one, was *Le pour et contre,* and the Hydra, that "aquatic serpent that has heads succeeding one after the other," was Desfontaines. Du Châtelet was named as "the aurora that always follows the sun," and the Café Gradot was called Akousmates, the god of "noise and tumultuous voices."[201]

Works such as these only enflamed the polemical energies unleashed by Voltaire and Desfontaines themselves, yet others attempted to dampen them with a tone of judiciousness and *honnêteté.* The *Nouveaux amusements du coeur et de l'esprit* published a sharp critique of the *Nouvelle astronomie du Parnasse,* calling it a "miserable work" and a "defamatory libel" that possessed "neither gaiety nor finesse." "It excites the indignation of *honnête gens,*" the editor wrote.[202] More sympathetic to Voltaire was a work called "Le Mediateur" published in the same periodical. It used similar language to attack *La Voltairomanie,* but ultimately called for peace as the only appropriate response to this unfortunate civil war within Parnassus.[203] Even more neutral was the "Disinterested Judgment on the Controversy That Erupted between M. Voltaire and abbé Desfontaines." "Do you think the public is satisfied to see itself inundated by your miserable libels," the author chastised? "You are the heroes of Parnassus, yet you do not even blush as you deliver yourselves into a war that would bring shame upon even the lowliest of men." The author nevertheless tried to find substance in the dispute, praising the qualities of the two men while also validating the criticisms of each. Voltaire was supported in his view that Desfontaines "had perhaps overestimated the public's confidence in his abilities," while Desfontaines was supported in his view that Voltaire "shows insufficient discretion regarding religion, materialism, deism, and even athe-

200. Josephe de Neufville de Montodor, *La nouvelle astronomie du Par'asse français, ou L'apothéose des écrivains vivans dans la présente année 1740* (n.d., 1740).

201. Ibid.

202. Review of Neufville de Mohtodor, *La nouvelle astronomie du Parnasse français,* in *Nouveaux amusemens du coeur et de l'esprit* 5 (n.d.): 99–101.

203. "La mediateur," *Nouveaux amusemens du coeur et de l'esprit* 3 (n.d.): 345–59.

ism." Voltaire needed to continue writing poetry, history, and even philosophy, this self-appointed judge concluded, and he also needed to cede to Desfontaines the right to criticize his works. "This being accomplished, Sirs, the indulgent public will disregard all of your past follies." [204]

Peace was also in the interest of the two combatants since no one gained any advantage by continuing such a vindictive dispute. The authorities were also troubled by this warfare, and since Voltaire's political position was more precarious than Desfontaines', and since he had instigated the public feud in the first place, he was particularly vulnerable in this respect. As the controversy heated up in early 1739, he confessed to Madame de Graffigny an impulse to absent himself in Holland. [205] Instead, he worked behind the scenes to negotiate a resolution. In May, he and Desfontaines each signed separate disavowals with the royal police officer Hérault. This officially ended their open civil war, but it in no way terminated the wider struggle. [206] In June, Voltaire's other enemy, Jean-Baptiste Rousseau, arranged for yet another edition of *La Voltairomanie* to be published in Holland. The journals in France and throughout the Republic of Letters also continued to echo the dispute for many years to come. Similarly, since Desfontaines continued to publish his own journal until well into 1743, he retained his platform for continuing his strong, but nonlibelous, criticism of Voltaire's Newtonian philosophical positions.

For his part, Voltaire was also moved to disassociate his name from the base, personal invective that had characterized his *Préservatif.* Since he also aspired much more than Desfontaines to be viewed as a serious and *honnête philosophe,* this agenda was particularly crucial for him. Attempting to restore his own intellectual authority in the minds of the serious savants who found his behavior disgraceful, and especially attempting to rehabilitate his Newtonianism from the insinuation that it derived from nothing more than libertine, materialist impiety, Voltaire worked to position himself more respectably. Defending his precise Newtonian position directly, he issued a rebuttal to his critics in the form of a pamphlet responding to various precise objections. Here he responded directly, and in detail, to the technical objections raised by Banières, Desfontaines, Regnault, and others without indulging in any sarcastic wit or satirical repartee whatsoever. [207] Indeed, he did not even name his critics personally,

---

204. "Jugement désinteressé du demelée qui s'est elevé entre M. de Voltaire et l'abbé Desfontaines," *Nouveaux amusemens du coeur et de l'esprit* 3 (n.d.): 246–56.

205. Vaillot, *Avec Madame du Châtelet,* 103.

206. On the resolution, see ibid., 112–115.

207. Voltaire, *Réponse à toutes les objections principales qu'on a faites en France contre la philosophie de Neuton* (Paris, 1739). This pamphlet is reprinted in *Œuvres com-*

focusing instead on the intellectual arguments they had made and his rebuttals of them.

Voltaire also reasserted his *honnête* critical style by publishing, first in the *Mercure de France* in June 1739, and then in the *Nouvelle bibliothèque, ou Histoire littéraire* the following month, an anonymous celebration of du Châtelet's essay on the nature of fire, recently published by the academy.[208] In the text he praised the essay as a victory for savant women everywhere. He also made clear that her views were founded "on the ideas of the great Newton" and on the experiments of his celebrated followers 'sGravesande and Boerhaave. This associated du Châtelet's work with Voltaire's own essay on fire published in the same volume, despite the many differences between them. It also worked to align and sanction their mutual embrace of Newtonian science with the authority of the Royal Academy. In his *Réponse aux objections principales,* Voltaire made this association even more explicit. Here he wrote that "the truths [of Newtonian science] have penetrated the Academy of Sciences, despite its dominant taste for Cartesian philosophy." "They were first proposed there by a great mathematician who has since, by his measurements taken at the polar circle, found and determined the figure that Newton and Huyghens attributed to the earth."[209] When this pamphlet appeared, Cassini de Thury had not yet conceded this debate to Maupertuis, but he soon would, further cementing the alliance between Voltaire, Newton, Maupertuis, and the Royal Academy implied in this passage. All of this worked to distance Voltaire, the aspiring Newtonian savant and philosophe, from Voltaire, the libertine poet and contentious public critic of Desfontaines. This was precisely the point.

These associations were further cemented, at least in Voltaire's mind, in early 1739. During the bitterest period of his controversies with Desfontaines, one of his letters, a long missive to Maupertuis, was published in the *Bibliothèque française.*[210] The richly detailed scientific text reiterated many of the same arguments made in his other apologia, but in this one his views were directly associated with those articulated and defended by Maupertuis. Especially powerful was Voltaire's discussion of God's power to grant an attractive property to matter. Here he countered those who denied this possibility by challenging them to read and refute the brilliant defense of this position offered in Maupertuis' *Discours sur les différentes figures des astres.* Voltaire similarly

---

plètes de Voltaire (ed. Besterman), 15: 724–750, and my citations will be drawn from this version.

208. Reprinted in *Oeuvres complètes de Voltaire* (ed. Beuchot), 23: 65–69.

209. Voltaire, *Réponse à toutes les objections,* 15: 697–718.

210. Voltaire, "Lettre de M. de Voltaire à M. de Maupertuis," 15: 698–718.

associated his own views on the figure of the earth with Maupertuis', also offering readers an image of a philosophical partnership between the two men, one marked by judicious reflection, moderation, and zeal for the truth. The letter thus worked to implicitly reflect Maupertuis' *honnêteté* and authority onto that of the scandal-prone writer. Voltaire insinuated himself with Dortous de Mairan in the same way. Without naming the academician precisely, he described how "one of the most estimable philosophes of our time, and one of your friends as well as mine, has honored me by writing and expressing his critical views about Newtonian attraction."[211] This gave Voltaire the occasion to further show the respect accorded to him by some of the leading academicians, while also demonstrating his comfort with modest, academic-style debate.

The letter ended with an appeal for peace in the Republic of Letters. Since the true man of letters abhors the idea that the concept of *odium theologicum* will translate into an *odium philosophicum*, Voltaire asserted that "a Newtonian can love a Cartesian." "For a long time I have said that all those who sincerely love the arts should be friends. This truth is worth more than a geometrical demonstration."[212] Yet the 1730s had been anything but a decade of friendship among French philosophers, and as the 1740s opened, the possibility of peace seemed remote. Voltaire had established his presence as a philosophical contender, but he had also generated a number of controversies challenging this very position. Maupertuis also remained distant from Voltaire, even if their respective career trajectories were beginning to converge. Du Châtelet, closely attached to both, further complicated the situation, especially as she began to grow more assertive in her own philosophical convictions. The 1740s opened, therefore, with the Newton wars still raging, yet by the end of the decade the situation had radically changed. Ironically, it was the injection of another "ism"—Leibnizianism—that catalyzed these changes, and the result was the field that allowed the philosophe movement, and thus the Enlightenment, to solidify in France.

211. Ibid., 702–703.
212. Ibid., 718.

7

Leibnizianism and the Solidification of the French Enlightenment

By 1740, the French Newton wars were in full swing. Voltaire was at the center of the struggle, and one theater of combat was the campaign, waged by him and the Marquise du Châtelet, to respond to his critics while stabilizing his public perception as a philosophe. Du Châtelet's agendas, however, did not always conform with Voltaire's, especially as she began to chart her own philosophical career. Against each as well stood a host of different opponents. These included Cartesians inside and outside the Paris Academy, and journalists who saw in these battles anything but a serious scientific debate. Voltaire also had enemies in the wider Republic of Letters, but he also had friends as well, including Maupertuis, who became a more active ally in his public Newtonian campaigns once the figure of the earth debate had been resolved on favorable terms. He and his Newtonian allies helped to give academic credibility to Voltaire's efforts, but they also complicated matters since they did not always share their more controversial colleague's precise agendas. Du Châtelet's own and often independent minded initiatives added further complexity, and Maupertuis also pursued some radical agendas of his own. These sometimes reinforced Voltaire's efforts while fighting against them at other times. Further complicating matters was Castel and the French Jesuits, who sided with Voltaire, Maupertuis, and the Newtonians on certain occasions, and with Fontenelle, Dortous de Mairan, and the Cartesians on others. Moreover, whatever their precise commitments, the French Jesuits never relented in making their precise approach to natural philosophy a force to be reckoned with in the wider public sphere.

The public, therefore, had much to consider as the battle raged into the 1740s. Voltaire continued his efforts to find a place among the table of serious, officially sanctioned philosophers, and the arrival of another Newtonian academician in the 1730s created an occasion for further initiatives in this regard.[1] The academician in question was Georges-Louis Le Clerc, the future Comte

---

1. Most helpful in my understanding of Buffon's career are Lesley Hanks, *Buffon avant l'Histoire naturelle* (Paris, 1966); and Jacques Roger, *Buffon* (Paris, 1989).

de Buffon, and the work that launched his public Newtonian debut was his translation of *Vegetable staticks, or An account of some statical experiments on the sap of plants,* which the English chemist and fellow of the Royal Society Stephen Hales published in London in 1727.[2] The work was a thoroughgoing illustration of Newtonian experimental philosophy, and Buffon published a French translation of the work in 1735. Hales offered what he claimed to be an experimental proof of gravitational attraction by using Newtonian theory to explain the apparent rise of fluids inside capillary tubes. Savants throughout the Republic of Letters were soon discussing the work, and the *Bibliothèque britannique* fueled interest by publishing three long reviews in its first issues of 1733 and 1734.[3] These were the same issues that reviewed Voltaire's *Letters on England*, and in this way French savants were drawn into the discussion of this English work at the same time as they begin to witness the first skirmishes of the Newton wars.

In the context of this interest, Buffon undertook his French translation, which further catalyzed the dialogue.[4] His translation of *Vegetable staticks* also marked his public, if not his academic, debut, and the paths that led him to this moment are revealing of the changing environment of French science during these years. Born to a family of Burgundian notables in 1707, Georges-Louis Le Clerc was targeted at birth for a career in law, perhaps even a judgeship in the Parlement of Dijon. At the young age of ten, however, he received all of his considerable inheritance, and this permitted him to lead what increasingly became an independent, and even rebellious, life devoted to intellectual sociability. Especially influential in Buffon's youth was the circle of the President Bouhier, a parlementary judge and leading Dijon savant. Bouhier's library contained over thirty-five thousand volumes and another two thousand manuscripts, and it was here that Buffon began to cut his intellectual teeth.

Launched in some measure by Bouhier, Buffon had by 1726 both graduated with a degree in law and oriented himself toward a career in the sciences. His first passion in these years was mathematics, especially the new eighteenth-century analytical mathematics. In 1727, he pursued this interest through a correspondence with Gabriel Cramer, an important mathematician in Geneva. He also left Dijon in 1728 to study mathematics at the Oratorian college of Angers.

2. Stephen Hales, *Vegetable staticks, or, An account of some statical experiments on the sap in vegetables* (London, 1727).

3. Review of Stephen Hales, *Vegetable staticks,* in *Bibliothèque britannique* 1: 219–236, 2: 272–293, 3: 222–234.

4. Stephen Hales, *La statique des végétaux, et l'analyse de l'air* (Paris, 1735).

Once the intellectual home of Lamy, Jaquemet, and Reyneau, Angers had been instrumental in the great efflorescence of Oratorian mathematics after 1690, even if it had lost much of its intellectual glow by the late 1720s.[5] Among the chaired professors when Buffon arrived, however, was Father Jean-Simon Mazières, the man who had won the 1726 Paris Academy prize on the laws of impact among elastic and inelastic bodies.[6] Mazières was also the author of *Treatise on the Small Vortices of Subtle Matter* published in 1727.[7] Writing about this book in the *Journal des savants*, a reviewer noted that the author "criticized the prevailing methods for explaining the *pesanteur* and elasticity of bodies" while also asserting that "to attribute these effects to an occult quality is to not explain them at all."[8] Buffon no doubt absorbed these and other Cartesian viewpoints during his time in Angers along with a lot of the latest analytical mathematics.

Analytical mathematics was particularly important for Buffon, for during these years he studied l'Hôpital's *Analyse des infiniment petits*, Reyneau's *Analyse demontrée*, Remond de Montmort's *Essai sur les jeux d'hazard*, and Fontenelle's recently published *Eléments de la géometrie de l'infini*. Given this reading list, it is also highly likely that he came into contact with the mathematical work of Newton and the Newtonians during these years as well. Since none of Buffon's mathematical work or correspondence from this period survives, it is impossible to trace with any certainty his actual intellectual development. Nevertheless, he clearly became an accomplished mathematician by 1730, one worthy of admission to the Royal Academy. His actual admission, however, occurred in other ways. Returning to Burgundy in early 1731, after a series of travels that took him to England among other places, where he was made a fellow of the Royal Society, Buffon took possession of an estate at Montbard. The next year, he also established a residence for himself in Paris at the home of the royal apothecary and academician Boulduc. Thus, by 1732 he had positioned

5. See Pierre Costabel, "L'Oratoire de France et ses colleges," in *Enseignement et diffusion des sciences en France au XVIIIe siècle*, ed. René Taton (Paris, 1964), 66–100. On the wider cultural significance of Oratorian mathematical work, see J. B. Shank, " 'There Was No Such Thing as the "Newtonian Revolution," and the French Initiated It': Eighteenth-Century Mechanics in France before Maupertuis," *Early Science and Medicine* 9 (August 2004): 257–292.

6. E. Bonnardet, *Les Fondations de prix à l'Académie des sciences* (Paris, 1881), 17.

7. Jean-Simon Mazières, *Traité des petits tourbillons de la matière subtile* (Paris, 1727).

8. Review of Jean-Simon Mazières, *Traité des petits tourbillons*, in *Journal des savants* (1727): 457–458.

himself to pursue what would become his characteristic living arrangement, splitting his time between the scientific, and eventually academic, sociability of Paris and his life and work at his provincial home near Dijon.[9]

Admission to the Royal Academy came soon after. Bouhier thought that his acceptance was long in coming and that it had been prepared for many years before the actual invitation arrived.[10] Whatever the behind-the-scenes machinations, the historical record offers the following account. Continuing his mathematical correspondence with Cramer during his travels, Buffon produced work that Cramer believed worthy of the attention of Clairaut in Paris. His research involved an obscure game of chance called "*le franc-jeu,*" and his innovation rested in the incorporation of geometric methods into the otherwise analytic science of mathematical probability.[11] In April 1733 Maupertuis and Clairaut approved Buffon's paper on behalf of the Royal Academy, praising in particular the geometrical erudition and "invention" of the author.[12] In a rare honor, Clairaut also read the paper before the entire assembly in order to alert them to this important work.[13]

Admission to the academy, however, did not follow directly. Maurepas played the decisive role. In 1731, the minister had approached the academy regarding ways to understand and improve the strength of wood. This issue had particular relevance to ship-building, which was within Maurepas' portfolio as naval secretary, yet the academy found it hard to satisfy the minister's request in this area. How Buffon was made aware of the minister's interest is unclear, but in May 1733 Buffon returned to Montbard to begin a series of experiments using the forests of his *seigneurie* as his laboratory.[14] In the fall, he returned to Paris, where he was invited to read another mathematical paper before the assembly, and where he no doubt reported to Maurepas about his other work as well.[15] In December, he was among the nominees for a vacant seat in the academy as an *associé astronome*, but this was an awkward fit for Buffon in terms of both field and rank.[16] Maurepas rejected this nomination, appointing Maraldi instead.

9. Hanks, *Buffon avant l'*Histoire naturelle, 20–36; Roger, *Buffon,* 27–34.

10. Roger, *Buffon,* 37, 44.

11. Hanks offers a full account of the problem and Buffon's solution, *Buffon avant l'*Histoire naturelle, 35–61.

12. PVARS, April 25, 1733.

13. Ibid., May 6, 1733.

14. Roger offers the most economical account (*Buffon,* 40–45). Hanks analyzes Buffon's wood science in more detail (*Buffon avant l'*Histoire naturelle, 141–213).

15. PVARS, November 25, 1733.

16. Ibid., December 20, 23, 1733.

But he also shifted Grandjean de Fouchy from his *adjoint mécanicien* chair into the class of astronomers. This move better reflected Grandjean's work, and it also created a seat for Buffon, which was offered to him in January 1734.[17] "One has granted me a thousand times more honor than I merit," the future count wrote disingenuously to Bouhier afterwards. Whatever humility he may have actually felt, he thereafter exercised it with extreme vigor as he became a very active academician.[18]

As an *adjoint mécanicien,* Buffon was positioned in the academy in ways that could have encouraged his ongoing mathematical work. Once a member of the company, however, his intellectual orientation instead began to shift. Already in 1731 Buffon had shown signs of dissatisfaction with mathematics, writing to Cramer with respect to the famous St. Petersburg problem in probability that the mathematical concepts of number and quantitative value do not agree with the empirical measures of experience and common sense.[19] Whether driven by these growing epistemological convictions or by some other motivation, Buffon ceased to pursue serious mathematical research, writing to Cramer in February 1736 that "I am going to deliver myself entirely from my taste for mathematics."[20] This departure did not imply a complete disappearance from the field of mathematical debate, but Buffon did stop producing original mathematical work. Instead, he began to cultivate a different orientation rooted in the empirical and experimental sciences, especially natural history. He continued to pursue his research on wood and trees at Montbard, further intensifying his divided life between Paris and the provinces. Through this work, he also strengthened his ties with Maurepas as well.[21]

A letter from 1737 describes Buffon as a chemist, yet he was neither this, nor a mathematician, nor a *mécanicien,* nor any other single type of savant in this period. He was rather an ambitious and talented academician who, like many before him, sampled various scientific identities without pursuing any one exclusively. For this very reason, then, his decision to make a translation of Hales's *Vegetable staticks* the vehicle for his public print debut in 1735 was deeply significant. In the preface to the work, which celebrated Hales's achievements, Buffon positioned himself squarely on the side of English experimental philosophy, defending not only the epistemology of inductive reasoning from experience and experiment but also the precise physics of gravitational attrac-

17. Ibid., January 9, 1734.
18. Cited in Roger, *Buffon,* 44.
19. See ibid., 31, 60–65.
20. Cited in ibid., 60
21. Ibid., 46–47.

tion that the English Newtonians defended. Ironically, this both reinforced and complicated his ties to the other French Newtonians inside the Royal Academy. Maupertuis and Clairaut, for example, were anything but English-style experimental philosophers, yet Buffon's willingness to publicly defend Newtonianism nevertheless made him allies with them. Similarly, while Buffon's growing distaste for mathematics and the mathematicization of the sciences placed him against the grain of these and other French Newtonians, his commitment to the physics of attraction and the epistemology of empiricism and experimentalism, combined with his hostility toward Cartesian rationalism and vortical mechanics, brought him back together with them in other ways.

Here one sees how Newtonianism in France functioned less as a coherent and consistently defended set of scientific ideas than as a political position that united savants with very different intellectual agendas. The same was true of Cartesianism after 1730, and as the war between these two self-conscious factions began to rage, Buffon's arrival on the scene marked an important development. The French public now had another French Newtonian to watch, and one who defended—from his seat in the Royal Academy, no less!—the more aggressive Anglo-Dutch form of attractionist Newtonianism that Voltaire had articulated in his *Lettres philosophiques*. In practice, Buffon carried himself much more like Maupertuis than Voltaire, and like his senior academic colleague he also showed no interest in allying with Voltaire's libertarian campaigns on behalf of Newton after 1734. In a letter from 1738, Buffon mentioned a possible visit to Cirey, but the visit never occurred.[22] Buffon and Voltaire did share many acquaintances in common, especially Helvétius, who was a frequent visitor to Montbard. In 1739, Voltaire wrote to his friend praising Buffon's character. "If I were not with Madame du Châtelet I would like to be at Montbard with you and Buffon."[23] He also published a tract in 1739 that explicitly made Buffon's translation of *Vegetable staticks* the most important event after Maupertuis' *Discours* in the "penetration of the Newtonian truths inside the Paris academy despite the taste for Cartesianism that still dominates there."[24]

Buffon, however, appears not to have wholly reciprocated in these sentiments. "M. Voltaire's commentary on Newton was wholly unsuccessful," the academician wrote with respect to Voltaire's *Eléments de la philosophie de Newton*.[25] Furthermore, in the Parisian salons and cafés where Voltaire's work was

22. Ibid., 57–58.

23. Cited in Hanks, *Buffon avant l'*Histoire naturelle, 91.

24. Voltaire, *Réponse aux objections principales qu'on a faites en France contre la philosophie de Newton* (Paris, 1739), in *Oeuvres complètes de Voltaire* (ed. Beuchot), 23: 71–72.

25. Cited in Roger, *Buffon*, 58.

read more sympathetically, Buffon was a nonentity. He preferred his rural life at Montbard to the distractions of *mondain* Paris. Marivaux, he once wrote, was best suited to "small and precious minds." This meant that Buffon, unlike Maupertuis, rarely crossed paths socially with Voltaire in wider Parisian society.[26] Yet the intellectual similarities between the two men were real and substantial, at least with respect to their understanding of Newtonian philosophy. This meant that Buffon's presence inside the Royal Academy, and in the wider public sphere of French science after 1735, greatly aided Voltaire's position. Indeed, from the perspective of ideas alone, Buffon was far more symbiotic with Voltaire than was Maupertuis.

Yet in a manner similar to Voltaire's relationship with Maupertuis before 1740, his intellectual affinities with Buffon did not translate into an identity transforming alliance. Lamenting these difficulties in 1738, Voltaire confessed that he was "a child lost to a party that has M. de Buffon as its leader." "I am like the soldiers who fight with brave hearts without fully understanding the interests of their prince."[27] These laments spoke to the actual social relationship that Voltaire enjoyed with the senior academicians to which he was intellectually sympathetic. Completely outside the orbit of the establishment networks that brought new members into the Royal Academy, and not fitting the profile of a would-be royal academician in any case, Voltaire was nevertheless tied to figures like Buffon and Maupertuis through the wider social networks that linked learned men to the state, each other, and the wider Republic of Letters. Yet even here, Voltaire was distanced from official science in ways that remained problematic for him.

Buffon, by contrast, was at the center of the nexus of learned society, the academy, and the Crown that brought favor and rewards to aspiring savants. His diligence in this arena, like Maupertuis', was also rewarded, for in the summer of 1739, as the Voltaire-Desfontaines dispute continued to rage, Maurepas named Buffon *superindendent* of the Jardin du Roi, further adding a lucrative annual pension of three thousand livres to the deal as proof of his favor. Buffon, therefore, had no professional reason to align himself too closely with Voltaire and his Newtonian campaigns no matter what philosophical convictions the two men may have shared. Certainly, the veneer of aristocratic libertinism and spirited, intelligent free thought was one that Buffon cultivated, as did many other men of his stature in this period. Yet the exchange between Buffon and Voltaire was not reciprocal since Voltaire, an unestablished writer, critic, and

26. Ibid., 59.

27. Voltaire to Helvétius, 3 October 1739, in Hanks, *Buffon avant* l'Histoire naturelle, 258–259.

would-be savant, did not receive the same social return from critical independence and libertinism as the fully established Buffon.

The same was true of Voltaire's relationship with Maupertuis, even if the latter, unlike Buffon, aspired to be something of a libertine philosophe himself. Maupertuis, unlike Buffon, had visited Cirey in September 1738 during his own period of self-described rebellion against the academy.[28] In the intervening weeks, however, the trajectories of the two men again parted. Voltaire became embroiled in his bitter and scandalous public dispute with Desfontaines, while Maupertuis' struggles with the defenders of the elongated earth moved in the opposite direction. While Desfontaines and Voltaire were negotiating their joint disavowals with the royal authorities, Cassini de Thury was preparing to announce his concession to Maupertuis on the figure of the earth. The concession came the following year, and in the interim Maupertuis had been awarded his lucrative royal pension as a reward for his service to the Crown and the public.[29]

No wonder, then, that Voltaire found Maupertuis avoiding him in September 1739 when he and du Châtelet were visiting Paris under the protection of the Duc and Duchesse de Richelieu. The senior academician was, in fact, in the city at the time, and Voltaire claimed to have presented himself "twenty times at his door" without receiving any recognition. "Why do you not come to see me and Madame du Châtelet?" he asked.[30] Any number of factors might explain the snub (if, in fact, it was a snub and not just a simple miscommunication), for it was certainly the case that Voltaire needed his association with Maupertuis far more in this period than the senior academician needed his. The exchange of cultural credit between the two men was never entirely one way, however, and even if the economy was tilted heavily against Voltaire in late 1739 (his *Siècle de Louis XIV* was condemned by the Paris Parlement in December), it began to readjust itself as Maupertuis grew comfortable with his new security inside the academy and with the favor showed to him by the royal authorities, especially by Maurepas.[31] After 1740, in fact, Maupertuis increasingly found himself drawn toward Voltaire's identity as part of his own maneuverings in and around the Paris Academy, and this subtle shift in orientation played an important role in Voltaire's own project of philosophical self-fashioning as well.

28. Voltaire to Maupertuis, 1 October 1738, in Voltaire, *Corr.,* 89: D1622.

29. See Terrall, *Man Who Flattened the Earth,* 149–154.

30. Voltaire to Maupertuis, 22 September 1739, Voltaire, *Corr.,* 90: D2073. The visit to Paris is discussed in Vaillot, *Avec Madame du Châtelet,* 120–121.

31. On the condemnation of *Siècle de Louis XIV,* see Vaillot, *Avec Madame du Châtelet,* 126.

Maupertuis' New Public Radicalism

Maupertuis had always fashioned himself a serious, academic savant and a libertine *esprit fort,* but after 1740 his willingness to present the latter face before the wider public began to increase. He continued to carefully guard his public reputation as a serious, judicious savant, yet a change in orientation also occurred, and it can be illustrated by looking at Maupertuis' ongoing efforts after 1740 to advance his position with respect to the figure of the earth debate. Up until Cassini de Thury's public concession in April 1740, Maupertuis had pursued a carefully restrained program of intellectual contestation. In the wider public, the shape of the earth controversy had become an integral part of the broader contestation that pitted defenders of English philosophy, particularly Newtonianism, against the increasingly patriotic defenders of French Cartesianism. In this space, Voltaire in particular had also used connections between Newtonian physics and the oblate earth to align his own Newtonianism with Maupertuis' position on this question. The academician, by contrast, had not cultivated the same associations, at least in his public presentations and writing. Yet after 1740, this separation began to erode. The change is evident in a series of works that Maupertuis published after his own status had been secured and his own position ratified by Cassini de Thury and others inside the Royal Academy.

Perhaps the first to appear was an anonymous book, published in very small numbers by an unnamed press, *Letter from an English Clockmaker to an Astronomer from Peking. Translated by M****.[32] One of the very few extant copies of this work is now housed in the Bibliothèque de l'Arsenal in Paris, the former library of the Paulmy family.[33] One of the eighteenth-century scions of this family was the Marquis d'Argenson, an intellectual compatriot of both Maupertuis and Voltaire. It is possible that the book found its way into his family library through his personal acquisition. No matter how the book was acquired, the volume, while short (less than sixty-five pages), was an elegantly and expensively bound tome that bore all the traces of a work printed at personal expense for circulation among friends. The text also appears to have attracted no public

---

32. *Lettre d'un horloger anglais à un astronome de Pekin. Traduite par M**** (n.p., 1740). The text is also reprinted with an accompanying critical analysis in David Beeson, "Lettre d'un horloger à un astronome de Pékin," *Studies in Voltaire and the Eighteenth Century* 230 (1985): 189–222; I will draw my citations from this edition. Terrall includes this text in her list of Maupertuis' published writings, but she makes only a passing reference to the text in her biography. *Maupertuis,* 160, n. 104.

33. Bibliothèque de l'Arsenal, Paris, Reserve 8° S 13886.

notice whatsoever, for despite its rather provocative content—Maupertuis told the astronomer Lalande that he ultimately deemed the work inappropriate and worked to suppress it—no periodicals or contemporary correspondence from the period make reference to it.[34]

For precisely this reason, however, it offers an interesting indication of Maupertuis' public strategizing in the early 1740s. The book carried the date "1740" on its title page and in the opening letter it makes reference to Cassini de Thury's concession regarding the shape of the earth at the academy's public assembly of April 27 of that year. This dates the work fairly precisely as an immediate postmortem on the controversy itself. True to form, the book offers a thorough recapitulation of the history of the debate, spiced throughout with witty and sarcastic critical commentary. The conceit of the book—an Englishman writing letters to a Chinese astronomer—allowed Maupertuis to position himself as an outside observer sympathetic to his own position. Voltaire's voice in the *Lettres philosophiques* was of course implied in this presentation, and from this perspective the Englishman/Maupertuis narrated the history of the shape of the earth controversy while offering a relentless critique of the Cassinis, their science, and the "fraud" that placed them at the pinnacle of French scientific authority.

As the Englishman tells it, the question of the true shape of earth is as old as the Babylonian caliphate, but only under the supportive reign of the great king Louis XIV did the resources become available to resolve the matter once and for all. The Sun King brought the greatest minds in Europe together at his newly founded Academy of Sciences and then turned them toward questions of scientific and public utility. One such savant was Jacques Cassini II, who took charge of the royal effort to map the earth in 1701. Based on this work, Cassini II began to declare that the earth was elongated at the poles. Yet the irony that the Englishman emphasized was that despite the erroneous nature of his claim, it took the Cassinis and the French Academy another forty years to discover and admit their errors.[35]

The letters document in great detail what the Englishman describes as a roughly half-century-long history of error and folly driven by the French Royal Observatory. The great expense of the operations is stressed; the time consumed by the expeditions and the elaborate labors recounted; and the apparent

---

34. Lalande's report is cited in Beeson, "Lettre," 205. Beeson discusses the attribution of the book to Maupertuis and its printing in ibid., 204–207. Terrall claims that four copies of the book were printed. *Man Who Flattened the Earth,* 373.

35. Beeson, "Lettre," 211–221.

meticulousness of the work itself emphasized. Yet while Newton found the true shape of the earth using only a pendulum, a pencil, and a few mathematical formulae, the Cassinis spent a king's ransom and four decades of labor only to reconfirm over and over again the same error. This irony also drives the wit of Maupertuis' tract. In a letter entitled "A very simple instrument that one should have used to determine the shape of the earth," the Englishman draws on Abraham de Moivre's new probability mathematics to suggest that a die would have been a more cost-effective instrument for resolving the question. By rolling the die to see if the earth was flattened or elongated at the poles, "you would at least have had an equal probability of finding the correct answer, whereas the odds are 63 to 1 that by this method you would have found, like Cassini, the wrong answer six times in a row."[36]

Maupertuis' Englishman also stressed the haughty arrogance of the Cassinis and their obstinacy in the face of criticism to their work. He further implicated Fontenelle in what he presented as a kind of conspiracy inside the academy by pointing to the repetition of Cassini's errors in the annual academy *histoires* despite the patent contradictions that this forced him to utter. For the Englishman, this was evidence that the academy secretary was following the authority of the Cassinis more than his own reason.[37] The text particularly indicted the suspicious conformity of the Cartesian theory of *pesanteur* with Cassini's theory of the elongated earth, a connection that implicated Dortous de Mairan in the conspiracy as well.[38]

Maupertuis' text echoed Voltaire in many respects, not least in its biting wit and satirical tone. Yet while the poet-cum-philosophe made the broader public the audience for his critical, philosophical assertions, becoming an outlaw as a result, Maupertuis restricted his sarcasm to a very small and closed circle of elite friends. The same guardedness about embracing the public radicalism of Voltaire's philosophe persona is evidenced in a second shape of the earth work that Maupertuis published in 1740, the anonymous *A Disinterested Examination of the Different Works That Have Been Done to Determine the Figure of the Earth*.[39] The text on its surface was what its title suggested—a disinterested examination of the figure of the earth debate. Yet one part joke and another part

36. Ibid., 213.

37. Ibid., 214–215.

38. Ibid.

39. [Maupertuis], *Examen désintéressé des différens ouvrages qui ont étés faits pour déterminer la figure de la terre* (Paris, 1738 [1740]). For a broader discussion of this work and its reception, see Terrall, *Man Who Flattened the Earth*, 154–160.

brilliant literary ruse, the book was also one of the publishing sensations of the 1740s. The anonymity of the author was a key to the book's success, and Montesquieu described the book's character well in a letter to the Englishman Martin Folkes. "A very well-written book has just appeared here, *A Disinterested examination*," he wrote. "The author seems to be a wise and reserved man who does not utter inanities."[40] Dortous de Mairan also praised the book in similar terms. "Whoever the author may be, this book is a credit to whoever wrote it. It is in the hands of everyone in Paris, including the ladies, who also dabble in wanting to know the shape of the planet they walk upon."[41] Yet Maupertuis refused to reveal himself to even his closest friends, and in this way he created a public literary phenomena that was at first completely detached from his name.

The book was indeed a sensation, and all the leading periodicals of the day ran reviews of the text while struggling to resolve the mystery about its author.[42] Who wrote such a learned, meticulous, and judiciously argued text, and what was the ultimate thesis that this sage wanted to impart to the public? No clear answer was possible in 1740, but by 1743, when a second edition of the work had been published, Maupertuis' authorship had been revealed. What, then, did he accomplish by creating and sustaining what du Châtelet called "a joke pushed too far?"[43] Two things primarily. First, between 1740 and 1743, when the book's anonymity was still secure, the text's technical mastery of the scientific questions, combined with its judicious and thoroughly balanced presentation of the controversy, effectively distanced the technical scientific debate between Maupertuis and the Cassinis from the often bawdy polemics about Newtonianism and Cartesianism that were coloring its wider public perception. This allowed Maupertuis to accomplish two things. First, it created a vehicle for transferring the technical debate as it had transpired inside the academy into the wider public sphere, while, second (and in direct contrast to the approach of the *Lettre d'un horloger anglais*) offering a perspective for judging the merits of each position grounded only in *honnête* and disinterested objectivity. Since each antagonist in the dispute received the same respectful and seemingly unbiased treatment, the text ultimately demonstrated Maupertuis' authenticity as a thoroughly *honnête homme* despite his partisanship in the debate itself.

As it became clear that he was the author of the work, a number of positive valuations further accrued to his name. In addition to winning praise as the

40. Cited in Terrall, *Man Who Flattened the Earth,* 155.

41. Ibid., 155–156.

42. For a sampling, see the review of *Examen désintéressé* in *Journal des savants* (1740): 153–158; and in *Observations sur les écrits modernes* 21 (n.d.): 25–43.

43. Du Châtelet to J. Bernoulli, 30 June 1740, in Voltaire, *Corr.,* 91: D2254.

author of such a learned and apparently *honnête* tome, his views with respect to the shape of the earth were also given greater credit since they were now seen, in retrospect, as the ones most in agreement with the objective and disinterested stance of the book. The indictment of the Cassinis was heightened, by contrast, since the book's thoroughly *honnête* depiction of their work and conclusions was now read as a brilliant and ruthlessly subtle satire. In the *Examen*, Maupertuis had ostensibly defended the Cassinis by asserting that their calculations, which differed greatly from his own, were certainly not arbitrary "any more than 2 times 2 equals 5." Given the disparity in the numbers, therefore, only two conclusions were possible: either the Cassinis were right, a conclusion that was left standing in the neutral presentation of the text, or "M. Cassini is the most clumsy and unfortunate astronomer in the world," a conclusion that the anonymous author declared absurd. By 1743, however, after the concession of the Cassinis on the shape of the earth and the revelation of Maupertuis' authorship, these same sentences were now read as they were intended in the first place, as a scathing parody of Cassini and his work.[44]

The entire *Examen* operated in the same way, and as its authorship became clear, the nature of its satirical intent also became public knowledge as well. Yet this brand of satire was significantly different from the crude and open libels that Voltaire and Desfontaines had exchanged little more than a year earlier. It was also much different than the privately addressed bon mots of the *Lettre d'un horloger anglais*. The anonymous nature of the text was key to this result since it at once distanced Maupertuis the man from the critical venom the book contained while elevating Maupertuis the writer to the status of a brilliant wit. Indeed, while the joke of the text was certainly vicious and acknowledged as such by its audience, it was also brilliantly artful, so much so, in fact, that its subtlety could be appreciated in ways that raised Maupertuis' status as an *honnête homme* even as it exposed him as the public enemy of the Cassinis. Furthermore, while Voltaire continued to struggle to assert his own Newtonian convictions in a way that projected a similar intellectual character for himself, Maupertuis began to tap into the wider public vogue for Newtonianism that Voltaire had catalyzed by defining himself more and more as an assertive, and even radical, Newtonian.

One component of this effort was a portrait that Maupertuis commissioned in October 1739 just before he received news of his pension.[45] In the portrait, he allowed his scientific image to be celebrated in a bold and some would say shocking way. The portrait itself showed Maupertuis, dressed in an elegant fur

44. Cited in Terrall, *Man Who Flattened the Earth*, 159–160.
45. Terrall offers a full account of the history of the portrait in ibid., 160–164.

hat and coat from Lapland, confidently staring at the viewer as one of his hands compressed a globe, as if to flatten it at the pole. The other hand gestured out through a window to a scene of the Lapland countryside outside. The image was unusual, both in its monumental celebration of a man of science (the portraitist's other clients were leading aristocrats, royal ministers, and military officers) and also in its manner of declaring its glory. When the portrait was shown at the biennial Parisian art salon of 1741, it thus generated much public conversation and criticism. The abbé Nollet found the image provocative, writing to a correspondent in Geneva that "Maupertuis . . . is declaring open war on anyone who does not proclaim at the top of his lungs, 'The earth is flattened!' Never did Don Quixote make such a row to defend his Dulcinea." Nollet also lamented Maupertuis' decision to single himself out from the work of his team, noting that the almost imperceptible appearance of the other members of the polar expedition in the background of the painting "makes it look like they had little to do with the work that initiated [Maupertuis' fame]." [46]

Even more strident in its self-promotion was the engraving of the portrait that Maupertuis' friend, the Marquis de Locmaria, commissioned in 1741 and which Maupertuis actively circulated in the subsequent years. The engraving, like the original portrait, was displayed publicly at the Parisian art salon, and in it a flattering quatrain, written and signed by Voltaire, was added that celebrated Maupertuis' achievements. "This poorly understood globe, which he knew how to measure, became the monument upon which his glory was founded. His role was to fix the shape of the earth, but also to please and enlighten it." [47] These verses closely tied Maupertuis to Voltaire not only in name and through their shared Newtonianism but also through the combination of science and pleasure that Voltaire already personified. Maupertuis soon began to publicly associate himself with this identity as well. His work during this period also encouraged these associations, further bringing his public persona closer to that of Voltaire.

The intense public interest in a comet visible from Paris in March 1742 gave Maupertuis one occasion to act. He exploited the opportunity by penning a *mondain* letter about the phenomenon to a fictional learned lady.[48] Published anonymously as *Lettre sur la comète*, the text allowed Maupertuis to

46. Cited in ibid., 163.

47. Other engraved portraits of Maupertuis, like the bust image contained in the Bibliothèque publique et universitaire, Neuchâtel, often contained the same quatrain, and they also circulated via the periodical press as well.

48. *Lettre sur la comète* (Paris, 1742). See Terrall, *Man Who Flattened the Earth*, 192–195.

FIGURE 22. *Maupertuis flattening the earth.*
Courtesy of the Bibliothèque nationale de France.

demonstrate his skill with the subtly erotic, belletrist style of public, scientific exposition that Fontenelle had pioneered in his *Conversations on the Plurality of Worlds*. Algarotti had also used this genre to great effect in his *Newtonianism for Ladies*, and at one level Maupertuis' text offered itself as a companion piece to Algarotti's dialogue. Like its Italian counterpart, the text offered a thorough account of Newtonian celestial mechanics as part of its pleasant exposition of the comet. Voltaire, seeking to break away from any association with this light, poetic, and feminine approach to scientific writing, had instead adopted clear and direct philosophical prose when composing his treatments of Newtonianism. Yet it was a sign of the changing climate that Maupertuis, admittedly already in possession of the serious scientific authority that Voltaire so craved, now began to associate his name with the learned wits and feminized *bel esprits* that Voltaire had previously rejected.

Although published anonymously, the letter's author was quickly revealed in the many reviews of the work that were published after its appearance.[49] Since these reviews were also very positive on the whole, they worked to bring acclaim to Maupertuis in his new identity as *mondain* writer and *bel esprit*. The *Journal des savants* spoke for many in praising "the illustrious academician" and author for "exposing what we know about comets with a great deal of clarity and *agrément*." "The work can be regarded as the relaxation of a philosophe who is known for being less an *homme d'esprit* and more a mathematician."[50] This was exactly the translation Maupertuis was intending to effect, and other journals expressed similar praise for the effort. Despite the Newtonianism that anchored the entire text, Desfontaines thought that "nothing is more clear or more solid than the work by M. de Maupertuis on the [nature of comets]."[51] He took issue with Maupertuis' articulation of the Newtonian theory that comets are eventually absorbed into the sun, rejuvenating its solar fire, calling it "an idea that is a little strange," and one that is grounded on no evidence."[52] He also offered a history of comet theory that climaxed with Cassini's Cartesian theories articulated two decades earlier.[53] Overall, however, Desfontaines praised the book and its author, suggesting, as he had with Algarotti's work, that this fusion of science with *bel esprit* was only to be encouraged.

49. Review of *Lettre sur la comète*, in *Journal des savants* (1742): 345–351; *OEM*, 24 (n.d.): 25–42; *Bibliothèque raisonnée des ouvrages des savants de l'Europe*, 52 vols. (Amsterdam, 1728–1753), 29: 429–435.

50. Review of *Lettre sur la comète*, in *Journal des savants* (1742): 351.

51. Desfontaines, review of *Lettre*, *Observations sur les écrits modernes* 29 (n.d.): 30.

52. Ibid., 40.

53. Ibid., 25–30.

Not everyone loved the book, however, even if the criticism it received served to promote rather than retard Maupertuis' growing public perception as a *mondain* wit. Troubled in equal measure by what he saw as the crude deployment of Fontenelle's artful style of *mondain* philosophy and by the prominence of English science, especially Newtonianism, in the book, a rather obscure Jesuit named Gilles Basset des Rosiers published a pamphlet criticizing Maupertuis' *Lettre*.[54] Adopting the pedagogical genre of an advice letter to a nine-year-old girl, Basset des Rosiers harshly criticized Maupertuis for distorting Fontenelle's elegant style. He also insinuated that his misuse of this genre made the work unfit for the chaste innocence of a young girl. He further saw English science at the root of both the aesthetic and moral failings of the work, and in his pamphlet he therefore connected Maupertuis directly to the unholy trinity, often associated with Voltaire, of English philosophy, immorality, and sexual impropriety. The discourse of other Jesuits, such as Castel and Regnault, was also deployed to drive the message home. "The address to a lady is the only thing that links [this text] to *des Mondes*," he wrote. Basset des Rosiers echoed Castel's mathematical discourse, while giving it a moral spin as well, when he wrote that "nothing but cones, ovals, parabolas, and conic sections" are to be found here. "It is frightening. I don't believe there is in France a single woman tough enough to read this *Letter* and learned enough to understand it, unless it be Madame la marquise du Châtelet."[55]

Maupertuis' own sexual relationship with du Châtelet was probably too little known to have served as an actual reference here, yet the improprieties widely ascribed to her relationship with Voltaire were enough to signal a bad stew of Newtonianism and sexual libertinism simmering in Maupertuis' *Letter*. By 1742, accusations of this sort had become standard fare, and in the wider public sphere such criticisms created as much interest as indignation. When the *Nouveaux amusments du coeur et de l'esprit* published *Newtonian Epistle on the Proper Genre of Philosophie to Bring Happiness* at about the same time, readers were not scandalized by the image it presented of a lovelorn woman, alone in a pastoral landscape, asking Newton to help her to use the stars to find the meaning of life. They were rather titillated, a pleasure that was heightened by the aura of theological and moral danger that still hovered closely over all discussions of Newtonian philosophy.[56] Maupertuis' *Lettre sur la comète* ad-

54. Gilles Basset des Rosiers, *Critique de la letter sur la comète, ou letter d'un philosophe à une demoiselle âgée de 9 ans* (Paris, 1742).

55. Ibid.

56. *Épître Newtonienne sur la genre de la philosophie proper à rendre heureux,* in *Nouveaux amusments du coeur et de l'esprit,* 9: 415–425.

mirably exploited this charge, and thus Madame de Graffigny responded in a manner typical of many when she sent a copy of Maupertuis' text, along with the Jesuit's response, to her friend Devaux so that he could have the pleasure of reading them together. A copy of Maupertuis' *Lettre sur la Comète* currently housed in the *Bibliothèque de l'Institut de France* in Paris, in fact, is bound together in the same eighteenth-century volume with a copy of Bassier to des Rosiers' critique. Such juxtapositions, which appear to have been commonplace, ultimately advanced Maupertuis' ambitions as an *esprit fort* more than they hindered them.[57]

His publishers also served his agendas by participating in what Desfontaines called an "ingenious joke." Having sold out of the first edition in a matter of weeks, they reprinted the text with a new "Avertissement" disclaiming it. "We are obliged to avow," the editors wrote with mock solemnity, "that while some found in this little work several curious and passably written things, other people of good taste were shocked by a kind of tasteless gallantry that they found in several spots. We pleaded with the author to retract these things, but he found it impossible to comply, granting us only permission to declare that it is in spite of us that we have reprinted the work as it is. Far from ceding to our advice, the author remains persuaded that the *Lettre* would have been even a bit better had he added more spice to it, since the less one is attentive, interested, and familiar with a subject, the more certain these things are to please. He also charged us with announcing to the public another work that he is currently working on about nebulae. It will also be addressed to Mademoiselle de . . . , since he has found an ingenious allusion between these stars and this charming person, and he promises not to omit any of the graces that conform to such an agreeable subject, and that tie together so well geometry, astronomy, and algebra."[58]

Desfontaines saw through this ruse immediately, calling it "certain to shock some" but "full of salt" nevertheless. Drawing upon his Anglophilia, he described it as "in the true taste of Dr. Swift."[59] The playful production of shock and scandal in this way was the calling card of the libertine wit in eighteenth-century France, and in this case the joke, and its further provocations, sustained the public buzz that Maupertuis hoped to cultivate. He also channeled his more recognizably scientific works toward the same agendas after 1740. Most illus-

57. See Terrall, *Man Who Flattened the Earth*, 195, n. 74.
58. "Avertissement" to the second edition of Maupertuis, *Lettre sur la comète*, in *Observations sur les écrits modernes* 29 (n.d.): 40–42.
59. Ibid., 42.

trative was the new edition of his *Discours sur les différentes figures des astres* published in 1742.[60] The first edition had opened the doors to the Newton wars by offering a cautious and still very tentative set of arguments supporting the possibility of the Newtonian theory of universal attraction. It also initiated the focus on the shape of the earth as the focal point of this question. A decade later, the book had become a historical monument, continually referenced whenever anyone, especially Voltaire, narrated the progress of Newton's reception in France. In these narrations, Maupertuis was offered as the first Frenchman to defy the Cartesian biases of the Paris Academy and to defend Newton. Yet the book itself defied this characterization in many respects. First of all, it was not a defense of Newtonianism but a defense of the possibility of such a defense. Its articulation of Newtonian physics also emphasized the mathematical nature of Newton's concept of force, not its physical and empirical materiality. The book was likewise heavily oriented toward the technical question of the shape of the earth and not at all a manifesto of Newtonian philosophy in general.

In short, what Maupertuis thus did in 1742 was to revise his text in a way that made it better conform with the perception of the text and its author that the Newton wars had created over the previous decade. The original text had posed the Newtonian and Cartesian "systems" side by side for comparison, but in the new text he radically reduced the Cartesian content and relegated it to a single "historical" chapter. He also added a new frontispiece that sought to displace Descartes' famous image of a comet moving through the celestial vortices with an image of an empty cosmos containing a comet moving effortlessly through space tracing a perfect mathematical oval. In keeping with his new assertiveness about the physical truth of Newtonian physics, Maupertuis also changed his presentation of gravitational attraction, replacing the Malebranchian mathematical phenomenalism of the 1732 edition with the assertion that Newton's achievement "was to have discovered an attractive force spread throughout all the particles of matter, which acts in inverse proportion to the square of their distances."[61] This was the physicalist understanding of gravity that Voltaire and Buffon articulated, but it was the first time Maupertuis had so squarely and publicly aligned himself with this understanding.

Framing himself and his Newtonianism this way aligned Maupertuis more closely with Voltaire, but it also associated him with the discourses of

60. Maupertuis, *Discours sur les différentes figures des astres, avec une exposition des systèmes de MM. Descartes et Newton* (Paris, 1742). This new edition is discussed in Terrall, *Man Who Flattened the Earth*, 190–192; and Beeson, *Maupertuis*, 147–150.

61. Cited in Terrall, *Man Who Flattened the Earth*, 190.

FIGURE 23. *Frontispiece from Maupertuis,* Discours sur les différentes figures des astres *(Paris, 1742). Courtesy of the Bibliothèque nationale de France.*

materialism and atheism that hovered around the poet's defense of this precise philosophy. Once again, however, Maupertuis toyed with rather than fled from these connections. His maneuverings can be illustrated by his relationship to a metaphysical text that he wrote sometime after 1740.[62] Its arguments would eventually appear in print in 1750 as *Essay on Cosmology,* but as early as 1741 he made mention of the work in a letter to Algarotti, saying that he had written a book "on the world and its fortunes" that he intended to publish as soon as possible. He did not in fact publish it, however, and in the same letter he gave an indication why. "I am a bit afraid for the success of this book," he wrote, "for it is just the thing to scandalize the weak."[63]

The final text of *Essay in Cosmology* offered an account of the metaphysical necessity of God and nature rooted in the mathematical principle of least action.[64] It is hard to know if this was the argument that his manuscript contained, but whatever the details the text certainly offered metaphysical arguments regarding the relationship between God and nature. These were precisely the kind of arguments that most scandalized the political and religious authorities in France, and accordingly the details of his text are probably less important than its overall character. He most likely circulated the text through clandestine channels to a few close friends, for through these or other networks du Châtelet caught wind of it, writing to him in August 1741 that "if I was ever curious about anything it is your cosmology." "The parallax of the moon is more interesting to astronomers," she added with reference to another, more technical work that Maupertuis had recently published, "but as for us *mondain* people, I would like the cosmology much more, and I am indignant about not seeing it."[65] Voltaire also requested a copy, but Maupertuis worried that the "novelty of my ideas may attract criticism." It was better, he thought, to leave small minds "mired in scandal" than to "take the trouble to see them well refuted."[66] In the end, no text appeared.

Evidence of its circulation, however, and of Maupertuis' wisdom in choosing to repress it, appeared soon after, during what was arguably Maupertuis' apotheosis as a worldly man of letters. In May 1743, the abbé de Saint-Pierre died, vacating one of the forty seats in the Académie française. Voltaire had contested unsuccessfully for the last vacant seat, which was awarded to Dor-

62. The history of this text is analyzed in Terrall, *Man Who Flattened the Earth,* 195–196, 280–286, 352–354.

63. Cited in ibid., 195.

64. Maupertuis, *Essai de cosmologie* (Berlin, 1750).

65. Du Chatelet to Maupertuis, 8 August 1741, in Voltaire, *Corr.,* 92: D2522.

66. Cited in Terrall, *Man Who Flattened the Earth,* 196.

tous de Mairan only a few months earlier.[67] Given Maupertuis' rapid ascent into the ranks of the lettered, many thought that he made an ideal candidate. The favor of Maurepas was crucial for admission, and Maupertuis knew that he could count on his old patron for that. He also had the support of his old Café Gradot companion, the abbé Terrasson, who, along with Dortous de Mairan and Fontenelle (who was still vigorous at ninety years of age), remained the only French savants with dual membership in the Royal Academy of Sciences and the Académie française. Montesquieu and President Hènault were also inclined to support him, but when he visited the bishop of Mirepoix, another of the forty immortals, seeking his support, he found that he had been denounced before his arrival as "a deist" and "the author of a certain manuscript *Cosmologie* where the existence of God is demonstrated algebraically."[68] Writing to the younger Johann Bernoulli after the visit, Maupertuis raged against the "indignities" sustained against him during his campaign. "They want to make people believe I had written books against religion," he fumed, "[but] you know very well that most people call books opposed to religion books that are opposed to them."[69]

In fact, as the campaign went on, Maupertuis learned who his real opponents were: the French Cartesians, especially Dortous de Mairan and Fontenelle. The reasons for their opposition also became clear to him. It was the *Examen désinterressé* that was the trigger, for, as he wrote to Bernoulli, "several of these Mssrs. have been ridiculed [*persifflés*]. They invented this wonderful pretext of religion, which covers them in shame. It was Fontenelle and Dortous de Mairan who were the leaders of the cabal." Graffigny, who was close to the maneuverings, confirmed this assessment, writing at the time that "old Fontenelle opposes [Maupertuis' nomination] with all the strength his age allows." She also alleged that he and his allies were "writing anonymous letters against him." According to the Duc de Luynes, in fact, it was only after Maupertuis and Fontenelle reconciled that his admission was secured. "The friends of M. de Maupertuis brought Fontenelle to a reconciliation, which took place before the meeting." Afterward Maupertuis was admitted.[70]

Maupertuis' triumph was also an implicit victory for Voltaire, even if he still felt that he should have been named to one of the seats given to these French mathematicians. The constellation of forces that propelled Maupertuis' ascent

67. See Vaillot, *Avec Madame du Châtelet*, 172–176. He notes that "the great obstacle to his election was his reputation for impiety and irreligion."

68. Cited in Terrall, *Man Who Flattened the Earth*, 197.

69. Ibid.

70. Ibid.

were nevertheless conducive to Voltaire's success as well, and Maupertuis' re-configuration, therefore, as both an authoritative savant and as a worldly *bel esprit* ultimately worked, albeit implicitly, to further sanction Voltaire's attempt to establish himself as a serious philosophe and *honnête homme*. Moreover, as their intellectual and social positions converged, Voltaire was able to reap the benefits, both intellectual and social, of Maupertuis' successful Newtonian campaigns. For Voltaire, Maupertuis' success, and the wider acceptance of their shared Newtonian positions, translated into increasing public acceptance and support. The actual intellectual unity of this fledgling *parti Newtonien,* however, was more apparent than real, and the fractures within it were well illustrated by the emergence in precisely this period of a third, complicating position: that of Leibnizianism. This label emerged after 1740 as an alternative to both Newtonianism and Cartesianism, yet its coherence (or lack thereof) was a product of the same contested and prismatic reception that had produced the other two philosophical "-isms" during the previous decades. The emergence of this third position also worked to complicate and to clarify the Newtonian/Cartesian binary in influential ways.

Enlightenment Leibnizianism in France

The relationship between Leibniz and Leibnizianism in eighteenth-century France mirrored the relation between Newton and Newtonianism that had evolved during the same period.[71] Like Newton, Leibniz was a major presence in French intellectual life from the 1680s onward, but only in the 1720s did something called "Leibnizianism" begin to cohere. At first, in fact, Leibniz was associated with Cartesianism far more than with his own very un-Cartesian philosophy. For example, the Leibniz that appeared to counter Samuel Clarke's Newtonianism so powerfully and confidently in the pages of Desmaizeaux's *Recueil* was not at all identical with the philosopher, mathematician, and *physicien géomètre* who had developed his own system of metaphysical physics after 1680. Moreover, since Leibniz's principle argument in his debate with Clarke—namely, that the theory of universal gravitation lacked rationalist, mechanical rigor and was not, therefore, a valid physical explanation—became a pillar of the self-proclaimed Cartesian attack against Newtonian physics in France, Leibniz became synonymous with Cartesianism despite the important philosophical differences that existed between them. The Marquise du Châtelet

71. A different view is offered in W. H. Barber, *Leibniz in France from Arnauld to Voltaire: A Study in French Reactions to Leibnizianism, 1670–1760* (Oxford, 1955).

pointed to this commonplace French conflation in a letter to Maupertuis when she described a certain "Mr. Fuller" (she had Euler in mind) as "a Leibnizian, and therefore a Cartesian." She also noted "the spirit of party" that was shaping philosophical discourse in France," and because of the unmistakable power of these pressures, Leibniz's own philosophy, which constituted either a hybrid of, or an alternative to, the philosophies of Descartes or Newton, was forced to enter France caught between the polarities of the Newton wars.[72]

One area in which Leibniz's actual philosophy was directly confronted was in the recurring disputes over *vis viva*. These debates forced thinkers to directly confront Leibniz's very different conception of material bodies, mechanical forces, and the rationality of physical causality. They also disrupted the clarity of the Newtonian/Cartesian opposition that did so much to shape philosophical debate in this period. In particular, while opposition to *vis viva* led both Cartesians and Newtonians to clarify, and then more vigorously defend, their view that matter was an inert, forceless entity devoid of any intrinsic active principle, supporters of *vis viva* were likewise led to clarify the nature of the "living force" that Leibniz posited at the center of mechanical action in bodies. Each side was also pushed to construct ever more vigorous arguments that distanced their views from the specter of Spinozist materialism. The *vis viva* debate further focused attention on epistemology since the evidentiary power of the various arguments were often as much at issue as the particular philosophical positions themselves. Thus, the more rationalist Cartesians, who adhered to a strict belief in the superiority of synthetic, deductive reasoning, often found themselves joining with the Leibnizians, who argued for the necessity of the "principle of sufficient reason," while the more empirical and philosophically skeptical Cartesians often found themselves aligned with the experimental Newtonians instead. In this way, Leibnizianism complicated rather than clarified the Newtonian/Cartesian battles in France.

The *vis viva* debates nevertheless introduced the core principles of Leibnizian physics and metaphysics into French public debate, even if they did not produce a self-proclaimed group of Leibnizians who singled out this "philosophical system" for defense against those of the Newtonians and Cartesians. It was the battle between the latter two that dominated public discussion in France, and not accidentally perhaps the same dynamics that gave rise to this struggle also catalyzed the emergence of eighteenth-century "Leibnizianism." The calculus priority dispute, which did so much to trigger the Newton wars, also played a key role in triggering this outcome. Called to defend Leibniz against the ongoing attacks of the English and Dutch Newtonians, Leibniz's

72. Du Châtelet to Maupertuis, 21 May 1738, in Voltaire, *Corr.*, 89: D1507.

self-appointed successor and protégé Christian Wolff began to publish a series of works articulating and defending his master's philosophy.[73] These defenses, published in Latin and then translated into German with unfriendly titles such as *Logic* and *Metaphysics,* integrated Leibniz's new immaterialist metaphysics of the monads, which he had developed in his last years, with the core conceptions of body, force, mechanics, and rationalist rigor that were central to his earlier philosophy.[74] He then turned all these ideas into a complete philosophy of nature that challenged both Newtonianism and Cartesianism simultaneously. Wolff also fought battles against the religious authorities that found Leibniz's rationalism, determinism, and ontology of active material forces just one more form of Spinozistic atheism. Wolff was, in fact, removed from his university post at Halle in 1723 as a result of these controversies, but he prospered in exile, attracting further followers to the fledgling Leibnizian cause through his charismatic teaching and advocacy.[75]

By 1740, Leibnizianism had become a vigorous, self-conscious movement in Germany, and it reached one plateau when it found its own Fontenelle/Algarotti to sing the praises of this German philosophy to the wider, feminized public. The appearance of Samuel Formey's *La belle Wolffienne* in 1741, therefore, marked the maturation of Leibnizianism into a philosophical contender within the Republic of Letters.[76] Its status was further improved when the young king of Prussia, Frederick II "the Great," reformed his Berlin Academy of Sciences in 1744. Frederick named Formey his Fontenelle as the institution's new perpetual secretary, and like his French counterpart, Formey came to manage a public outreach program that included a large correspondence network and

73. See Catherine Wilson, "The Reception of Leibniz in the Eighteenth Century," in *The Cambridge Companion to Leibniz,* ed. Nicolas Jolley (Cambridge, 1995), 442–474; Jonathan Israel, *Radical Enlightenment,* 541–562; and Manfred Kuehn, "The German Aufklärung and British Philosophy," in *British Philosophy and the Age of Enlightenment. Routledge History of Philosophy,* ed. Stuart Brown (New York, 1996), 5: 309–331.

74. Christian Wolff, *Oratio de Sinarum philosophia* (Halle, 1726); *Philosophia rationalis sive logica methodo scientifica pertracta* (Frankfurt and Leipzig, 1728); *Philosophia prima sive ontologia* (Frankfurt and Leipsic, 1730). German titles include *Vernünftige Gedanken von den Kräften des menschlichen Verstandes* (Halle, 1712); *Vernünft. Gedanken von Gott, der Welt u. der Seele des Menschen* (Frankfurt and Leipzig, 1719); *Vernünft. Gedanken v. den Menschen Thun u. Lassen zur Beförder ihrer Glückseligkeit* (Halle, 1720); *Vernünft. Gedanken v. dem Gesellsch. Leben der Menschen* (Halle, 1721); and *Vernünft. Gedanken v. d. Absichten der natürl. Dinge* (Frankfurt, 1723).

75. Israel, *Radical Enlightenment,* 544–552.

76. Jean Louis Samuel Formey, *La belle Wolfienne,* 6 vols. (The Hague, 1741–1753). For a review, see *Bibliothèque raisonnée* 27 (1741): 267–274.

an annual volume of *Histoire et mémoires* modeled on the one published by the Paris Academy since 1699.[77] These Francophone works connected Formey and the Berlin Academy to the discourses of *honnête,* cosmopolitan science throughout the Republic of Letters, discourses that made partisan advocacy for Wolff or Leibniz inappropriate. Frederick encouraged this disinterested objectivity, since it served his agendas for his royal academy, yet since he also filled his academy with many of the leading Wolffians and Leibnizians in the German lands, and since he also created a different academy structure that included a class of "speculative philosophers," not to mention "experimental philosophers" as well, his reform supported the perception that a nationalist link tied German culture, the Prussian Crown, and the "German philosophy" of Leibniz together.[78]

In France, these developments exerted little influence at first. Readers of the learned periodicals could read about Wolff's works and the Leibnizian philosophy they espoused.[79] They could also read about the various charges of Spinozistic atheism and materialism leveled against each.[80] This discourse, like that which swirled around radical Newtonians such as Toland, was omnipresent is these periodicals, yet while radical Newtonianism circulated via libertine tracts defending free thought and deist natural religion, Wolff's Leibnizianism was presented in dense Latin tomes. Accordingly, the latter exerted little noticeable influence on French discussions despite the radicalism often associated with it.[81] Leibniz's own writings were even more obscure, and since many of his philosophical positions, especially his rationalist epistemology, had been

77. Christian Bartholmess, *Histoire philosophique de l'Académie de Prusse depuis Leibniz jusqu'a Schelling,* 2 vols. (Paris, 1850). This history is also summarized in Terrall, *Man Who Flattened the Earth,* 236–240. Formey was admitted to the academy in 1744, but only assumed the secretary's duties in 1748.

78. The fact that the academy also contained a class devoted to "German language and literature" furthered this nationalist conception.

79. A French translation of Wolff's *Philosophia rationalis sive Logica* was published as *La logique, ou Réflexions sur la force de l'entendement humain et sur leur legitime usage dans la connaissance de la verité* (Berlin, 1736) and reviewed in *Journal des savants* (1738): 113–119. Desfontaines reviewed the same work in *OEM* 12 (n.d.): 81–96.

80. Israel notes the impact of Wolff's controversial reputation in France, *Radical Enlightenment,* 544. For a sampling, see *Bibliothèque raisonnée* 20 (1738): 284–304.

81. An exception that proves the rule was a radical tract entitled *La découverte de la verité et le monde détrompé à l'egard de la philosophie et de la réligion,* which the *Bibliothèque raisonnée* reviewed in 1746. The journal linked this treatise to the philosophy of Leibniz and Wolff, prompting the Leibnizian Samuel König to issue a letter denouncing this linkage. The journal published König's rebuttal, which stated that "the French already

absorbed entirely into French Cartesianism, the German exerted little direct philosophical influence himself.

The defenders of *vis viva,* in fact, were the only ones in France to stake out and defend anything like a Leibnizian philosophy before 1740. Yet even here their defense was restricted entirely to the conception of body and force that he employed and to the precise conception of mechanics that he developed in conjunction with this philosophy. Furthermore, since none of these defenders was a Leibnizian in any deeper, metaphysical sense, they often took pains to restrict their use of Leibnizian philosophy to this mechanics (or as he called it, "dynamics") alone. Isolating one from the other, however, was often easier said than done, and as the broader Leibnizian system that Wolff articulated began to gain German adherents, it also began to attract interested students in France.

Most intriguing in this respect, and unfortunately most mysterious as well, was Maupertuis, who first came to the attention of Johann Bernoulli and other senior *géomètres* in the 1720s as an alleged defender of Leibnizian *vis viva.* He never publicly professed these convictions, and throughout his life he maneuvered in and around the principles that defined the German's wider philosophy without ever expressing any singular attachment to them. Certainly Maupertuis never professed himself a "Leibnizian," even after he moved to the Berlin Academy in 1746, where such a declaration would have earned him enormous cultural capital. Instead, he remained a "Newtonian" by self-definition, when he chose to define himself at all. His relationship with this Newtonian identity, however, was complex and ever changing. We have already seen his shift from a stance of cautious defense of the philosophical possibility of Newtonian attraction in the 1732 edition of his *Discours* to the more aggressive assertions about the physical reality of attractive force in his revised 1742 edition. Other aspects of Maupertuis' Newtonianism were similarly fluid. He was never a defender of Newtonian experimental philosophy, for example, yet his friend and Lapland colleague Le Monnier nevertheless dedicated his French translation of Côtes' *Leçons de physique experimentales* to him.

However imprecise, such associations reinforced the clarity of Maupertuis' public Newtonian persona even as the non-Newtonian influences in his work were also becoming more pronounced. He did not resume his defense of *vis viva,* even though controversy around that topic again flared up in 1740. Instead, he returned to his previous work in mathematical mechanics, bringing to it a new non-Newtonian sensibility that had many affinities with Leibnizianism. In March 1739, Maupertuis initiated his last mathematical exchange with

---

have a confused idea of Wolff's philosophy, so these associations must be disavowed." *Bibliothèque raisonnée* 35 (1745): 367–397 and 38 (1747): 238.

Johann Bernoulli, the doyen of mathematical mechanics, sending him a paper that connected his newer interest in the shape of the earth with his older interest in the mechanics of moving bodies in fluid media. The paper focused on the forces of attraction operative within the oceans, and while Bernoulli found the work prolix and poorly developed, Maupertuis used it as the springboard for writing one of his very few published academic papers from this period.[82] During the same years, du Châtelet also initiated a correspondence with him as she studied Leibniz as part of her own research in physics and metaphysics. Maupertuis' letters do not survive, but we can glean from hers that they discussed collision, forces, conservation, and the laws of motion, topics that were central to Leibnizian dynamics.[83] Du Châtelet was particularly interested in Maupertuis' claim, made in his 1734 academy paper on attraction, that he had a proof for why God preferred the inverse square law of attraction over all others. Maupertuis does not appear to have satisfied the marquise's request, but his own work reveals that metaphysical principles of this sort—the kind that Leibniz made his stock-in-trade—were increasingly on his mind as well.[84]

Maupertuis' own use of metaphysical principles first appeared in an academic paper he presented in late 1740 based on his correspondence with Bernoulli.[85] His topic had changed, for the new paper focused on equilibrium principles operative among attractive bodies at rest. Bernoulli had done work on similar problems, but the real target for Maupertuis was Varignon, who had used "87 pages in quarto," the paper argued, "to demonstrate several useful and curious propositions that are nothing but very particular cases of those we shall give here." In the published version that appeared in the academy's *mémoires* for 1740 (without commentary in the *histoire*), Varignon's name was suppressed while the general point was reiterated. "One acquires immediately by this theorem the solution of several problems of mechanics that in the past arrested able mathematicians," the paper asserted.[86] At one level, this paper offered what was by then a four-decade-old celebration of universality and generality as the hallmark of French analytical mechanics. Yet Maupertuis also of-

82. See Terrall, *Man Who Flattened the Earth*, 173–174; HARS-Mem. (1740): 170–176.

83. Du Châtelet to Maupertuis, letters from 1738 dated 2 February, 10 February, 30 April, 9 May, 21 May, 22 May, 21 June, 7 July, 1 September, 3 September, and 20 January 1739, in Voltaire, *Corr.*, 89: D1442, D1448, D1486, D1496, D1507, D1509, D1528, D1548, D1606, D1610, and 90: 1804.

84. Du Châtelet to Maupertuis, 9 May 1738, in *Corr.*, 89:

85. Maupertuis, "Loi du repos des corps," PVARS, February 20, 1740 and *HARS-Mem.* (1740): 170–176.

86. Cited in Terrall, *Man Who Flattened the Earth*, 174.

fered a slightly different defense of this virtue by making general principles not just the marker of, but also the conceptual foundation for, good mechanics.

For Varignon, Fontenelle, and the other Malebranchian analysts of earlier in the century, generality was a virtue to be aspired to and a sign, once achieved, that the science developed was close to the truth. Maupertuis suggested something different: that *physiciens géomètres* should seek out general principles directly and then use them to guide their research into the particular truths of nature. This was a less analytical and more synthetical approach to the role of general principles in mechanics, and it revealed Maupertuis' emerging interest in Leibniz's more rationalist and synthetic approach to this science.

These tendencies were fully displayed in Maupertuis' subsequent, and ultimately last, academic paper delivered first to the members and then to the public at the spring assembly in April 1744.[87] In the years between these two academic interventions, Maupertuis had effected his public reconfiguration as a libertine wit and *esprit fort,* yet these associations were nowhere manifest in this thoroughly abstruse piece of French academic science. The argument of the paper, however, was nevertheless provocative in its own way even if it appears to have generated no response whatsoever among its intended audience. Terrall ably explores the technical aspects of the work, which focused on a new topic for Maupertuis: the refraction of light through dense media.[88] They can be economically summarized by describing the project as an effort to more fully explain the behavior of light conceived according to the Newtonian model of active, material forces. By conceiving of light this way, Maupertuis was able to treat it as a particular example of the much more general phenomena of material bodies moving according to the laws of Newtonian physics, especially the law of universal gravitation. The real significance of his paper, however, emerged not from its application of this physics to this precise empirical case, but rather from the very un-Newtonian use of this physical phenomenon to claim that metaphysical principles were foundational to the whole physics of nature.

In short, to quote Terrall, Maupertuis took the Newtonian corpuscular theory of light and built out of it "a decidedly non-Newtonian mechanics," one in which Leibniz played a decisive role.[89] Deploying Leibniz's principle of sufficient reason, Maupertuis explored the reasons why light behaved the way it did in passing through dense media. His point was that a foundational metaphysical principle, the principle of least action, as he called it, was a fun-

---

87. "Accord de différentes loix de la nature qui avaient jusqu'ici paru incompatibles," PVARS, April 15 and 18, 1744.
88. Terrall, *Man Who Flattened the Earth,* 176–180.
89. Ibid., 177.

damental causal factor in the changes themselves. In making this argument, he also invoked God, saying, along with Leibniz, that the laws of mechanics must also be consonant with the rational perfection of the deity who created them. As he wrote in a richly Leibnizian passage, "There is no doubt that all things are regulated by a supreme being who, even as he imprinted on matter the forces that denote his power, destined it to execute effects that mark his wisdom." Moreover, since the harmony between the material principles of nature and the wisdom of God's rational plan must be perfect, "all natural effects could be deduced from each taken separately." "If our mind were vast enough," he continued, "it would see the causes of physical effects equally by calculating the properties of bodies and by investigating that which is most fitting to carry out these effects."[90] In sum, empirical, mathematical mechanics and philosophical reasoning about necessary causes were each valid ways to arrive at the true constitution of nature since each was but one side of a larger, more perfect whole.

Ultimately, Maupertuis argued that physics should follow both paths simultaneously, recognizing that descriptive mathematical mechanics alone is often limited while philosophical rationalism is often misleading. "Let us calculate the motions of bodies," he concluded, "but let us also consult the designs of the intelligence that cause them to move."[91] Nevertheless, even this moderate inclusion of divine necessity as a principle of mathematical physics had no precedent in the annals of French academic mathematics. The methodological Cartesians who criticized Newtonian physics on philosophical grounds came closest to this position, but even they restricted themselves to epistemological arguments about scientific rigor while avoiding altogether the deep, causal, and metaphysical determinism that Maupertuis offered in his paper. In part, they did so because their philosophical masters, Descartes and Malebranche, had done so as well, seeing the heresies implicit in accounts of nature that treated God's creation in terms of necessary, rational, and deterministic principles. Leibniz had seen things differently. Furthermore, in the hands of Maupertuis, the German's conception of a divinely created world operating according to deterministic principles of force, motion, action, and conservation was turned into a foundation for a wholly new approach to physics and mechanics, one admittedly grounded in the very different principles of Newtonian physics as well.

As abstruse as it was, Maupertuis' turn toward Leibnizian metaphysical physics was not at all unconnected to his other transformations in the 1740s,

90. Cited in ibid., 179.
91. Cited in ibid., 180.

namely, his increasing comfort with radical thought and the social identities associated with it. His particular deployment of Leibnizian philosophy, in fact, attached him to the discourses of radical materialism and atheism in ways that his Newtonianism never really did. For many who understood Leibnizian metaphysics in detail, his philosophy rather than Newtonian physics was the real source of irreligion and atheism. Both could be imagined as offering a philosophy of nature that made innate material forces in bodies the causal agent in a deterministic universe governed by impersonal, mathematical laws. But while Newton, and especially his Newtonian followers, had worked hard to distance their conception of Newtonianism from precisely these Spinozistic and pantheistic notions of a necessary and deterministic world machine, Leibniz, and then Wolff even more emphatically, had celebrated the rational necessity and mathematical certainty of the Leibnizian system as proof of its scientific power. Consequently, links between Leibnizianism and Spinozism became even more commonly drawn in the Republic of Letters after 1740 than those between Newtonianism and the same. This was especially true given the way that the priority dispute gave added motivation for English and Dutch Newtonians to insist upon these very associations. No hint of these suspect affinities was present in Maupertuis' actual academic paper, yet these associations were definitely in the background as Maupertuis discussed the rational necessity of nature before the assembled public at the Louvre in April 1744.

Further supporting this context were other works by Maupertuis published in the same period that reinforced these radical associations even if the senior academician continued to use anonymous publication to shield himself from direct connection with them. The manuscript version of the future *Essay on Cosmology* was circulating in Paris from as early as 1741, and since it offered an account of God and nature rooted in active matter physics and rationalistic metaphysics, it was in many respects the philosophical companion to the mathematical and mechanical account of "least action" delivered to the academy. The bishop of Mirepoix's reaction to this brand of scandalous mathematical thinking has already been noted, and since these reactions were commonplace in 1740s France, they can be posited as a vital context for the reception of Maupertuis' other works as well. Especially provocative was another short, anonymous work published in this same period, first entitled *Physical Dissertation Occasioned by the White Negro* and then *Physical Venus*. Terrall describes the book as Maupertuis' *entrée* into the new research areas of biological generation and the sciences of life, which indeed it was. She also notes the reception of these books as scandalous libertine tracts. They were that too. Yet these texts also suggested a linkage between Maupertuis' scientific and philosophical Leibnizianism and his *mondain*, libertine persona cultivated in the same period.[92]

The "occasion" for the book, as Maupertuis called it, was an albino African slave from South America that some enterprising soul, well attuned to the spectacle-loving nature of the Parisian public, had brought to the city in 1744. The white-skinned Negro was displayed in aristocratic homes, salons, and at the Royal Academy of Sciences, and he created a public sensation that drew in equal parts upon the public's general love for curiosities of nature and its particular fascination with "monsters" of biological generation.[93] Trembley's apparently self-regenerating fresh water polyp was another, similar scientific spectacle, and in 1741 Réaumur presented this wonder to "the Academy, the court, and the city, which, in our enlightened century, hardly differ from savants."[94] These were Dortous de Mairan's words, published in the academy's *histoire* for that year, and they spoke to the way that science was inseparably linked to the Parisian public in eighteenth-century France. Attuned to this fascination with scientific wonders, while also eager to exploit it for his own purposes, Maupertuis therefore penned a quickly written text in late 1744 addressing some of the questions raised by the *nègre blanc*.

As both Desfontaines and Graffigny noted, however, the book was really not about the albino Negro at all.[95] Maupertuis quickly left behind the details of this precise curiosity, preferring instead to develop a more general argument about life and its origins. He framed the text in the idioms of *le beau monde*, claiming that a lady that he "could not refuse" had compelled him to write the text. This situated the book as a piece of *mondain* philosophy, and Maupertuis heightened the erotic charge implicit in the genre by pursuing an aggressively licentious agenda in the text. The book ultimately offered an account of the biological origins of life, but opening it he addressed his female interlocutor with the following caution: "Do not be annoyed if I tell you that you were once a worm, or an egg, or a kind of mud. But do not think either that all is lost when . . . that form that you have now, . . . that body that charms

---

92. For a full account of these books and their reception, see Terrall, *Man Who Flattened the Earth*, 207–226.

93. The visit to the academy is noted in PVARS, January 8, 1744.

94. Dortous de Mairan had succeeded Fontenelle as the academy's perpetual secretary in January 1741. These passages appeared in *HARS-Hist* (1741): 33–34, and are cited in Terrall, *Man Who Flattened the Earth*, 205–206.

95. Desfontaines ran a review of Maupertuis' text in his new, but short-lived *Jugemens sur quelques ouvrages nouveaux*, 11 vols. (Paris, 1744–1746), 9: 216–231.

everyone, will be reduced to dust."[96] Continuing in this vein, Maupertuis went on to teach his charming companion the mysteries of generation while making it clear throughout that he was struggling to resist the allure of her bodily attractions. When the sexual nature of biological procreation was treated, the erotic character of the text was intensified as Maupertuis indulged in a vividly pornographic treatment of animal sexuality. In his text, the queen bee was described as maintaining a "harem of lovers" that she fully satisfies "in the most unrestrained debauch," while the "impetuous bull, proud of his strength" is described as an animal that "doesn't fool around with caresses." "He throws himself immediately on the heifer, penetrates deep into her entrails, and spills out in a great flood the liquor that makes her fertile."[97]

Amidst this symphony of witty and learned smut, however, was a serious philosophical argument, and it was here that Maupertuis' libertinism came together with his Leibnizianism in important ways. Generational theory at the time held that reproduction was a one-way street, with either the female seed being activated by male semen or the male seed being fertilized by its implantation in the female womb. In each case, no biological mixing between the male and the female was thought to occur. This was precisely what Maupertuis challenged. He developed instead what historians of biology today call an "epigenetic" theory of reproduction by positing life as the joint biological product of males and females together.[98] Maupertuis offered a host of different arguments in support of this idea, emphasizing in particular how the epigenetic theory helped to explain the way that children often resemble both parents.

More radical was Maupertuis' attempt to offer a physical and mechanical account of this epigenetic transformation. Here he drew upon Newtonian affinity chemistry, among other things, to explain the mixing of elements that produced life. Writing in a passage that strongly echoed Newtonian chemists such as Geoffroy, he described how "these forces and these *rapports* are nothing other than what other, more daring philosophers call attraction." Getting in a jab at the opponents of such forces in the Paris Academy, he further added that "this ancient term, revived in our times, at first shocked those *physiciens* who thought they could explain everything with out it." Nevertheless, as "in-

96. Cited in Terrall, *Man Who Flattened the Earth,* 211.

97. Cited in ibid., 219.

98. On this history, see Jacques Roger, *Les sciences de la vie dans la pensée française du XVIIIe siécle; la génération des animaux de Descartes à l'Encyclopédie* (Paris, 1963); and Clara Pinto-Correia, *The Ovary of Eve: Egg and Sperm and Preformation* (Chicago, 1997).

comprehensible as they are, these forces seem to have penetrated even into the Academy of Sciences, where they weigh new opinions carefully." Maupertuis had no compunction accepting and deploying these material forces in scientific explanation, and he thus asked, in a setup to all that followed: "Why, if this force exists in nature, might it not play a role in the formation of bodies?" [99]

The text answered this question with a thoroughgoing physical and material account of the origins of life. Pierre Brunet sees in this effort a Newtonian attempt to reduce all of nature to the physics of universal gravitation, but Terrall rightly calls this characterization one-dimensional.[100] "Maupertuis used the viability of gravitational attraction to justify other forces that act selectively and teleologically," she writes, and in formulating things this way, Terrall implicitly position's Maupertuis' science of life alongside his new interest in metaphysical rationalism and the immaterial teleology of final causes. She also points to the influence of Leibniz in his thinking even if she does not stress this connection.[101] His science of life as it is developed in the dissertation on the *nègre blanc*, and especially *Vénus physique* is in fact very Leibnizian, although it is probably best described as a hybrid and personal fusion of several different philosophical discourses simultaneously.[102]

Maupertuis sees matter composed of tiny particles possessing inherent properties that make them inclined to organize into rational, ordered systems. Certainly Newtonian affinity chemistry, with its systematic science of attractive and repulsive forces, could be deployed to construct a materialism of this sort. Far closer to Maupertuis' conception, however, was Leibniz's metaphysics, with its fundamental monad that contains within its entelechy the rational principles that govern the entire cosmos. Especially Leibnizian was the rationalist determinism that Maupertuis employed, since even materialist Newtonianism was anchored in the antirationalist notion of a voluntarist creator ordering nature in whatever way he saw fit. In this respect, the immanent teleology that Maupertuis located at the very center of matter's corpuscular nature, combined with the deterministic rationalism that he saw governing the whole of nature,

---

99. Cited in Terrall, *Man Who Flattened the Earth*, 216.

100. Pierre Brunet, *Maupertuis, l'oeuvre et sa place dans la pensée scientifique et philosophique du XVIIIe siècle* (Paris, 1929), 318.

101. Terrall, *Man Who Flattened the Earth*, 216–217.

102. A recent book that connects Leibniz to eighteenth-century materialist discourses, including that of Maupertuis, is Claire Fauvergue, *Diderot, lecteur et interprète de Leibniz* (Paris, 2006).

placed his materialist philosophy of life at the nexus between Leibnizianism and radical Spinozism much more than with Newtonianism.[103]

Maupertuis again protected himself from the provocative *Claims of Vénus physique* by publishing the text anonymously and by refusing to claim the book as his own even after its publication. Many attributed the book to him, although Voltaire was also named more than once as the probable author.[104] Other possible authors were also suggested, and precisely because of the mystery of the book's authorship and the scandalous, titillating nature of its content, it sold very well. Desfontaines noted that the first edition of the dissertation on the *nègre blanc* was "much sought after," and it was quickly reprinted twice before the end of 1744.[105] *Vénus physique* appeared soon after, and it continued to sustain the buzz, earning reviews in Desfontaines' new *Jugemens sur quelques ouvrages nouveaux*, Elie Fréron's *Lettres sur quelques écrits de ce tems*, and the *Bibliothèque raisonée des ouvrages des savants de l'Europe*.[106] It also generated critical pamphlets, including *L'anti-Vénus physique*, an attack by the same Jesuit, Basset des Rosiers, who had written against Maupertuis' *Lettre sur la comète*.[107]

Most pointed was the *Bibliothèque raisonnée*, which minced no words in connecting the book directly to materialism and atheism. Attributing the book either to Maupertuis or Voltaire, the reviewer wrote: "I hardly need to point out to the reader that this system leads naturally to pantheism. It gives to matter all the energy, all the power, that Spinozists, naturalists, and materialists attribute to it." It also places "the true cause" of the motions and organization of animals and plants "in matter, which does everything by instinct, by a force that does not come from outside, and that it exists in and of itself."[108] Attentive readers of this journal would not have missed the association between this philosophy

103. On the connection between Leibnizian philosophy and eighteenth-century theories of generation, see Justin Smith, "Leibniz's Hylomorphic Monad," *History of Philosophy Quarterly* 19, no. 1 (2002): 21–42; Ohad Nachtomy, Ayelet Shavit, and Justin Smith, "Leibnizian Organisms, Nested Individuals, and Units of Selection," *Theory in Biosciences* 121 (2002): 205–230.

104. It was surmised in *Bibliothèque raisonnée* 35 (1745): 312, that either Voltaire or Maupertuis had written the text.

105. Review of *Dissertation physique à l'occasion du negro blanc*, in *Jugemens sur quelques ouvrages nouveaux* 9 (1744): 217.

106. The publishing history and reception of the text is chronicled in Terrall, *Man Who Flattened the Earth*, 208–209, 223–226.

107. Gilles Basset des Rosiers, *L'anti-Vénus physique* (n.p., 1746).

108. Review of *Vénus physique*, in *Bibliothèque raisonné*, 35: 312.

and Wolff's, which the journal had earlier characterized as a metaphysics in which "all simple beings have an active force that is essential to them," and where God is that force and thus also "lives in nature." For Wolff, the journal declared, "the writings of John 5:17 are no different than reason," which is why "the philosophy of Leibniz and Wolff that is so in vogue in Germany" has nothing to ground it but "their fertile imaginations."[109] By implication, the same was true for Maupertuis' *Vénus physique* as well.

More important than the precisely Leibnizian or Newtonian nature of this philosophical radicalism, however, was the immorality and impiety associated with it. Basset des Rosiers drove this point home in his pamphlet *L'anti-Vénus physique* without bothering to pin down the precise philosophical origins of the heresies. He charged the anonymous author with subverting the moral order by seducing readers into his prurient philosophical mindset. He also complained about the shameless mixing of genres and discourses in the text, calling such textual subversion an affront to the moral order. "The fashion [today] is to write in all genres and to succeed or to excel in none." For the Jesuit this genre and stylistic destructiveness was part and parcel of Maupertuis' degenerate desire to tear down all that was good, true, and virtuous.[110] Others certainly found the work just as scandalous, but many also found it brilliantly provocative. Elie Fréron, the future enemy of the *Encyclopédistes,* thought that the work illustrated how "an anatomical dissertation can become pleasant reading." The abbé Raynal, writing in Melchior Grimm's new *Correspondence littéraire,* called the book "clearly, vigorously, ardently, and elegantly written." "It is from the hand of a master," he declared, "and our ladies have abandoned their novels to read it."[111] Desfontaines also attributed *esprit* to the writing, while worrying about the licentiousness of its content. When Maupertuis drew upon Réaumur's work on the mating habits of insects, Desfontaines expressed a grossly understated regret that "he did not borrow the gravity of his style."[112] Graffigny claimed to have spent an entire day in bed reading the book with her lover, but she nevertheless thought the text made the author look "somewhat ridiculous."[113]

To the extent that Maupertuis was actively cultivating an identity as an *esprit fort,* this reception must have pleased him, especially since the buzz circulated in a way that allowed him to reap all of its benefits without suffering any

109. Ibid., 20: 301, 34: 84.

110. Cited in Terrall, *Man Who Flattened the Earth,* 225.

111. Cited in ibid., 225–226.

112. Review of *Vénus physique,* in *Jugemens sur quelques ouvrages nouveaux* 2 (n.d.): 251.

113. Terrall, *Man Who Flattened the Earth,* 224.

of the persecution that often came with it. Voltaire benefited as well, since his name was continually associated with Maupertuis' in this context (Basset des Rosiers likened both to "poets who write geometry").[114] In an important twist, however, this was a philosophical libertinism that no longer had Newtonianism as its precise and obvious linchpin. The radical philosophy of *Vénus physique* was associated more with Leibnizianism than with Newtonianism, and the appearance of this text as the philosophical cause célèbre of the moment created a new dynamic between them.

This dynamic was further catalyzed by the direct intrusion of Leibnizianism into Voltaire's philosophical life through the independent intellectual work of the Marquise du Châtelet. Du Châtelet's role in either stimulating or feeding upon Maupertuis' turn toward rationalist metaphysics and immaterialist determinism after 1738 has already been noted. Yet their correspondence also led to a series of initiatives on her part that implicated Voltaire with Leibnizianism in new and important ways. As early as 1737, the couple read Wolff during their studies at Cirey, and evidence of this study was apparent in du Châtelet's academy essay on fire, which she submitted at the end of that year.[115] While Voltaire manifested his debt to Anglo-Dutch experimental philosophy in his submission, offering the academy a set of physical theories grounded in Newton-inspired inductions from experimental observations, du Châtelet's essay offered a much more rationalist argument, which established the nature of fire through deductions from necessary material principles.[116] Furthermore, while she was writing the defense of Voltaire's *Eléments* that would appear in the *Journal des savants* in September 1738, she also engaged in her correspondence with Maupertuis about force, bodies, action, and the laws of collision. In March 1739, this thinking was further intensified when, at the suggestion of Maupertuis and Bernoulli, she invited the German Wolffian Samuel König to come to Cirey as her tutor. This was the period of Voltaire's scandalous battle with Desfontaines, yet while the couple stayed mobile and Voltaire fled the consequences of his libelous pamphlet, König began to teach du Châtelet Wolff's Leibnizian philosophy. His method of instruction was a key component of the philosophical teachings themselves. He presented the Leibnizian system to her deductively, insisting that all doubt be removed about each demonstration

114. *L'Anti-Vénus physique*, 201.

115. See Vaillot, *Avec Madame du Châtelet*, 76–77, 81–84.

116. Voltaire, "Essai sur la nature du feu et sur sa propagation"; and Emilie de Breteuil, marquise du Châtelet, "Dissertation sur la nature et la propgation du feu" in *Recueil des Pièces qui ont remporté les Prix de l'Académie des Sciences* 4 (1739). Voltaire's essay is reprinted in *Oeuvres complètes de Voltaire* (ed. Beuchot), 22: 279–326.

before he would move on to the next proposition. In this way, du Châtelet acquired both an understanding of Leibnizian natural philosophy and an appreciation for the rational necessity that was seen to anchor the whole system.[117]

As the months went on, and for reasons that are unclear, König grew more and more frustrated with his situation. In December, he returned to Germany, but in many ways his work was already done. By the time of his departure, du Châtelet was already at work composing her own work of metaphysical physics, which she published in 1740 as *Institutions de physique*.[118] The appearance of the work was a major event in the Republic of Letters, not least because of the sex of its author. Addressed to serious philosophers and the philosophically literate public, the *Institutions de physique* was du Châtelet's assertion of her own identity as a sophisticated and independent philosophical thinker. In this respect, it was not unlike Voltaire's *Eléments*, even if the particular social, gender, and discursive characteristics of her effort were very different. Like the work of her male companion, du Châtelet's treatise was also rewarded with long and largely favorable reviews in all the major learned periodicals, including the *Journal des savants*, the *Journal de Trévoux*, and the new *Bibliothèque raisonnée*. It also captured a notice in the organ of Parisian society news, the *Mercure de France*.[119] Yet while Voltaire used Newton to launch his serious scientific debut, du Châtelet launched hers using a synthesis of Newton's physics with what she argued was the superior mechanics, metaphysics and methodology of Leibnizian philosophy.[120]

The French reception of the *Institutions de physiques* can be charted by comparing two of its principle reviews. In late 1740 and early 1741, the *Journal des savants* devoted two long and richly detailed articles to du Châtelet's treatise. Adopting the judicious neutrality that was the journal's hallmark, the reviews repeated, and thus reinforced, most of du Châtelet's central claims. Indeed, for

117. Vaillot, *Avec Madame du Châtelet*, 126–127.

118. Emilie de Breteuil, marquise du Châtelet, *Institutions de physique* (Paris, 1740).

119. Review of Emilie du Châtelet, *Institutions de physique*, in *Journal des savants* (1740): 737–755; (1741): 135–153; *Journal de Trévoux* (1741) 894–927; *Bibliothèque raisonnée* (1741): 27: 433–464; and *Mercure de France* (1741): 336–345.

120. A thorough discussion of du Châtelet's treatise is found in Linda Gardiner Janik, "Searching for the Metaphysics of Science: The Structure and Composition of Mme. Du Châtelet's *Institutions de physique*, 1737–1740," *Studies in Voltaire and the Eighteenth Century* 201 (1982): 85–113. On du Châtelet's intellectual style more generally, see Judith Zinsser, "Entrepreneur of the Republic of Letters: Emilie de Breteuil, Marquise du Châtelet and Bernard Mandeville's *Fable of the Bees*," *French Historical Studies* 25 (Fall 2002): 595–624.

AEMILIA DE BRETEUIL
CONJUX MARCHIONIS DU CHATELLET

FIGURE 24. *Portrait of Emilie du Châtelet, from Jacob Brucker,* Pinacotheca
scriptorum nostra aetate literis illustrium: Émilie du Châtelet *(Augsburg: J. Haid-
ium, 1741–1755). Courtesy of the Herzog August Bibliothek, Wolfenbüttel, Germany.*

many, the reviews served as a convenient précis of the book, making study of the dense treatise in its entirety unnecessary. However, in condensing the book with its characteristic judiciousness, the journal also shaped its meaning in important ways. Most significant was the way that its discussion absorbed du Châtelet's Leibnizianism into the discourses of Cartesian anti-Newtonianism, discourses that were widely prevalent in the journal during these years. Her defense of the Leibnizian system of the monads; her presentation of his doctrine of the preestablished harmony; her analysis of his principles of continuity and diversity: each was fully articulated but each was also absorbed into the still-pervasive binary Newtonian/Cartesian framework.

This framing appeared very early as the journal discussed du Châtelet's overall approach in the treatise. Her work was offered as a pedagogical manual addressed to her son, and in her preface du Châtelet defended this approach by arguing that deductions from evident first principles were the appropriate foundation from which to build a clear understanding of natural philosophy. The journal quoted these statements verbatim, joining with du Châtelet in praising their correctness. As the reviewer declared, "this method is very advantageous from the point of view of solidity, and there is no other that leads to the truth with greater surety."[121] In this way, the journal linked du Châtelet's use of the Euclidean, deductive method developed by Wolff and taught to her by König, to the other Cartesian treatises that made synthetic, deductive demonstration the model for clear and rigorous science. In describing du Châtelet's presentation of the Leibnizian principle of sufficient reason, they further directed this conviction against the unrigorous theory of Newtonian attraction. Quoting du Châtelet, the journal declared that "to deny this principle is to fall into strange contradictions," for without it "one cannot say that the universe, so tied together in all its parts, is produced by a Supreme Wisdom. For saying that something happens without sufficient reason is to say that it simply happens by chance, which is to say nothing." Illustrating the point, they noted that this principle had led to the "banishment from philosophy of all the causes that are nothing more than words without meaning," causes such as the Aristotelian occult properties. It had also led to the notion of a vegetative soul in plants, and "in our own days, attraction, if one alleges it to be a physical cause."[122]

The rest of the journal's presentation continued in this Cartesian vein. When Leibniz's more peculiar and specific theories were discussed, the journal merely recounted du Châtelet's presentation while praising her for making these strange ideas as clear as possible. When, by contrast, her arguments broached

121. Review of du Châtelet, *Institutions de physique*, in *Journal des savants* (1740): 740.
122. Ibid., 741.

the philosophical rationalism central to the Cartesian rejection of Newtonianism, the journal reinforced her arguments in an aggressive way. Analyzing her discussion of the nature of matter, for example, the reviewer stressed how her Leibnizianism reinforced the errors of Lockean materialism.[123] Lockean-thinking matter-materialism and Newtonian gravitation were twin components of the post-Voltaire conception of Newtonianism in France, and in its review of the *Institutions de physique,* the *Journal des savants* was making du Châtelet's Leibnizianism an important antidote to it. They did the same in their account of du Châtelet's discussion of space and time. Repeating Leibniz's arguments against Clarke from their famous correspondence and then turning du Châtelet's treatise into a contemporary French reinstatement of these anti-Newtonian critiques, the journal explained that "God acts for a reason." "Consequently, since the reason for the place of the universe in space and the limits of extension are neither in these things themselves, nor in the will of God, one must conclude that the hypothesis of the void is false."[124] Newton, of course, defended the opposite position, and here du Châtelet was positioned most strongly as an anti-Newtonian.

Du Châtelet's treatise, however, was not really an anti-Newtonian tract, even if she did offer the critiques of Newtonian philosophy that the journal made sure to reinforce. Instead, her project centered on integrating the recognized achievements of Newtonian physics with what to her mind were the superior metaphysics and mechanics of Leibniz. In this area the journal had a more difficult time merging her thought with Cartesianism. Du Châtelet devoted three chapters (13–15) to the problem of *pesanteur,* and this discussion ended with a discussion of the Newtonian theory of gravity. She also devoted another chapter (16) to the theory of universal attraction itself. In its review, the journal had no problem reinforcing du Châtelet's critique of Newtonianism, especially her conclusion that the English were "too zealous for attraction" while forgetful of the need for a sufficient reason to believe in it. Especially prudent, the reviewer thought, was her reminder that the Newtonians "too quickly disregard the mechanical explanations that have been offered for the phenomena [of weight]." They attribute it too quickly to attraction, she emphasized, yet whatever insufficiencies the former may have they in no way constitute an argument for the latter.[125]

These critiques of Newtonianism, however, were attached to a conception of mechanical force and action drawn from Leibniz. These arguments were

123. Ibid., 744–746.
124. Ibid., 747.
125. Review of du Châtelet, *Institutions de physique,* in *Journal des savants* (1741): 145–147.

far less compatible with Cartesianism. Using the Leibnizian language of living and dead forces, du Châtelet connected Newtonian planetary theory, namely, the inverse square law, together with Galileo's law of falling bodies, uniting each with a conception of mechanical action that agreed with God's supreme rationality. The result was a new synthesis of the prevailing natural philosophies of the day, yet all the journal could do was describe it while noting its disagreement with Dortous de Mairan's conception of *vis viva*.[126] The point to emphasize here is that du Châtelet's attempt to position Leibniz as a third way between Cartesianism and Newtonianism ultimately was flattened between the powerful pulls of the prevailing conceptual binary. Few if any Frenchman rallied to du Châtelet's cause, therefore, and Leibnizianism, so well articulated and defended by her, remained little more than an exotic philosophical monster for the French. Despite the notoriety of her work, they still thought in terms of a Manichean, and at times even patriotic, choice between Newton and Descartes. Yet her effort was not without some influence. The reconfigurations her work encouraged can be seen by considering a second review of the work, one published in the pro-Newtonian *Bibliothèque raisonnée des ouvrages des savants d'Europe*.[127] Some scholars have attributed this review to Voltaire even though it was published without an author's byline. Whether or not he was the author, it is clear that the review is the work of an avowed Newtonian.[128] Consequently, it reveals, in ways parallel to the Cartesian take on du Châtelet's work in the *Journal des savants*, how Leibnizianism became a wedge in the ongoing Newtonian-Cartesian polemics of the period.

The reviewer began, like his counterpart in Paris, with a celebration of du Châtelet's achievement. He particularly praised her admirable status as a *femme savante*. After these niceties, however, the review quickly took a different tone. Rather than judiciously summarizing the contents of the book, the reviewer instead engaged in a critical dialogue with it. The result in the end was a Newtonian critique of Leibnizianism that was witty at times, serious at others, and moderate throughout. Noting at the outset that Leibnizian philosophy was little known outside Germany, and that du Châtelet's work articulated the recent German synthesis of Leibniz's writings by Christian Wolff, the reviewer quickly launched into his critique. Regarding the principle of sufficient reason, he found du Châtelet's presentation to be "admirably brief and methodical."

126. Ibid., 148–153.

127. Review of *Institutions de physique*, in *Bibliothèque raisonnée* 27 (1741): 433–464.

128. Beuchot claims the review for Voltaire and reprints it in his collection of Voltaire's works, *Oeuvres complètes de Voltaire*, 23: 129–146. My citations will be to the version found in the *Bibliothèque raisonnée*.

Yet he wondered what was so innovative about it. No one should be surprised to learn that nothing happens without a reason. However, if Leibniz's position was that everything is as it is for a reason, and that reason alone can command certain propositions out of existence, then he was overstepping the boundaries of human thought. "There is no particular reason why the planets move from east to west," the reviewer wrote, and to claim otherwise is to assert human reason above the free will of the creator.[129] Other, familiar Newtonian arguments were offered to counter other aspects of du Châtelet's Leibnizian presentation, and while the rebuttal was often pointed, it was also lighthearted. The reviewer, for example, criticized du Châtelet for agreeing with Leibniz that God could only create the best of all possible worlds. He did so, however, while also conceding that the marquise offered proof that God did at least create many very excellent things.[130]

The review overall operated through a dichotomy that placed the judicious and philosophically modest empiricism and skepticism of the Newtonians against the hoary metaphysical subtleties of Leibniz and Wolff. Ultimately, the reviewer insisted, their natural philosophy was nothing more than "a complete fantasy" (*un roman complet*).[131] Their effort to differentiate between essential and accidental attributes in bodies, for example, led Leibniz and Wolff into "a labyrinth of subtleties," while Locke's empirical skepticism about such matters was far simpler.[132] Similarly, while Leibniz's claim that "there was a sufficient reason for everything that God did" claimed too much for human reason, the Newtonian focus on the empirical facts of matter and motion, by contrast, could not be refuted without refuting experience itself.[133] Wit also played a role, such as when the reviewer likened Leibniz's preestablished harmony between God's determinism and the freedom of the human will to "a burlesque sermon where one man preaches while the other makes the gestures."[134] The reviewer's befuddlement in the face of Leibniz's immaterial and extensionless monads acting as God's agent for material action in the world also led him to quip that "one must, it seems to me, have the mind of the person who wrote the *Institutions de physique* to bring clarity to such things."[135] Nationalist bias was also invoked as the writer

129. Review of du Châtelet, *Institutions de physique*, in *Bibliothèque raisonnée* 27 (1741): 435.

130. Ibid., 27: 439.

131. Ibid., 27: 447.

132. Ibid., 27: 440.

133. Ibid., 27: 451.

134. Ibid., 27: 447.

135. Ibid., 27: 449.

noted how no Frenchman, Englishman, or Italian had adopted this "strange system." Du Châtelet was thus trying to defend a German "romance" that even many Germans only regarded as "a game useful for exercising the mind."[136]

These formulations reinforced the association between Leibnizianism and the medieval philosophical systems of yore while making Newtonianism the philosophy of modern, judicious empiricists. This was precisely the point, and to the extent that the reviewer also emphasized Leibniz's conception of matter as an active agent driven by necessary metaphysical laws, he also reinforced the connections between this philosophical monster and Spinozistic materialism. Maupertuis' *Vénus physique* would further reinforce these connections three years later, and against both the *Bibliothèque raisonné* offered an image of Newtonianism rooted in philosophical skepticism about the metaphysical nature of matter, and modest empiricism as the only judicious stance for the science of physics. "When someone can offer a good mechanical account of [the phenomena of attraction], he will render a service to the public by publishing it," the journal declared. "However, since the greatest philosophes have been looking for over sixty years for such an account without finding it, let us just hold on to attraction as a fact proven by experience, at least until God reveals the sufficient reason for it to some Leibnizian."[137]

Voltaire may have been the author of this review, for the text certainly has a Voltairean flavor to it. However, even if he did not write it, he initiated his own intervention into the new vogue for metaphysics, especially German metaphysics, after 1740. He wrote a short work of his own called *The Metaphysics of Newton,* publishing it first as a stand-alone work, and arranging for it to appear simultaneously with du Châtelet's *Institutions de physique.*[138] Like the latter, Voltaire's work was a serious metaphysical exposition that entirely eschewed the idioms and tone of libertine philosophy. In subsequent years, Voltaire also appended his metaphysical exposition to the physical explanations of his *Eléments de la philosophie de Newton,* creating in this way a two-part metaphysics-physics structure for the text that mimicked exactly the organization of du Châtelet's *Institutions.* In 1744–1745, amidst the buzz created by Maupertuis' deeply metaphysical and libertine-materialist treatment of the science of life, Voltaire reissued this new and improved edition of the *Eléments.*[139] This al-

136. Ibid., 27: 447.

137. Ibid., 27: 452.

138. Voltaire, *La métaphysique de Newton* (Amsterdam, 1740).

139. *Les eléments de la philosophie de Newton,* 2nd ed. (London [La Haye], 1744). The details behind the subsequent editions of Voltaire's *Eléments* are recounted in Walters and Barber, introduction, 119–140.

lowed him to further define Newtonianism in ways that distanced it from both the retrograde philosophical system building of the Leibnizians and the *malhônnete* libertinism and irreligion of the radicals.

The Leibnizian context for the text is paramount, for rather than offering a thoroughgoing exposition of Newton's metaphysical views alone, Voltaire instead offered a Newtonian critique of the views of Leibniz (and at times Descartes and Malebranche as well). He ended his text by inviting his readers to decide for themselves between Newton and Leibniz, but his real point, implied throughout, was that Newton really had had no need for a metaphysics at all. Reason, experience, and good physics were all that one needed to understand nature's truths, or so the author suggested. Thus, in a series of chapters on topics ranging from God and liberty to the nature of space, extension, matter, forces, the animal soul, ideas, the monads, and natural religion, Voltaire worked his readers through a series of arguments that showed the validity of Newton's views and the errors that followed from the metaphysical approach of Leibniz and other philosophers.[140]

The result, therefore, is less a work of metaphysics per se than a defense of Newton's unmetaphysical approach to questions of philosophy. The latter offered a strong alternative to the deductive, metaphysical demonstrations of Wolff and du Châtelet. Throughout, humble respect for the mystery of God combined with disgust at the arrogance of philosophical hubris merges to assert the superiority of Newton's preference for mathematics and experiment over metaphysics. Also celebrated was Newton's stance of philosophical modesty in the face of egregious and dangerous metaphysical excess. "Is it really proven that everything is a plenum despite the mob [*foule*] of metaphysical and physical arguments that support the void," Voltaire asks of the Leibnizians? "Is it proven that your alleged monad must have the useless knowledge of everything that happens in this plenum? I appeal to your conscience: do you not see how such a system is nothing but pure imagination? Is a vow of human ignorance regarding the elements of matter not superior to such a vain science? What a misuse of logic and mathematics to allow them to lead one into such a labyrinth, walking down the road to error with the flame destined to enlighten us."[141]

Newton's philosophical modesty made his deployment of logic and mathematics far more trustworthy. Moreover, and without drawing the irreligious conclusions explicitly, Voltaire also pointed to the ways that Leibnizian meta-

140. My citations will be drawn from the edition found *Œuvres complètes de Voltaire* (ed. Besterman), 15: 195–252.

141. Ibid., 15: 244.

physical rationalism generated moral as well philosophical monsters. In his chapter on God, after reassuring his readers about Newton's piety and declaring that "all of Newtonian philosophy necessarily points to a Supreme being who freely created and arranged everything," Voltaire exposed the materialism inherent in the non-Newtonian approaches to the deity and his creation. Positing a questioning atheist as his interlocutor, Voltaire defined the correct Newtonian view while putting a number of standard materialist arguments into the mouth of his antagonist. Many of these resonated strongly with the metaphysics of Leibniz, and they especially echoed Maupertuis in *Vénus physique*. "There are no created beings, no distinct substances," the atheist posited. "The universe is one, necessarily existing and developing without end." "Rest is a fiction," he continued. "An infinitely subtle matter circulates everywhere and eternally through the pores of bodies," and there is "an equal quantity of motive force in nature," and this force moves bodies "in a necessary way."[142]

While Leibniz, of course, distanced his own philosophy from each of these positions, Maupertuis indulged in many of the claims in his *Vénus physique*. Yet he did so in an anonymous work that was anything but a systematic metaphysical treatise. The very subtlety of Leibniz's arguments, however, when combined with the resonances between them and Maupertuis' metaphysical materialism, proved Voltaire's point. "No one was a more zealous partisan of natural religion than Newton," Voltaire declared, but his genius rested in knowing how to restrict his understanding within the bounds of human knowledge.[143] He refrained from hubristic claims about God's enslavement to rational necessity, and as a result his philosophy was superior because it humbly restricted itself to calculations from facts. As Voltaire summed up, Newton knew the most profound approach of all: "He knew how to doubt."[144]

Voltaire used the familiar discourses of the Newton wars to make his points, yet by the 1740s these positions had taken on new meaning as a result of the emergence of a more aggressive Leibnizianism within the Republic of Letters. In this context, Leibnizianism, rather than Cartesianism, became the increasing focus of pro-Newtonian discourses. Claims to philosophical modesty, together with indictments of chimerical philosophical system-building, were likewise redirected against Leibniz's metaphysics of active forces and deterministic rationalism. Meanwhile, the Cartesian vortices, once the central target of these attacks, began to recede from the center of these struggles. Voltaire's re-

142. Ibid., 15: 198–199.
143. Ibid., 15: 219.
144. Ibid., 15: 232.

orientation of his Newtonianism in this direction in the revised and expanded *Eléments de la philosophie de Newton* of 1745 furthered this transformation, for it allowed him to realign the relationship between Newtonianism and Cartesianism in France at a time when the situation was ripe for change.

The Solidification of Enlightenment Newtonianism after 1745

These changes can be introduced by considering Voltaire's response to another consequence of du Châtelet's *Institutions de physique:* the final revival of the *vis viva* debate in France. In reviewing du Châtelet's *Institutions de physique,* the *Journal des savants* ended its discussion by summarizing her challenge to the opponents of *vis viva* in France. Most prominent among the opponents was Dortous de Mairan, who was just then beginning his brief tenure as the interim successor to Fontenelle as the academy's perpetual secretary.[145] Voltaire likewise ended his *Metaphysique de Newton* with a critique of the Leibnizian conception of *vis viva,* siding with Archimedes, Descartes, Newton, Varignon, Mariotte, and others in claiming matter's essential inertness and the correctness of mass times velocity (mv) as the correct measure of its motive force. To end the book this way was a bit odd, since he had cut short his metaphysical discussions in these chapters on more than one occasion by letting readers know that they raised physical questions that would be dealt with more appropriately (which is to say nonmetaphysically) in the physics sections of the text. The transition from the metaphysical discussion of "the active force that puts all in motion in the universe" (the title of Voltaire's final chapter) to the precise mechanics of *vis viva* could easily have triggered a similar deferral, but in this case Voltaire offered a thorough account of the controversy instead. He also did so while defending the joint position of the Newtonians and the Cartesians through a familiar presentation of the standard arguments against Leibniz's measure of motive force as mv^2.[146]

Framed by its presence in a text about metaphysics, therefore, the discussion implicated the precise measurement of force in mechanics with the wider problems and dangers of Leibnizian metaphysics. This in turn associated the defenders of mv^2 and Leibnizian mechanics, however narrow and technical

145. Review of Châtelet, *Institutions de physique,* in *Journal des savants* (1741): 148–153. Dortous de Mairan served as academy secretary from January 1741 to December 1742.

146. Voltaire, *Metaphysique de Newton,* in *Œuvres complètes de Voltaire* (ed. Besterman), 15: 245–252.

their reasoning, with philosophical monstrosity on the one hand and metaphysical materialism on the other. The argument also positioned Voltaire on the side of the French Cartesians (and the English Newtonians in this case), and opposite du Châtelet, who had defended the Leibnizian position in her *Institutions de physiques.* The debate that was thus triggered attracted a great deal of public notice since it pitted the current secretary of the Royal Academy of Sciences against a rare intellectual wonder, a learned marquise.

Dortous de Mairan initiated the exchange when he published a pamphlet defending his Cartesian measure of motive force in matter while challenging du Châtelet's Leibnizian conception of the same thing.[147] The marquise responded in kind, publishing a pamphlet of her own that not only reiterated her Leibnizian critique of Dortous de Mairan's arguments in favor of mv, but also took issue with what she alleged to be his condescending attitude toward her status as a woman.[148]

The debate made for great public theater, and Desfontaines spoke for many in calling Dortous de Mairan's position "the most commonly held view" while praising him for his "love of truth," his "depth of research," and for the neatness, precision, and politeness of his style and the solidity of his reasoning." He also praised du Châtelet as well, reaffirming his "admiration, in light of her rare genius and profound learning, of the great use to which she has put her mind and her leisure." [149] Voltaire also composed his own "Doubts on the measure of motive forces and their nature." He first submitted his text to the Paris Academy, and then published it two months later in the *Nouvelle bibliothèque, ou Histoire littéraire* with the academic approval of Pitot and Clairaut appended.[150] The brief essay contributed nothing original to the discussion, but it did confirm Voltaire's knowledge of the facts while showing his strong support for the Cartesian-Newtonian position in the dispute. It also gave him an opportunity to again associate the Leibnizian conception of force with Spinozistic materialism while distancing Newtonian gravitational attraction from the same

147. Jean-Jacques Dortous de Mairan, *Lettre de M. de Mairan, . . . à Madame*** [la marquise du Châtelet] sur la question des forces vives, en réponse aux objections qu'elle lui fait sur ce sujet dans ses* Institutions de Physique (Paris, 1741).

148. Gabrielle-Émilie Le Tonnelier de Breteuil, marquise du Châtelet, *Réponse de Madame***[Du Châtelet] à la lettre que M. de Mairan, secrétaire perpétuel de l'Académie des sciences, lui a écrite, le 18 février, 1741, sur la question des forces vives* (Brussels, 1741).

149. "Review of the *vis viva* Debate between Dortous de Marain and Emilie du Châtelet," *Observations sur les écrits modernes* 23 (n.d.): 336; 24 (n.d.): 137–141.

150. Voltaire, "Doutes sur la mésure des forces vives et leur nature," in *Oeuvres complètes de Voltaire* (ed. Beuchot), 23: 165–172.

danger. "Force is not a being apart, an internal principle, a substance which animates and distinguishes bodies, as some philosophes have argued.... [It] is only a property, subject to variations, like all the other modes of matter." [151]

The significance of this alignment became most clear in 1743, when, after almost a half-century of recurring argument, the controversy finally began to resolve itself. Combatants had repeated for decades that the dispute centered on a matter of words. Concessions that valid arguments existed on both sides had also become commonplace among the combatants. Desfontaines further noted the wider public appreciation for the futility of the controversy when he wrote, with respect to du Châtelet and Dortous de Mairan, that "everything that can be said on this question has now been said. There only remains, therefore, authority, well or poorly evaluated, at the root of each position." [152] The petty nationalism and "spirit of party" that many saw at the root of the dispute was also expressed by the *Journal de Trévoux* in its review of du Châtelet's *Institutions de physique*. "The doctrine of *vis viva* has divided and continues to divide learned Europeans. Germany and almost all of the north has adopted it. England rejects it, and battles against it with all its might. Can this diversity be explained by saying that Leibniz was German and completely anti-Newtonian? France, less against M. Leibniz than for M. Descartes, also fights against it, ... but in the end it is for the public to decide." [153]

Whatever the public frustration, savants nevertheless continued to energize themselves over this question in intense, public dispute. Then, in July 1743, a text appeared that quieted the matter and began the process of its ultimate resolution. The book was a sophisticated mathematical treatise entitled *Traité de dynamique* authored by the very young and largely unknown *adjoint astronome* at the Paris Academy Jean le Rond d'Alembert.[154] D'Alembert had been a member of the Academy for less than two years when his treatise appeared, and the text marked his public intellectual debut. It also intervened directly in the mechanics debates of the period in ways that were significant for the wider development of Enlightenment Newtonianism in France.

D'Alembert's biography is admirably recounted elsewhere, but it is important nevertheless to offer some background here in order to understand the

151. Ibid., 23: 171.

152. "Review of the *vis viva* Debate," *OEM* 24 (n.d.): 140.

153. Review of du Châtelet, *Institutions de physique,* in *Journal de Trévoux* (1741): 926–927.

154. Jean le Rond d'Alembert, *Traité de dynamique, dans lequel les loix de l'équilibre et du mouvement des corps sont réduites au plus petit nombre possible, et démontrées d'une manière nouvelle* (Paris, 1743)

FIGURE 25. *Maurice Quentin de La Tour*, Portrait de Jean Le Rond d'Alembert. *Louvre, Paris, France. Courtesy of Art Resource, NY.*

nature of d'Alembert's influence.[155] Born the bastard and unacknowledged son of one of Paris's most glittering *salonnières*, Madame de Tencin, the young d'Alembert came of age with a peculiar relationship to French society. Connected on the one hand to its very pinnacles, he grew up in one respect as a well-connected child of privilege. Yet left as a newborn in 1717 on the doorsteps of the Parisian church of St. Jean Le Rond, from which he took his name, d'Alembert was also an orphan who had to forge a life on his own through the exertions of his own will. As a child, he was sent to the foundling home but then quickly adopted and supported materially by his father, Louis-Camus Destouches, and adoptive mother, Madame Rousseau. D'Alembert lived most of his life with his adoptive mother, only "weaning himself from her" (his term) in 1765 at the age of forty-eight. He also appears to have suffered from neither material nor emotional want during his youth. At the same time, he came of age without any acknowledged connections within Parisian society, and thus he was forced to make his way in society as something of a parvenu even if he had a host of clandestine supporters behind the scenes at every step.

In the early 1730s, he attended college in Paris at Quatres Nations, discovering his aptitude for mathematics through the study of Varignon's lectures in analytical mechanics delivered at the Collège royale.[156] During this period, he produced a commentary, never published, on the works of Guisnée and l'Hôpital, and he further initiated himself into the tradition of French analytical mathematics through a program of reading and autodidactic mathematical instruction. At the same time, he also pursued a different career path, studying law and winning appointment to the bar in 1738. His interest in mathematics nevertheless remained keen, and starting in late 1739 he began to submit papers to the Royal Academy addressing current problems in mathematical research.

His first, which exposed some errors in Reyneau's *Analyse démontrée*, another of d'Alembert's textbooks, attracted the interest and praise of Clairaut.[157] A second paper, which treated the motion of a body in resisting fluids, was also praised by Dortous de Mairan.[158] It is perhaps worth noting that this second

155. The most complete biography of d'Alembert is Ronald Grimsley, *Jean d'Alembert, 1717-83* (Oxford, 1963). Most helpful for me in understanding d'Alembert's scientific thought were Thomas L. Hankins, *Jean d'Alembert: Science and the Enlightenment* (Oxford, 1970); Véronique Le Ru, *Jean le Rond d'Alembert philosophe* (Paris, 1994); and Keith M. Baker, *Condorcet: From Natural Philosophy to Social Mathematics* (Chicago, 1974), esp. 85–194.

156. The best account of d'Alembert's scientific formation is found in Hankins, *D'Alembert*, 19–22.

157. PVARS, July 29, 1739.

158. Ibid., February 6, 1740.

paper was submitted in February 1740, during the period when Maupertuis was discussing similar issues with Bernoulli in preparation for the development of his least action principle. There is no evidence of any connection between the two men during these years and no reason to think that the still unknown d'Alembert would have met the senior academician before this date. Their mathematical work, however, was very closely allied, and a year later d'Alembert sent a third paper to the academy, this time on the integration of rational fractions.[159] Since this work was similarly praised by Clairaut and Dortous de Mairan, the young lawyer soon became a candidate for admission into the academy. He was elected to fill a vacant *adjoint astronome* seat the following May, and he responded to the invitation by leaving his law career forever behind.[160]

The *Traité de dynamique* was the first fruit of these labors, and it established d'Alembert immediately and ever after as a leading member of the European mathematical community. From the perspective of France in the 1740s, however, several aspects of the book were particularly influential. First was the title, which explicitly engaged with Leibnizian mechanics in important ways. Leibniz had invented the term "dynamique" in the 1690s as a way of distinguishing his deeply metaphysical science of ontologically conceived bodies and forces from the standard "kinematic" *mécanique* of the day.[161] The Malebranchian analytical mechanics that d'Alembert learned from Varignon, Reyneau, Guisnée, l'Hôpital, and others was kinematic in this way in that it conceived of terms such as *force, body,* and *impact* in purely mathematical terms while eschewing any metaphysical understanding of their ontology. Newton's mechanics could also be interpreted in kinematic terms, too, even if those who conceived of universal gravitation as an empirically confirmed fact of physical nature challenged this understanding. Leibniz's mechanics, however, was aggressively and unequivocally metaphysical and ontological, and accordingly he

159. Ibid., February 18, 1741.

160. Ibid., May 17, 1741.

161. The key text was "Specimen Dynamicum," *Acta eruditorum* (1695). An English translation is found in *G. W. Leibniz. Philosophical Essays,* ed. Roger Ariew and Daniel Garber (Indianapolis, 1989), 117–138. On Leibniz's dynamics see Daniel Garber, "Leibniz: Physics and Philosophy," in *The Cambridge Companion to Leibniz,* ed. Nicolas Jolley (Cambridge, 1995), 270–352, idem, "Leibniz and the Foundations of Physics: The Middle Years," in *The Natural Philosophy of Leibniz,* ed. Kathleen Okruhlik and James Robert Brown (Dordrecht, 1985), 27–130; Pierre Costabel, *Leibniz and Dynamics* (Ithaca, 1973); and Richard Westfall, *Force in Newton's Physics: The Science of Dynamics in the Seventeenth Century* (New York, 1971).

coined the new term "dynamics" (*dynamique* in his French writings) to signify the new "science of powers or motive causes" that was to be distinguished from simple mechanics, or the "science of effects rather than causes."[162]

These last definitions were formulated by d'Alembert in the introduction to his *Traité de dynamique*, and they marked the text as a direct intervention into what had become a very hotly contested scientific battleground. In this arena, the most arcane questions of mathematical physics were linked to wider concerns about philosophy, religion, and moral order. Only a year after the appearance of d'Alembert's book, Maupertuis would deliver his paper to the Royal Academy defending the value of metaphysics for the practice of physics and arguing for the return of teleological principles in the science of mechanics. His manuscript connecting such thinking to cosmological questions about the presence and force of God in nature was already in clandestine circulation, and a year later his scandalous application of these ideas to the science of life would appear. All of this conspired to draw attention to a work with "dynamique" in the title, even if d'Alembert responded to these controversies with a meticulously argued mathematical treatise that showed little sympathy for the nonmathematical student of these questions.

As narrowly technical as it was, the text was nevertheless influential, especially in the way that d'Alembert synthesized his training in the Malebranchian (and by the 1740s Cartesian) understanding of forces and mechanics together with his fluency in the post-Maupertuis discourse of Newtonian forces and mechanics. The result was an approach to dynamics that appeared to both supersede Leibnizian dynamics and undermine the very claim of this science altogether.[163] In short, what d'Alembert articulated was a new conception of dynamics that was not dynamic in the Leibnizian sense at all. D'Alembert's conception of force, body, and motion was thoroughly mathematical, and he thus rejected altogether any metaphysical or ontological project for developing a science of forces and their action. Instead, through a systematic display of rigorous mathematical deductions from first principles, deductions that he presented using the language of Cartesian *évidence*, d'Alembert demonstrated how the laws of motion can be derived from a set of axiomatic principles without any recourse to metaphysics whatsoever. In other words, he exorcized the motive forces and metaphysical causes from Leibniz's dynamics while laying claim to this label in the name of pure mathematical mechanics.[164]

162. D'Alembert, *Traité de dynamique*, xxiii.

163. See the review of d'Alembert in *Journal des savants* (1743): 522–528.

164. See especially Hankins, *d'Alembert*, 151–169; Le Ru, *Jean le Rond d'Alembert philosophe*, 19–63.

It was in this context that he also claimed to resolve the *vis viva* controversy by showing that the dispute was rooted in the muddled metaphysical language of force. D'Alembert showed analytically how the dispute was rooted in a mere dispute of words by returning the question to pure mathematics. Thomas Hankins notes that d'Alembert's solution was not really a solution at all since his mathematical system still left certain aspects of the problem unclarified. More important than this, however, especially in the context of 1740s France, was the perception of a solution that the treatise offered. D'Alembert showed how mv was the appropriate measure under one set of circumstances, and mv^2 in another, and even if he left the precise details of the solution less than transparent, the perception took hold that d'Alembert had eliminated the problem by providing a more general framework that subsumed each side as a special case. Thus when he called the *vis viva* dispute "too undignified to occupy a philosophe any longer," many followed his lead.[165]

French Leibnizianism also went into retreat with the dissolution of the *vis viva* controversy, but even more important was how the factors that produced this precise turn of events also encouraged a new integration of Cartesianism and Newtonianism in France, one with important long-term consequences. Voltaire had already positioned himself alongside the empiricist and philosophically modest Cartesians by siding with them against Leibniz's metaphysical monsters. D'Alembert gave new impetus to this association by offering a new science of dynamics that agreed with their position and that also opened the door to a French Newtonian conceptualization of it. Maupertuis had laid the foundation for this conjunction when he had shown in the 1732 edition of his *Discour sur les différentes figure des astres* that the forces that the French had placed at the center of their mathematical mechanics for over four decades could be reinterpreted in terms of the language of Newtonian universal gravitation.[166] In a very real sense, d'Alembert took this formulation and turned it into a public, philosophical identity. He declared himself a Newtonian from very early on, and yet he did so while nevertheless employing what was in essence a very Cartesian, and especially Malebranchian, conception of bodies and forces in his science of mechanics. The reconfigurations triggered by Leibnizianism and Voltaire's response to it further pushed d'Alembert in this direction, as did Maupertuis' turn toward Leibnizian metaphysics and metaphysical libertinism after 1740. All of these shifts allowed Newtonianism to acquire a new

165. Hankins, "Eighteenth-Century Attempts to Resolve the *Vis Viva* Controversy," 281–282. See also Le Ru, *Jean le Rond d'Alembert philosophe,* 77–92.

166. Hankins notes the similarities between d'Alembert's conception of forces and Maupertuis' formulation in the *Discours* in *D'Alembert,* 159–160.

legitimacy in France. It increasingly ceded to other philosophies its status as a veiled form of Spinozistic materialism while increasingly acquiring a reputation as the foundation for sound empirical, mathematical, and philosophically modest science.

Central to this new configuration was d'Alembert's conception of "Newtonian mechanics" since it fused the skeptical phenomenalism of the older Malebranchian mathematical tradition together with the new language of empiricism and philosophical modesty central to eighteenth-century Newtonian discourse. This new alignment was especially important in bringing the Paris Academy fully into the camp of the Newtonians, and no one was more influential in this respect than Alexis-Claude Clairaut. Throughout the late 1730s and into the 1740s, Clairaut served as Maupertuis' more purely mathematical sidekick, producing work that supported his Newtonian conception of the shape of the earth while also supporting his turn toward the Newtonian theory of light and refraction. In the early 1740s, however, as Maupertuis began to turn toward a more metaphysical approach to mechanics and physics while also cultivating a new reputation as libertine *esprit fort,* the work of the two men began to diverge. While Maupertuis was addressing the feminized public about comets and playing witty games of authorial subterfuge with the defenders of the elongated earth, Clairaut was deepening the mathematical approach to Newtonian celestial mechanics that had once been Maupertuis' obsession as well. The result was the publication in early 1743 of Clairaut's *Theory of the Shape of the Earth, Pulled from the Principles of Hydrostatics,* a dauntingly dense and sophisticated application of differential and integral analysis to the question of the shape of the earth.[167] Keith Baker rightly calls this work the single clearest marker of what the science of Newton's *Principia* had become during its fifty-year gestation inside the Paris Academy.[168] Clairaut's treatise was of course too technically sophisticated to gain wide readership, but it was reviewed favorably in the learned press, with reviewers always noting the Newtonian physics that anchored Clairaut's mathematical approach.[169] D'Alembert's mathematical

---

167. Alexis-Claude Clairaut, *Théorie de la figure de la terre, tirée des principes de l'hydrostatique* (Paris, 1943). An analysis true to the technical difficulty of Clairaut's original is found in Greenberg, *Problem of the Earth's Shape,* 426–619.

168. Baker, *Condorcet,* 7–8.

169. See review of Clairaut, *Théorie de la figure de la terre,* in *Journal des savants* (1743): 463–469. Desfontaines admitted that the work exceeded his technical abilities, and he lamented that such an important book would not be read. He did recommend the introduction, however, calling it "a work written with clarity and precision." See the review of this work in *Observations sur les écrits modernes* 33 (n.d.): 216.

work, which also remained exceedingly technical and sophisticated throughout these years, received this same treatment. Together they reinforced the perception that Newton's theory of universal gravitation and the physics of attraction that anchored it were now the accepted science of the French Royal Academy.

The mathematical mechanics of Newton's *Principia* was indeed accepted within French mathematical mechanics, but this acceptance had occurred decades earlier, in the 1690s, when Varignon and others read Newton's language of forces in terms of Malebranchian mathematical phenomenalism and created French analytical mechanics.[170] What remained controversial was the claim that the forces of universal gravitation were physical entities constituted within matter itself. Yet here the brief explosion of discourse about Leibnizian metaphysics and dynamics provided the wedge necessary to alleviate this controversy in ways favorable to the Newtonian cause. D'Alembert's Newtonian mechanics treated forces and bodies as mathematical constructs susceptible to differential and integral analysis while questions of metaphysics and ontological dynamics were pushed entirely to the side. When called upon to defend the forces of gravity that he placed at the center of his science, he further appealed to experience in the manner of the Newtonians, arguing that gravity is a fact of nature proven by experience even if its cause remains unknown.

This had been Voltaire's position all along, even if he sometimes defended a more deeply physical conception of gravitational force since, he believed, the physical existence of these forces followed naturally from the stunning regularity of the empirical effects. Maupertuis, too, had defended this empirical conception of universal gravitation, even if he was increasingly oriented in other directions. It had also been the basis of Buffon's Newtonian pronouncements from as early as 1735. The more recent declarations of Le Monnier, La Condamine, and Bouguer, whose pendulum experiments demonstrating the effects of the gravitational pull of mountains were reported at the academy's spring public assembly of 1739, further confirmed it. Even more influential in supporting this view was Clairaut, who used dauntingly complex mathematics to predict, and thus confirm, the empirical effects that this force of gravity was believed to produce.

The overall result was the increasingly widespread acceptance in France after 1745 of the idea that universal gravitation was a fact of nature proven by experience even if its metaphysical nature and causal agency remained unknown. D'Alembert's French Newtonian stance regarding physics, with its ardent Cartesianism regarding the role of analytical mathematics and deductive

170. See Shank, "There Was No Such Thing."

mathematical demonstration in the development of mechanics, reinforced this acceptance. It especially allowed for the gradual conversion of the empirical Cartesians in and around the Paris Academy to the new Newtonian language of empiricist and antimetaphysical physics. Also required, however, was the departure of the methodological Cartesians who insisted upon rationalist standards for rigorous science. Savants such as Privat de Molières and Gamaches continued to challenge the philosophical absurdity of Newtonian attraction through appeals to synthetic demonstration and mechanical causality, and while d'Alembert called this group "a sect that is very much weakened" in 1743, their weakness was not apparent to everyone.

The year 1740 saw the appearance of Gamaches' thoroughly Cartesian *Physical Astronomy,* a work that was publicly celebrated by commentators as varied as Desfontaines, the Jesuit Castel, and the *Journal des savants.* Privat de Molières was also continuing to bask in the glow of his own celebrity with no indication of an end in sight. Prémontval's public lecture courses based on Privat de Molières' *Leçons de physique* began in early 1741, and inside the academy the aged yet increasingly outspoken Cartesian Fontenelle ended his forty-three-year tenure as perpetual secretary by ceding his office to an equally outspoken Cartesian, Dortous de Mairan. He did so, moreover, with the overwhelming support of the company. Dortous de Mairan only served thirty-three months as academy secretary, and he only prepared eulogies, therefore, for five public assemblies. In one, however, he attracted great public attention by celebrating Cardinal Polignac, an honorary academician, for his unpublished Latin poem *Anti-Lucrece*.[171]

The poem connected Epicurean materialism with active matter atomism in ways that associated Newtonian physics with these positions. Commentators like Desfontaines seized on the eulogy as an opportunity to celebrate both the piety of Polignac and the responsible Cartesianism of Dortous de Mairan while echoing the discourse that linked Newtonianism with materialism and irreligion. The acclaim triggered by the eulogy led to the posthumous publication of the poem in 1747, and this allowed periodicals like the *Journal des savants,* which offered several long reviews of the Latin text, to continue echoing these formulations throughout the remainder of the decade.[172] The discourse

171. The eulogy, which was considered a masterpiece by some, was published separately by Dortous de Mairan in 1742. It received reviews in *Journal des savants* (1742): 471–481; and *Observations sur les écrits modernes* 29 (n.d.): 145–155.

172. Review of Polignac, *Anti-Lucrece,* in *Journal des savants* (1747): 600–604, 669–676.

of Cartesian anti-Newtonianism was, in fact, never completely exterminated in eighteenth-century France, yet its authority began to wane as the 1740s progressed while other discourses ascended in its place.

One of the most vigorous challenges to this discourse after 1740 originated from one of the least likely of venues: the University of Paris. As Laurence Brockliss has shown, the Sorbonne remained throughout the seventeenth and eighteenth centuries a full generation behind the philosophical currents of the times. The university only embraced the new mechanics and physics of Descartes at the very end of the seventeenth century, and they likewise adopted Newton's revisions of Descartes very late as well.[173] Nevertheless, in the 1740s, when the debate over Newtonian attraction still remained hot, the cause of Cartesianism was set back by the public intervention of an otherwise obscure priest and university professor named Pierre Sigorgne. In March 1741, Sigorgne published a long pamphlet, *Examination and Refutation of the* Leçons de physique of *M. de Molières*, which systematically exposed all the technical and physical problems inherent in Privat de Molières' vortical account of celestial and terrestrial mechanics.[174] The text offered little that was original, simply repeating the many contradictions and inconsistencies that anti-Cartesians had long shown to be inherent to the vortical system. Appearing as it did, however, amidst the public disputes over these topics, and coming from an authority very different than those who were typically contesting these matters, the pamphlet was an important public intervention.

Despite his outspoken public support for Privat de Molières elsewhere in his journal, Desfontaines was quick to pick up on the criticism and to give Sigorgne a platform for airing his views. He called the "young ecclesiastic" a "writer as polite as his physics is profound," and he assured his readers that they could count on the ability and judiciousness of the author and the solidity of his work.[175] Privat de Molières published a response to Sigorgne's first pamphlet, and his adversary responded with a second reasserting his original critiques.[176] Reviewing the entire debate after all three pamphlets had appeared, Desfontaines had a hard time sustaining his earlier discourse in praise of Privat

173. L. W. B. Brockliss, *French Higher Education in the Seventeenth and Eighteenth Centuries: A Cultural History* (Oxford, 1987).

174. L'abbé Pierre Sigorgne, *Examen et réfutation des leçons de physique expliquées par M. de Moliéres au collège royal de France* (Paris, 1741).

175. Review of Pierre Sigorgne, *Examen et réfutation des leçons de physique*, in *Observations sur les écrits modernes* 24 (n.d.): 261–262; 25 (n.d.): 107–110.

176. Joseph Privat de Molières, *Réponse aux principaux objections contenues dean l'examen des leçons de physique de M. l'abbé de Molières* (Paris, 1741); Pierre Sigorgne,

de Molières. Noting that the replies had grown to enormous lengths, far exceeding the length of the original *Examen* that started the controversy, Desfontaines complained that such verbiage was proof that the debate itself had gone off track. "Far better to keep the criticisms to five or six sheets published Chez Clouzier," the journalist complained.[177] He further faulted Privat de Molières more than Sigorgne for the excess, implying that there was a connection between the academician's verbiage and his vortical system. The genius of the Cartesian vortices was its simplicity and clarity, or so its defenders had been saying for over half a century. Desfontaines shared this aesthetic love of the theory, and he thus joined with Sigorgne in lamenting how "the vortices lose their beauty as soon as one multiplies them too much like one is doing today."[178]

To account for all the empirical phenomena and maintain the compatibility between vortical mechanics and the quantitative laws of Newton's *Principia*, Privat de Molières deployed an elaborate system of small vortices nested within larger vortices in the manner of Malebranche. What success he had achieved came from this complexity, but it struck Sigorgne that this complexity was evidence that Privat de Molières had distorted the real truth inherent in Descartes' system. Gamaches had also followed Malebranche in this respect, going so far as to posit a vortex of infinite complexity as the real causal source of physical action in nature. More interested in the philosophical clarity of the whole system than in the ability of the vortices to offer actual empirical and mechanical accounts of particular phenomena, Gamaches' system was praised by Sigorgne as more coherent than Privat de Molières'. However, Sigorgne also implied that neither succeeded in actually accounting for the phenomena. Moreover, these systems failed, he added, while also turning the sublime clarity of Descartes' thought into a horribly complex and unwieldy system. This was an abomination to the true legacy of Descartes, or so Sigorgne argued, and in this way he worked from Descartes back against the eighteenth-century Cartesians to disassociate the system of the vortices from the real essence of Cartesianism.[179]

The real Cartesianism, Sigorgne argued, was rooted not in the failed systems of Privat de Molières and Gamaches but in the pellucid mathematical demonstration of simple and elegant truths of nature. Desfontaines largely followed Sigorgne in this critique, and in doing so he illustrated how diehard

---

Replique à M. de Molières ou démonstration physico-mathématique de l'impossibilité et de l'insuffisance des tourbillons (Paris, 1741).

177. Review of the Sigorgne-Privat de Molières debate in *Observations sur les écrits modernes* 25 (n.d.): 108.

178. Ibid., 110.

179. Sigorgne, *Examen et réfutation des leçons de physique*.

defenders of Cartesian mechanism could still be led to see the absurdity of the vortical system. Desfontaines' fairly late discussion of Gamaches' *Astronomie physique*, which actually came after his discussion of the Sigorgne–Privat de Molières dispute in his journal, also illustrated the same tendency. His review celebrated the conceptual clarity of Gamaches' work, noting in particular how his union of geometry with calculation had led him to see the invariable laws that governed all movement. This allowed Desfontaines to connect Gamaches' method with that of Newton's *Principia*, even if he also distanced each from those who founded systems based upon "the void of Epicurus" and the "occult qualities of Aristotle." This distinction between the sublime mathematical riches of Newton's *Principia* and the errors and absurdities of philosophical Newtonianism was never systematically developed or sustained, yet the distinction itself was crucial, for it was one of many that opened the door to the reconnection of methodological Cartesianism with Newtonian physics in ways that made Newton more palatable to French savants.[180]

Desfontaines was also led in this direction by a growing awareness, common to many in France after 1740, that the Cartesian vortical system defended by Gamaches, Privat de Molières, and others was indeed a Baroque monster crumbling at its foundations. Sigorgne's intervention had exposed these inadequacies for all to see, and when he then went on to declare himself a Newtonian, publishing a text called *Newtonian Foundations, or Introduction to the Philosophy of Newton* that he intended as a university textbook, he made manifest the proper conclusion to be drawn from this critique.[181]

The final years of Privat de Molières' life also confirmed this conclusion in ways that illustrated well the changing climate of the times. After his widely noticed public dispute with Sigorgne, Privat de Molières further confronted a challenge inside the academy. In the months preceding the appearance of Sigorgne's pamphlet, and motivated by reasons that are hard to discern, Privat de Molières offered a set of experimental demonstrations to the academy that allegedly documented the vortical account of *pesanteur*.[182] His demonstration was largely a reprise of the swirling water demonstrations that Huygens had pioneered in the 1660s and that Saulmon had revived in the 1710s. By 1741, however, they had a new context. In the intervening years, Anglo-Dutch experimental philosophy had infiltrated the French public sphere in the figure

180. Review of Gamaches, *Astronomie physique*, in *Observations sur les écrits modernes* 26 (n.d.): 49–67.

181. Pierre Sigorgne, *Institutions Newtoniennes, or Introduction s la philosophie de M. Newton* (Paris, 1747). This text was reviewed in *Journal des savants* (1748): 3–9.

182. PVARS, January 11 and 23, 1741.

of abbé Jean-Antoine Nollet, who was already famous in France for his public experimental demonstrations that he had begun offering in 1736.[183] Nollet was also a member of the Royal Academy by this time, and perhaps Privat de Molières was seeking to respond to experimental criticisms made of his work by Nollet in the academy or in one of his public courses. Whatever their inspiration, their outcome was anything but successful for him.

A full year passed between Privat de Molières' first demonstration and Nollet's response—the period in which he was intensely occupied with Sigorgne's challenge. But when Nollet's response came it was decisive. In March 1742 Nollet asked that academic commissioners be appointed to verify a set of experimental refutations he proposed against Privat de Molières' work.[184] Réaumur, Clairaut, d'Alembert, and Montigny (another young analytical mathematician) were chosen, a clear sign of the personnel shifts that were transforming the institution. A week later the committee received Nollet's report.[185] Three days later, Privat de Molières began reading his response, which was interrupted by the Easter vacation but continued again in April.[186] Not satisfied with it Nollet performed the experiments himself before the academy on May 5.[187] Privat de Molières prepared a second response, but at this time he fell deathly ill. In the end, all he could produce was a document, which Dortous de Mairan read to the company on May 9, that the membership deemed unworthy of transcription.[188]

Within a week, Privat de Molières was dead, and in the ultimate irony, his *associé mechanicien* seat, which he so hoped to vacate for a place among the *pensionnaires,* was awarded to Nollet.[189] The controversy was not over, however, for members of Privat de Molières' family petitioned the academy asking that his final papers on the experimental proof of the vortical system of *pesanteur* be published as an honor to his legacy.[190] This put the academy into an awkward position since, as the account in the academy's registers read, "the

183. On Nollet, see Jean Torlais, *L'abbé Nollet (1700–1770) et la physique experimentale au XVIIIe siècle* (Paris, 1954); and esp. Lynn, *Popular Science and Public Opinion.* Paola Bertucci has just published *Viaggio nel paese delle meraviglie: Scienza e curiosità nell'Italia del settecento* (Turin, 2007), a history of Nollet's travels in Italy, and I am grateful to her for her willingness to share her expertise on Nollet with me.

184. PVARS, March 7, 1742.

185. Ibid., March 14, 1742.

186. Ibid., March 17, April 11, 1742.

187. Ibid., May 5, 1742.

188. Ibid., May 9, 1742.

189. Ibid., 2, 27 June, 1742.

190. Ibid., December 7, 1742.

committee had not recommended publishing the work, and no member of the academy is allowed to put the name of the academy on a publication without this approval." It is likely that few in the company approved of the work, yet the membership was sympathetic to the family's appeal to the honor of their deceased colleague. In December, Clairaut read the resolution, announcing that Privat de Molières' work would be published.[191] In this way, Privat de Molières' final pronouncements on behalf of Cartesian vortical physics appeared with the imprimatur of the Royal Academy of Sciences even as a consensus within the company was beginning to form in opposition to this very physics.

The personnel shifts alluded to above also encouraged these trends in important ways. A changing of the academic guard in fact occurred during this period, helping to shift the balance of cultural authority in favor of the Newtonians. It began in earnest in December 1731 with the appointment of Maupertuis to Saurin's *pensionnaire* seat. After this date, the transformation only accelerated. Fontenelle retired from the Academy in December 1740, and his successor as perpetual secretary, Dortous de Mairan, followed him into semi-retirement less than three years later. Grandjean de Fouchy, a thoroughly competent yet prosaic astronomer, was elected to succeed the two titans of French Cartesianism in August 1743, and he came to personify the increasing professionalism of the company.

The three decades that separated Grandjean's appointment from that of his successor, the Marquis de Condorcet, witnessed a shift of power inside the institution toward those who called themselves Newtonians. The details of this shift can be briefly noted. Saurin died in 1737, and he was followed to the grave by Privat de Molières in 1742 and Johann I Bernoulli in 1748. Bernoulli had already begun to disappear intellectually by 1740, however, and in this respect his final years were similar to those of the abbé Bignon, who retired in 1736 and then lived quietly until his death in 1743. Nicole, meanwhile, having survived his years in the mathematical desert, acquired newfound academic authority in precisely these years. He served as one of the academy's directors in 1740 and 1741.[192] When he died in 1758, a year after Réaumur and a year before Maupertuis, it was the other Newtonians—Clairaut, La Condamine, Bouguer, Buffon, Nollet, d'Alembert, and Le Monnier—who dominated the royal institution of science in France.

191. Ibid. The work appeared in 1743 as Jean-Baptiste-Jacques Corgne de Launay and Joseph Privat de Molières, *Principes du système des petits tourbillons, mis à la parté de tout le mond et appliqués aux phénomènes les plus generaux* (Paris, 1743).

192. A useful chart listing the officers of the Royal Academy before 1750 is found in Sturdy, *Science and Social Status,* 421–423.

Not every one of these academicians was a polemical Newtonian in the manner of Maupertuis or Voltaire, yet each sided in one way or another with Newtonian science as it had come to be established in France after 1732. Meanwhile, self-proclaimed Cartesians became hard to find inside the institution after 1750. Indeed, as d'Alembert declared in the second edition of his *Traité de dynamique* published in 1758, "the Cartesian sect barely exists anymore."[193] A changing of the journalistic guard also occurred in these years, and this initiated a parallel transformation in the wider public sphere that reinforced the one at the Royal Academy. Bignon's retirement in 1736 brought an end to his thirty-five-year tenure as editor of the *Journal des savants,* yet in this case the change brought few noticeable changes to the pattern of judicious, neutral commentary that the journal had acquired under his stewardship. Similarly, the *Bibliothéque raisonnée* continued the venerable legacy of Bayle and Le Clerc until 1756, offering a platform for the views of 'sGravesande, Desaguliers, and Musschenbroek until their deaths in 1742, 1744, and 1761, respectively. The journal did the same for the next generation of Anglo-Dutch Newtonians while continuing to merge their work with the discourses of constructive philosophical skepticism and critical natural religion that had long been the hallmark of Protestant journalism in the Low Countries.

Other journalistic shifts, however, brought about more noticeable changes, and these reinforced the new cultural authority of the Newtonians in France. In December 1745, Desfontaines died, leaving his latest journalistic initiative, *Jugements sur quelques ouvrages nouveaux,* without an heir. Prévost remained active into the 1760s, and he was soon joined by other critical French journalists who practiced the same brand of weekly criticism and cultural commentary that these men had pioneered in the 1730s. Most influential was Guillaume Thomas Raynal, who began to write a weekly newssheet entitled *Nouvelles littéraires* in 1747 before becoming the principle editor of the *Mercure de France* in 1750. Raynal was a critical journalist in the manner of Desfontaines, yet unlike his predecessor he was a friend of Voltaire and a frequent guest at the salon of Madame Geoffrin where he, d'Alembert, and the other emerging philosophes congregated.[194] Desfontaines' invectives against Voltaire were ably continued by Elie Fréron, who began his anti-philosophe periodical *Année littéraire* in 1754.[195] But whereas Voltaire had struggled to establish himself as a credible, authoritative philosophe in the 1730s by battling against critical journalists like Prévost and Desfontaines, his cultural authority and that of his allies was sup-

193. D'Alembert, *Traité de dynamique,* 2nd ed. (Paris, 1758), vi.
194. On Raynal, see Sgard, *Dictionnaire des journalistes,* 2: 821–824.
195. See Jean Balcou, *Fréron contre les philosophes* (Geneva, 1975).

ported after 1750 through the writings their own, sympathetic critical journalists such as Raynal.

Equally crucial to this shift was the change that occurred in 1745 at the
Jesuit-edited *Journal de Trévoux*. Throughout the 1730s and early 1740s, Castel remained vigorous in his campaign to use the Jesuit journal as a bullhorn
for his own critical, philosophical discourse. In 1740, he initiated yet another
buzz in the French public sphere with the publication of his *The Optics of Colors, Founded on Simple Observations,* an anti-Newtonian work that anticipated
many of Goethe's more famous critiques of Newtonian light and color theory.[196]
The work was reviewed in several major journals, including the *Journal des
savants,* and it also triggered critical pamphlets that Desfontaines reviewed in
his journal and Castel responded to through the *Journal de Trévoux.*[197] In 1743,
he followed this work with a direct intervention into the Newton wars, publishing *The true system of physics of M. Isaac Newton, exposed and analyzed in tandem with that of M. Descartes.*[198] This work was widely reviewed as well, and
it offered Castel's idiosyncratic and Jesuit perspective on the public debates
then raging between the Newtonians and Cartesians in France.[199] Characteristically, Castel did not take sides, preferring to frame the "true Newton" as one
who conformed with his position outside and above the fray. He also described
him as somebody who ultimately agreed with Castel's own complex and hybrid
philosophical positions. Castel's tone throughout was vigorous, learned, witty,
and critical as it had been for over two decades, and in this respect the work
merged seamlessly with the many book reviews he published after 1732 during
the intense debates over Newtonianism.

In 1744, however, for reasons that remain unclear, he ran afoul of the Jesuit
authorities in charge of the journal, and especially with its rising young intellectual star, Father Guillaume François Berthier.[200] In an ascent reminiscent of

196. L.-B Castel, S.J., *L'optique des coleurs, fondée sur les simples observations* (Paris,
1740). For an analysis, see Michel Blay, "Castel critique de la théorie newtonienne des
couleurs," in *Autour du Pere Castel et du claevecin oculaire,* in *Études sur le XVIIIe siècle,*
ed. Roland Mortier and Hervé Hasquin (Brussels, 1995), 43–58.

197. Review of Castel, *L'optique des coleurs,* in *Journal des savants* (1740): 195–201; *OEM*
24 (n.d.): 10–22; *Journal de Trévoux* (1740): 1235–1263.

198. Louis-Bertrand Castel, S.J., *Le vrai système de physique génerale de M. Issac Newton, exposé et analysé en parallele avec celui de M. Descartes* (Paris, 1743).

199. Review of Castel, *Le vrai système,* in *Journal des savants* (1743): 643–654; *Mercure de France* (1743): 2444–2451; *Journal historique de Verdun* (September 1743): 163–167;
Journal de Trévoux (1743): 2548–2630, 2974–3020, (1744): 232–272.

200. For this history, see Schier, *Castel,* 44–45; and John N. Pappas, *Berthier's Journal
de Trévoux and the Philosophes* (Geneva, 1957), 18–35.

Castel's own rise two decades earlier, Berthier had come to Paris in 1742 as a talented young scholar and teacher who had previously distinguished himself as a philosophy professor at the Jesuit colleges of Rennes and Rouen. Charged with finishing *The History of the Galician Church* begun by the recently deceased Father Brumoy, Berthier demonstrated such assiduity and erudition in the completion of his task that he earned the notice of his Jesuit superiors in Paris.[201] According to John Pappas, they had come to see the *Journal de Trévoux* as an institution in decline. Its editors had "all the talents [and] all the qualities one might wish for in a scholar," a document from the period declared, but they were not "sufficiently industrious." These Jesuit authorities also expressed frustration with the excessively polemical and contentious nature of the journal, and seeking reform they appointed the forty-year-old Berthier editor-in-chief of the journal in January 1745.[202]

Castel appears to have found his new boss difficult and the new imperatives toward journalistic propriety contrary to his intellectual temperament. Accordingly, after quarreling with Berthier sometime in 1746, he either resigned from the journal or was forced into retirement. Ironically, only five years later Berthier would lead the *Journal de Trévoux* into arguably its most fierce polemical campaign when he made the periodical the organ of his vociferous opposition to Diderot and d'Alembert's *Encyclopédie* and the emerging French *parti philosophique* that supported it. Amidst these new battles, Castel remained a teacher and public intellectual at large, and given his reputation he was often dragged into the middle of Jesuit/philosophe wars as an arbiter. Voltaire in particular appealed to him on more than one occasion during his struggles with Berthier. Castel even contemplated (although never executed) a public attack on his Jesuit rival, and he corresponded with Diderot, writing a praiseworthy review of his scandalous *Lettre sur les aveugles* for the *Journal de Trévoux* that Berthier refused to publish.[203] He was also a frequent and sympathetic correspondent with Montesquieu, and in 1756 he joined with many of the philosophes, including Voltaire, in publicly criticizing Rousseau's *Discourse on the Origins of Inequality*.[204] Castel, however, also remained a Jesuit, offering the authority of the Old Testament and the Catholic Church Fathers as the antidote to Rousseau's deistic ideas about natural law and religion. Writing of Voltaire,

201. This history was eventually published as Jacques Longueval, Pierre Claude Fontenay, Pierre Brumoy, and Guillaume François Berthier, *Histoire de l'eglise gallicane* (Paris, 1825–1828).

202. Pappas, *Berthier and the Philosophes*, 17–18.

203. Schier, *Castel*, 48.

204. Ibid., 52–56.

he reiterated what had become a characteristic position when he noted that he "knows religion since he was raised right here [at Louis-le-Grand], but he is nevertheless furious against it. . . . He goes for systems, English systems, and they make him furious against our monarchy."[205]

Berthier, however, was far more inflexible in his opposition to Voltaire and the emerging philosophe movement, and in this respect the displacement of Castel at the *Journal de Trévoux* clarified things in ways that also solidified the cause of the Newtonians. This precise cultural battlefield, however, did not emerge all at once, and even after the Newton wars had apparently placed the new Newtonian philosophes led by Voltaire and his allies inside the Royal Academy into confrontation with the defenders of the old order, echoes from the past continued to be heard. In 1752, for example, after celebrating his ninety-fifth birthday, Fontenelle published his last piece of public scientific exposition, a defense of Cartesian physics called *The Theory of the Cartesian Vortices*.[206] It was neither a long nor a systematic work, and it responded to no noticeable public debate at the time. However, appearing as it did as the apparent coda to Fontenelle's then two-decade-old project to defend Cartesianism against the onslaught of Newtonianism, it soon came to define his legacy in French intellectual history. Despite the complexities of his philosophical labors throughout his first ninety years, and the debt, often acknowledged, that Voltaire and the other philosophes owed to his work and legacy, the publication of this final text placed Fontenelle in the unfortunate position of dying as an enemy of the moderns and as France's last unreconstructed Cartesian.[207]

This legacy still haunts Fontenelle's reputation among historians, and it also informs the reputation of Dortous de Mairan, who outlived Fontenelle by almost fifteen years but still saw his legacy colored by his vigorous commitments to Cartesianism. His last great intervention into a public scientific debate occurred in 1750, when he used the occasion of the Academy's reissue of his *Treatise on Ice*, a text that had first appeared in 1716, to launch a passionate defense of the text's method of systematic, hypothetico-deductive natural philosophy. By 1750, mechanical, hypothetico-deductive accounts of natural phenomena had become synonymous with the hoary "systems" that Newtonian empiricism and math-

205. Cited in ibid., 56.

206. Fontenelle, *La théorie des Tourbillons Cartesiennes* (Paris, 1752). The text is found in *Oeuvres de Fontenelle*, 7: 365–463.

207. For a further discussion of the long history of Fontenelle's relationship to Cartesianism, see J. B. Shank, "On the Alleged Cartesianism of Fontenelle," *Archives Internationales d'Histoire des Sciences* 53, nos. 150–151 (2003): 139–156.

ematical physics had made obsolete.[208] Thus, while periodicals like the *Journal des savants* reviewed Dortous de Mairan's work and honored the *honnêteté* of his career and convictions, they did so while ignoring the more common public perception that Dortous de Mairan was an old, genial, and thoroughly honorable man defending a dead natural philosophy.[209] Dortous de Mairan died a decade later having done nothing in the interim to change this perception.

Meanwhile, a younger cohort inside the academy rose to prominence through the aggressive, public practice of their newly defined and accepted Newtonian science. Especially successful in this respect were Clairaut and d'Alembert, who quickly established their reputations as leading lights in Paris and in the wider European community of mathematical physicists through their practice of analytical mechanics. The two men quickly became rivals as well, and while their work was far more similar than not, there were real intellectual differences that divided them. Most significantly, Clairaut was much more empirical and pragmatic in his approach to mathematical physics, working hard to fit equations together with empirical results while evincing little interest in d'Alembert's passion for Cartesian philosophical certainty and clarity. He was also more comfortable than d'Alembert with physicalist conceptions of force since, for him, mechanical forces were merely facts of nature confirmed by the agreement between mathematical prediction and empirical phenomena. In this way, they came to attach different meanings to the banner "Newton" that they each flew with honor.[210] Their different orientations also spoke to wider fissures that existed within the otherwise unified camp of the Newtonians in France, fissures that remained even as Newtonianism became a unifying label.

What unified Newtonians in France, in fact, was less agreement about the precise philosophy that should bear this name than solidarity regarding the rejection of those arguments that self-consciously opposed it. There was, of course, general agreement about the validity of the core principles of Newtonianism—the predictive truth of Newtonian mathematical physics, the validity of the law of universal gravitation, the necessity of mathematical empiricism and experimental demonstration in science, among other things. More powerful in unifying the movement, however, was its shared opposition to Cartesianism, Leibnizianism, and the other philosophical identities that had emerged in opposition to Newtonianism. Unity was further encouraged by the way that core elements

---

208. Jean-Jacques Dortous de Mairan, *Dissertation sur la glace*, new ed. (Paris, 1750).

209. Review of Dortous de Mairan, *Dissertation sur la glace*, in *Journal des savants* (1750): 298–303.

210. See Hankins, *d'Alembert*, 32–36.

of these rival positions—philosophical modesty and mathematicism with respect Cartesianism, and the language of forces with respect to Leibnizianism—could be appropriated to frame what were in fact original philosophical hybrids (d'Alembert's Cartesian Newtonianism, Maupertuis' Newtonian Leibnizianism, etc.). Framed this way, these positions did not become splintering heresies but orthodox variations on the core Newtonian theme. Through these hybrid stances and a strong consensus about what good philosophy must avoid, an image of philosophical coherence was forged among French Newtonians that masked a much more complex intellectual and historical reality.

Moreover, if the centrifugal tendencies within French Newtonianism were contained through the construction of a consensus about what Newtonianism was not (i.e., a philosophical system, a metaphysical account of forces, a reckless and libertine materialism, etc.), they were also made centripetal by the institutional realignments that moved Newtonianism so united to the center of the cultural field of French public science. Sigorgne's public defense of Newtonianism from his ecclesiastically sanctioned chair at the University of Paris was a telling indicator of the shifts that had occurred, for it showed that, contrary to the worries of Cardinal Polignac, Newtonian philosophy was not necessarily the enemy of orthodox religion and morality. Sigorgne's teaching at the Sorbonne was reinforced in 1748 through the publication of his Latin *Brief Account of Physical Astronomy according to Newtonian Principles,* a work that introduced Newtonian philosophy through the scholastic method still used at the University of Paris.[211] This text, as well as the discussion it received in the learned press, further demonstrated that the truths of Newtonian science could and should be taught at the leading centers of French higher education.[212] This conviction was further confirmed in 1753 when King Louis XV named the abbé Nollet to the first ever royally supported chair in experimental physics at the Collège de Navarre of the University of Paris. Philosophical radicalism and libertinism, of course, remained ever present dangers, as Berthier would begin arguing with new vehemence in his attacks against the *Encyclopédie.* However, by the date of this text's appearance in 1751, these dangers had been disassociated with the precise physics and philosophy of Newton in such a way as to allow for the apotheosis of the latter and the relocation of the former into other corners of the cultural field.

A number of other, and often ironic, developments also worked to reinforce this realignment. One interesting change involved the revival of quasi-vortical accounts of gravity and magnetism through the new Newtonian conception of

211. Pierre Sigorgne, *Astronomiae physicae juxta Newtonis principia breviarum* (Paris, 1748).

212. See the review of Sigorgne's work in *Journal des savants* (1750): 47–52.

the mechanical ether. Philosophically speaking, Newtonianism and Cartesianism had come to a truce in the 1740s by agreeing that the facts of nature demonstrated the truth of Newton's theory of universal gravitation even if the actual cause of attraction remained to be discovered. This consensus was also built upon a rejection of physical claims about the power of active forces in matter to effect bodily action at a distance through empty space. Such propositions were deemed philosophically absurd at best and dangerously materialist at worst, and Newtonian attraction was thus accepted as an empirical fact so long as it was not claimed as a physical, active principle. Given this understanding, the challenge was then issued to physicists to find the mechanical cause of gravity that everyone knew must exist.

Newton had even speculated about such a cause, first at the end of his "General Scholium" published in the 1713 edition of the *Principia,* and then in the queries that he appended to the 1717 edition of the *Opticks.*[213] In these texts, Newton explained, albeit speculatively, how the action of an all-pervasive but infinitely subtle fluid, or ether, may be the mechanical cause of physical action in the world. This ether obviously had much in common with Descartes' subtle fluid, yet Newton never posited vortical action as the central feature of this fluid, nor did he systematically account for how this ether might cause physical effects. These writings nevertheless suggested that his ultimate conception of gravity rested with some sort of direct, point-contact, materio-mechanical agent as the cause behind the apparent phenomenon of action at a distance. Thus, as the discourse of Cartesian vortical mechanics waned in credibility and that of Newtonian gravitational theory grew, many mechanically minded philosophers found themselves grabbing hold of the Newtonian ether as the substitute they needed for causal mechanistic physics.

This transformation was encouraged in 1745 when the *Bibliothèque raisonnée* published a letter from Newton to Robert Boyle written at the end of the seventeenth century in which he categorically denied that bodies attract one another across empty space. The letter also suggested that the real cause behind this phenomena was an all-pervasive mechanical ether.[214] At one level this letter merely reinforced the ongoing project of distancing Newtonianism from any association with active matter materialism. Yet it did provoke another question: what exactly was the nature of this ether that Newton often used, and could Newton's ether be developed as an explanatory framework for accounting for the facts we see?

213. *Principia*, 943–944; "Query 18-24," *Opticks*, 348–354.
214. Newton to Robert Boyle, letter published in *Bibliothèque raisonnée* 35 (1745): 154–173.

Eighteenth-century electrical research, which exploded after 1720, often did just that. Savants as widely dispersed as John Gray in London, the abbé Nollet in Paris, and Benjamin Franklin in Philadelphia all dressed their work in the clothing of Newtonian experimental physics, yet they all employed some variety of subtle effluvia when explaining the mechanical cause of the apparent attraction of magnetically charged particles across seemingly empty space.[215] In 1748, this reconfiguration of Newtonianism reached one plateau with the publication of Colin Maclaurin's *An Account of Sir Isaac Newton's Philosophical Discoveries*.[216] This would-be successor to Pemberton and Voltaire was a man with impeccable Newtonian credentials, and thus when he argued, as he did in his text, that Newton was at root an ether mechanist when it came to the theory of universal gravitation, Maclaurin's judgment carried great authority. The *Journal des savants* ran a long and celebratory review of the French translation of Maclaurin's work in early 1750 emphasizing the articulation of Newtonian ether mechanics that it offered. They also noted the book's dedication to the Cartesian vorticist Dortous de Mairan.[217] Such pronouncements did nothing to derail the nonmechanical use of Newton's theory of universal gravitation in mathematical physics, nor did they really generate a revival of attempts to physically account for natural phenomena through recourse to ether mechanisms, even if some did attempt to realize this project.[218] Instead, the revelation that Newton was allegedly an ether mechanist at heart worked to further reassure those who needed reassurance that Newtonianism was anything but a veiled form of pantheist materialism.

This reassurance was still necessary for many in the 1740s, a fact that helps to explain why Maupertuis' permanent move to Berlin in early 1745 further solidified the cultural authority of the Newtonians in France. By this date, Mauper-

215. See John L. Heilbron, *Electricity in the Seventeenth and Eighteenth Centuries: A Study of Early Modern Physics* (Berkeley, 1979). Patricia Fara also notes the pervasiveness of ether mechanism in eighteenth-century accounts of magnetism. See *Sympathetic Attractions: Magnetic Practices, Beliefs, and Symbolism in Eighteenth-Century England* (Princeton, 1997).

216. Colin Maclaurin, *An Account of Sir Isaac Newton's Philosophical Discoveries* (London, 1748); *Exposition des découvertes philosophiques de M. le Chevalier. Ouvrage traduit de l'Anglois par M. Lavirotte* (Paris, 1749).

217. Review of Colin Maclaurin, *Exposition des découvertes philosophiques de M. le Chevalier*, in *Journal des savants* (1750): 3–13.

218. Perhaps the most prolific and famous of the eighteenth-century ether mechanists was the Genevan George-Louis Le Sage. For a recent assessment of his mechanical theories of gravity, see the essays collected in Matthew Edwards, ed., *Pushing Gravity: New Perspectives on Le Sage's Theory of Gravitation* (Montreal, 2002).

tuis' philosophical work was moving in new directions. His interest in cultivating a reputation as a libertine man of letters and *esprit fort* had led him to author a number of aggressive and philosophically radical texts. With the appearance of his *Vénus physique* in 1745, this identity transformation made it possible to directly associate Maupertuis' libertinism with his other, recently manifested interest in Leibnizian metaphysical physics. These perceptions were further encouraged when the academician, rather surprisingly, announced that he was leaving Paris to move to Berlin, the heart of Leibnizian Europe.

He had been drawn east by Frederick the Great's lucrative offer to become the head of his newly reformed Berlin Academy of Sciences.[219] This institution was already noteworthy for its class of speculative philosophers, a group that had no counterpart at either the Paris Academy or the Royal Society of London. It was also perceived to be a bastion of Wolffian Leibnizianism. In this environment, the metaphysical leanings already apparent in Maupertuis' final papers in Paris were reinforced, and as the president of the Berlin Academy he further encouraged these tendencies within the institution.[220] In 1746, the Berlin Academy made the question of the Leibnizian monads the basis for its prize essay contest (the abbé Condillac submitted an entry), and even if no evidence exists that Maupertuis had a stake in this contest, he did write to Johann II Bernoulli at the time asking that he submit an entry that would "revive the honor of metaphysics."[221] In 1750 Maupertuis also published his *Essai de cosmologie,* the work that contained the geometrical and metaphysical conception of God that had so scandalized the bishop of Mirepoix in 1743.[222] While Maupertuis was moving in these directions, Newtonians in Paris were actively distancing their conception of forces and mechanical action from the metaphysical and materialist theories that were becoming more and more central to Maupertuis' work. Consequently, the former darling of French Newtonianism increasingly found his science alienated from that practiced by his old allies inside the Paris Academy, a distance that was physically reinforced by his residence in Berlin.

This separation worked to solidify French Newtonianism. The public refashioning of Maupertuis as a quixotic and reckless metaphysician and libertine was already underway before he left; however, once departed (to the very homeland of Teutonic metaphysics, no less!) it became all the easier to

219. See Terrall, *Man Who Flattened the Earth,* 226–234.

220. Terrall discusses the new metaphysical work of Maupertuis at the Berlin Academy in ibid., 265–269.

221. Cited in ibid., 258. The text of Condillac's submission exists in a modern, critical edition, Condillac, *Les monades,* ed. L. L. Bongie (Paris, 1995).

222. Maupertuis, *Essai de cosmologie* (Amsterdam, 1750).

define French Newtonianism in terms other than his. Maupertuis also made this reconfiguration easier through his conduct in Berlin. A series of scandals rocked his term as academy president, and these further suggested to many that his oft-noted ambitiousness had spilled over into a haughty and *malhonnête* arrogance. Most damaging was his handling of what came to be called the "König affair."[223] Briefly summarized, the controversy began in 1751 when du Châtelet's teacher and Maupertuis' former friend, Samuel König, published a paper linking Maupertuis' least action principle to several statements by Leibniz made half a century earlier. Maupertuis saw in this paper both an accusation of plagiarism and a challenge to his honor. He, therefore, deployed his full authority at the Berlin Academy to punish König for this alleged attack. Many saw in Maupertuis' conduct an egregious and despotic misuse of his authority. And since the matter also centered upon Maupertuis' relationship with Leibnizian philosophy, the associations between it and philosophical despotism more generally became easy to draw. No one saw more opportunity in this controversy than Voltaire. Following the death of du Châtelet in 1749, and after a very uncomfortable period during which he and Maupertuis attempted to share the intellectual spotlight at Frederick's court in Berlin, he used the König affair as the pretext for publicly turning against his former ally and confidant.[224]

In April 1752, the Berlin Academy issued its ruling against König, bringing the full weight of its authority down upon him and his charges. In December, the *Journal des savants* contributed to the growing international controversy by describing the nature of the dispute and its resolution. A few months later, it also reviewed, objectively yet sympathetically, König's *Appeal to the Public* that defended his position in the dispute.[225] Dozens of other periodicals did the same, and soon the same journals were discussing pamphlets critical of the action with titles such as *Letter from an Academician from Berlin to One in Paris* and *Letter from a Cosmopolitan to One in London*. Overall, the trial and condemnation of König triggered an international debate about the conduct and character of the Berlin Academy and its president, and Voltaire played a key role in fueling the dispute.[226] The first pamphlet noted above may have been written by him, and he certainly authored the critical review of the *Oeuvres de*

223. The details of this controversy are recounted in Terrall, *Man Who Flattened the Earth*, 292–309.

224. For Voltaire's role in these episodes, see Pomeau and Mervaud, *De la cour au jardin, 1750–1759*, 21–102.

225. Review of Samual König, *Appel aux publique*, in *Journal des savants* (1752): 818–821; (1753): 225–230.

226. See Pomeau and Mervaud, *De la cour au jardin, 1750–1759*, 108–111.

M. de Maupertuis, published at Dresden, that the *Bibliothèque raisonnée* published in September 1752.[227] In this review, Voltaire made explicit the reckless nature of Maupertuis' metaphysical leanings while mocking his thought overall. The review was also followed in the same issue by a tract defending König's position in his dispute with the Berlin Academy.[228]

Voltaire further linked Maupertuis' intellectual failings with his conduct as academy president by publishing a scathing satire of Maupertuis entitled *Diatribe du Doktor Akakia*.[229] The satire combined scathing ridicule of Maupertuis' recent excursions into metaphysics and the science of life with equally barbed indictments of his despotic and vainglorious character. Indeed, each side of this satire worked to reinforce the sting of the other as Voltaire presented Maupertuis as a pedantic and hopelessly pompous metaphysician (the title "Dr." was deployed as an epithet) whose vanity and arrogance was served by despotically expounding his absurd philosophy. Voltaire would employ many of the same tropes more famously at the end of the decade in the construction of his character Dr. Pangloss for *Candide*.

Voltaire was still in Berlin at the court of Frederick when the first edition of *Dr. Akakia* appeared. The text soon forced his departure. Frederick sided with Maupertuis and demanded that Voltaire either sign a retraction or leave. The author decided to depart soon after, but not before arranging to have his satire published in Leiden, adding to it a burlesque piece called *The Memorable Session* that travestied Maupertuis' conduct inside the Berlin Academy.[230] These works triggered the usual scandals, with counter pamphlets, charges of *malhonnêteté*, and a cacophony of recriminations all around. But from the perspective of Paris, which had last seen Maupertuis in 1745, the scandal served Voltaire's interests far more than his rival's. Maupertuis' increasingly radical published works, his growing reputation as an arrogant and vainglorious climber, and the reporting regarding the König affair in the learned press all gave Voltaire's satire the ring of truth. It, therefore, worked to further solidify the gap that had been forming for almost a decade between the perceived New-

227. Review of *Oeuvres de M. de Maupertuis*, in *Bibliothèque raisonnée* 49 (1752): 158–172. The text is reprinted in *Oeuvres de Voltaire* (ed. Beuchot), 23: 535–545. An equally scathing review of Maupertuis' published letters appeared in the same volume, 431–439.

228. "Sur le *Jugement de l'Académie* et *l'Appel* de König," *Bibliothèque raisonnée* 49 (1752): 173–209.

229. The publishing history of this work, which was really a collection of brief satirical pieces, is explained in Pomeau and Mervaud, *De la cour au jardin, 1750–1759*, 113, n. 65.

230. Ibid., 115–119. The full text, titled *Histoire de Dr. Akakia et du natif de Saint-Malo*, is found in *Oeuvres complètes de Voltaire* (ed. Beuchot), 23: 559–585.

tonian philosophical culture of France and that of its much more metaphysical rival in Germany. It also reinforced the perception that French savants were more reasonable and moderate as a result of their Newtonian philosophical orientation, while German metaphysicians, even French-born ones, were pompous, reckless buffoons. It finally showed how much Voltaire's cultural authority had grown since his scandals with Desfontaines a decade earlier, for while his wit and taste for satirical venom was in no way moderated in these texts, the cultural authority that his discourse carried was much more potent.

The *Diatribe du Doktor Akakia,* in fact, marked the end of what had been a relatively scandal-free decade for Voltaire, a decade that allowed him to transform himself from a would-be philosophe struggling to wash himself of the taint of dishonorable libertinism into a serious and increasingly respected philosophical voice. By 1752, his position at the center of the newly established Newtonian community in France reinforced the esteem of each, and while he still had his critics, and while his return to scandal mongering at the expense of Maupertuis challenged again his reputation as an *honnête homme,* by this date he was secure enough to weather the storm. This newfound strength was partially a result of his own efforts at rehabilitation, for after his reconciliation with Desfontaines in 1739 Voltaire worked hard to fully establish his status as a respected and *honnête* savant who deserved the accolades accorded learned men by the French authorities. Works such as *The Metaphysics of Newton* and the newly revised 1745 edition of the *Eléments de la philosophie de Newton* were a key part of this program, as was Voltaire's public stance in support of Dortous de Mairan in his debate over *vis viva* with du Châtelet. Equally important was his opposition to the Leibnizianism that she and Maupertuis began to promote.

The success of his history *Le siècle de Louis XIV* also played a role, for after escaping from the scandals that had clouded its initial reception, the book had become a recognized masterpiece of historical literature. Its celebration of the Sun King through praise of his efforts on behalf of the arts, letters, and science also sent the right message to the reigning sovereign, who was also aspiring to rule over a glorious and culturally vibrant kingdom. In 1744, these forces converged in the appointment of Voltaire as the royal historian of France. This appointment demonstrated his newfound favor at the very center of the French establishment, and a year earlier he had been made a fellow of the Royal Society of London, an honor that he called a just reward for his suffering on behalf of Newton in France.[231] He also launched a serious, if unsuccessful, campaign to join the forty immortals in the Académie française during these years, ul-

231. Voltaire to Martin Folkes, November 1743, Voltaire, *Corr.,* 93: D2890.

timately losing out to Dortous de Mairan and Maupertuis in these elections. These establishmentarian aspirations required him to appear at Versailles and to conduct himself as courtier, work he performed adequately if begrudgingly. As a result of these efforts, he also grew closer to Maurepas, conducting diplomatic missions for him and his longtime friend, the Marquis d'Argenson. To further secure this courtly favor, he even agreed to a plan that would have sent him to Rome in early 1745 to receive a papal benediction. His allies at Versailles thought that an expression of favor from Pope Benedict would further convince *dévots* at court of his moral and intellectual propriety. The plan was never executed, but Voltaire did begin to compose a flattering verse portrait of the pope that would have been his calling card had he actually knelt before the pontiff to kiss his ring.[232]

In 1746, his official authorization as an establishment intellectual was furthered by his admission to the Académie française. This located him at the very pinnacle of French letters, yet he was only admitted after a hotly contested election that had forced him, like Maupertuis, to demonstrate his probity to the bishop of Mirepoix and the other conservatives among the forty immortals.[233] During the period of the election, the fiery Jansenist periodical *Nouvelles ecclésiastiques* attempted to use Voltaire's candidacy to further their own program of exposing the unholy anti-Jansenist alliance between the pope and the French Crown. "This author, I say, is in dialogue with the pope while [Jansenist] bishops, priests, and the faithful are treated as the excommunicated," the journal fumed. "This is the author of the *Lettres philosophiques,* burned by the hand of the hangman, letters that appalled those who still have any religion. . . . Yet the Holy Father gives this poet gold medals as marks of his esteem. . . . Is there still any faith left on earth? Or must one concede that truth has retired and abandoned us?"[234]

Fortunately for Voltaire, Jansenist criticism such as this was the same as praise within the ecclesiastical and political hierarchies that ruled France. His appointment to the French Academy was secured, therefore, despite this Jansenist opposition. He followed it with a thoroughly formal and official *Pangyrique de Louis XIV* published in 1748, a work produced in his capacity as royal historian and French academician. In perhaps the apotheosis of this decade-long effort at establishmentarian self-fashioning, Voltaire was further chosen in August 1749 to read a pangyrique to Saint Louis in the Royal Chapel

232. On Voltaire's life between 1743–1745, see Vaillot, *Avec Madame du Châtelet,* 193–234.

233. Ibid., 248–258.

234. Cited in ibid., 249.

of the Louvre as part of the annual celebration of this patron saint of France. The oration was attended by the king, the royal family, his court, the leading members of the ecclesiasty, and a throng of blue-blooded dignitaries.[235]

By 1749, therefore, Voltaire had succeeded in establishing himself as a thoroughly respectable savant and a royally sanctioned man of letters. At one level, he had achieved this success by carefully avoiding the more radical philosophical discourses and positions that had characterized his work in the 1730s. In other ways, however, his new authority in officialdom still served these causes. For one, it allowed Voltaire's Newtonianism, now carefully shorn of its more dangerous materialist and irreligious associations, to move to the center of the French establishment with him. It would be just as accurate, however, to say that Newton's newly acquired authority and status in France drew Voltaire into the mainstream with it. Whatever the precise direction of influence, each was reinforcing the authority of the other by 1750 while earlier they had fought against one another. Voltaire's reputation as a scandalous libertine and freethinker certainly was not erased by the honors he began to receive. Instead, it was moderated by his new reputation as an authoritative and respected spokesperson for philosophical truth.

So why, in the midst of this growing official security, did Voltaire seemingly surrender all these gains and instead launch another scandalous public attack on Maupertuis? His present-day legacy as the personification—perhaps the earliest—of critical, aggressive, libertarian intellectualism makes the choice appear natural and overdetermined. Historically speaking, however, Voltaire's return to his more provocative intellectual practices of the 1730s was anything but self-evident. It was, in fact, a historically contingent choice. Nevertheless, beginning in 1749, and continuing through 1752 with his critique of Maupertuis, and then on into the subsequent decades, Voltaire reclaimed his stance as a detached public critic, the one that had been thrust on him by the scandal of the *Lettres philosophiques*. He also resumed his characteristic practice, begun then as well, of critical, public philosophizing. The philosophe movement, led by him, followed in the wake of this resumption, and with it the wider French Enlightenment that the newly formed *parti philosophique*, united around Voltaire, came to personify.

Yet here again this crystallization was anything but inevitable. It took one final event to provide the climactic and defining cause that unified all the others and transformed them into a self-conscious movement. Prior to 1749, Newtonianism had provided the glue that unified this inchoate philosophe movement, channeling its adherents toward the critical, public philosophizing that would

235. Ibid., 396–397.

become the hallmark of the French Enlightenment. Newtonian philosophy continued to play a unifying role in the later Enlightenment as well; however, it was neither a single philosophy, nor even a single philosopher, that catalyzed the full-fledged initiation of this movement. Instead it was a book: Diderot and d'Alembert's *Encyclopédie, ou Dictionnaire raisonnée des arts et sciences*. The prospectus for this work appeared in 1749, and the first volume was released in June 1751. Within nine months, the text was banned amidst a scandal that rocked both the French public sphere and the wider Republic of Letters. Both the text and the scandal it triggered ushered in the self-conscious program of public philosophic Enlightenment in France, along with the social movement that made this cause its rallying cry. Newtonianism played a key role in this process, but it was not its guiding force. The final link between Newton and Enlightenment in France was forged rather by this monumental book and its impact, and it is thus with an account of this linkage that this genealogy of Newton's role in provoking the French Enlightenment should end.

{ *Instituting the*
French Enlightenment
Diderot and d'Alembert's
Encyclopédie

Too many historians have recognized Diderot and d'Alembert's *Encyclopédie* as the defining text of the French Enlightenment to warrant a justification of that claim here. Just as many have argued that the project's initial publication, scandalous reception, and eventual prohibition were the key catalysts in the formation of the Enlightenment as a French social movement. Given this, I further take this claim for granted. What has not often been stressed, however, is that these undeniably transformative events also completed a half-century process of intellectual and social change in France, changes that are chronicled in the previous chapters of this book. What I propose here, therefore, by way of coda, is not a reinterpretation of the history of the *Encyclopédie,* so completely and ably told by others, but rather a retelling of this familiar story in ways that highlight how it served both as the climax of Newton's reception in eighteenth-century France and, seamlessly, as the final stitch between this reception and the beginning of the French Enlightenment.

Many of the principle protagonists in this story have already been introduced, but one has yet to appear, the character who in fact played the most decisive role in it. Denis Diderot was born in 1713 to humble artisans in Langres.[1] His parents were worldly and affluent enough to send him to the Jesuit college down the street, where he distinguished himself as a student of rhetoric and mathematics. They were not, however, equipped to offer him entrée into the world of eighteenth-century letters and science. As a young man, there-

---

1. My account of Diderot's early years is drawn primarily from "Part I: The Testing Years," in Arthur M. Wilson, *Diderot* (Oxford, 1972). Other works that have been important in shaping my understanding of Diderot and his work are Robert Darnton, *The Business of Enlightenment: The Publishing History of the Encyclopédie, 1775–1800* (Cambridge, 1979); Lynne Dixon, *Diderot, Philosopher of Energy* (Oxford, 1988); Jacques Proust, *Diderot et l'Encyclopédie* (Paris, 1962); Paul Vernière, *Spinoza et la pensée française avant la Révolution* (Paris, 1954), esp. 528–611; and Franco Venturi, *Jeunesse de Diderot (1713–1753)*, trans. by Juliette Bertrand (Geneva, 1967).

fore, Diderot was compelled to live the life of a poor, independent writer feeding himself through his wit and his pen. This was a lifestyle that d'Alembert romanticized as the ideal one for a philosophe in his widely influential *Essay on the Society of the Lettered [gens des lettres]* first published in 1753.[2] Neither he nor any of his associates at the salon of Madame Geoffrin or inside the royal academies actually lived this way, however. Diderot did. Consequently, his ascent to intellectual prominence reveals much about the interplay between social reality and cultural ideals in the making of the Enlightenment philosophes.

Diderot's success at his Jesuit college, which produced what he called his father's proudest moment when his son arrived home at the end of his studies with a prize, allowed him to win parental approval for his move to Paris to continue his studies. They hoped he would enter one of the learned professions, perhaps the law, but when Diderot instead found himself attracted to the thriving life of science and letters in the city, conflicts between he and his family began to emerge. He studied philosophy at the University of Paris, taking a master's degree in 1732. He also honored his father's wishes by enrolling in law school and apprenticing with a Parisian lawyer and family friend. He found the arrangement unpalatable, however, and after an even more distasteful stint as an apprentice in his father's cutlery shop in Langres, Diderot returned to Paris around 1735, this time as a rebel and outcast from his family.[3]

His activities during the next decade are notoriously hard to document. He considered entering the clergy, going so far as to accept the tonsure, but he stopped short of full ordination. He was also bitterly poor since he received only sporadic parental support during these years, a condition that led him to vow in his later and more affluent years to never refuse the entreaties of hungry beggars. Despite much hardship, he appears to have eked out a basic living by working as a writer, translator, and occasional journalist. He was known to have written some pieces for Desfontaines' *Observations sur les écrits modernes,* which began publication in 1735. Given his strong command of English, a skill whose origin is unclear, he also received payment for a number of translation projects. His first was a French edition of the Englishman Temple Stanyan's *Grecian History,* a staid work of erudite scholarship. This won him his first public recognition, a brief mention in the *Journal des savants* as the

2. *Essai sur la société des gens de lettres et des Grands, sur la reputation, sur les mécènes, et sur les recompenses litteeraires,* in *Oeuvres de d'Alembert,* 5 vols. (Paris, 1821–1822), 4: 335–373.

3. Wilson, *Diderot,* 15–28.

FIGURE 26. *Louis-Michel van Loo*, Portrait de Denis Diderot. *Louvre, Paris, France. Courtesy of Art Resource, NY.*

book's "rather negligent" translator.[4] It also earned him the paltry payment of three hundred francs.[5]

Diderot's next project brought him more acclaim but only slightly more money. He undertook a translation of *An Inquiry concerning Virtue and Merit* by Anthony Ashley Cooper, the third Earl of Shaftesbury. Shaftesbury was, among other things, John Locke's patron and an important English moral philosopher.[6] Diderot took liberties with his text, turning his translation into something of a statement of his own philosophy regarding natural religion, morality, and the value of religious tolerance. Yet since Shaftesbury had also defended these views, Diderot's work on another level was simply a gloss of the joint philosophical positions that he and the earl shared. The resulting text, which appeared in 1745, produced a modest reception in the learned press.[7] Desfontaines attested to his personal acquaintance with the anonymous author while also offering the book a favorable review.[8] The *Journal des savants* gave the work a fairly thorough and unobjectionable discussion.[9] Cutting more to the heart of the provocative nature of Diderot's text, the *Journal de Trévoux* told readers on page one of its February 1746 issue that they should "imagine Mr. Locke discoursing on morality."[10] This, not inappropriately, raised the specter of English deism with respect to the text, yet the reviewer found no reason to exercise himself over this matter. As a result, his review was ultimately rather prosaic, and neither it nor the text as a whole generated much wider controversy.

After his translation of Shaftesbury's *Inquiry*, Diderot returned to his less philosophical yet more financially lucrative translation projects. He produced in particular a translation of Dr. R. James's six-volume *Medicinal Dictionary* between 1746 and 1748, a text that a century later Mark Twain called "a majestic literary fossil."[11] It was through a commission to translate another English mastodon, Ephraim Chamber's *Cyclopedia,* that Diderot would be set to work

4. Review of *Histoire de Grèce, traduite de l'anglais de M. Temple Stanyan,* in *Journal des savants* (1743): 451–462, (1745): 547–555, (1746): 231–238. Diderot's translation skills are noted in the last review, 238.

5. Wilson, *Diderot,* 47–50.

6. See Lawrence Klein, *Shaftesbury and the Culture of Politeness: Moral Discourse and Cultural Politics in Early Eighteenth-Century England* (Cambridge, 1996).

7. Diderot, *Principes de la philosophie morale; ou Essai de M.S*** sur le mérite et la vertu. Avec réflexions* (Amsterdam, 1745).

8. Review of *Principes de la philosophie morale, ou Essais de M. S.***,* in *Jugemens sur quelques ouvrages nouveaux* 8: 86–87.

9. Review of *Principes de la philosophie morale,* in *Journal des savants* (1746): 210–219.

10. Review of *Principes de la philosophie morale,* in *Journal de Trévoux* (1746): 198–220.

11. Cited in Wilson, *Diderot,* 53.

on his more famous *Encyclopédie*. During this period translation projects such as these allowed him to earn a meager living at a time when he had few other means of material support. At least in the case of his Shaftesbury translation, they also gave him an opportunity, however constrained, to exercise his own muscles as a critical writer and thinker. In many respects, his lifestyle during these lean years matched quite well Marivaux's image of the *l'indigent philosophe* even if Diderot had not yet been given the opportunity (at least publicly) to demonstrate his aptitude for immorality and impiety.

During these years, however, Voltaire was also redefining the identity of the philosophe in France, infusing the much older tradition of aristocratic libertinism and intellectual independence with the new discourses of philosophical and political liberty coming out of England and the Low Countries. He was also trying to stabilize the radical nature of this position with public claims upon the *honnête* values of moderation, utility, and service to truth. Negotiating a balance between critical intellectual liberty and *honnêteté* in the pursuit of philosophical truth would become the central ethical project of the Enlightenment philosophes after 1750, and when Voltaire begin to define this new identity in the 1730s and 1740s he at least had the advantages of a sizeable fortune and the support of established and politically powerful patrons. He also had a foothold in the Republic of Letters secured through his reputation as a playwright and poet. Diderot, by contrast, had no similar status within the corporate structures that still ruled the arts and sciences in ancien régime France. Accordingly, when he began to position himself after 1745 as a philosophe in the manner of Voltaire, the hardships he faced were much more severe.

One commonality he shared with Voltaire was an interest in science, particularly Newtonian science, and for Diderot these philosophical discourses likewise played a key role in his youthful philosophic self-fashioning. He had excelled at mathematics as a student, and given his skills as a student of English, it is likely that he read many of the leading works in mathematical science in this period, including the major works in English. In one of the rare documentary glimpses into his life in the 1730s, he reports that he had begun to work on a French commentary of Newton's *Principia*, only to have abandoned the project when the Minim friars Le Suer and Jacquier published their commentary on the treatise in March 1738. If Diderot's recollections were accurate, then it means that the young and still thoroughly unknown writer was actively participating in the Newton wars even if his presence in these struggles never made it into the public eye.[12] In 1740, he also authored a text (now lost) called "Astronomical and Physical Description of the World," and in 1748 he further

---

12. Wilson, *Diderot*, 30.

published a short work of mathematical *mémoires* that revealed in more detail
the nature of his scientific work in this period.[13] The five short pieces manifest
a familiarity with the analytical mathematics that was so influential in France,
along with a knowledge of the different English methods that were important
across the Channel. The papers also show a knowledge of the new mathematics of probability that was important in each of these venues as well.[14] Diderot
pursued his mathematical work independent of any institution, and he appears
to have enjoyed no contact whatsoever with either the Paris Academy or any
of its members before 1746. Yet in this characteristically isolated and marginal
way, he was nevertheless at the cutting edge of the scientific, and especially the
mathematical, currents of the day.

That he could sustain a viable intellectual life at all was a testimony to what
the institutions of public science had wrought since their full-fledged emergence in France in the 1690s. By the late 1730s, when Diderot was cutting his
philosophical teeth, the print public sphere had expanded enormously, with
ever more books available and ever more mechanisms for their access close at
hand. The learned press was also exploding in quantity and quality, giving independent savants like Diderot virtual access to innumerable books, discoveries, and scientific debates. These journals also conveyed a new intellectual style
that made anonymous readers and writers into authoritative critics, warranted
in their judgments by the wider public to which they addressed themselves.
Diderot's position as a struggling writer and journalist living on the margins of
the print public sphere in Paris gave him ready access to all this material, and
by all accounts he voraciously consumed every bit of it. He also absorbed the
new intellectual style that this expanding print culture fostered, along with the
new social possibilities that it suggested as well.[15] He further immersed himself in the scientific sociability that a city like Paris offered, attending Nollet's

13. The first is discussed in J. Th. De Bouy, "Denis Diderot. Ecrits inconnus de jeunesse," *Studies on Voltaire and the Eighteenth Century* 119 (1974): 217–230; also see Denis
Diderot, *Mémoires sur différens sujets de mathématiques* (Paris, 1748), reprinted in J. Assézat and M. Tourneux, eds., *Oeuvres complètes de Diderot,* 20 vols. (Paris, 1875–1877), 9:
73–182. For a review of the mathematical papers, see *Journal des savants* (1749): 3–8.

14. For an analysis of Diderot's mathematical work, see Lester Gilbert Krakeur and Raymond Leslie Krueger, "The Mathematical Writings of Diderot," *Isis* 33, no. 2 (June 1941):
219–232.

15. For an incisive analysis of the connection between the expanding print culture of the
eighteenth century and the new critical spirit of the period, see Chartier, "Do Books Make
Revolutions?"

experimental physics courses begun in 1735 and the public chemistry lectures that Venel gave at the *Jardin du roi*. At the latter, he met the vitalist doctors who would eventually contribute the articles on chemistry and physiology to his *Encyclopédie*.[16] Cafés, theater, and a host of other sociable activities were also a part of his routine as well even if no precise documentary record of his activities in the 1730s exists.[17] Through this urban circulation, both material and virtual, he became socialized into the culture of French public science and the wider Republic of Letters even if he still remained anonymous within it.

These sociable networks also brought him together with many of his future colleagues even if few of them were aristocrats or members of the French intellectual establishment. Jean-Jacques Rousseau reports meeting Diderot almost immediately after he arrived in Paris in August 1742.[18] This young Genevan, seeking fame and fortune as an *homme des lettres,* quickly became one of Diderot's closest friends, and soon they were joined in their socializing by another would-be philosophe and future *Encyclopédiste*, the abbé Condillac. D'Alembert also joined this circle soon after as he began to establish a new presence for himself in Parisian society. His *Traité de dynamique* appeared in 1743, and riding the wave of its public acclaim, not to mention his implicit, though often unacknowledged, blood ties to Parisian high society, he was introduced to the gatherings hosted by Madame du Deffand. D'Alembert's charm was infectious, and his talent for mimicry was particularly loved by salon guests. As a result, he was soon after invited to the salon of Madame Geoffrin, and by 1746 he was also dining each week with Condillac, Rousseau, and Diderot at the Hôtel du Panier Fleuri near the Palais royal.[19] Rousseau immortalized these dinners in his *Confessions*, making them central in forming the intimate ties that joined the philosophe movement in the subsequent years.[20]

These ties undoubtedly united the movement but they were nonetheless complex. D'Alembert was far and way the most accomplished and established of the bunch. He continued to increase his stature as a mathematician and physicist throughout these years, winning, among other things, the 1746

16. Elizabeth Williams, *A Cultural History of Medical Vitalism in Enlightenment Montpellier* (Aldershot, 2003), 119–123.

17. Wilson, *Diderot*, 30–32.

18. Maurice Cranston, *Jean-Jacques. The Early Life and Works of Jean-Jacques Rousseau, 1712–1754* (Chicago, 1982), 162–163; Wilson, *Diderot*, 45–46.

19. Grimsley, *D'Alembert*, 7–9.

20. Jean-Jacques Rousseau, *Confessions*, in Hachette, ed., *Oeuvres complètes de Jean-Jacques Rousseau*, 13 vols. (Paris, 1885–1905), 8: 246–247.

Berlin Academy prize for his mathematical treatment of the phenomenon of the wind.[21] He sent a copy of this text to Voltaire, initiating a correspondence with him that would grow more vigorous and intimate as the years progressed.[22] He also crossed paths with the new royal historian and French academician in the Parisian salons that he frequented. D'Alembert's ties to these social circles and to the Royal Academy of Sciences did not confer access to his friends, however. Diderot and Condillac were restricted, therefore, to secondhand accounts of d'Alembert's wanderings through their dinner conversations with their much better connected friend. Rousseau, by contrast, formed his own social ties, bringing a new system of musical notation with him from Geneva that both excited Diderot's own, budding musical sensibilities and the interest of the Royal Academy of Sciences.[23]

Music and mathematics were still closely allied inside the Parisan academy, and soon after Rousseau's arrival in August 1742 he was granted an audience with several royal mathematicians and invited to attend one of the academy's sessions.[24] His musical system was ultimately rejected, by no less than Dortous de Mairan, but he published it anyway, first in the *Mercure* and then in a short book entitled *Dissertation on Modern Music*. These texts defended his own system of musical notation while also criticizing the one accepted by Dortous de Mairan and the Royal Academy.[25] Rousseau also enjoyed success in the patronage game so crucial to aspiring writers and savants. He attracted the support of the Duc and Duchesse de Richelieu, important patrons of Voltaire as well, and through their introductions, he entered Parisian society, proving, like d'Alembert, to be a successful salon guest.[26] In December 1745, his connections with the Richelieus in particular also brought him into correspondence with Voltaire.[27] Diderot, by contrast, remained an outsider to all of this Parisian so-

21. Hankins, *D'Alembert*, 45–46.

22. Voltaire to d'Alembert, 13 December 1746, Voltaire, *Corr.*, 94: D3484. See also John N. Pappas, *Voltaire and d'Alembert* (Bloomington, 1962), 4.

23. Cranston, *Jean-Jacques*, 157–158; Wilson, *Diderot*, 45.

24. Rousseau claimed to have attended and read a paper at the academy's meeting of August 22, 1742, although no record of the visit appears in PVARS. See Cranston, *Jean-Jacques*, 157, n. 4. On music and mathematics in the Paris Academy, see Albert Cohen, *Music in the French Royal Academy of Sciences: A Study in the Evolution of Musical Thought* (Princeton, 1981).

25. Rousseau, *Dissertation sur la musique moderne* (Paris, 1743). See also PVARS, September 5, 1742; review of Rousseau, *Dissertation sur la musique moderne*, in *Mercure de France* (February 1743); and Cranston, *Jean-Jacques*, 157–158.

26. See Cranston, *Jean-Jacques*, 201–216.

27. Voltaire to Rousseau, 15 December 1745, in Voltaire, *Corr.*, 93: D3270.

ciety until much later even if he had virtual access to it through the companion-ship of his friends.

The sociable ties that would constitute the core of the philosophe move-ment were thus in place by 1746, but the union was entirely personal at this point, formed through idiosyncratic friendships rather than social, political, or institutional alliances. The intellectual coherence of the group was even less developed, and while science, particularly Newtonian science, played a role in the intellectual orientation of each man, it was in no way the defining creed that brought them all together. If an ideology did unite them, it was the values of the Republic of Letters, and especially the new identity of the critical, public philosophe that Voltaire had begun defining with his Newtonian camapaigns during the previous decade. The Newton wars had solidified this conception, and the cause of Newtonianism had, therefore, become synonymous with the cause of liberty and *philosophie* in general as a result of these public battles. Dumarsais further articulated this union, minus the Newtonianism, in his ar-ticle "Philosophe" published in 1742 in the widely read libertine tract *Nouvelles libertées de pensée.*[28] Diderot would eventually let this essay define the word "philosophe" in his own *Encyclopédie,* and in the 1740s, the future editor, along with his would-be philosophe companions, began to chart out, each from their own social position, an intellectual career consonant with their understanding of this identity.

For d'Alembert, the model of Maupertuis proved most powerful, as he com-bined serious mathematical research and intense professional labor inside the Royal Academy with a parallel life as a witty and affable socialite. Maupertuis also infused this combination with an orientation toward libertinism and phil-osophical radicalism in the manner of Voltaire's philosophical self-fashioning of the 1730s, yet d'Alembert eschewed any associations with philosophical rad-icalism of this sort. It was rather the more moderate and conformist Voltaire of the 1740s that d'Alembert modeled himself after, and given these cultural sym-pathies and their many common social circles, the two men drew closer together during these years. Ironically, perhaps given their social differences, Diderot found himself far more closely aligned with Maupertuis' libertine materialism than d'Alembert, despite the absence of any personal or professional contact between them. Indeed, while Diderot began to cultivate his own identity as a libertine, materialist philosopher, drawing upon many of the same influences as Maupertuis in this effort, d'Alembert instead embraced an antimaterialist Cartesian Newtonianism, a stance appropriate to his growing authority inside the Paris Academy.

28. "Philosophe," in *Nouvelles libertés de pensée* (Amsterdam, 1743): 173–204.

Locke's empirical and sensate metaphysics was also central to his philosophical outlook, and in this respect Condillac's *Essay on the Origins of Human Consciousness* published in 1746 also echoed d'Alembert's own budding philosophical sensibilities through its commitment to metaphysical modesty, empiricism, and rational, mathematical analysis in scientific epistemology.[29] Neither thinker had any sympathy for the active matter materialism that was also associated with Locke through the legacy of Toland, Collins, and others, nor with active matter Newtonian attractionist physics that was also associated with these English deists. Voltaire had also begun to publicly distance himself from such discourses by the 1740s, for he increasingly defined and defended Newtonian-Lockean philosophy by contrasting it starkly and publicly with both the active force metaphysical physics of du Châtelet and the Leibnizians and the rationalist, teleological materialism of Maupertuis' science of life. Accordingly, as the 1740s came to a close, he too found himself comfortably sharing the same intellectual bed as d'Alembert and Condillac.

Diderot was of a different sensibility entirely, and beginning in 1746, with the publication of his first, fully independent work, the anonymous and illegally published *Pensées philosophiques,* he began to articulate what would become one of eighteenth-century France's most radical and materialist philosophies.[30] He appropriated the aphoristic structure of Pascal's revered and pious *Pensées* for his work, and even invoked the name of the stern Jansenist directly on one occasion; however, his "thoughts" were turned toward anything but Christian apologetics. "Superstition is more injurious to God than atheism," he wrote in a typically provocative aphorism.[31] In another, he declared that "skepticism is the first step toward truth."[32] "People have the right to demand of me that I seek the truth, but not that I find it," he asserted in *pensée* 29.[33] Overall, his work honored this principle by using the aphoristic structure to assert provocative claims that challenged existing pieties while never developing any systematic philosophical position. The value of natural religion; the prejudice and superstition of established church authority; the naturalness of the passions and the unnaturalness of moral strictures against bodily pleasure

---

29. E.-.B., abbé de Condillac, *L'essai sur les origines des connaissances humaines* (1946; Paris,2002). See also Le Ru, *D'Alembert,* 127–186.

30. Diderot, *Pensées philosophiques* (La Haye, 1746), reprinted in *Oeuvres completes de Diderot,* 1: 123–170.

31. Ibid., 1: 130.

32. Ibid., 1: 140.

33. Ibid.

and desire: these and other challenges to the established order were offered by Diderot in a style that was simultaneously sharp, incisive, aggressive, and elegant.

Diderot also took on the established structures of philosophy, launching aphoristic bombs against scholasticism and theological tyranny but also against the lay natural philosophy of Descartes and Malebranche. In the end, he produced a masterpiece of philosophical skepticism, leaving no edifice unscathed while also demonstrating his keen intellect at every turn. The book appears to have circulated very widely despite its scandalous content and lack of official sanction. On July 7, 1746, the Paris Parlement further assured the book's success by awarding Diderot the highest honor that a book of this nature could achieve: it burned the *Pensées philosophiques* on the front steps of its chambers while publicly declaring that the book "presents to restless and reckless spirits the venom of the most criminal and absurd opinions that the depravity of human reason is capable of." [34]

Despite its anonymous publication and because of its scandalous reception, many came to know Diderot as the author of the text. For those who did, the *Pensées philosophiques* established him as the very personification of the immoral libertine, skeptic, and irreligious freethinker so often conjured in the conservative intellectual discourse of the period. His work also brought him to the attention of the Paris authorities, even if he was never legally tied to the book. A police file was opened with his name on it, and roughly a year later someone named Perrault reported to Berryer, the lieutenant general of the Paris police, that "this miserable Diderot" was "a very dangerous man who speaks of the holy mysteries of our religion with contempt." [35] Even more damning was the report of his parish priest. He described Diderot as "a young man who passed his early life in debauchery," and he further condemned his current unorthodox living arrangements. [36] The priest told Berryer that Diderot uttered "blasphemies against Jesus Christ and the Holy Virgin that I would not dare to put in writing." "I have never spoken to this young man, [and] I do not know him personally," he continued, "but I am told that he has a great deal of wit and that his conversation is most amusing. In one he confessed to being the author of one of the two works condemned by the Parlement and burned, [and] I have

34. Wilson, *Diderot*, 55–56.

35. Ibid., 61.

36. Diderot had married a woman against the will of his family, and as a result his marriage was not legally recognized. He and his wife nevertheless lived together in Paris, but she did not use Diderot's name.

been assured that he has been working for more than a year on another work still more dangerous to religion." [37]

Indeed, despite the growing surveillance of Diderot by the Parisian authorities, the would-be philosophe continued to illegally publish and circulate libertine tracts. The next to appear was *La promenade du sceptique,* a dialogue of sorts that invited readers to travel down the roads to Enlightenment suggested by the leading philosophical sects of the day.[38] These included the Pyrrhonians, the Spinozists, the idealists (i.e., Berkeley), the deists, and of course, the atheists. No adjudication between these positions was offered, but Diderot again leveled the playing field between these secular natural philosophies and those offered by the Christian church. Safer in some respects, since it was clearly a work of pornographic fiction, but more scandalous in others, since it used a licentious romance to articulate a set of libertine philosophical critiques, was Diderot's *Bijoux indiscrets,* which appeared, again without legal sanction, in January 1748.[39] This novel used the convention of a fantastic voyage to offer a set of witty and raunchy philosophical investigations rooted in contemporaty cultural commentary. One chapter indulged in a satire of both sides of the Newtonian/Cartesian debate, and as a whole the work strongly associated Diderot with libertine free thought.[40] He tempered this identity, however, by also bringing out his mathematical *mémoires* in the same year and by completing his translation of James's thoroughly prosaic *Medicinal Dictionary* in 1748. In fact, when his entire *oeuvre* during this period is viewed together, it becomes clear that he was not pursuing a radical intellectual project to the exclusion of all others, but rather, in the manner of Voltaire, Maupertuis, and others, a philosophical project in the new fashion, one that artfully navigated between the conflicting pulls of liberty, philosophical inquiry, *honnête* decorum, and public service.

Diderot, of course, charted this course from a very different social position than did his other, more comfortably established contemporaries. In 1749 the consequences of this difference caught up with him. In June, he published another brilliant and subtle philosophical book, *Lettre sur les aveugles,* a work that Arthur Wilson aptly describes as "disarming." [41] In it, Diderot took a fairly

37. Ibid., 61–62.

38. *Oeuvres complètes de Diderot,* 1: 171–257.

39. Diderot, *Les bijoux indiscrets* (n.p., 1748).

40. Chapter 9, "L'état de l'Académie de Banza." Throughout the text "Banza" was his figure for Paris, and in this chapter Newton appeared as "Circino" and Descartes as "Olibri."

41. *Lettre sur les aveugles, à l'usage de ceux qui voient* (Pondon, 1749), reprinted in *Oeuvres complètes de Diderot,* 1: 265–342.

mundane, if philosophically intriguing, fact of the day—the existence of a blind mathematician in England named Nicholas Saunderson—and weaved out of his story a text "written with the seeming artlessness of someone idly improvising on a musical instrument." "One subject suggests another," Wilson writes, and soon the reader, "led on and on through a sort of steeplechase over most of the various metaphysical jumps, finally gets himself soaked in the water-hole called 'Does God exist?'" [42] The work is indeed brilliantly evocative, but it contains no definitive demonstration of any kind. Quite the contrary, it is an open-ended exploration of possibilities that raises questions rather than answers them. Nevertheless, the work, when read alongside Diderot's previous record, became the occasion for his official prosecution. A royal *lettre de cachet* was issued for his arrest, and early on the morning of July 24, 1749 two officers of the Paris Police woke him from his bed to escort him to the royal prison at Vincennes. [43]

The order to arrest him had come from the Comte d'Argenson, the director of the Paris police, who was also the official in charge of policing the Paris book trade. Diderot was kept locked up for over three months. Voltaire had written his first letter to Diderot a month earlier, expressing his pleasure with *Lettre sur les aveugles* and inviting him to join him for a "philosophical repast." [44] The Paris police cancelled this date. But since the warden at Vincennes was, coincidently, François-Bernard du Châtelet, a relative of the marquise with whom Voltaire still cohabitated (she died during childbirth four months later), he implored her to intervene on Diderot's behalf. He also contacted his friend, the Marquis d'Argenson, the brother of the police commissioner, hoping that he might intervene as well. These initiatives did help to shorten Diderot's incarceration somewhat, but more importantly they cemented a relationship between Diderot and Voltaire as philosophical combatants fighting for liberty against the multi-headed monster of superstition, clerical authority, and political despotism. [45]

This alliance was reinforced by Diderot's activities after his release from prison in November 1749. Four years earlier, Durand, the publisher responsible for all of Diderot's important works and the major source of his income, joined a consortium of other booksellers in the release of a prospectus for a new translation of Ephraim Chambers's successful English work *Cyclopedia*. Desfontaines spoke highly of both the original (he had read the English text) and the new translation project, and the *Journal de Trévoux* did the same, saying

42. Wilson, *Diderot,* 97.
43. Ibid., 103.
44. Voltaire to Diderot, 10 June 1749, in Voltaire, *Corr.,* 95: D3940.
45. Wilson, *Diderot,* 103–116; Vaillot, *Avec Madame du Châtelet,* 384–386.

of the project that "nothing is more useful, more abundant, better analyzed, better related, in a word more perfect and finer than this [proposed] *Diction-naire*."[46] Based on this interest, the publishers began to recruit writers and editors appropriate to the project. Much of the work was of a highly technical, scientific nature, and thus contacts were made early on with members of the Royal Academy of Sciences. The first editor-in-chief, in fact, was a rather obscure and undistinguished member of the class of *géometres* named Gua de Malves. He had inexplicably resigned from the academy in June 1745, just a month after the prospectus had appeared, and soon he was managing the projected translation of Chambers's compendium.[47]

Condorcet, in his eulogy of Gua de Malves, claimed that the latter recruited Diderot to the project, but the business records suggest that d'Alembert arrived first, a fact that may be attributed to his academic connections with the editor. Diderot did not appear on the same register until February 1746, three months after d'Alembert, although it is possible that he was involved with the project earlier. No matter how Diderot and d'Alembert initially came to the project, they became its principle coeditors in October 1747. At this point, the project was still conceived as a translation of Chambers's English volume, but over the next three years the work grew into something far more ambitious.[48] In the process Diderot also made two other acquaintances of lasting significance. The first was the journalist Raynal, who was winding down the publication of his weekly newssheet *Nouvelles littéraires* while preparing to begin his appointment as editor of the *Mercure de France*. The second was Friedrich Melchoir Grimm, a German-born savant residing in Paris who would take over and expand Raynal's periodical through his own, even more widely influential *Correspondence littéraire* begun in 1754. Diderot and Raynal assisted Grimm in the production of this journal, and together they turned it into yet another important organ of the philosophe movement.[49]

In November 1750, Diderot's growing network of learned companions was further activated by the publication of the prospectus for the forthcoming encyclopedia that he and d'Alembert were developing.[50] Chambers's *Cyclopedia*

---

46. Review of a prospectus for a translation of Ephraim Chambers's *Cyclopedia*, in *Juge-mens sur quelques ouvrages nouveaux*, 8: 70–72; *Journal de Trévoux* (1745): 934–939.

47. Wilson, *Diderot*, 75–78.

48. Ibid., 78–82.

49. Ibid., 119–120, 221–222.

50. Ibid., 120–121. Denis Diderot, *Prospectus de l'Encyclopédie, ou dictionnaire raisonné des sciences des arts et des metiers* (Paris, 1750). My citations will be drawn from ARTFL

still remained central to the project, but gone was the exclusive focus on translating this work alone. Instead, Diderot proposed a universal *Encyclopédie, ou dictionnaire raisonnée des sciences, des arts, et des métiers* that would be framed by a taxonomy of all knowledge drawn from the work of Francis Bacon. The prospectus described the new conception in detail, and it included, as an appendix, the tree of knowledge to be used (Diderot called it a *"système des connaissance humaines"*). The text also made clear the collaborative nature of the undertaking, describing how collective production was central to both the epistemological and social character of the work. "No man can treat all the sciences and arts by himself," Diderot wrote. Even if one had the mind and the leisure to undertake the task, there is no way a single individual could amass all the experience necessary in his lifetime. Accordingly, Diderot's *Encyclopédie* would be produced through a division of labor and expertise. "Mathematics will be treated by mathematicians, fortifications by engineers, chemistry by chemists," and so on, Diderot explained. While he admitted that this "reduces the worth of the editor somewhat," he was convinced that it would "add enormously to perfecting the work." [51]

In the first volume of the *Encyclopédie*, which appeared six months after the prospectus in June 1751, the collaborative nature of the project was given further material and social coloration. [52] The title page announced that it was authored not by a single author but by "un Société des gens des lettres" unified by the editors to complete the task. The work was indeed the product of dozens of savants joined to the cause through invitations from Diderot and d'Alembert to write articles. The initial prospectus, authored by Diderot, gave the work an overarching ideology, and the two editors also contributed numerous articles to the work itself. In a very real sense, however, it was a collectively authored work, sustained by the particular contributions of dozens of *gens des lettres* simultaneously. Voltaire contributed some articles, including (perhaps) the scandalous article on the soul that generated so much controversy. [53] In this way the project reinforced his ties with the editors. Rousseau also contributed articles, as did Condillac. And d'Alembert recruited royal academicians, such as Buffon, to contribute as well. Maupertuis, by contrast, was not invited to participate. Dozens of others also played a role, some famous, others anonymous or

---

Encyclopedia Project's electronic edition, available at http://www.lib.uchicago.edu/efts/ARTFL/projects/encyc/prospectus.html.

51. "Prospectus," in "ARTFL Encyclopedia."

52. For a review of the first volume, see *Journal des savants* (1751): 617–627.

53. See Pomeau and Mervaud, *De la cour au jardin, 1750–1759*, 207–208.

ENCYCLOPÉDIE,

OU

DICTIONNAIRE RAISONNÉ

DES SCIENCES,

DES ARTS ET DES MÉTIERS,

PAR UNE SOCIÉTÉ DE GENS DE LETTRES.

Mis en ordre & publié par M. *DIDEROT*, de l'Académie Royale des Sciences & des Belles-Lettres de Pruſſe ; & quant à la PARTIE MATHÉMATIQUE, par M. *D'ALEMBERT*, de l'Académie Royale des Sciences de Paris, de celle de Pruſſe, & de la Société Royale de Londres.

Tantùm ſeries juncturaque pollet,
Tantùm de medio ſumptis accedit honoris ! HORAT.

TOME PREMIER.

A PARIS,

Chez
{
BRIASSON, *rue Saint Jacques, à la Science.*
DAVID l'aîné, *rue Saint Jacques, à la Plume d'or.*
LE BRETON, Imprimeur ordinaire du Roy, *rue de la Harpe.*
DURAND, *rue Saint Jacques, à Saint Landry, & au Griffon.*
}

M. DCC. LI.

AVEC APPROBATION ET PRIVILEGE DU ROY.

FIGURE 27. *Title page,* Encyclopédie, ou Dictionnaire raisonée des sciences et des arts, *vol. 1 (Paris, 1751). Pierpont Morgan Library. Courtesy of Art Resource, NY.*

obscure, and while the group was diverse, a commitment as self-identified *gens des letters* to serve the public good through the unification and articulation of learning united them.[54]

In this way, the group came to acquire a collective identity as *Encyclopédistes*, a badge that many wore with honor. Others slung the label at them as an epithet, and increasingly the term became entangled with others—the label *philosophe* most frequently—as observers began to describe the new union of critical, philosophically minded *gens des lettres* that emerged in France after 1751. The *Encyclopédie* project galvanized this group. It also attached them in a new way to the values circulating within the Republic of Letters regarding the proper role of learning and learned men in the making of a modern and civilized society. The result was a new program of Enlightenment in France, one that was about collectively serving the public through philosophical inquiry and the wide dissemination of useful knowledge. The *Encyclopédie* embodied these values on every page, and the scientific articles were one area where these values shone most clearly. The articles on the physical sciences, many of which were written by d'Alembert himself, also articulated the Newtonian consensus that had been established in France over the previous decade. Accordingly, the volumes worked to connect Newtonianism and the new project of Enlightenment at every turn.

"Action," "analyse," "astronomie," "attraction": these were some of the many articles in the first volume where the new consensus about Newtonianism theory and its epistemological anchors empiricism, experimentation, mathematical analysis, and philosophical skepticism were expressed. In the second volume, it was the articles "calcul" and "cause" that accomplished the same goal. All of this was further reinforced by d'Alembert's *Discours préliminaire*, the text that introduced the first volume, and through it, the *Encyclopédie* as a whole. In part 1 of this two-part work, d'Alembert allied himself with Condillac and the Lockean sensationalism that they both embraced by laying out a philosophical understanding of human consciousness and knowledge that also drew heavily on their shared commitment to mathematical analysis.[55] The work was quickly recognized as one of the great philosophical summations of the century, for it articulated well the French Newtonian fusion of Cartesian mathematicism and skeptical English empiricism that had occurred over the previous three decades. It also celebrated the scientific epistemology of the mathematicians inside the Paris Academy, and as such it anchored the technical mathemati-

---

54. Frank A. Kafker and Serena L. Kafker, *The Encyclopedists as Individuals: A Biographical Dictionary of the Authors of the Encyclopédie* (Oxford, 1988).

55. D'Alembert, *Discours préliminaire*, 3–59.

cal and physical articles of the *Encyclopédie* itself. This further linked French Newtonianism with the wider philosophical project of the *Encyclopédie* and its supporters.

French Newtonianism, however, had not emerged as a unifying philosophy in France through general acclamation and acceptance. It had been established through a series of hard-fought, public struggles. The text of the *Encyclopédie* articulated the new intellectual consensus that this victory had wrought, yet it did so while also linking the Enlightenment it exemplified with the public contestation that had created this consensus during the preceding decades. In his article "Encyclopedia," published after the project had been officially condemned, Diderot articulated the link between Enlightenment and the preceding Newton wars by noting that his universal dictionary was not just a compendium of knowledge but in addition a tool for "changing the common way of thinking."[56] In this article, he also reiterated the importance of collective authorship by independent *gens des lettres,* stressing that "the work is to be carried out by a dispersed society of men of letters and artists, each concerned with his own division, and joined together only by the general interest of humankind and by a feeling of mutual beneficence."[57] Elsewhere in the same article he described the glue that bound the *Encyclopédistes* together: a common "zeal for the best interests of the human race and a feeling of mutual goodwill."[58] These pronouncements and others like them strongly echoed the ethic of service and sociability central to Dumarsais' description of the "philosophe." However, by making critical intellectual change central to this vision as well, the article also connected this ethic to Voltaire's history as a combative advocate for philosophical truth and his redefinition of the philosophe identity that his campaigns on behalf of Newton had wrought.

The text of the *Encyclopédie* itself also engaged in the kind of libertarian public criticism that Voltaire and others had begun to associate with the philosophe persona. Because of the scandals it triggered, the work also posed in a new and more collective way the same dilemmas that had characterized Voltaire's public philosophical campaigns of the 1730s (e.g., independence vs. obligation, criticism vs. service, liberty vs. *honnête* restraint). However, it did so in ways that also confirmed the realignment between Newtonianism, Voltaire's example, and critical philosophical liberty that had occurred over the preced-

56. Diderot, "Encyclopedia," vol. 5. My citations will be drawn from the English translation of this article found in Diderot, *Rameau's Nephew and Other Works,* ed. and trans. by Jarques Barzun and Ralph H. Bowen (Indianapolis, 2001): 277–307.

57. Diderot, "Encyclopedia," 283.

58. Ibid., 284.

ing decades as well. True to his words, Diderot's *Encyclopédie* did provoke critical responses while challenging the common way of thinking. In doing so, his work also carried on and solidified the critical legacy of the Newton wars even if Newtonian philosophy was rarely the topic that provoked these controversies. Instead, critical friction was generated through the deployment of critical, libertine philosophy itself while Newtonianism, once conceived as the philosophical father of such tendencies, was increasingly detached from them. In this way, the *Encyclopédie* was able to perform a double duty in France. It first connected Newtonian science with Enlightenment by tying the movement to philosophical veracity through its invocation of the new consensus about Newton's valid science of nature. But then second, it also channeled the critical libertarian intellectualism that the battles to win its approval in France had spawned toward the future project of ongoing philosophical advancement. Indeed, it was not the articulation and dissemination of Newtonian science in the *Encyclopédie* so much as the way that the text channeled and focused the antiestablishment intellectual energies of the Newton wars into the 1750s that allowed it to trigger the French Enlightenment. The project accomplished this goal, moreover, not by forging among its participants a single, shared philosophy or science but rather by galvanizing the wider community of *gens des letters* around the cause of Enlightenment defined in terms of a united *parti philosophique* fighting to defend philosophical liberty and to change the common way of thinking.

Not every *Encyclopédiste* was equally zealous in this cause, and d'Alembert in particular resigned as editor amidst the growing scandals surrounding the work.[59] Diderot, by contrast, was steadfast in this effort, and in this way he became the key force in turning this publishing project into the cause célèbre that would ever after define Enlightenment in France. His efforts in this respect began early on. Father Berthier, Castel's successor at the *Journal de Trévoux*, began what would become the defining crusade for himself, his journal, and his order in Enlightenment France when he published a very critical review of Diderot's initial prospectus in the *Journal de Trévoux* for February 1751.[60] Diderot responded by publishing a critical and fairly satirical rebuttal soon after, and this established a pattern.[61] As the subsequent volumes appeared, the Jesuit became more and more impassioned in his attacks against the text. He

59. Wilson, *Diderot*, 287–290; Hankins, *D'Alembert*, 95–103.

60. Review of Diderot, *Prospectus de l' Encyclopédie*, in *Journal de Trévoux* (1751): 841–863, 1677–1697.

61. *Lettre de M. Diderot au R. P. Berthier* (s.l., 1751), reprinted in *Oeuvres complètes de Diderot*, 13: 165–168.

further used the Jesuit journal as the organ for his views. This aligned the entire Society of Jesus in France against the editors of the *Encyclopédie,* and while contestation like this was nothing new for the Jesuit periodical, a new dynamic nevertheless developed, since unlike any previous institution, the *Encyclopédie* truly had no status in the corporate hierarchy of Old Regime France. It offered itself from one perspective as a rival academy, conceived just like other official societies as an organized union of independent *gens des lettres* devoted to science and the public good. What it lacked was recognized political sanction for its activities, and given this independence critics like Berthier came to see the *Encyclopédistes* as a rogue assembly that authorized libertines and freethinkers to subvert the established order. Accordingly, while earlier incidents of public intellectual warfare in France, including those triggered by the Jesuits at the *Journal de Trévoux,* had resulted in pledges of *honnête* reconciliation among the partisans or a peace brokered by the political authorities, the battle between Berthier and the *Encyclopédie* instead resulted in an escalating and irreconcilable war. The battles ultimately produced the battle lines of a new and more openly combative conception of Enlightenment.[62]

To be sure, Berthier had much to complain about with respect to Diderot and his text. The first volume contained the radically materialist discussion of the human soul that, while anonymous, may have been written by Voltaire. If so, then it constituted a direct link between the materialist discourses of Lockean and Newtonian philosophy articulated by Voltaire in the 1730s and the new philosophical radicalism of the 1750s. Also provocative was the article "authority," which largely discredited its value, even in theological matters.[63] Diderot's repeated deployment of his critical, libertine wit in the construction of the text further stirred the fires of controversy. Typical of these tendencies was the article "anthropophagie," or cannibalism, that appeared in the first volume. In it, Diderot, writing anonymously, laid out the conventional understanding of the practice by drawing heavily from travel narratives and other works of protoanthropology that had been widely available in France for almost two centuries. There was, in fact, nothing at all scandalous about Diderot's presentation until he used the structures of cross-referencing that he had celebrated in his initial prospectus to give the text a witty, anticlerical twist.[64] The system for cross-referencing articles that Diderot included throughout the *Encyclopédie* was intended to help readers find other articles of related interest. At the end of

62. On Berthier and the *Encyclopédie,* see Pappas, *Berthier and the Philosophes.*
63. Wilson, *Diderot,* 154.
64. The system of cross-referencing is discussed in his article "encyclopedia" as well. Diderot, *Rameau's Nephew and Other Works,* 294–296.

the discussion of cannibalism, Diderot referred readers to the article "Eucharist" for further clarification. Here was a witty yet subversive act of libertine anticlericalism that marked the work as anything but an erudite compendium. Moreover, since such critical witticisms were willfully intended and carefully chosen for effect, their mere presence marked the work as a dangerous tract and its contributors as subversive freethinkers eager to subvert the established order.

Most damning of all perhaps was the famous "Prades Affair" that followed in the wake of the appearance of volume 1.[65] This scandal appeared, at least to Berthier and his allies, to offer direct proof that the *Encyclopédie* had grown out of a conspiracy of libertine freethinkers and atheists hell-bent on disrupting the established structures of religion and morality.[66] The controversy began when a little-known abbé, Prades, defended a doctoral thesis at the Sorbonne in November 1751. The first volume of the *Encyclopédie,* with its scandalous article "ame," had already been circulating for over five months when the doctors of theology in Paris approved the abbé's thesis. Prades' work became the focus of a scandal when it was pointed out that the text defended many of the same positions found in the first volume of the *Encyclopédie.* Diderot and the other *Encyclopédists* quickly seized on this as an example of the disingenuous hypocrisy of official theology. Critics like Berthier, however, accused the *Encyclopédistes* of staging a plot to subvert the University of Paris and the established order more generally. One outcome of the controversy was stronger ties between Voltaire and d'Alembert, as the latter used the former's contacts in Berlin (*Dr. Akakia* was still months away) to get Prades admitted into the Berlin Academy.[67] Diderot had been admitted earlier that year as a result of the same connections, and this worked to sanction the position of each as they faced off against their clerical opponents in France. In January 1752, amid the scandals, volume 2 of the *Encyclopédie* appeared, but by February the entire project had been deemed disruptive to political and moral order and condemned by the royal authorities.[68]

Importantly, though, the *Encyclopédie* did not die with this suppression. The publishers merely shifted their operations across the border into Switzerland, changing the precise nature of their illegal operation in ways that permitted the continued publication and circulation of the books in France.[69] The produc-

65. Wilson discusses this scandal at length, *Diderot,* 154–158.

66. Pappas, *Berthier and the Philosophes,* 72–73.

67. Pappas, *Voltaire and d'Alembert,* 6–7.

68. Wilson, *Diderot,* 159–160, 161–169.

69. See Darnton, *Business of Enlightenment.*

tion of the volumes thus continued apace even if the controversy surrounding the text only increased as a result of its persecution. The scandal surrounding the *Encyclopédie,* in fact, never died away; it rather intensified throughout the 1750s, becoming on the one hand the rallying point for the newly aggressive philosophe movement and, on the other, the focus of the newly energized critics of philosophers and *philosophie* that Darrin MacMahon has recently exposed.[70] D'Alembert resigned from his post as editor amid these controversies, but he in no way repudiated the project. Moreover, even without his personal involvement in the project, the texts that he had already produced continued to contribute to the movement. The second part of his *Discours préliminaire,* in fact, narrated a history of human knowledge that gave important ideological sustenance to the *Encyclopédistes.* D'Alembert's history of modern thought gave successive authority to Bacon, Descartes, Newton, and the eighteenth-century Newtonians, especially those in France, in the creation of modern civilization. Accordingly, his narrative became the genealogical mythology invoked by *Encyclopédistes* and philosophes alike as they united in opposition to obscurantism and tyranny. D'Alembert took up this call as well, even if he did so in ways that also conformed to the obligations of his academic and other duties. He also suffered for his choices, not winning promotion inside the institution for two decades despite a record of continual scientific achievements that warranted professional advancement.[71]

The Royal Academy of Sciences in fact found itself in an uncomfortable position with respect to the new encyclopedic movement, and this too placed *Encyclopédistes* like d'Alembert into an awkward position. On the one hand, the science articulated by the text was by and large the academy's science as well. Also familiar was Diderot's principle of eschewing any one philosophical system while amassing particular knowledge through the individual contributions of disciplined experts. This was, in fact, little more than a reformulation of the ideology articulated continuously by the Royal Academy of Sciences for over half a century.[72] In many respects, therefore, not least in their shared attitudes toward Newtonian science, the *Encyclopédie* and the Paris Academy were brothers in arms in the new project of Enlightenment. The Academy of Sciences would always remain the institution most consonant with the ideals and values of the philosophes (Voltaire had celebrated it as well in his *Lettres philosophiques*), yet the antiestablishment currents that also defined the move-

70. McMahon, *Enemies of Enlightenment.*

71. See Hankins, *d'Alembert,* 43–44.

72. See Fontenelle's 1699 "Preface" to the first volume of *Histoire et mémoires de l'Académie royale des sciences,* in *Oeuvres de Fontenelle,* 6: 37–50.

ment cut against this alliance as well. One source of very public tension involved the mechanical arts. The *Encylopédie* in effect realized the dream, announced a century earlier by Colbert and Louis XIV, of producing a set of volumes describing the industrial and mechanical arts in the kingdom. Despite repeated efforts to accomplish this goal, the academy had failed in this mission. Diderot further accentuated this failure by publishing, for the first time, a set of engravings produced in conjunction with these ill-fated royal volumes. For critics of the *Encyclopédie,* like Elie Fréron, this act of "plagiarism" (it was actually an act of business savvy) demonstrated the corruption at the heart of the Encyclopedic movement. For others, however, it reinforced the perception, first raised during the fierce academic opposition to Newtonianism, that the Royal Academy was a bastion of traditionalism in need of reinvigoration by independent savants like those active with the *Encyclopédie.*[73]

Voltaire and d'Alembert would soon conclude something similar, launching their own program to advance Enlightenment by stacking the French academic establishment with *Encyclopédistes* and others loyal to their *parti philosophique.*[74] Voltaire's public attacks on Maupertuis, which began in the immediate wake of the condemnation of the *Encyclopédie* in France, further returned him to his previous critical tendencies, and these in turn placed him at the center of the new philosophe movement triggered by the cause of the *Encyclopédie.* By 1757, the anointed leader of the *parti philosophique* in France was complaining to d'Alembert that the articles in the *Encyclopédie* had become too orthodox while also strategizing with him ways to get his loyalists into positions of intellectual prominence in France. In this way, as John Pappas writes, "[the *Encyclopédie*] served to make [Voltaire] aware of his raison d'être." It "imbued him with a sense of mission and opened before him the vision of his role as leader and defender of the philosophes."[75] The campaign to *l'ecrasez l'infâme* had begun.

The *Encyclopédie* did indeed launch Voltaire and the philosophe movement in the way that Pappas describes, but it was a beginning that was prepared by over a half-century of cultural and institutional changes in France. It was also conditioned by the particular reception of Isaac Newton's science in France. This reception triggered changes in the way that the French thought about the natural world, but it also, and perhaps more importantly, changed the way they understood the role of the philosopher in society. Looking backward in

73. On this controversy and the mechanical arts initiatives within the *Encyclopédie* more generally, see Proust, *Diderot et l'Encyclopédie,* 189–232.

74. This plan is explored most fully in Pappas, *Voltaire and d'Alembert.*

75. Ibid., 12.

1787, two years before the fall of the Bastille would make his pronouncements prophetic, the French savant Rulhière told his colleagues in the Académie française that 1749 marked a watershed in the history of France. It was a turning point provoked by a change in the character of French learning. The year 1749 saw an explosion in the number of philosophical works, he explained, and these books provoked "a succession of unfortunate events that little by little and from day to day stripped from the government that public approbation that up until that time it had enjoyed." The nation passed from "the love of belles lettres to the love philosophy," he mourned, and the result was a new era of complaints, perpetual remonstrances, and disrespectful clamoring regarding religion and morals.

At the center of this decay was the new "empire of public opinion." "Men of letters immediately had the ambition to be its organs, and almost its arbiters," Rulhière sighed, and this led to a new desire to "instruct rather than to please." The phrase *dignity of men of letters* "quickly became an approved expression, and one in common use." [76] Noting the same shifts a century later, the editor of the diaries of Barbier, one of France's most active and engaging eighteenth-century courtiers, wrote that 1749 was a remarkable year in the literary history of France. "It was on this date that writings hostile to religion appeared and multiplied, and that war broke out between skepticism and faith." Barbier's diary attested to the shift, for in this year, almost at once, the courtier stopped speaking of "ballad writers and poets" and started speaking of philosophes." "It is in this year that the real eighteenth century begins." [77]

By the real eighteenth century, Barbier's editor meant the century of Enlightenment, the Enlightenment described by Rulhière as a self-conscious movement among men of letters to use philosophy to challenge and change the established order of things. At one level this book agrees with this assessment. The Enlightenment defined as a self-conscious movement led by *gens des lettres* to defend intellectual liberty and the philosophical positions it produced against the authority of clerics, censors, and royal officials did commence after 1749 with the launching of the *Encyclopédie*. However, the *Encyclopédie* was not sui generis, and the Enlightenment that it spawned was likewise not without earlier beginnings as well. In particular, it took changes within the institutional character of French learning over the course of the early eighteenth century to make these initiatives possible.

Here the transformations within the international Republic of Letters triggered by the new structures of publicity that opened up after 1690 are most

76. Cited in Wilson, *Diderot*, 94–95.
77. Ibid., 94.

important. Also crucial was the particular way that these structures were negotiated within Old Regime France. Important as well was the legacy of Isaac Newton, a legacy in one part intended and scientific, but in another and equally important part unintended and cultural. Newton's scientific work did occasion a host of cultural shifts driven by their own intellectual logic. However, since his science quickly fragmented into a variety of competing Newtonianisms, and since each of these conceptions played a different role in the scientific and cultural changes that transpired in different places at different times, it is impossible to make Newton the agent of his own destiny. Rather, his legacy was made throughout the Republic of Letters by savants, caught in their own web of interests and ideals, as they deployed and debated his ideas in the service of their own agendas.

In France, this negotiation produced a variety of outcomes, some more directly connected to the French Enlightenment than others. Least connected was the tradition of analytical mechanics, pioneered in the 1690s by l'Hôpital and Varignon while working in the orbit of Malebranche, Leibniz, Johann Bernoulli, and eventually Newton. By the 1750s this highly mathematicized approach to mechanics had come to be called Newtonian mechanics, a fact that allowed its practitioners, whatever their commitment to the critical intellectual program associated with this label, to be associated with Enlightenment through the veneration of what they perceived to be Newtonian science. However, the science of the *Principia* was not identical with the analytical mechanics practiced in its name after 1750, and similarly the link between Enlightenment and this highly mathematical brand of physics—the science of Clairaut and d'Alembert, and in subsequent decades Lagrange and Laplace—was not a natural bond but a particular and contingent historical fusion. Laplace was not an Enlightenment philosophe because he practiced Newtonian science, and likewise Newtonian science was not the engine of Enlightenment because it became the dominant physics of the period. Rather, each possessed its own history, and the particular historical contingencies that united them are the real source of their connection.

Nevertheless, Newtonian science played a crucial role in the beginning of the French Enlightenment even it was not because of Newton's direct influence in the creation of eighteenth-century mathematical physics. Instead, the changes were first initiated by the way that his peculiar reception within the political climate of England and the Low Countries after 1690 conspired to create a new field of discourse around his name and work. This generated a new field of political contestation within the Republic of Letters that was then catalyzed by the accidents of the priority dispute and the new social possibilities created by the expanding print public sphere and its new and more openly critical style of discourse. This style then found its way into France through these same channels, bringing with it the attachments to Newton's name forged in these contexts. Further contingencies then conspired to align large sectors of the French intellectual establishment against this new cultural formation, and in this way the field was prepared for Voltaire's decisive intervention. He triggered the opening of the Enlightenment by using Newton to initiate controversies that worked to define a new critical intellectual style in France. This style was then adopted by Diderot and the *Encyclopédistes*, among others, and as the French learned world found itself engaged after 1750 in an open, public, and aggressively critical exchange over the nature of *philosophie*, and its place in the moral and political order, the French Enlightenment was underway. Newtonianism remained the banner under which the philosophe movement grew, but it was less the science of Newton itself than the ideal of critically defending *philosophie* that sustained this identity.

Because Newtonianism had arrived in France as an outlaw and only became accepted through the sustained antiestablishment efforts of its defenders, it became the creed of the *parti philosophique*. In narrating their own history, however, the philosophes rarely paid attention to the accidents and contingencies that had actually shaped this history. Like all political movements, they constructed a mythology of origins consonant with their own interests, and to their credit their story has proven so powerful that it has since been accepted uncritically by historians and modern Enlightenment ideologues ever since. Speaking from a position somewhere outside the Enlightenment, this book has offered a different and more complex account of this history, one that has tried to show the actual historical linkages that tied Newton to Enlightenment in France. Newton's solitary genius is still offered far too often as the singular reason for his status as the father of modern physics, while the French Enlightenment continues to be celebrated too frequently as Newton's natural and unmediated offspring. Each of these mythologies needs to be scrutinized, and if this book strikes a heavy blow against this overly pervasive and persuasive edifice, it will have accomplished its goal.

Bibliography

MANUSCRIPT COLLECTIONS

Archives de l'Académie des Sciences, Paris, France
Collection des réglemens et déliberations de l'académie royale des sciences par ordre
de matière, avec table des matiéres. MSS 1, J 1.
Dossiers of individual academicians: Jean Le Rond d'Alembert, Guillaume
Amontons, Abbé Jean-Paul Bignon, ? Bomie, Pierre Bouguer, Charles Étienne
Louis Camus, François-Joseph de Camus, Louis Carré, Alexis-Claude Clairaut,
Guillaume Delisle, Joseph-Nicolas Delisle, Gilles Filleau Des Billettes,
Jean-Jacques Dortous de Mairan, Jean-Paul Grandjean de Fouchy, Jean Gallois,
Étienne-Simon de Gamaches, Étienne François Geoffroy, Louis Godin, Thomas
Gouye, ? Guisnée, Nicolas Louis de La Caille, Charles Marie de La Condamine,
Phillipe de La Hire, Gabriel-Philippe de La Hire, Joseph-Jérôme Lefrançois
de Lalande, Pierre Le Monnier, Pierre-Charles Le Monnier, Jacques-Eugène
d'Allonville de Louville, Pierre-Louis Moreau de Maupertuis, Joseph Privat
de Molières, Pierre Rémond de Montmort, François Nicole, René Antoine de
Réaumur, ? Saulmon, Joseph Saurin, Pierre Sigorgne, Jean Baptiste Terrasson,
and Pierre Varignon
Lalande, Jerôme. "Annotated Reglements of the Paris Academy of Sciences."
Microfilm copy of MS in the Biblioteque Laurenziana, Florence.
"Rapport des ouvrages qui ont été lus dans les assemblés de l'Académie des sciences
depuis la rentrée de la saint Martin {...} jusques à celle de Pasques {...}, suivi du
Rapport des ouvrages lus dans les assemblées de l'Académie des sciences depuis la
rentrée de Paques {...} jusques à celle de la Saint Martin de la meme année, par M.
de Montigny." 9 vols. MSS 1 J 23.
Registres manuscrits des Procès-verbaux des séances de l'Académie Royale des
Sciences de Paris.

Archives Nationales, Paris, France
Correspondence of J.-N. Delisle. Archive de la Marine, 2JJ 60–69.
"Pièces justicatives" pertaining to the "affair of the couplets," X2B 925–927.

Archives de l'Observatoire, Paris, France
Bigourdan, Guillaume. "Catelogue de la correspondence de J.-N. Delisle." MSS 1029a.
Correspondence of J.-N. Delisle. B1 -8.

Bibliothèque de l'Arsenal, Paris, France
Dossiers of Charles Marie La Condamine. MSS 11830; 12493; 12,495; 12, 498–499.
Jean-Paul Grandjean de Fouchy. "Académie Royale des Sciences. Collections de ses
Réglements et Déliberations par ordre de Matieres." MS 4624.

Information concerning Maupertuis's early years is found in the dossiers of Laurent
  Angliviel de La Beaumelle: 11,830, 12,493, 12,495, 12,498, and 12,499.
"Mémoire où l'on examine s'il est avantageux que l'Académie des sciences qui depuis
  sa fondation n'a eu qu'un sécrétaire, en ait deux aujourd'hui, l'un pour la partie
  physique, l'autre pour la partie mathématique." MS 7464, ff. 48–53
"Observations diverses, mémoires et pièces detachés conçernant les arts et sciences."
  MS 6130.
Papers allegedly given to the Socièṭè des arts are found in the dossier of La
  Beaumelle: 11,830, 12,493, 12,495, 12,498, and 12,499.
"Réglement de la Société des Arts (Paris 1730)." MS 12,727, ff. 981–991.

Bibliothèque de l'Institut, Paris, France
D'Alembert's lecture notes from the Collège royale are found in MS 2469.

Bibliotèque nationale de France, Paris, France
Correspondence of La Condamine. MSS N.A. f. fr. 21015.
Diary of Claude Bourdelin. MS N.A. f. fr., 5148.
Diary of J.-J. Lalande. MS f. fr. 12274.
"Observations et correspondances astronomiques, XVIIe et XVIIIe S." MS N.A. fr.
  6197.
Papers of the abbé Jean-Paul Bignon. MSS f. fr. 22225–22236.
Papers of Emilie du Châtelet. MSS f. fr. 12265–12269.
Papers of J.-N. Delisle. MSS f. fr 9678.
Papers of Malesherbes. MSS f. fr. 22133.

Bibliothèque universitaire et cantonale, Lausanne, Switzerland
Fonds Jean-Pierre de Crousaz.

Stockholm National Museum, Stockhom, Sweden
A list of members of the Société des Arts is found in C.C. 354.

Universitaetsbibliothek, Basel, Switzerland
Papers and correspondence of the Bernoulli family.

LEARNED PERIODICALS FROM THE
SEVENTEENTH AND EIGHTEENTH CENTURIES

Bibliothèque ancienne et moderne. 28 vols. Amsterdam, 1714–27.
Bibliothèque anglaise. 22 vols. Amsterdam, 1717–1728.
Bibliothèque britannique, ou Histoire des ouvrages des savans de la Grand-Bretagne.
  25 vols. La Haye, 1733–1747.
Bibliothèque choisie. 22 vols. Amsterdam, 1703–1713.
Bibliothèque germanique. 20 vols. Berlin, 1720–1740.
Bibliothèque raisonnée des ouvrages des savants de l'Europe. 52 vols. Amsterdam,
  1728–1753.
Bibliothèque universelle et historique. 16 vols. Amsterdam, 1686–1693.

L'Europe savante. 12 vols. The Hague, 1718–1719.
Histoire de l'Académie royale des sciences. Avec les mémoires de mathématique et de physique . . . tirés des registres de cette académie. 92 vols. Paris, 1702–1797.
Journal des savants. 127 vols. Paris, 1665–1792.
Journal historique de la république des lettres. 3 vols. Leiden, 1732–1733.
Journal littéraire de la Haye. 23 vols. La Haye, 1713–22, 1728–37.
Jugemens sur quelques ouvrages nouveaux. 11 vols. Avignon, 1744–1746.
La nouvelliste du Parnasse, ou Réfléxions sur les ouvrages nouveaux. 4 vols. Paris, 1731–1732.
Mémoires pour servir à l'histoire des sciences et des beaux arts. 67 vols. Trévoux, 1701–1767.
Mémoires secrets de la république des lettres. 7 vols. Amsterdam, 1744–1759.
Mercure galant. 45 volumes. Paris, 1672–1714.
Mercure de France. 148 volumes. Paris, 1724–1820.
Nouveaux amusemens du coeur et de l'esprit. 15 vols. Paris, 1737–1745.
Observations sur les écrits modernes. 31 vols. Paris, 1735–1742.
Le pour et contre, ouvrage périodique d'un goût nouveau. 20 vols. Paris, 1733–1740.
Réflexions sur les ouvrages de literature. 12 vols. Paris, 1737–1741.

PUBLISHED WORKS

Acerra, Martine. *Les marines de guerres européennes XVIIe–XVIIIe siècles.* Edited by Martine Acerra, José Merino, and Jean Meyer. Paris: Presses de l'Université de Paris-Sorbonne, 1985.
———. *Rochefort et la construction navale française, 1661–1815.* Paris: Libr. de l'Inde, 1993.
Adams, D. J. *Bibliographie d'ouvrages français en forme de dialogue, 1700–1750.* Oxford: Voltaire Foundation, 1992.
Adams, Robert. *Leibniz: Determinist, Theist, Idealist.* Oxford: Oxford University Press, 1994.
Aiton, E. J. "The Application of the Infinitesimal Calculus to Some Physical Problems by Leibniz and His Friends." *Studia Leibnitiana* 14 (1985): 133–143.
———. "The Celestial Mechanics of Leibniz." *Annals of Science* 16 (1960): 65–82.
———. "The Celestial Mechanics of Leibniz: A New Interpretation." *Annals of Science* 20 (1964): 111–123.
———. "The Celestial Mechanics of Leibniz in Light of Newtonian Criticism." *Annals of Science* 18 (1962): 31–41.
———. "The Inverse Problem of Central Forces." *Annals of Science* 20 (1964): 81–99.
———. "An Imaginary Error in the Celestial Mechanics of Leibniz." *Annals of Science* 21 (1965): 169–173.
———. "The Mathematical Basis of Leibniz's Theory of Planetary Motion." *Studia Leibnitiana* 13 (1984): 209–225.
———. *The Vortex Theory of Planetary Motions.* New York: American Elsevier, 1972.

D'Alembert, Jean le Rond. "Essai sur la société des gens de lettres et des Grands, sur la reputation, sur les mécènes, et sur les récompenses littéraire." In *Oeuvres complètes de d'Alembert.* Edited by Bossange. 5 vols. Paris: Belin, 1821–1822.

———. *Mélanges de littérature, d'histoire et de philosophie.* 4th ed. 5 vols. Amsterdam: Z. Chatelain, 1767.

———. *Preliminary Discourse to the Encyclopedia of Diderot.* Translated by Richard N. Schwab. 2d ed. Chicago: University of Chicago Press, 1995.

———. *Traité de dynamique.* Paris: Chez David, 1743.

Algarotti, Francesco. *Il newtonianismo per le dame ovvero dialoghi sopra la luce e i colori.* Naples: Giambatista Pasquale, 1737.

Andrewes, William J. H., ed. *The Quest for Longitude.* Cambridge: Collection of Historical Scientific Instruments, Harvard University, 1996.

Antoine, Michel. *Louis XV.* Paris: Fayard, 1989.

———. *Le conseil du roi sous le regne de Louis XIV.* Geneva: Droz, 1970.

Argenson, René-Louis de Voyer, marquis d'. *Considérations sur le gouvernement ancien et présent de la France.* Amsterdam: Chez Marc-Michel Rey, 1764.

Arnauld, Antoine, and Pierre Nicole. *Logic or Art of Thinking.* Translated and edited by Jill Vance Buroker. Cambridge: Cambridge University Press, 1996.

———. *La logique, ou L'art de penser.* Edited by Pierre Clair and François Girbal. Paris: Presses universitaires de France, 1965.

Artz, Frederick B. *The Development of Technical Education in France, 1500–1789.* Cambridge: MIT Press, 1966.

Aucoc, Leon, ed. *L'Institut de France. Lois status, et règlements conçernant les anciennes academies et l'Institut de France depuis 1635 jusqu'à 1889.* Paris: Imprimerie nationale, 1889.

Azouvi, François. *Descartes et la France: Histoire d'une passion nationale.* Paris: Fayard, 2002.

Bachelard, Suzanne. *Les polémiques conçernant le principe de moindre action au XVIIIe siècle.* Paris: Université de Paris, Palais de la découverte, 1961.

Badinter, Elizabeth. *Émilie, Émilie, l'ambition féminine au XVIIIe siècle.* Paris: Flammarion, 1983.

———. *Les passions intellectuelles.* 3 vols. Paris: Fayard, 1999–2007.

Bailly, Jean Sylvain. *Histoire de l'astronomie moderne depuis la fondation de l'école d'Alexandrie, jusqu'à l'epoque de M.D.CC.LXXXII.* 3 vols. Paris: Freres De Bure, 1779–1782.

Baker, Keith M. *Condorcet: From Natural Philosophy to Social Mathematics.* Chicago: University of Chicago Press, 1974.

———. "Enlightenment and the Institution of Society: Notes for a Conceptual History." In *Main Trends in Cultural History.* Edited by Willem Melching and Wyger Velema. The Hague: Rodopi, 1995.

———. "A Foulcauldian French Revolution?" In *Foucault and the Writing of History.* Edited by Jan Goldstein. Oxford: Oxford University Press, 1994.

———. *Inventing the French Revolution: Essays on French Political Culture in the Eighteenth Century.* Cambridge: Cambridge University Press, 1990.

Balcou, Jean. *Fréron contre les philosophes.* Geneva: Droz, 1975.

Ballantyne, Archibald. *Voltaire's Visit to England, 1726–1729.* London: Murray, 1919.

Banières, Jean. *Examen et réfutation des Elémens de la philosophie de Neuton de M. Voltaire, avec une dissertation sur la réflexion & la réfraction de la lumière.* Paris: Lambert, Durand, 1739.

Banks, Kenneth. *Chasing Empire across the Sea: Communications and the State in the French Atlantic, 1713–1763.* Montreal: McGill–Queen's University Press, 2002.

Barber, W. H. *Leibniz in France from Arnauld to Voltaire: A Study in French Reactions to Leibnizianism, 1670–1760.* Oxford: Clarendon Press, 1955.

———. "Mme de Châtelet and Leibnizianism: The Genesis of the *Institutions de Physique.*" In *The Age of Enlightenment: Studies Presented to Theodore Besterman.* Edited by William Henry Barber. Edinburg and London: Oliver and Boyd, 1967.

———. "Voltaire and Samuel Clarke." In *Studies on Voltaire and the Eighteenth Century,* no. 179. Edited by Theodore Besterman. Geneva: Institut et musée Voltaire: 1979.

Barnes, Annie. *Jean le Clerc et la république des lettres.* Paris: E. Droz, 1938.

Barrell, Rex A. *Bolingbroke and France.* Lanham: University Press of America, 1988.

Barrière, Pierre. *L'Académie de Bordeaux: Centre de culture internationale au XVIIIe siècle (1712–1792).* Bordeaux: Éditions Bière, 1951.

Barruel, Augustin. *Les helviennes, ou Lettres provinciales philosophiques.* 4 vols. Paris: La Société typographique de Méquignon Fils Aîné, et Boiste Père, Imprimerie d'A. Clo, 1830.

Bartholmess, Christian. *Histoire philosophique de l'Académie de Prusse depuis Leibniz jusqu'a Schelling.* 2 vols. Paris: Franck, 1850.

Beer, Peter, ed. *Newton and the Enlightenment: Proceedings of an International Symposium, Held at Cagliari, Italy, on 3–5 October 1977.* New York: Pergamon Press, 1979.

Beeson, David *Maupertuis: An Intellectual Biography.* Oxford: Voltaire Foundation at the Taylor Institution, 1992.

Belin, Jean Paul. *Le commerce des livres prohibés à Paris de 1750 à 1789.* New York: B. Franklin, 1967.

———. *Le mouvement philosophique de 1748 à 1789: Étude sur la diffusion des idées des philosophes à Paris d'après les documents concernant l'histoire de la librairie.* Paris: Belin frères, 1913.

Berkeley, George, bishop of Cloyne. *The Analyst; or A Discourse Addressed to an Infidel Mathematician.* London: Printed for J. Tonson, 1734.

Bernoulli, [Johann] I. *Der Briefwechsel von Johann I. Bernoulli.* Edited by O. Spiess, Pierre Costabel, and Jeanne Peiffer. 3 vols. Basel: Birkhäuser Verlag, 1955, 1988, and 1992.

———. "Essai d'une nouvelle physique céleste." In *Recueil des pièces qui ont remporté les prix de l'Académie Royale des Sciences*, 3: 5–88. Paris: Académie des sciences, 1720–1772.

———. *Nouvelles pensées sur le système de M. Descartes*. Paris: Chez Claude Jombert, 1730.

Bertoloni-Meli, Domenico. *Equivalence and Priority: Newton versus Leibniz*. Oxford: Clarendon Press, 1993.

Bertrand, Joseph. *L'Académie des sciences et les académiciens de 1666 à 1793*. Paris: J. Hetzel, 1869.

Bertucci, Paola. "Back from Wonderland: Jean Antoine Nollet's Italian Tour (1749)." In *Curiosity and Wonder from the Renaissance to the Enlightenment*. Edited by R. J. W. Evans and Alexander Marr. New York: Ashgate, 2006.

———. *Sparking Controversy: Jean Antoine Nollet and Medical Electricity South of the Alps*. Florence: Leo S. Olschiki, 2005.

Biagioli, Mario. *Galileo, Courtier: The Practice of Science in the Culture of Absolutism*. Chicago: University of Chicago Press, 1993.

———. *Galileo's Instruments of Credit: Telescopes, Images, Secrecy*. Chicago: University of Chicago Press, 2006.

———. "Etiquette, Interdependence, and Sociability in Seventeenth-Century Science." *Critical Inquiry* 22 (1996): 193–238.

Bigourdan, Guillaume. *Histoire de l'astronomie d'observation et des observatoires en France*. 2 vols. Paris: Gauthier-Villar, 1930.

Bila, C. *La croyance à la magie au XVIIIe siècle en France*. Paris: J. Gamber, 1925.

Birembaut, Arthur. "Les liens de famille entre Réamur et Brisson, son dernier élève." *Revue d'histoire des sciences* 11 (1958): 167–169.

Birn, Raymond. "*Le Journal des savants* sous l'ancien régime." *Journal des savants* (January–March 1965): 15–35.

Blanchard, Anne. *Dictionnaire des ingénieurs militaires, 1691–1791*. Montpellier: Université Paul-Valéry, 1979.

———. *Les ingénieurs du "roy" de Louis XIV à Louis XVI*. Montpellier: Université Paul-Valéry, 1979.

Blay, Michel. "Castel critique de la théorie newtonienne des couleurs." In *Autour du Pere Castel et du claevecin oculaire*. Edited by Roland Mortier and Hervé Hasquin. Études sur le XVIIIe siècle, 27. Brussells: Éditions de l'Université de Bruxelles, 1995.

———. "L'introduction du calcul différentiel en dynamique: L'example des forces centrales dans les mémoires de Varignon en 1700." *Sciences et Techniques en Perspective* 10 (1985–1986): 157–190.

———. "Quatre mémoires inédites de Pierre Varignon consacré à la science du mouvement." *Archives internationales d'histoire des sciences* (1989): 218–248.

———. *La naissance de la mécanique analytique: La science du mouvement au tournant des xviie et xviiie siècles*. Paris: Presses universitaires de France, 1992.

———. *Les raisons de l'infini: Du monde clos à l'univers mathématique.* Paris: Gallimard, 1993. Translated as *Reasoning with the Infinite: From the Closed World to the Mathematical Universe* by M. B. DeBevoise. Chicago: University of Chicago Press, 1998.

Bléchet, Françoise. "Un précurseur de l'*Encyclopédie* au service de l'état: L'abbé Bignon." In *L'encyclopédisme: Actes du colloque de Caen, 12–16 janvier 1987.* Paris: Klincksieck, 1991.

———. "Le rôle de l'abbé Bignon dans l'activité des sociétés savantes au XVIIIe siècle." *Actes du 100e Congrès National des Sociétés Savantes, Section d'histoire moderne et contemporaine et Commission d'histoire des sciences et des techniques,* 31–41. Paris: Bibliothèque Nationale 1979.

———. "Fontenelle et l'abbé Bignon, du president de l'Academie royale des Sciences au secretaire perpétuel: Quelques lettres de l'abbé Bignon à Fontenelle." *Corpus, revue de philosophie* 13 (1990): 51–61.

Bluche, François. *Louis XIV.* Translated by Mark Greengrass. Oxford: Blackwell, 1990.

Boisnard, Luc. *Les Phélypeaux: Un famille des ministres sous l'ancien régime.* Paris: Sedopols, 1986.

Boissier, Gaston. *L'Académie française sous l'ancien regime.* Paris: Hachette, 1909.

Bonnardet, E. *Les fondations de prix à l'Académie des sciences.* Paris: Gauthier-Villars, 1881.

Bonnassieux, Pierre, and Eugène Lelong, eds. *Conseil de Commerce et Bureau de Commerce, 1700–1791: Inventaire analytique des procès-verbaux.* Paris: Imprimerie National, 1900.

Bonney, Richard, ed. *Economic Systems and State Finance.* Oxford: Clarendon Press, 1995.

———. "What's New about the New French Fiscal History?" *Journal of Modern History* 70 (1998): 639–667.

Bos, Henk J. M. "Differentials, Higher-Order Differentials and the Derivative in the Leibnizian Calculus." *Archive for the History of the Exact Sciences* 14 (1974–1975): 1–90.

———. *Redefining Geometrical Exactness: Descartes' Transformation of the Early Modern Concept of Construction.* New York: Springer-Verlag, 2000.

Boss, Valentin. *Newton and Russia: The Early Influence, 1698–1796.* Cambridge: Harvard University Press, 1972.

Bouguer, Pierre. "Entretiens sur la cause de l'inclinaison des planètes." In *Recueil des pièces qui ont remporté les prix de l'Académie Royale des Sciences.* Paris: C. A. Jombert, 1752.

Bouhours, Dominique. *Les entretiens d'Ariste et Eugène.* Paris: S. Mabre-Cramoisy, 1671.

———. *La manière de bien penser dans les ouvrages de l'esprit.* Paris: F. Delaulne 1687.

Bouhier, Jean. *Correspondance littéraire.* Edited by Henri Duranton. Lyons: Centre de Sant-Étienne, 1974–1988.

Bouillier, Francisque. "Divers projets de réorganisation des anciennes académies." *Séances et travaux de l'Académie des Sciences Morales et Politiques* 115 (1881): 636–673.

———. "Une parfaite académie d'après Bacon et Leibniz." *Revue des deux mondes* 29 (1878): 673–697.

———. *Histoire de la philosophie cartésienne*. 2 vols. Paris: C. Delagrave et cie, 1854.

Boullée, Étienne. *Treatise on Architecture*. Edited by Helen Roussenau. London: A. Trianti, 1953.

Bourde, Andre J. *Agronomie et agronomes en France au XVIIIe siècle*. Paris: S.E.V.P.E.N, 1967.

———. *The Influence of England on the French Agronomes, 1750–1789*. Cambridge: Cambridge University Press, 1953.

Bourdieu, Pierre. *Distinction: A Social Critique of the Judgment of Taste*. Translated by Richard Nice. Cambridge: Harvard University Press, 1987.

———. *Outline of a Theory of Practice*. Translated by Richard Nice. Cambridge: Cambridge University Press, 1977.

Bouy, J. Th. de. "Denis Diderot. Ecrits inconnus de jeunesse." *Studies on Voltaire and the Eighteenth Century*, vol. 119. Oxford: Voltaire Foundation, 1974.

Boyer, Carl B. "The First Calculus Textbooks." *Mathematics Teacher* 39 (1946): 159–167.

———. *The History of the Calculus and Its Conceptual Development*. New York: Dover, 1959.

———. "The New Math of the 1740s in England and France." In *XIIe Congrès International d'Histoire des Sciences*. Paris: A. Blanchard, 1970–1971.

Brennan, Thomas. *Public Drinking and Popular Culture in Eighteenth-Century Paris*. Princeton: Princeton University Press, 1988.

Brian, Eric. *Mesure de l'état: Administrateurs et geometres au XVIIIe siècle*. Paris: Albin, Michel, 1994.

Briggs, E. R. "The Political Academies of France in the Early Eighteenth Century with Special Reference to the Club de l'Entresol and Its Founder, the Abbé Pierre Joseph Alary." Ph.D. dissertation, Cambridge University, 1931.

Briggs, J. Morton. "Aurora and Enlightenment." *Isis* 58 (1967): 491–503.

Brockliss, Laurence. *Calvet's Web: Enlightenment and the Republic of Letters in Eighteenth-Century France*. Oxford: Oxford University Press, 2002.

———. *French Higher Education in the Seventeenth and Eighteenth Centuries: A Cultural History*. Oxford: Oxford University Press, 1987.

Brockliss, Laurence, and Colin Jones. *The Medical World of Early Modern France*. Oxford: Clarendon Press, 1997.

Bromley, John S. "The Loan of French Naval Vessels to Privateering Enterprises (1688–1713)." In *Les marines de guerres européennes XVIIe–XVIIIe siècles*. Edited by Martine Acerra et al., 65–90. Paris: SEDES, 1985.

Brown, Harcourt. "From London to Lapland: Maupertuis, Johann Bernoulli, and *La terre applatie, 1728–1738*." In *Literature and History in the Age of Ideas*. Edited by

Charles G. S. Williams, 65–90. Columbus: Ohio State University Press, 1975.

———. *Scientific Organizations in Seventeenth-Century France, 1620–1680.* New York: Russell and Russell, 1967.

———. *Voltaire and the Royal Society of London.* Toronto: University of Toronto Quarterly, 1947.

———. "The Composition of the Letters concerning the English Nation." In *The Age of Enlightenment: Studies Presented to Theodore Besterman.* London: Oliver & Boyd, 1967.

Brumfitt, J. H. *The French Enlightenment.* Cambridge: Schenkman Pub. Co., 1973.

———. "History as Propaganda in Voltaire." *Studies on Voltaire and the Eighteenth Century* 24 (1963): 271–287.

———. *Voltaire: Historian.* London: Oxford University Press, 1970.

Brunel, Lucien. *Les philosophes et l'Académie française au dix-huitième siècle.* Geneva: Slatkine Reprints, 1967.

Brunet, Pierre. *L'introduction des théories de Newton en France au XVIIIe siècle.* Paris: A. Blanchard, 1931.

———. *La vie et l'oeuvre de Clairaut (1713–1765).* Paris: Presses Universitaires de France, 1952.

———. *Les physiciens hollandais et la méthode expérimentale en France au XVIIIe siècle.* Paris: A. Blanchard, 1926.

———. *Maupertuis, étude biographique.* Paris: A. Blanchard, 1929.

———. *Maupertuis, l'oeuvre et sa place dans la pensée scientifique et philosophique du XVIIIe siècle.* Paris: A. Blanchard, 1929.

Brunschvig, Léon. *Les étapes de la philosophie mathématique.* Paris: F. Alcan, 1912.

Buffier, Claude. *Traité de la société civile et du moyen de se rendre heureux.* Paris: Chez Pierre-François Giffart, 1726.

Buffon, George-Louis Leclerc, Comte de. *Histoire naturelle, génerale et particulière, avec la description du Cabinet du Roy.* 15 vols. Paris: l'Imprimerie royale, 1749–1767.

———. *La statique des végétaux, et l'analyse de l'air.* Translated by Stephen Hales. Paris: Chez J. Vincent, 1735.

Burtt, E. A. *The Metaphysical Foundations of Modern Science.* London: Routledge and K. Paul, 1932.

Bush, Newell Richard. *The Marquis d'Argens and His Philosophical Correspondence.* Ann Arbor: Edwards Brothers, 1953.

Callot, Emile. *Maupertuis: Le savant et le philosophe.* Paris: M. Rivière, 1964.

———. *Six philosophes français du XVIIIe siècle: La vie, l'ouvre et la doctrine de Diderot, Fontenelle, Maupertuis, La Mettrie, d'Holbach, Rivarol.* Annecy: Gardet, 1963.

Campbell, Peter. *Power and Politics in Old Regime France, 1720–1745.* London and New York: Routledge, 1996.

Casaban, Bernard de. "Joseph Saurin, member de l'Académie royale des sciences de Paris (1655–1737)." In *Mémoires de l'Académie de Vaucluse*, 6e série, Tome I, 2eme semester. Avignon: Seguin, 1968.

Casini, Paolo. "Briarée en miniature: Voltaire et Newton." *Studies on Voltaire and the Eighteenth Century* 179 (1979): 70–72.

———. "Le 'newtonianisme' au siècle des lumières. Recherches et perspectives." *Dix-huitième siècle* 1 (1979): 139–159.

Cassini, Jean Dominique. *Réponse à la dissertation de M. Celsius*. Paris: Imprimerie royale, 1738.

———. *Traité de la grandeur et de la figure de la terre*. Paris: Chez Pierre de Coup, 1723.

Cassirer, Ernst. *The Philosophy of the Enlightenment*. Translated by Fritz C. A. Koelln and James P. Pettegrove. Princeton: Princeton University Press, 1951.

Castel, R. P. Louis-Bertrand de. *Mathématique universelle, abregée à la portée de tout le monde*. Paris: Chez Pierre Simon, 1728.

———. *La géometrie naturelle en dialogues*. In *Nouveaux amusemens du coeur et d'esprit* 1 (1741): 80–93; 185–202.

———. *L'optique des coleurs, fondée sur les simples observations*. Paris: Briasson, 1740.

———. *Le vrai systeme de physique génerale de M. Issac Newton, exposé et analysé en parallele avec celui de M. Descartes*. Paris: Chez Claude-François Simon, Fils, Imprimeur-Libraire . . ., avec approbation et privilege du Roi, 1743.

———. *Traité physique sur la pesanteur universelle des corps*. Paris: A. Cailleau 1724.

Chakrabarty, Dipesh. *Provincializing Europe: Postcolonial Thought and Historical Difference*. Princeton: Princeton University Press, 2000.

Chandler, Philip. "Clairaut's Critique of Newtonian Attraction: Some Insight into His Philosophy of Science." *Annals of Science* 32 (1975): 369–378.

Chandrasekhar, S. *Newton's* Principia *for the Common Reader*. Oxford: Clarendon Press, 1995.

Chapin, Seymour L. "The Academy of Sciences during the Eighteenth Century: An Astronomical Appraisal." *French Historical Studies* 5 (1968): 371–404.

———. "Astronomy and the Paris Academy of Sciences during the Eighteenth Century." Ph.D. dissertation, University of California, Los Angeles, 1964.

Chapman, Sara E. *Privat Ambition and Political Alliances: The Phélypeaux de Pontchartrain Family and Louis XIV's Government, 1650–1715*. Rochester: University of Rochester Press, 2004.

Charliat, P.-J. "L'Académie royale de marine et la révolution nautique au XVIIIe siècle." *Thalès* 1 (1934): 71–82.

Chartier, Roger. *The Cultural Origins of the French Revolution*. Translated by Lydia G. Cochrane. Durham: Duke University Press, 1991.

———. *The Cultural Uses of Print in Early Modern France*. Translated by Lydia G. Cochrane. Princeton: Princeton University Press, 1987.

———. *Lectures et lecteurs dans la France d'ancien regime*. Paris: Éditions du Seuil, 1987.

———. *Cultural History: Between Practices and Representations.* Translated by Lydia G. Cochrane. Ithaca: Cornell University Press, 1988.

Chartier, Roger, ed. *Passions of the Renaissance.* Translated by Arthur Goldhammer. Vol. 3 of *A History of Private Life.* Edited by Philippe Ariès and Georges Duby. 5 vols. Cambridge: Belknap Press, 1989.

Chartier, Roger, Dominique Julia, and Marie-Madeleine Compere. *L'éducation en France du XVIe au XVIIIe siècle.* Paris: Société d'édition d'enseignement supérieur, 1976.

Du Châtelet, Gabrielle Emilie Le Tonnelier de Breteuil, Marquise. *Dissertation sur la nature et la propagation du feu.* Paris: Prault, 1744.

———. *Institutions de physique.* Paris: Prault, 1740.

———. "Institutions physique: Nouvelle edition (1742)." In Christiaan Wolff, *Gesammelte Werke.* New edition. Hildesheim: Georg Olms, 1988.

———. *Réponse de Madame***[Du Châtelet] à la lettre que M. de Mairan, secrétaire perpétuel de l'Académie des sciences, lui a écrite, le 18 février, 1741, sur la question des forces vives.* Brussels: Chez Foppens, 1741.

Childs, Nick. *A Political Academy in Paris, 1724–1731: The Entresol and Its Members.* Oxford: Oxford University Press, 2000.

Clairaut, Alexis-Claude. *Théorie de la figure de la terre, tirée des principes de l'hydrostatique.* Paris: David, 1743.

Clark, William, Jan Golinski, and Simon Schaffer, eds. *The Sciences in Enlightened Europe.* Chicago: University of Chicago Press, 1999.

Clarke, Samuel. *De l'existence et des attributs de Dieu, des devoirs de la réligion naturelle, et de la verité de la religion chrétienne.* Translated by J.-Fr. Bernard. Amsterdam, 1717.

———. *A Collection of Papers which Passed between the Late Learned Mr. Leibnitz and Dr. Clarke in the Years 1715 and 1716 relating to the Principles of Natural Philosophy and Religion.* London: J. Knapton, 1717.

Clarke, Jack A. "Abbé Jean-Paul Bignon: 'Moderator of the Academies' and Royal Librarian." *French Historical Studies* 8 (Fall 1973): 213–235.

Clément, Pierre. "Les successeurs de Colbert: Pontchartrain." *Revue des deux mondes* 46 (1863): 918–919.

Cochet, Jean-Baptiste. *Eléments de mathématiques de Monsieur Varignon.* Paris: P. M. Brunet, 1731.

Cohen, H. Floris. *The Scientific Revolution: A Historiographical Inquiry.* Chicago: University of Chicago Press, 1994.

Cohen, I. Bernard. *The Newtonian Revolution, with Illustrations of the Transformation of Scientific Ideas.* Cambridge: Cambridge University Press, 1980.

———. *Introduction to Newton's* Principia. Cambridge: Cambridge University Press, 1978.

———. "Isaac Newton, Hans Sloane, and the Académie Royale des Sciences." In *Mélanges Alexandre Koyré.* Edited by I. B. Cohen and René Taton. 2 vols. Paris: Hermann, 1964.

Collins, Anthony. *A Discourse of Free-thinking.* London, 1713.

——. *Discours sur la liberté de penser, écrit à l'occasion d'une nouvelle secte d'esprits forts, ou de gens qui pensent librement.* London, 1714.

——. *A Philosophical Enquiry concerning Human Liberty.* London: R. Robinson, 1717.

Collins, J. Churton. *Voltaire, Montesquieu, and Rousseau in England.* London: E. Nash, 1908.

Collins, James. *The State in Early Modern France.* Cambridge: Cambridge University Press, 1995.

Colombey, Emile. *Ruelles, salons, et cabarets.* 2 vols. Paris: E. Dentu, 1892.

Compère, Marie-Madeleine. *Du collège au lycée (1500–1850): Généalogie de l'enseignement secondaire français.* Paris: Gallimard, 1985.

Condillac, abbé de. *L'essai sur les origines des connaissances humaines.* Paris: 1746.

Condorcet, Jean-Antoine-Nicolas de Caritat. *Esquisse d'un tableau historique des progrès de l'esprit humain.* Edited by O. H. Prior. Paris: Boivin et cle, 1933.

——. *Vie de Voltaire.* With a foreword by Elisabeth Badinter. Paris: Quai Voltaire, 1994.

Conlon, Pierre M. *Voltaire's Literary Career from 1728 to 1750.* Geneva: Institut et musee Voltaire, 1961.

Correspondance mathématique et physique de quelques célèbres géométres du XVIIIe siècle. Edited by P. H. Fuss. Vol. 2. New York: Johnson Reprint, 1968.

Costabel, Pierre. Introduction to *Mathematica. Oeuvres complètes de Malebranche.* Edited by André Robinet, vol. 17, part 2. Paris: Vrin, 1968.

——. *Leibniz and Dynamics.* Ithaca: Cornell University Press, 1973.

——. "L'Oratoire de France et ses colleges." In *Enseignement et diffusion des sciences en France au XVIIIe siècle.* Edited by René Taton. Paris: Hermann, 1964.

——. "La mécanique dans *l'Encyclopédie.*" In *L'Encyclopédie et le progrès des sciences et techniques.* Paris: Presses Universitaires de France, 1952.

——. *La signification d'un debat sur trente ans (1728–1758): La question des forces vives.* Paris: Societé Française d'Histoire des Sciences et des Techniques, 1983.

——. *Pierre Varignon (1654–1722) et la diffusion en France du calcul differentiel et intégral.* Paris: Palais de la découverte, 1965.

Cotter, Charles H. *A History of Nautical Astronomy.* New York: American Elsevier Pub. Co., 1968.

Cottret, Bernard. *Bolingbroke: Exil et écriture au siècle des lumières.* Paris: Klincksieck, 1992.

Couturat, Louis. *La logique de Leibniz d'après des documents inédits.* Paris: F. Alcan, 1901.

Cranston, Maurice. *Jean-Jacques. The Early Life and Works of Jean-Jacques Rousseau, 1712–1754.* Chicago: University of Chicago Press, 1982.

Crocker, Lester G. *Diderot the Embattled Philosopher.* New York: Free Press, 1966.

Crosland, Maurice. "The Development of a Professional Career in Science in France." *Minerva* 13 (1975): 38–57.

Crousaz, Jean-Pierre de. *Commentaire sur l'analyse des infiniment petit.* Paris: Montalant, 1722.

———. *Essay sur le mouvement, ou l'on traite de sa nature, de son origine, de sa communication en general.* Groninge: J. Coste, 1726.

Crow, Thomas E. *Painters and Public Life in Eighteenth-Century Paris.* New Haven: Yale University Press, 1985.

Daniel, Gabriel. *Voyage du monde de Descartes.* Paris: Vve de S Bénard, 1690.

Daniel, Stephen Hartley. *John Toland, His Methods, Manners, and Mind.* Kingston: McGill-Queen's University Press, 1984.

Danville, François de. "L'enseignement scientifique dans les collèges des Jésuites." In *Enseignement et diffusion des sciences en France au dix-huitième siècle.* Edited by René Taton. Paris: Hermann, 1986.

Darnton, Robert. *The Business of Enlightenment: The Publishing History of the Encyclopédie, 1775–1800.* Cambridge: Harvard University Press, 1979.

———. *The Kiss of Lamourette: Reflections in Cultural History.* New York: W. W. Norton, 1990.

———. *The Literary Underground of the Old Regime.* Cambridge: Harvard University Press, 1982.

———. *Mesmerism and the End of the Enlightenment in France.* Cambridge: Harvard University Press, 1970.

Daston, Lorraine. *Classical Probability in the Enlightenment.* Princeton: Princeton University Press, 1988.

———. "The Ideal and the Reality of the Republic of Letters in the Enlightenment." *Science in Context* 4 (1991): 367–386.

Daston, Lorraine, and Katherine Park. *Wonders and the Order of Nature, 1150–1750.* Cambridge: MIT Press, 1998.

Daumas, Maurice. "La chimie dans l'Encyclopédie." *Revue d'histoire des sciences* 2 (1951): 334–343.

———. *Les instruments scientifiques aux XVIIe et XVIIIe siècles.* Paris: Presses Universitaires de France, 1953.

De Beer, Sir Gavin, and André Michel Rousseau. *Voltaire's British Visitors.* Geneva: Institut et Musée Voltaire, 1967.

Dear, Peter. *Discipline and Experience: The Mathematical Way in the Scientific Revolution.* Chicago: University of Chicago Press, 1995.

———. *Revolutionizing the Sciences: European Knowledge and Its Ambitions, 1500–1700.* Princeton: Princeton University Press, 2001.

Delorme, Suzanne. "Académies et salons." *Revue de Synthèse* 26 (1950): 115–154.

———. "'La Géometrie de l'infini' et ses commentateurs de Jean Bernoulli à Monsieur de Cury." *Revue d'histoire des sciences* 10 (1957): 339–359.

———. "La vie scientifique à l'époque de Fontenelle d'après les 'Eloges des Savants.'" *Archeion* 19 (1951): 217–235.

Delorme, Suzanne, ed. *Fontenelle, sa vie et son oeuvre, 1657–1757. Journées Fontenelle, organisées au Centre international de synthèse, Salon de Madame de Lambert, les 6, 7, 9, et 10 mai 1957.* Paris: Librarie Droz, 1961.

Densmore, Dana. *Newton's Principia: The Central Argument.* Santa Fe: Green Lion Press, 1995.

Derham, George. *Theologie-physique.* Translated by J. Lufneu. Rotterdam: Chez Jean Daniel Beman, 1726.

———. *Théologie astronomique, ou démonstration de l'existence et des attributs de Dieu, par l'examen et la déscription des cieux.* Translated by Fr. Bellanger. Paris: Chez Chaubert, 1729.

Desaguliers, John Theophilus. *A Course of Experimental Philosophy.* 2 vols. London: Printed for John Senex, W. Innys and Richard Manby, and John Osborne and Thomas Longman, 1734–1744.

———. *A System of Experimental Philosophy.* London: Printed for W. Mears . . . B. Creake . . . and J. Sackfield . . ., 1719.

Desautels, Alfred R. *Les Memoires de Trévoux et le mouvement des idees au XVIIIe siècle.* Rome: Institutum Historicum S.I., 1956.

Descartes, René. *Oeuvres.* Edited by Charles Adam and Paul Tannery. 12 vols. Paris: Vrin, 1897–1957.

Desfontaines, Pierre-François Guyot, abbé de. *La voltairomanie.* Edited by Mark Waddicor. Exeter: Exeter University Press, 1983.

Desfontaines, Pierre-François Guyot, abbé de, Voltaire, and Sr. Francoise Jore. *La voltairomanie avec le préservatif et Le factum de Sr. Claude Francois Jore.* London: 1739.

Desmaizeaux, Pierre. *Recueil de diverses pièces, sur la philosophie, la réligion naturelle, l'histoire, les mathématiques, &c par Mrs Leibniz, Clarke, Newton, et autres auteurs celèbres.* Amsterdam: Chez H. Du Sauzet, 1720.

Dhombres, Jean. "Quelques rencontres de Diderot avec les mathématiques." In *Denis Diderot, 1713–1784. Actes du colloque internationale de Diderot.* Edited by Anne-Marie Chouillet. Paris: Amateurs de Livres, 1985.

Dibon, Paul. "Communication in the Respublica Literaria of the Seventeenth Century." *Res Publica Literarum* 1 (1978): 42–55.

Dickinson, H. T. *Bolingbroke.* London, Constable, 1970.

Dictionnaire de l'Académie française. Paris, 1694.

Diderot, Denis. *Les bijoux indiscrets.* Paris: Au Monomotapa, 1748.

———. *Rameau's Nephew and Other Works.* Edited and translated by Jacques Barzun and Ralph H. Bowen. Indianapolis: Hackett Publishing Co., 2001.

———. *Lettre de M. Diderot au R. P. Berthier.* N.p., 1751.

———. *Lettres sur les aveugles, à l'usage de ceux qui voient.* London, 1749.

———. "Mémoires sur différents sujets de mathématiques." Paris: Chez Durand, Pissot, 1748. Reprinted in J. Assézat and M. Tourneux, eds., *Oeuvres complètes de Diderot.* 20 vols. Paris: Garnier frères, 1875–1877.

———. *Pensées philosophiques.* La Haye: Aux dépens de la Compagnie, 1746.

———. "Prospectus." The ARTFL Encyclopedia Project, http://www.lib.uchicago
.edu/efts/ARTFL/projects/encyc/prospectus.html.

Dieckmann, Herbert. *Le Philosophe: Texts and Interpretations.* Washington
University Studies. New series, Language and Literature, no. 18. St. Louis:
Washington University Press, 1943.

Dijk, Suzanna van. *Traces des Femmes: Présence féminine dans le journalisme français
du XVIIIe siècle.* Amsterdam: APA-Holland University Press, 1988.

Dijksterhuis, E. J. *The Mechanization of the World Picture.* Oxford: Clarendon Press,
1961.

Dixon, Lynne. *Diderot, Philosopher of Energy: The Development of His Concept
of Physical Energy.* Studies on Voltaire and the Eighteenth Century, vol. 255.
Oxford: Voltaire Foundation, 1988.

Dobbs, Betty Jo Teeter. *Alchemical Death and Resurrection: The Significance
of Alchemy in the Age of Newton.* Washington, D.C.: Smithsonian Institution
Libraries, 1990.

———. *The Foundations of Newton's Alchemy, or "The Hunting of the Greene Lyon."*
Cambridge: Cambridge University Press, 1975.

———. *The Janus Face of Genius: The Role of Alchemy in Newton's Thought.*
Cambridge: Cambridge University Press, 1991.

Dobbs, Betty Jo Teeter, and Margaret C. Jacob. *Newton and the Culture of
Newtonianism.* Atlantic Highlands: Humanities Press, 1995.

Doyon, André, and Lucien Liaigre. *Jacques Vaucanson, mécanicien de génie.* Paris:
Presses Universitaires de France, 1966.

Drouet, Joseph. *L'abbé de Saint-Pierre: L'homme et l'oeuvre.* Paris: Librairie Ancienne
Honore Champion, 1912.

Duchet, Michèle. *Anthropologie et histoire au siècle des lumières Buffon, Voltaire,
Rousseau, Helvetius, Diderot.* Paris: F. Maspero, 1971.

Dugas, René. *A History of Mechanics.* Translated J. R. Maddox. New York: Dover
Publication, 1988.

Dumarsais, César Chesneau. "Philosophe." In *Nouvelles Libertés de penser.*
Amsterdam, 1743.

Dumas, Gustave. *Histoire du Journal de Trévoux depuis 1701 jusqu'en 1762.* Paris:
Bouvin et cie, 1936.

Du Plan, Alfred Doneaud. *Histoire de l'Académie royale de la marine.* Paris:
Berger-Levranlt, 1878–1882.

Easton, Patricia. *Bibliographia Malebranchiana: A Critical Guide to the Malebranche
Literature into 1989.* Carbondale: Southern Illinois University Press, 1991.

Edwards, Matthew, ed. *Pushing Gravity: New Perspectives on Le Sage's Theory of
Gravitation.* Montreal: Apeiron, 2002.

Ehrard, Jean. *L'idée de la nature en France dans le premier moitié du dix-huitième
siècle.* Paris: Klincksieck, 1964.

Ehrard, Jean, and Jacques Roger. "Deux périodiques français du 18è siècle:
'Le Journal des Savants' et 'Les Mémoires de Trévoux.' Essai d'une étude

quantitative." In *Livre et société dans la France du XVIIIè siècle*. Paris: Mouton et cie, 1965.

Ehrman, Esther. *Mme. du Châtelet: Scientist, Philosopher and Feminist of the Enlightenment*. Leamington [Spa]: Berg, 1986.

Eisenstein, Elizabeth. *The Printing Press as an Agent of Change: Communication and Cultural Transformation in Early Modern Europe*. 2 vols. Cambridge: Cambridge University Press, 1979.

Elias, Norbert. *The Civilizing Process: The Development of Manners*. Translated by Edmund Jephcott. New York: Urizen Books, 1978.

———. *The Court Society*. Translated by Edmund Jephcott. Oxford: Blackwell, 1983.

Elkana, Yehuda. "Newtonianism in the Eighteenth Century." *British Journal for the Philosophy of Science* 22 (1971): 297–306.

Ellis, Harold A. *Boulainvilliers and the French Monarchy: Aristocratic Politics in Early Eighteenth-Century France*. Ithaca: Cornell University Press, 1988.

Evans, James. "Fraud and Illusion in the Anti-Newtonian Rear Guard: The Courtaud-Mercier Affair." *Isis* 87 (1996): 74–108.

Fara, Patricia. *Fatal Attraction: Magnetic Mysteries of the Enlightenment*. London: Icon Books, 2005

———. *Newton: The Making of a Genius*. New York: Columbia University Press, 2002.

———. *Sympathetic Attractions: Magnetic Practices, Beliefs, and Symbolism in Eighteenth-Century England*. Princeton: Princeton University Press, 1996.

Feingold, Mordechai. *The Newtonian Moment: Newtonian Science and the Making of Modern Culture*. Oxford: Oxford University Press, 2004.

Feingold, Mordechai, ed. *Jesuit Science and the Republic of Letters*. Cambridge: MIT Press, 2003.

Ferrone, Vincenzo. *The Intellectual Roots of the Italian Enlightenment: Newtonian Science, Religion, and Politics in the Early Eighteenth Century*. Translated by Sue Brotherton. Atlantic Highlands: Humanities Press, 1995.

———. *Scienza natura religione: Mondo newtoniano e cultura italiana nel primo Settecento*. Naples: Jovene, 1982.

Fletcher, Dennis J. "Bolingbroke and the Diffusion of Newtonianism in France." *Studies on Voltaire and the Eighteenth Century* 53 (1967): 29–46.

———. "The Fortunes of Bolingbroke in France in the Eighteenth Century." *Studies on Voltaire and the Eighteenth Century* 47 (1966): 207–232.

Florida, R. E. *Voltaire and the Socinians*. Banbury: Voltaire Foundation, 1974.

Fontenelle, Bernard le Bovier de. *Eléments de la géometrie de l'infini*. Edited by Michel Blay and Alain Niderst. Paris: Klinckseik, 1995.

———. *Histoire et mémoires de l'Académie royale des sciences, depuis son établissement en 1666 jusqu'à 1699*. 11 vols. Paris: Charles-Estienne Hochereau, J. Boudot, and others, 1729–1733.

———. *Oeuvres complètes de Fontenelle*. Edited by Alain Niderst. 9 vols. Paris: Fayard, 1990–2001.

Formey, Jean Louis Samuel. *La belle Wolfienne*. 6 vols. La Haye: Chez la veuve de Charles le Vier, 1741–1753.

Force, James E. *William Whiston, Honest Newtonian*. Cambridge: Cambridge University Press, 1985.

Foucault, Michel. *The Archaeology of Knowledge*. Translated by A. M. Sheridan. New York: Pantheon Books, 1972.

———. *The Birth of the Clinic: An Archaeology of Medical Perception*. Translated by A. M. Sheridan. New York, Pantheon Books, 1973.

———. *Discipline and Punish: The Birth of the Prison*. Translated by A. M. Sheridan New York, Pantheon Books, 1977.

———. *The Order of Things: An Archaeology of the Human Science*. New York: Pantheon Books, 1970.

———. "Nietzsche, Genealogy, History." In *The Foucault Reader*. Edited by Paul Rabinow. New York: Pantheon Books, 1984.

Foucault, Michel, Graham Burchell, Colin Gordon, and Peter Miller. *The Foucault Effect: Studies in Governmentality: with Two Lectures by and an Interview with Michel Foucault*. Chicago: University of Chicago Press, 1991.

Frängsmyr, Tore, J. L Heilbron, and Robin E. Rider, eds. *The Quantifying Spirit in the Eighteenth Century*. Berkeley: University of California Press, 1990.

Fraser, Craig G. "The Calculus as Algebraic Analysis: Some Observations on Mathematical Analysis in the 18th Century." *Archive for the History of the Exact Sciences* 39 (1989): 317–335.

Fréret, Nicolas. *Nouvelles Libertés de penser*. Paris, 1742.

Freudenthal, Gad. "Early Electricity between Chemistry and Physics: The Simultaneous Itineraries of Francis Hauksbee, Samuel Wall, and Pierre Polinière." *Historical Studies in the Physical Sciences* 11 (1981): 203–229.

———. "Littérature et sciences de la nature en France au début du XVIIIe siècle." *Revue de Synthèse* 100 (1980): 267–295.

Frostin, Charles. "Le chancelier de France Louis de Pontchartrain, 'ses' premiers présidents et la discipline des cours souveraines (1699–1714)." *Cahiers d'histoire* 27 (1982): 9–34.

———. "La famille ministérielle des Phélypeaux; esquisse d'un profil Pontchartrain (XVIe–XVIIIe siècles)." *Annales de Bretagne* 86 (1979): 117–140.

———. "L'organisation ministérielle sous Louis XIV: Cumul d'attributions et situations conflictuelles (1690–1715)." *Revue historique de droit français et étranger* 58 (1980): 201–226.

———. "Les Pontchartrain et la pénétration commerciale française en Amérique espagnole (1690–1715)." *Revue historique* 498 (1971): 307–336.

———. "Pouvoir ministériel, *voies ordinaires de la justice* et *voies de l'autorité* sous Louis XIV: Le chancelier Louis de Pontchartrain et le secrétaire d'état Jerôme de Pontchartrain (1699–1715)." In *107e Congrès national des sociétés savants*, 1: 7–29. 3 vols. Brest, 1982.

Fumaroli, Marc. *L'âge de l'éloquence: Rhétorique et "res literaria" de la Renaissance au seuil de l'époque classique.* Paris: Champion, 1980.

Funkenstein, Amos. *Theology and the Scientific Imagination from the Middle Ages to the Seventeenth Century.* Princeton: Princeton University Press, 1986.

Furbank, Philip Nicholas. *Diderot: A Critical Biography.* London: A. A. Knopf, 1992.

Furetière, Antoine. *Dictionnaire universel.* La Haye: Leers, 1690.

Gabbey, Alan. "Newton's 'Mathematical principles of natural philosophy': A treatise on mechanics?" In *The Investigation of Difficult Things: Essays on Newton and the History of the Exact Sciences in Honour of D. T. Whiteside.* Edited by P. M. Harman and Alan E. Shapiro. Cambridge: Cambridge University Press, 1992.

Galison, Peter. *Image and Logic: A Material Culture of Microphysics.* Chicago: University of Chicago Press, 1997.

Gamaches, Étienne-Simon de. *Les agrémens du langage réduits à leurs principes.* Paris: Chez Guillaume Cavelier, Jacques Estienne, Guillaume Cavelier fils, 1718.

———. *Astronomie physique, ou Principes généraux de la nature appliqués au mécanisme astronomique et comparés aux principes de la philosophie de M. Newton.* Paris: C.-A. Jomber, 1740.

———. *Sisteme du movement.* Paris: Chez Jean-Michel Garnie, 1721.

———. *Système du coeur, ou Conjectures sur la manière dont naissent les différentes affections de l'âme.* Paris: Chez Denys Dupuis, 1704.

Gandt, François de. *Force and Geometry in Newton's Principia.* Translated by Curtis Wilson. Princeton: Princeton University Press, 1995.

Garber, Daniel. *Descartes' Metaphysical Physics.* Chicago: University of Chicago Press, 1992.

———. "Leibniz and the Foundations of Physics: The Middle Years." In *The Natural Philosophy of Leibniz.* Edited by Kathleen Okruhlik and James Robert Brown. University of Western Ontario Series in Philosophy of Science, 2. Boston: D. Reidel Pub. Co., 1985.

———. "Leibniz: Physics and Philosophy." In *The Cambridge Companion to Leibniz.* Edited by Nicolas Jolley. Cambridge: Cambridge University Press, 1995.

Gardiner Janik, Linda. "Searching for the Metaphysics of Science: The Structure and Composition of Mme. Du Châtelet's *Institutions de physique,* 1737–1740." *Studies on Voltaire and the Eighteenth Century* 201 (1982): 85–113.

Garrioch, David. *Neighborhood and Community in Paris, 1740–1790.* Cambridge: Cambridge University Press, 1986.

Gay, Peter. *The Enlightenment: An Interpretation.* Vol. 1, *The Science of Freedom.* New York: Knopf, 1969.

———. *The Enlightenment: An Interpretation.* Vol. 2, *The Rise of Modern Paganism.* New York: Knopf, 1977.

———. *The Party of Humanity: Essays in the French Enlightenment.* New York: Knopf, 1954.

———. *Voltaire's Politics: The Poet as Realist.* New Haven: Yale University Press, 1988.

Gearhart, Suzanne. "Rationality and the Text: A Study of Voltaire's Historiography."
Studies on Voltaire and the Eighteenth Century 140 (1975): 21–43.

Gillispie, Charles Coulston, ed. *Dictionary of Scientific Biography.* 18 vols.
New York: Scribner, 1970–1990.

———. *The Edge of Objectivity: An Essay in the History of Scientific Ideas.* Princeton:
Princeton University Press, 1960.

———. "Fontenelle and Newton." In *Isaac Newton's Papers and Letters on Natural
Philosophy.* Edited by I. B. Cohen. Cambridge: Harvard University Press,
1978.

———. *Pierre-Simon Laplace (1749–1827): A Life in Exact Science.* Princeton:
Princeton University Press, 1997.

———. *Science and Polity in France at the End of the Old Regime.* Princeton: Princeton
University Press, 1980.

———. *Science and Polity in France: The Revolutionary and Napoleonic Years.*
Princeton: Princeton University Press, 2005.

Gingras, Yves. "What Did Mathematics Do to Physics?" *History of Science* 39 (2001):
383–416.

Girbal, Francois. *Bernard Lamy (1640–1715).* Paris: Presses Universitaires de France,
1964.

Glotz, Marguerite. *Salons du XVIIIe siècle.* Paris: Nouvelles Éditions latines, 1949.

Goldgar, Anne. *Impolite Learning: Conduct and Community in the Republic of
Letters, 1680–1750.* New Haven: Yale University Press, 1995.

Golinski, Jan. *Science as Public Culture: Chemistry and Enlightenment in Britain,
1760–1820.* Cambridge: Cambridge University Press, 1992.

Goodman, Dena. "Public Sphere and Private Life: Toward a Synthesis of Current
Historiographical Approaches to the Old Regime." *History and Theory* 31 (1992):
1–20.

———. *The Republic of Letters: A Cultural History of the French Enlightenment.*
Ithaca: Cornell University Press, 1994.

Gordon, Daniel. *Citizens without Sovereignty: Equality and Sociability in French
Thought, 1670–1789.* Princeton: Princeton University Press, 1994.

Gowing, Ronald. *Roger Côtes: Natural Philosopher.* Cambridge: Cambridge
University Press, 1983.

Grafton, Anthony. *Cardano's Cosmos: The Worlds and Work of a Renaissance
Astrologer.* Cambridge: Harvard University Press, 1999.

———. *Defenders of the Text: The Traditions of Humanism in an Age of Science,
1450–1800.* Cambridge: Harvard University Press, 1991.

———. *The Footnote: A Curious History.* Cambridge: Harvard University Press, 1997.

Gravesande, Willem Jakob Strom van 's. *Eléments de physique, ou Introduction à la
philosophie de Newton.* Translated by C.-F. Roland de Virloys. 2 vols. Paris: Chez
Charles-Antoine Jombert, 1742.

———. *Mathematical Elements of Natural Philosophy, Confirmed by Experiments: or,
An Introduction to Sir Isaac Newton's Philosophy.* Translated by J. T. Desaguliers.

6th ed. 2 vols. London: Ed. for W. Innys, T. Longman and T. Shewell, C. Hitch, in Pater-Noster-Row; and M. Senex, in Fleet-Street 1747.

———. *Oeuvres philosophiques et mathématiques.* Edited by J. N. S Allamand. 2 vols. Amsterdam: M. M. Rey, 1774.

Greenberg, John. "Geodesy in Paris in the 1730s and the Paduan Connection." *Historical Studies in the Physical Sciences* 13 (1983): 239–260.

———. *The Problem of the Earth's Shape from Newton to Clairaut: The Rise of Mathematical Science in Eighteenth-Century Paris and the Fall of "Normal" Science.* Cambridge: Cambridge University Press, 1995.

Grimsley, Ronald. *Jean d'Alembert (1717–1783).* Oxford: Oxford University Press, 1963.

Guicciardini, Niccolò. *The Development of the Newtonian Calculus in Britain, 1700–1800.* Cambridge: Cambridge University Press, 1989.

———. "Dot-Age: Newton's Mathematical Legacy in the Eighteenth Century." *Early Science and Medicine* 9 (2004): 218–256.

———. *Reading the Principia: The Debate on Newton's Mathematical Methods for Natural Philosophy from 1687–1736.* Cambridge: Cambridge University Press, 1999.

Guerlac, Henry. *Newton on the Continent.* Ithaca: Cornell University Press, 1981.

———. "Newton's Changing Reputation in the Eighteenth Century." In *Carl Becker's Heavenly City Revisited.* Edited by Raymond O. Rockwood. Ithaca: Cornell University Press, 1958.

———. "Where the Statue Stood: Divergent Loyalties to Newton in the Eighteenth Century." In *Aspects of the Eighteenth Century.* Edited by Earl R. Wasserman. Baltimore: Johns Hopkins Press, 1965.

Habermas, Jürgen. *The Stuctural Transformation of the Public Sphere: An Inquiry into a Category of Bourgeois Society.* Translated by Thomas Burger. Cambridge: MIT Press, 1989.

Hacking, Ian. *The Taming of Chance.* Cambridge: Cambridge University Press, 1990.

Hahn, Roger. *The Anatomy of a Scientific Institution: The Paris Academy of Sciences, 1666–1803.* Berkeley: University of California Press, 1971.

———. "The Application of Science to Society: The Societies of Arts." *Studies on Voltaire and the Eighteenth Century* 25 (1963): 829–836.

———. "Changing Patterns for the Support of Scientists from Louis XIV to Napoleon." *History and Technology* 4 (1987): 353–371.

———. *Laplace as a Newtonian Scientist.* Los Angeles: University of California, William Andrews Clark Memorial Library, 1967.

———. *Pierre Simon Laplace, 1749–1827: Determined Scientist.* Cambridge: Harvard University Press, 2005.

———. "Scientific Careers in Eighteenth-Century France." In *The Emergence of Science in Western Europe.* Edited by Maurice Crosland. London: Macmillan Press, 1975.

———. "Scientific Research as an Occupation in Eighteenth-Century Paris." *Minerva* 13 (1975): 501–513.

———, ed. *L'hydrodynamique au XVIIIe siècle; aspects scientifiques et sociologiques.* Conferences du Palais de la decouverte, Ser. D, Histoire des sciences. Paris: Université de Paris, 1964.

Hales, Stephen. *La Statique des végétaux, et l'analyse de l'air.* Translated by Georges Louis Leclerc de Buffon. Paris: chez Debure l'Aîné, 1735.

———. *Vegetable staticks, or, An Account of Some Statical Experiments on the Sap in Vegetables.* London: W. and J. Innys, and T. Woodward, 1727.

Hall, A. R. *From Galileo to Newton, 1630–1720.* New York: Dover, 1963.

———. "Newton in France: A New View." *History of Science* 13 (1975): 233–250.

———. *Philosophers at War: The Quarrel between Newton and Leibniz.* Cambridge: Cambridge University Press, 1980.

———. *The Revolution in Science, 1500–1750.* London: Longman, 1983.

———. *The Scientific Revolution, 1500–1800: The Formation of the Modern Scientific Attitude.* London: Longman, 1954. Second ed., London, 1960.

Hankins, Thomas L. "Eighteenth-Century Attempts to Resolve the *Vis Viva* Controversy." *Isis* 56 (1965): 281–297.

——— "The Influence of Malebranche on the Science of Mechanics during the Eighteenth Century." *Journal of the History of Ideas* 28 (1967): 193–210.

———. *Jean d'Alembert: Science and the Enlightenment.* Oxford: Clarendon Press, 1970.

———. "The Reception of Newton's Second Law of Motion in the Eighteenth Century." *Archive international d'histoire de science* 20 (1967): 43–65.

———. *Science and the Enlightenment.* Cambridge: Cambridge University Press, 1985.

Hanks, Lesley. *Buffon avant l' "Histoire naturelle."* Paris: Presses Universitaires de France, 1966.

Hanley, William. "The abbé Rothelin and the *Lettres philosophiques.*" *Romance Notes* 23 (1983): 1–16.

Hanna, Blake T. "Polinière and the Teaching of Physics at Paris: 1700–1730." In *Eighteenth-Century Studies Presented to Arthur M. Wilson.* Edited by Peter Gay, 13–39. Hanover: University Press of New England, 1972.

Harisse, Henry. *L'abbé Prévost: Histoire de sa vie et de ses oeuvres d'après des documents bnouveaux.* Paris: Lévy, 1896.

Harrison, Peter. "Newtonian Science, Miracles, and the Laws of Nature." *Journal of the History of Ideas* 56 (1995): 531–554.

Harth, Erica. *Cartesian Women: Versions and Subversions of Rational Discourse in the Old Regime.* Ithaca: Cornell University Press, 1992.

———. *Ideology and Culture in Seventeenth-Century France.* Ithaca: Cornell University Press, 1992.

Hartsoeker, Nicolas. *Conjectures physiques.* Amsterdam: H. Desbordes, 1706.

———. *Cours de physique.* La Haye: Chez Jean Swart, 1730.

———. *Recueil de plusieurs pièces de physique, où l'on fait principalement voir l'invalidité des systèmes de M. Newton.* Utrecht: Veuve de G. Broedelet, 1722.

Hauksbee, Francis. *Physico-Mathematical Experiments on Various Subjects.* London: 1719.

Havard, Jean-Alexandre, ed. *Voltaire et Mme du Châtelet, révélations d'un serviteur attaché à leurs personnes.* Paris: E. Dentu, 1863.

Hazard, Paul. *La crise de la conscience européenne (1680–1715).* Paris: Boivin, 1935.

———. *La pensée europeenne au XVIIIe siècle, de Montesquieu à Lessing.* Paris: Fayard, 1963.

Healy, George Robert. "Mechanistic Science and the French Jesuits: A Study of the Responses of the *Journal de Trévoux* (1701–1762) to Descartes and Newton." Ph.D. dissertation, University of Minnesota, 1956.

Heilbron, John L. *Electricity in the Seventeenth and Eighteenth Centuries: A Study of Early Modern Physics.* Berkeley: University of California Press, 1979.

———. *Elements of Early Modern Physics.* Berkeley: University of California Press, 1982.

———. *The Sun in the Church: Cathedrals as Solar Observatories.* Cambridge: Harvard University Press, 1999.

Heimann, P. "Ether and Imponderables." In *Conceptions of Ether: Studies in the History of Ether Theories.* Edited by G. N. Cantor and J N. S. Hodge. New York: Cambridge University Press, 1981.

———. "Nature as Perpetual Worker: Newton's Aether and Eighteenth-Century Natural Philosophy." *Ambix* 20 (1973): 1–25.

———. "Science and the English Enlightenment." *History of Science* 26 (1978): 143–151.

———. "Voluntarism and Immanence: Conceptions of Nature in Eighteenth-Century Thought." *Journal of the History of Ideas* 39 (1978): 271–283.

Heimann, P., and J. E. McGuire. "Newtonian Forces and Lockean Powers." *Historical Studies in Physical Science* 3 (1971): 233–306.

Hesse, Mary B. *Forces and Fields: The Concept of Action at a Distance in the History of Physics.* New York: Philosophical Library, 1961.

Hirschfield, John M. "The Académie Royale des Sciences (1666–1683): Inauguration and Initial Problems of Method." Ph.D. dissertation, University of Chicago, 1957.

Hirschman, Albert O. *The Passions and the Interests: Political Arguments for Capitalism before Its Triumph.* Princeton: Princeton University Press, 1977.

Hobart, Michael E. *Science and Religion in the Thought of Nicolas Malebranche.* Chapel Hill: University of North Carolina Press, 1982.

Hôpital, Marquis Guillaume de l'. *Analyse des infiniment petits pour l'intelligence des lignes courbes.* Paris: Imprimerie royale, 1696.

Hoskin, M. A. "Mining All Within: Clarke's Notes to Rohault's *Traité de physique*." *Thomist* 24 (1962): 353–371.

Houtteville, abbé Claude François. *La Réligion chrétienne prouvée par les faits.* Paris: G. Depuis, 1722.

Husserl, Edmund. *The Crisis of European Sciences and Transcendental Phenomenology: An Introduction to Phenomenological Philosophy.* Translated and edited by David Carr. Evanston: Northwestern University Press, 1970.

Huygens, Christiaan. *Discours de la cause de la pesanteur.* Leiden: Chez Pierre van der Aa, marchand libraire, 1690.

Hutchison, Ross. *Locke in France: 1688–1734.* Oxford: Voltaire Foundation at the Taylor Institution, 1991.

Iltis, Carolyn. "The Decline of Cartesianism in Mechanics: The Leibnizian-Cartesian Debates." *Isis* 64 (1973): 356–373.

———. "The Leibnizian-Newtonian Debates: Natural Philosophy and Social Psychology." *British Journal for the History of Science* 6 (1973): 343–377.

———. "Madame du Châtelet's Metaphysics and Mechanics." *Studies in the History and Philosopy of Science* 8 (1977): 29–48.

Isherwood, Robert M. *Farce and Fantasy: Popular Entertainment in Eighteenth-Century Paris.* Oxford: Oxford University Press, 1986.

Ishiguro, Hidé. *Leibniz's Philosophy of Logic and Language.* Cambridge: Cambridge University Press, 1990.

Israel, Jonathan. *Enlightenment Contested: Philosophy, Modernity, and the Emancipation of Man, 1670–1752.* Oxford: Oxford University Press, 2006.

———. *The Radical Enlightenment: Philosophy and the Making of Modernity, 1650–1750.* Oxford: Oxford University Press, 2000.

Jacob, Margaret. *Living the Enlightenment: Freemasonry and Politics in Eighteenth-Century Europe.* New York: Oxford University Press, 1991.

———. *The Newtonians and the English Revolution, 1689–1720.* Ithaca: Cornell University Press, 1976.

———. *The Radical Enlightenment: Pantheists, Freemasons, and Republicans.* London: Cornerstone Book Publishers, 1981.

Jammer, Max. *Concepts of Force: A Study in the Foundations of Dynamics.* Cambridge: Harvard University Press, 1957.

———. *Concepts of Space.* Cambridge: Harvard University Press, 1957.

Jesseph, Douglas M. *Berkeley's Philosophy of Mathematics.* Chicago: University of Chicago Press, 1993.

———. *Squaring the Circle. The War between Hobbes and Wallis.* Chicago: University of Chicago Press, 1999.

Johns, Adrian. *The Nature of the Book: Print and Knowledge in the Making.* Chicago: University of Chicago Press, 1998.

Jolley, Nicholas, ed. *The Cambridge Companion to Leibniz.* Cambridge: Cambridge University Press, 1995.

———. "Leibniz and Phenomenalism." *Studia Leibnitiana* 18 (1986): 38–51.

——— "The Reception of Descartes' Thought." In *The Cambridge Companion to Descartes.* Edited by John Cottingham. Cambridge: Cambridge University Press, 1992.

Jones, Colin. *The Great Nation: France from Louis XV to Napoleon.* London: Allen Lane, 2003.

Jones, Richard Foster. *Ancients and Moderns: A Study of the Rise of the Scientific Movement in Seventeenth-Century England.* Berkeley: University of California Press, 1965.

Jourdain, Charles. *Histoire de l'Universite de Paris.* Paris: L. Hachette et cie, 1867.

Julia, Dominique, Jacques Revel, and Roger Chartier. *Les universités européennes du XVIe au XVIIIe siècle: Histoire sociale des populations étudiantes.* Paris: Éditions de l'Ecole des hautes études en sciences socials, 1986.

Kafker, Frank, and Serena Kafker. *The Encyclopedists as Individuals: A Biographical Dictionary of the Authors of the Encyclopédie.* Studies on Voltaire and the Eighteenth Century, vol. 257. Oxford: Voltaire Foundation, 1988.

Kaiser, Thomas E. "The Abbé de Saint-Pierre, Public Opinion, and the Reconstitution of the French Monarchy." *Journal of Modern History* 55, no. 4 (December 1983): 618–643.

Kellman, Jordan. "Discovery and Enlightenment at Sea: Maritime Exploration and Observation in the Eighteenth-Century French Scientific Community." Ph.D. dissertation, Princeton University, 1998.

Keohane, Nannerl. *Philosophy and the State in France.* Princeton: Princeton University Press, 1980.

Kiernan, Colm. *Science and the Enlightenment in Eighteenth-Century France.* Geneva,1968.

King, J. E. *Science and Rationalism in the Government of Louis XIV.* Baltimore: Johns Hopkins Press, 1949.

Klein, Lawrence. *Shaftesbury and the Culture of Politeness: Moral Discourse and Cultural Politics in Early Eighteenth-Century England.* Cambridge: Cambridge University Press, 1994.

Kleinbaum, Abby Rose. "Jean Jacques Dortous de Mairan (1678–1771): A Study of an Enlightenment Scientist." Ph.D.dissertation, Columbia University, 1970.

Kleinert, Andreas. "La vulgarisation de la physique au siècle des lumières." *Francia* 10 (1982): 303–312.

Kline, Morris. *Mathematical Thought from Ancient to Modern Times.* 2 vols. Oxford: Oxford University Press, 1990.

———. *Mathematics and the Physical World.* New York: Cromwell, 1959.

———. *Mathematics: The Loss of Certainty.* Oxford: Oxford University Press, 1980.

Knight, Isabel F. *The Geometric Spirit: The Abbé de Condillac and the French Enlightenment.* New Haven: Yale University Press, 1968.

Konvitz, Josef. *Cartography in France, 1660–1848: Science, Engineering and Statecraft.* Chicago: University of Chicago Press, 1987.

Kors, Alan Charles. *Atheism in France, 1650–1729.* Princeton: Princeton University Press, 1990.

———. *D'Holbach's Coterie: An Enlightenment in Paris.* Princeton: Princeton University Press, 1976.

Koselleck, Reinhart. *Critique and Crisis: Enlightenment and the Pathogenesis of Modern Society*. Translated by Victor Gourevitch. Cambridge: MIT Press, 1988.

Koyré, Alexandre. *Études galiléennes*. Paris: Hermann, 1939.

———. *From the Closed World to the Infinite Universe*. Baltimore: Johns Hopkins University Press, 1957.

———. *Newtonian Studies*. Chicago: University of Chicago Press, 1965.

Krakeur, L. G. "The Mathematical Writings of Diderot." *Isis* 33 (1941): 219–232.

Kramnick, Isaac. *Bolingbroke and His Circle: The Politics of Nostalgia in the Age of Walpole*. Cambridge: Harvard University Press, 1968.

Kreiser, Robert B. *Miracles, Convulsions, and Ecclesiastical Politics in Early Eighteenth-Century Paris*. Princeton: Princeton University Press, 1978.

Kronick, David A. *A History of Scientific and Technical Periodicals*. Metuchen: Scarecrow Press, 1976.

Kuehn, Manfred. "The German Aufklärung and British Philosophy." In Stuart Brown, ed. *British Philosophy and the Age of Enlightenment*. Routledge History of Philosophy, vol. 5. New York: Routledge, 1996.

Kuhn, Thomas. *The Structure of Scientific Revolutions*. Chicago: University of Chicago Press, 1970.

La Beaumelle, Laurent de. *La vie de Maupertuis*. Paris: Ledoyen, 1856.

Labriolle-Rutherford, Marie-Rose de. *Le pour et contre et son temps*. Oxford: Institut et Musée Voltaire, 1965.

Lacombe, Henri, and Costabel, Pierre, eds. *La figure de la terre du XVIIIe siècle à l'ère spatiale*. Paris: Gauthier-Villars, 1988.

Laeven, H. *The Acta Eruditorum under the Editorship of Otto Mencke (1644–1707)*. Amsterdam: APA-Holland University Press, 1990.

Lafuente, Antonio, and Jose L. Peset. "La question de la figure de la terre. L'agonie d'un débat scientifique au XVIIIe siècle." *Revue d'Histoire des Sciences* 37 (1984): 235–254.

Lagarrigue, B. P. L. *Un temple de la culture européen (1728–1753). L'histoire externe de la Bibliothèque raisonée des ouvrages des savants de l'Europe*. Nijmegen: Katholieke Universiteit Nijmegen, 1993.

Lallemand, Paul. *Histoire de l'éducation dans l'ancien Oratoire de France*. Paris: E. Thorin, 1888.

Lamy, Bernard. *Entretiens sur les sciences*. Edited by François Girbal. Paris: Presses Universitaires de France, 1966.

Landes, Joan B. *Women and the Public Sphere in the Age of the French Revolution*. Ithaca: Cornell University Press, 1988.

Lanson, Gustave, ed. *Lettres philosophiques. Édition critique avec une introduction et un commentaire*. 2 vols. Paris: É. Cornély, 1909.

Larkin, Steve, ed. *Le pour et contre. Nos 1–60*. Oxford: Voltaire Foundation, 1993.

Lathuillère, Roger. *La préciosité: Étude historique et linguistique*. Geneva: Droz, 1966.

Latour, Bruno. *Pandora's Hope: Essays on the Reality of Science Studies.* Cambridge: Harvard University Press, 1999.

———. *Science in Action: How to Follow Scientists and Engineers through Society.* Cambridge: Harvard University Press, 1987.

———. *We Have Never Been Modern.* Translated by Catherine Porter. Cambridge: Harvard University Press, 1991.

Launay, Jean-Baptiste-Gabriel de Corgne de. *Principes du système des petits tourbillons, ou abrégé de la physique de l'abbé de Molières.* Paris: Jombert, 1743.

Laval. *Voyage à la Louisianne fait par ordre du roy en l'année mil sept cent vingt: Dans lequel sont traitées diverses matieres de physique, astronomie, geographie, & marine. et des reflexions sur quelques points du sisteme de M. Newton.* Paris: Jean Mariette, 1728.

Lefranc, Abel. *Histoire du Collège de France.* Paris: Hachette, 1893.

Legendre, Gilbert-Charles, Marquis de St. Aubin. *Traité historique et critique de l'opinion, ou Mémoires pour servir à l'histoire de l'esprit humain.* Paris: Chez Barbou, 1765.

Leibniz, Gottfried Wilhelm. *Leibnizens mathematische Schriften.* Edited by C. J. Gerhardt. 7 vols. Hildesheim: Olms, 1962.

———. *Philosophical Essays.* Translated and edited by Roger Ariew and Daniel Garber. Indianapolis: Hackett Pub. Co., 1989.

———. *Philosophical Papers and Letters.* A selection translated and edited by Leroy E. Loemker. 2 vols. Chicago: University of Chicago Press, 1956.

———. *Die philosophichen Schriften von Gottfried Wilhelm Leibniz.* Edited by C. J. Gerhardt. 7 vols. Hildesheim: Olms, 1971.

Lennon, Thomas M. *The Battle of the Gods and Giants: The Legacies of Descartes and Gassendi, 1655–1715.* Princeton: Princeton University Press, 1993.

Leouzon Le Duc, Louis Antoine. *Voltaire et la police; dossier recueilli à Saint-Petersbourg, parmi les manuscrits français originaux enlevés à la Bastille en 1789.* Paris: A. Bray, 1867.

Le Roy Ladurie, Emmanuel. *The Ancien Régime: A History of France, 1610–1774.* Translated by Mark Greengrass. Oxford: Blackwell, 1996.

Le Ru, Véronique. *Jean le Rond d'Alembert philosophe.* Paris: J. Vrin, 1994.

Lesage, George. *Cours abregé de physique.* Geneva: Chez Fabri & Barillot, 1732.

Le Seur, Thomas, and Francois Jacquier. *Philosophiae naturalis principia mathematica. Auctore Issaco Newtono.* 3 vols. Geneva: typis Barrillot & filii, bibliop. & typogr., 1739–1742.

Lettres critiques écrites d'Angleterre au R. P. Castel de la Compagnie de Jésus, sur trois articles importans et surprenans de son nouveau Système de la pesanteur universelle. Par le Chevalier*** de la Société Royale de Londres. Paris, 1725.

Libby, Margaret Sherwood. *The Attitude of Voltaire to Magic and the Sciences.* New York: Columbia University Press, 1935.

Licoppe, Christian. *La formation de la pratique scientifique. Le discours de l'experience en France et en l'Angleterre (1630–1820).* Paris: La Découverte, 1996.

Lindberg, David C., and Robert S. Westman. *Reappraisals of the Scientific Revolution*. Cambridge: Cambridge University Press, 1990.

Lindeboom, Gerrit Arie. *Herman Boerhaave: The Man and His Work*. London: Methuen, 1968.

Lindquist, Svante. "The Spectacle of Science: An Experiment in 1744 concerning the Aurora Borealis." *Configurations* 1 (1992): 57–94.

Lister, Martin. *A Journey to Paris in the Year 1698*. London: Printed for Jacob Tonson . . ., 1699.

Lougee, Carolyn C. *Le paradis des femmes: Women, Salons, and Social Stratification in Seventeenth-Century France*. Princeton: Princeton University Press, 1976.

Lough, John. ed. *The "Encyclopédie" of Diderot and d'Alembert. Selected Articles*. Cambridge: Cambridge University Press, 1969.

Lynn, Michael. *Popular Science and Public Opinion in Eighteenth-Century France*. Manchester: Manchester University Press, 2006.

Mach, Ernst. *The Science of Mechanics: A Critical and Historical Study of Its Development*. Translated by Thomas J. McCormack. La Salle: Open Court Publishing, 1989.

Maclaurin, Colin. *An Account of Sir Isaac Newton's Philosophical Discoveries*. London: Printed for the author's children: and sold by A. Millar and J. Norse [etc.], 1748.

———. *Exposition des découvertes philosophiques de M. le chevalier Newton*. Translated by L.-A. Lavirotte. Paris: Pissot, 1749.

Mahoney, Michael S. "Changing Canons of Mathematical and Physical Intelligibility in the Late Seventeenth Century." *Historia Mathematica* 11 (1984): 417–423.

———. "Christian Huygens: The Measurement of Time and of Longitude at Sea." In *Studies on Christian Huygens*. Edited by H. J. M Bos et. al. Lisse: Swets & Zeitlinger B.V., 1980.

———. "Infintessimals and Transcendant Relations: The Mathematics of Motion in the Late Seventeenth Century." In *Reappraisals of the Scientific Revolution*. Edited by David C. Lindberg and Westman, Robert S., 461–491. Cambridge: Cambridge University Press, 1990.

———. *The Mathematical Career of Pierre de Fermat, 1601–1665*. Princeton: Princeton University Press, 1994.

———. "On Differential Calculuses." *Isis* 75 (1984): 366–372.

———. "Pierre Varignon and the Calculus of Motion." Unpublished manuscript.

Maindron, Ernest. *Les fondations de prix à l'Académie des sciences*. Paris: Gauthier-Villars, 1881.

Mairan, Jean-Jacques Dortous de. *Dissertation sur la glace, ou Explication physique de la formation de la glace de ses divers phénomenes*. Bordeaux: Chez R. Brun, 1716. 2nd ed. Paris: Imprimerie royale, 1749.

———. In *Malebranche's First and Last Critics: Simon Foucher and Dortous de Mairan*. Translated and edited by Richard A. Watson and Marjorie Greene. Carbondale and Edwardsville: St. Augustine's Press, 1995.

————. *Lettre de M. de Mairan, . . . à Madame *** [la marquise du Châtelet] sur la question des forces vives, en réponse aux objections qu'elle lui fait sur ce sujet dans ses institutions de physique.* Paris: Chez Charles-Antoine Jombert, 1741.

————. *Traité physique et l'historique de l'Aurore Boréale.* Paris: Imp. Royale, 1733.

Maire, Catherine. *Les convulsionnaires de St. Médard: Miracles, convulsions, et prophéties à Paris au XVIIIe siècle.* Paris: H. Champion: Libraire de la Société de l'Histoire de Paris, 1985.

Malebranche, Nicolas. *Correspondence avec J.-J. Dortous de Mairan.* Edited by Joseph Moreau. Paris: J. Vrin, 1947.

————. *Oeuvres complètes.* Edited by André Robinet. 20 vols. Paris: J. Vrin, 1958–1970.

Maluf, Ramez Bahige. "Jean Antoine Nollet and Experimental Natural Philosophy in Eighteenth-Century France." Ph.D. dissertation, University of Oklahoma, 1985.

Mancosu, Paolo. "The Metaphysics of the Calculus: A Foundational Debate in the Paris Academy of Sciences." *Historia Mathematica* 16 (1989): 224–248.

————. *The Philosophy of Mathematics and Mathematical Practice in the Seventeenth Century.* Oxford: Oxford University Press, 1996.

Manuel, Frank, and Fritzie Manuel. *Utopian Thought in the Western World.* Oxford: Blackwell, 1979.

Marchal, Roger. *Madame Lambert et son milieu.* Oxford: Voltaire Foundation, 1991.

Marguet, F. *Histoire de la longitude à la mer au XVIIIe siècle en France.* Paris, 1917.

Marin, Louis. *Portrait of the King.* Translated by Martha M. Houle. Minneapolis: University of Minnesota Press, 1988.

Marion, Marcel. *Dictionnaire des institutions de la France aux XVIIe et XVIIIe siècles.* Paris: A. Picard, 1923.

Marsak, Leonard M. *Bernard de Fontenelle: The Idea of Science in the French Elightenment.* Philadelphia: American Philosophical Society, 1959.

Mason, Amelia Gere. *The Women of the French Salons.* New York, 1891.

Masseau, Didier. *L'invention de l'intellectual dans l'Europe du XVIIIe siècle.* Paris: Puf, 1994.

Masson, Pierre-Maurice. *Madame de Tencin (1682–1749), une vie de femme de XVIIIe siècle.* Paris: Hachette, 1910.

Maupertuis, Pierre-Louis Moreau de. *Discours sur les différentes figures des astres avec une exposition des systèmes de MM. Descartes et Newton. In Du Maupertuis.* 4 vols. Lyon: Chez Jean-Marie Bruyset, 1756.

————. *La figure de la terre déterminée par les observations de MM. de Maupertuis, Clairaut, Le Monnier, Outhier, Celsius au cercle polaire.* Paris: Imprimerie royale, 1738.

————. *Essai de cosmologie.* Basel: Johann Bernoulli, 1750.

————. *Examen désintéressé des différens ouvrages qui ont étés faits pour déterminer la figure de la terre.* Oldenbourg: T. Bachmuller, 1738.

———. *Lettre d'un horloger anglais à un astronome de Pekin. Traduite par M****. N.p., 1740.

———. *Oeuvres*. 4 vols. Lyon: Bruyset, 1756.

Maury, Alfred. *Les académies d'autrefois: L'ancienne Académie des inscriptions et des belles-lettres*. Paris: Didier et cie, 1864.

———. *L'ancienne Académie des sciences*. Paris: Didier et Cie, 1864.

Mauzi, Robert. *L'idée du bonheur dans la litterature et la pensée françaises au XVIIIe siècle*. Paris: A. Colin, 1960.

Mazières, Jean-Simon. *Traité des petits tourbillons de la matière subtile*. Paris: C. Jombert et Pissot, 1727.

Mazzotti, Massimo. "Maria Gaetana Agnesi: Mathematics and the Making of the Catholic Enlightenment." *Isis* 92 (2001): 657–683.

———. "Newton for Ladies: Gentility, Gender and Radical Culture." *British Journal for the History of Science* 27, no. 2 (June 2004): 119–146.

McClellan, James E. "The Académie Royale des Sciences: A Statistical Portrait." *Isis* 72 (1981): 541–567.

———. *Science Reorganized: Scientific Societies in the Eighteenth Century*. New York: Columbia University Press, 1985.

McGuire, J. E., and P. M. Rattansi. "Newton and the Pipes of Pan." *Notes and Records of the Royal Society* 21 (1966): 108–157.

McKenna, Antony. *De Pascal a Voltaire: Le rôle des "Pensées" de Pascal dans l'histoire des idées entre 1670 et 1734*. Oxford: Voltaire Foundation at the Taylor Institution, 1990.

McMahon, Darrin M. *Enemies of the Enlightenment: The French Counter-enlightenment and the Making of Modernity*. New York: Oxford University Press, 2001.

McMillan, Cynthia Anne. "The Concept of the Mathematical Infinite in French Thought, 1670–1760." Ph.D. dissertation, University of Virginia, 1970.

Melton, James Van Horn. *The Rise of the Public in Enlightenment Europe*. Cambridge: Cambridge University Press, 2001.

Metzger, Hélène. *Attraction universelle et religion naturelle chez quelques commentateurs Anglais de Newton*. 3 vols. Paris, 1938.

———. *Les doctrines chimiques en France du début du XVIIe à la fin du XVIIIe siècle*. Paris, 1923.

———. *Newton, Stahl, Boerhaave et la doctrine chimique*. Paris, 1930.

Miller, Peter. "Friendship and Conversation in Seventeenth-Century Venice." *Journal of Modern History* 73 (2001): 1–31.

———. *Peiresc's Europe: Learning and Virtue in the Seventeenth Century*. New Haven: Yale University Press, 2000.

Molières, Joseph Privat de. *Leçons de mathématiques nécessaires pour l'intelligence des Principes de Physiques, qui s'enseignent actuellement au Collège Royale*. Paris: Thiboust, 1725.

———. *Leçons de physique, contenant les eléments de la physique déterminées par les seuls lois de mécanique, expliqués au Collège Royale.* 4 vols. Paris: Veuve Brocas, 1734–1738.

Monod-Cassidy, Hélène. *Un voyageur-philosophe au XVIIIe siècle: L'abbé Jean-Bernard Le Blanc.* Cambridge: Harvard University Press, 1941.

Montgomery, Frances K. *La vie et l'oeuvre du Père Buffier.* Paris: Association du doctorat, 1930.

Montucla, Jean-Étienne. *Histoire des mathématiques.* 2 vols. Paris, 1758;

Morize, André. *L'apologie du luxe au XVIIIe Siècle: "Le Mondain" et ses sources.* Paris: H. Didier 1909.

Mornet, Daniel. "Les enseignements des bibliothèques privées (1750–1780)." *Revue d'histoire littéraire de la France* 17 (1910): 449–496.

———. *Les Sciences de la nature en France au XVIIIe siècle.* Paris: A. Colin, 1911.

Morris, Thelma. *L'abbé Desfontaines et son rôle dans la literature de son temps.* Geneva: Institut et musée Voltaire, 1961.

Mortureux, Marie-Francoise. *La formation et le fonctionnement d'un discours de la vulgarisation scientifique au XVIII eme siècle à travers l'oeuvre de Fontenelle.* Paris, 1983.

Moureau, Francois. *Le* Mercure galant *de Dufresny (1710–1714), ou journalisme à la mode.* Oxford: Voltaire Foundation at the Taylor Institution, 1982.

Mousnier, Roland E. *The Institutions of France under the Absolute Monarchy, 1598–1789.* Translated by Brian Pearce. 2 vols. Chicago: University of Chicago Press, 1974–1979.

———, ed. *Un Nouveau Colbert.* Paris: Éditions SEDES/CDU, 1985.

Mouy, Paul. *Le développement de la physique cartésienne, 1646–1712.* Paris: J. Vrin, 1934.

Nachtomy, Ohad, Ayelet Shavit, and Justin Smith. "Leibnizian Organisms, Nested Individuals, and Units of Selection." *Theory in Biosciences* 121 (2002): 205–230.

Nadler, Steven M. *Arnauld and the Cartesian Philosophy of Ideas.* Princeton: Princeton University Press, 1989.

———. "Choosing a Theodicy: The Leibniz-Malebranche-Arnauld Connection." *Journal the History of Ideas* 55 (October 1994): 573–590.

Le Nécrologe des hommes célèbres de France, par une société de gens de lettres. 18 vols. Paris: Chez J. Dufou, 1766–1784.

Neveu, Bruno. *Erudition et religion aux XVIIIe et XVIIIe siècles.* Paris: Albin Michel, 1994.

Newton, Isaac. *The Method of Fluxions and infinite series; with its Application to the Geometry of Curved-lines.* Translated from the Latin by John Colson. London, 1736.

———. *La méthode des fluxions et des suites infinies. By Isaac Newton.* Translated by George-Louis Le Clerc. Paris: Chez De Bure l'aîné, 1740.

————. *Principes mathématiques de la philosophie naturelle.* Translated by
Gabrielle-Emilie de Breuteil, Marquise du Châtelet. Paris: 1756–59.
————. *The Principia. Mathematical Principles of Natural Philosophy.* Translated
by I. Bernard Cohen and Anne Whitman, assisted by Julia Budenz. Berkeley:
University of California Press, 1999.
————. *Opticks, or A Treatise of the Reflections, Refractions, Inflections and Coulours
of Light.* Edited by I. B. Cohen. New York: Dover, 1979.
————. *Traité d'Optique.* Translated by Pierre Coste. Amsterdam, 1720.
Niderst, Alain. *Fontenelle.* Paris: Plon, 1991.
————. *Fontenelle à la recherche de lui-même.* Paris: A.-G. Nizet, 1972.
————, ed. *Fontenelle: Actes du colloque tenu à Rouen de 6 au 10 octobre 1987.* Paris:
Presses universitaires de France, 1989.
Northeast, Catherine M. *The Parisian Jesuits and the Enlightenment, 1700–1762.*
Oxford: Voltaire Foundation at the Taylor Institution, 1991.
Okruhlik, Kathleen, and James Robert Brown. *The Natural Philosophy of Leibniz.*
Dordecht: D. Reidel Pub. Co., 1985.
Olmstead, John W. "The Scientific Expedition of Jean Richer to Cayenne." *Isis*
34 (1942): 117–128.
————. "The Voyage of Jean Richer to Arcadia in 1670: A Study in the Relations of
Science and Navigation under Colbert." *Proceedings of the American Philosophical
Society* 104 (1960): 612–634.
Osler, Margaret J., ed. *Rethinking the Scientific Revolution.* Cambridge: Cambridge
University Press, 2000.
Ozanam, Jacques. *Dictionnaire mathématique ou idée générale des mathématiques.*
Paris: Chez Estienne Michallet . . ., 1691.
————. *Récreations mathématiques et physiques,* 2 vols. Paris: J. Jombert, 1694.
Palmer, Robert R. *Catholics and Unbelievers in Eighteenth-Century France.*
Princeton: Princeton University Press, 1939.
Pannekoek, A. *A History of Astronomy.* New York: Dover, 1961.
Pappas, John N. *Berthier's Journal de Trévoux and the Philosophes.* Studies on
Voltaire and the Eighteenth Century, vol. 3. Geneva: Institut and musée Voltaire,
1957.
————. *Voltaire and d'Alembert.* Bloomington: Indiana University Press, 1962.
Passeron, Irène. "Clairaut et la figure de la terre au XVIIIe siècle: Cristallisation d'un
nouveau style autour d'une pratique physico-mathématique." Thèse de Doctorat,
Université de Paris VII—Denis Diderot, 1994.
Pater, C. de. "Petrus van Musschenbroek (1692–1761): A Dutch Newtonian." *Janus*
64 (1977): 77–87.
Paul, Charles B. *Science and Immortality. The Éloges of the Paris Academy of Sciences
(1699–1791).* Berkeley: University of California Press, 1980.
Pelletier, Monique. *Les cartes de Cassini: La science au service de l'état et des regions.*
Paris: CTHS, 2002.

Pemberton, Henry. *A View of Sir Isaac Newton's Philosophy*. London: Printed by
S. Palmer, 1728.

Perkins, Merle L. *The Moral and Political Philosophy of the Abbe de Saint-Pierre*.
Geneva: E. Droz, 1959.

Perry, Norma. *Sir Everard Fawkener, Friend and Correspondent of Voltaire*. Banbury:
Voltaire Foundation, 1975.

Pettit, Alexander. *Illusory Consensus: Bolingbroke and the Polemical Responses to
Walpole, 1730–1737*. Newark: University of Delaware Press, 1997.

Picard, Roger. *Les salons littéraires et la société française, 1610–1789*. New York:
Brentano's, 1943.

Picon, Antoine. *L'Invention de l'ingénieur moderne: l'Ecole des Ponts et Chaussées,
1747–1851*. Toulouse: Presses de l'Ecole nationale des ponts et chausses, 1992.

Pintard, Rene. *Le libertinage érudit dans la première moitié du XVIIe siècle*. Paris:
Slatkine, 1943.

Pinto-Correia, Clara. *The Ovary of Eve: Egg and Sperm and Preformation*. Chicago:
University of Chicago Press, 1997.

Pluche, Noël Antoine. *La spectacle de la nature, ou Entretiens sur les particularités de
l'histoire naturelle*. 8 vols. La Haye: J. Neaulme, 1739.

Pocock, J. G. A. *Barbarism and Religion. The Enlightenment of Edward Gibbon,
1737–1764*. 2 vols. Cambridge: Cambridge University Press, 1999.

———. "The Varieties of Whiggism from Exclusion to Reform: A History of Ideology
and Discourse." in *Virtue, Commerce History: Essays on Political Thought Mostly
in the Eighteenth Century*. Cambridge: Cambridge University Press, 1981.

Pomeau, René. *D'Arouet à Voltaire*. Vol. 1 of *Voltaire en son temps*. Oxford: Voltaire
Foundation, Taylor Institution, 1985.

———. *La réligion de Voltaire*. Paris: Nizet, 1969.

———. *Voltaire en son temps*. Edited by René Pomeau. 5 vols. Oxford: Voltaire
Foundation, 1985–1994.

Polinière, Pierre. *Experiences de physique*. 8 vols. 5th ed. Paris: Chez Clouzier,
Bordelet, David, Ganeau, 1741.

Popkin, Richard Henry. *The History of Scepticism from Erasmus to Spinoza*.
Berkeley: University of California Press, 1979.

Porter, Roy. *The Creation of the Modern World: The Untold Story of the British
Enlightenment*. New York: Norton, 2000.

———, ed. *The Scientific Revolution in National Context*. Cambridge University
Press, 1992.

Porter, Roy, and Mikulas Teich, eds. *The Enlightenment in National Context*.
Cambridge: Cambridge University Press, 1981.

Prémontval. *Discours sur la nature des quantités que les mathémathiques ont pour
objet*. Paris: Chez Antoine-Urbain Coutelier, 1742.

———. *Discours sur l'utilité des mathématiques. Pronocée par Monsieur de
Prémontval, à l'ouverture de ses conférences*. Paris: Chez Antoine-Urbain
Coutelier, 1742.

Proust, Jacques. *Diderot et l'Encyclopédie.* Paris: Slatkine, 1962.

Rappaport, Rhoda. "Government Patronage of Science in Eighteenth-Century France." *History of Science* 8 (1969): 119–136.

———. "The Liberties of the Paris Academy of Sciences." In *The Analytic Spirit: Essays in the History of Science in Honor of Henry Guerlac.* Edited by Harry Woolf. Ithaca: Cornel University Press, 1981.

Rathéry, E. J. B., ed. *Journal et mémoires inédits du Marquis d'Argenson,* 9 vols. Paris: E. J. B. Rathery, 1859–1867.

Rattansi, P. M. "Voltaire and the Enlightenment Image of Newton." In *History and Imagination: Essays in Honor of H. R. Trevor-Roper.* Edited by Hugh Lloyd-Jones, Valerie Pearl, and Blair Warden. New York: Holmes & Meier Publishers, 1981.

Ravaisson, François, ed. *Archives de la Bastille: Documents inédits.* 13 vols. Paris: A. Durand et Pedone-Lauriel, 1881.

Ravel, Jeffrey. *The Contested Parterre: Public Theater and French Political Culture, 1680–1791.* Ithaca: Cornell University Press, 1999.

Ravetz, J. R. "The Representation of Physical Quantities in Eighteenth-Century Mathematical Physics." *Isis* 52 (1961): 7–20.

Réaumur, René Antoine de. *L'art de convertir le fer forgé en acier et l'art d'adoucir le fer fondu ou de faire des ouvrages de fer fondu aussis finis que de fer forge.* Paris: Chez M. Brunet, 1722.

———. *Mémoires pour servir à l'Histoire des insectes.* Paris: Imprimerie royale, 1734–1740.

Recueil des pièces qui ont remporté les Prix de l'Académie des Sciences. 9 vols. Paris: Gabriel Martin, 1752–1777.

Regnault, Noël. *Les Entretiens physiques d'Ariste et d'Eudoxe, ou Physique nouvelle en dialogues. Qui renferment précisément ce qui s'est découverte de plus curieux et de plus utile dans la nature.* 4 vols. Paris: David, Durand, 1729–1734.

———. *Lettre d'un physicien sur la philosophie de Neuton, mise a la portée de tout le monde par M. Voltaire.* N.p., 1738.

———. *L'origine ancienne de la physique nouvelle, où l'on voit dans des entretiens par lettres.* Paris: Chez J. Clousier, 1734.

Regourd, François, and James McClellan III. "French Science and Colonization in the Ancien Régime: The 'Machine coloniale.'" *Osiris* 15 (2000): 31–50.

Rémond de Montmort, Pierre. *Essay d'analyse dur les jeux de hazard.* Paris: J. Quillan, 1708; 2nd ed., 1713.

Revel, Jacques. "Forms of Expertise: Intellectuals and 'Popular' Culture in France (1650–1800)." In *Understanding Popular Culture: Europe from the Middle Ages to the Nineteenth Century.* Edited by Steven L. Kaplan. New York: Mouton, 1984.

Reyneau, Charles. *Analyse démontrée, ou la méthode de résoudre les problèmes des mathématiques.* 2 vols. Paris: Chez Jacque Quillau, 1708.

————. *La science du calcul des grandeurs en générale, ou les éléments des mathématiques*. Paris: G. Quillau, 1714.

Riskin, Jessica. *Science in the Age of Sensibility: The Sentimental Empiricists of the French Enlightenment*. Chicago: University of Chicago Press, 2002.

————. "The Defecating Duck, or The Ambiguous Origins of Artificial Life." *Critical Inquiry* 29, no. 4 (2003): 599–633.

Robinet, A. "Les Académiciens des Sciences Malebranchistes." In vol. 3, *Oeuvres complètes de Malebranche*. Edited by Désiré Roustan and Paul Schrecker. Paris: Boivin, 1938–.

————. "Le groupe malebranchiste introducteur du calcul infinitesimal en France." *Revue d'histoire des sciences* 13 (1960): 287–308.

————. *Malebranche, de l'Académie des sciences. L'oeuvre scientifique, 1674–1715*. Paris: Vrin, 1970.

————. *Système et existence dans l'oeuvre de Malebranche*. Paris: J. Vrin, 1965.

————. *Malebranche et Leibniz relations personelles*. Paris: J. Vrin, 1955.

Roche, Daniel. *France in the Enlightenment*. Translated by Arthur Goldhammer. Cambridge: Harvard University Press, 1998.

————. *The People of Paris: An Essay in Popular Culture in the Eighteenth Century*. Berkeley: University of California Press, 1987.

————. "Un savant et sa bibliothèque. Les livres de Jean-Jacques Dortous de Mairan." *Dix-Huitieme Siècle* 1 (1969): 47–88.

————. *Le siècle des lumiéres en province: Académies et académiciens provinciaux, 1680–1789*. 2 vols. Paris: Mouton, 1978.

Rodis-Lewis, Geneviève. *Nicolas Malebranche*. Paris: Presses universitaires de France, 1963.

Roger, Jacques. *Buffon: A Life in Natural History*. Ithaca, N.Y.: Cornell University Press, 1997.

————. *Les sciences de la vie dans la pensée française du XVIII siècle*. Paris: A. Colin, 1963.

Roger, Jacques, ed. *Jean d'Alembert, savant et philosophe; portrait à plusieurs voix. Actes du Colloque organisé par Centre international de synthèse, Fondation pour la science, Paris, 15–18 juin 1983*. Paris: Éditions des archives contemporaines, 1989.

Rohault, Jacques. *Traité de physique*. 2 vols. Paris: G. Desprez, 1683.

Roncière, Charles de la. *Histoire de la marine française*. 6 vols. Paris: E. Plon, Nourrit et cie, 1899–1932.

Rosneau, Helen. *Boullée and Visionary Architecture*. London: Academy Editions, 1976.

Rorty, Richard. *Philosophy and the Mirror of Nature*. Princeton: Princeton University Press, 1979.

Rothkrug, Lionel. *Opposition to Louis XIV: The Political and Social Origins of the French Enlightenment*. Princeton: Princeton University Press, 1965.

Rousseau, André Michel. *L'Angleterre et Voltaire (1718–1789)*. Oxford: Voltaire Foundation at the Taylor Institution, 1976.

———. "Naissance d'un livre et d'un texte: Les *Letters concerning the English nation.*" Studies on Voltaire, 179. Oxford: Voltaire Foundation at the Taylor Institution, 1979.

Rousseau, G. S., and Roy Porter, eds. *The Ferment of Knowledge: Studies in the Historiography of Eighteenth-Century Science.* Cambridge: Cambridge University Press, 1980.

Rousseau, Jean-Jacques. "Confessions." In vol. 8 of *Oeuvres complètes de Jean-Jacques Rousseau.* Paris: Hachette, 1885–1905.

Ruestow, Edward Grant. *Physics at Seventeenth and Eighteenth-Century Leiden: Philosophy and the New Science in the University.* The Hague: Nijhoff, 1973.

Rule, John C. "Jerôme Phelypeaux, Comte de Pontchartrain, and the Establishment of Louisiana, 1696–1715." In *Frenchmen and French Ways in the Mississippi Valley.* Edited by John F. McDermotte. Urbana: University of Illinois Press, 1979.

Russell, Betrand. *A Critical Exposition of the Philosophy of Leibniz.* London: Geo. Allen & Unwin Ltd., 1900.

Said, Edward W. *Beginnings: Intention and Method.* New York: Basic Books, 1975.

Saigey, Émile. *Les sciences au XVIIIe siècle: La physique de Voltaire.* Paris: G. Baillière, 1873.

Saint-Pierre, Charles Irénée de Castel, abbé de. *Discours sur la polysynodie: Où l'on démontre que la polysynodie, ou pluralité des conseils, est la forme de ministère plus avantageuse pour un roi, & pour son royaume.* Amsterdam: Du Villard et Changuion, 1719.

———. *Ouvrajes politiques de Mr. L'Abbé de St. Pierre, Charles Irenée Castel.* 10 vols. Rotterdam: Jean Daniel Berman, 1734.

Saunders, Elmo Stewart. "The Decline and Reform of the Académie des Sciences à Paris, 1676–1699." Ph.D. dissertation, Ohio State University, 1980.

Saverien, A. *Dictionnaire universel de mathématique et de physique.* 2 vols. Paris: Charles-Antoine Jombert, 1753.

———. *Histoire des philosophes modernes,* 8 vols. Paris: Brunet, 1761–1767.

Schaffer, Simon. "Natural Philosophy." In *The Ferment of Knowledge: Studies in the Historiography of Eighteenth-Century Science.* Edited by G. S. Rousseau and Roy Porter. Cambridge: Cambridge University Press, 1980.

———. "Natural Philosophy and Public Spectacle in the Eighteenth Century." *History of Science* 21 (1983): 1–43.

———. "Newtonianism." In *Companion to the History of Modern Science.* Edited by R.C. Olby et al. New York: Routledge, 1990.

Schaffer, Simon, and Steven Shapin. *Leviathan and the Air Pump: Hobbes, Boyle and the Experimental Life.* Princeton: Princeton University Press, 1987.

Schiebinger, Londa. *Nature's Body: Gender in the Making of Modern Science.* Boston: Beacon Press, 1993.

———. *The Mind Has No Sex? Women in the Origins of Modern Science.* Cambridge: Harvard University Press, 1989.

Schier, Donald S. *Louis Bertrand Castel: Anti-Newtonian Scientist.* Cedar Rapids: Torch Press, 1941.

Schmaltz, Tad. *Radical Cartesianism: The French Reception of Descartes.* Cambridge: Cambridge University Press, 2002.

Schofield, Robert E. "An Evolutionary Taxonomy of Eighteenth-Century Newtonianism." *Studies in Eighteenth-Century Culture* 7 (1978): 175–192.

———. *Mechanism and Materialism: British Natural Philosophy in an Age of Reason.* Princeton: Princeton University Press, 1970.

Schosler, Jorn. *La Bibliothèque Raisonnée (1728–1753): Les réactions d'un périodique français à la philosophie de Locke.* Svendborg: Odense University Press, 1985.

Schwab, Richard N., Walter E. Rex, and John Lough. *Inventory of Diderot's Encyclopédie.* Geneva: Institut et Musée Voltaire, 1971.

Sédillot, Louis. *Les professeurs de mathématiques et de physique générale au College de France.* Rome: Imprimerie des Sciences Mathématiques et Physiques, 1869.

Sergescu, Pierre. "La littérature mathématique dans la première période (1655–1701) du Journal des Savants." *Archives internationales d'histoire des sciences* 1 (1947): 60–99.

Sgard, Jean. *Le Pour et contre de Prévost.* Paris: A. G. Nizet, 1969.

Sgard, Jean, ed.. *Dictionnaire des journaux 1600–1789.* 2 vols. Paris: Universitas, 1991.

———. *Dictionnaire des journalistes.* Oxford: Voltaire Foundation, 1999.

Shackleton, Robert. *Montesquieu: A Critical Biography.* Oxford: Oxford University Press, 1961.

Shaftesbury, Anthony Ashley Cooper. *Principes de la philosophie morale; ou Essai de M.S*** sur mle mérite et la vertu. Avec réflexions.* Translated by Denis Diderot. Amsterdam: Chez Zacharie Chatelain, 1745.

Shank, J. B. "The Abbé de Saint-Pierre and the Emergence of the 'Quantifying Spirit' in French Enlightenment Thought." In *A Vast and Useful Art: The Gustave Gimon Collection on French Political Economy.* Edited by Mary Jane Parrine. Stanford: Stanford University Press, 2004.

———. "Before Voltaire: Newtonianism and the Origins of the Enlightenment in France, 1687–1734." Ph.D. dissertation, Stanford University, 2000.

———. "Neither Natural Philosophy, nor Science, nor Literature: Gender and Natural Knowledge in Fontenelle's *Entretiens sur la pluralité des mondes.*" In *Men, Women, and the Birthing of Modern Science.* Edited by Judith Zinsser. DeKalb: Northern Illinois University Press, 2005.

———. "On the Alleged Cartesianism of Fontenelle." *Archives Internationales d'Histoire des Sciences* 53, nos. 150–151 (2003): 139–156.

———. "'There Was No Such Thing as the "Newtonian Revolution," and the French Instituted It': Eighteenth-Century Mechanics in France before Maupertuis." *Early Science and Medicine* 9 (August 2004): 257–292.

Shapin, Steven. "Of Gods and Kings: Natural Philosophy and Politics in the Leibniz-Clarke Disputes." *Isis* 72 (1981): 187–215.

———. *A Social History of Truth: Civility and Science in Seventeenth-Century England.* Chicago: University of Chicago Press, 1994.

———. *The Scientific Revolution.* Chicago: University of Chicago Press, 1996.

Sigorgne, Pierre. *Astronomiae physicae juxta Newtonis principia breviarum.* Paris, 1748.

———. *Examen et réfutation des leçons de physique expliquées par M. de Moliéres au collège royal de France.* Paris: Chez Jacques Clousier, 1741.

———. *Institutions Newtoniennes; ou Introduction à la philosophie de M. Neuton.* 2 vols. Paris: J.-F. Quillau, 1747.

———. *Réplique à M de Molières, ou démonstration physico-mathématique de l'insuffisance et de l'impossibilité des tourbillons.* Paris: Chez Jacques Clousier, 1741.

Simon, Renee. *Nicolas Fréret, academicien.* Geneva: Institut et musée Voltaire, 1961.

Smith, David Kammerling. "*'Au Bien du Commerce'*: Economic Discourse and Visions of Society in France." Ph.D. dissertation, University of Pennsylvania, 1995.

———. "Structuring Politics in Early Eighteenth-Century France: The Political Innovations of the French Council of Commerce." *Journal of Modern History* 74 (2002): 490–537.

Smith, Jay M. *The Culture of Merit: Nobility, Royal Service, and the Making of Absolute Monarchy in France, 1660–1789.* Ann Arbor: University of Michigan Press, 1996.

Smith, Justin. "Leibniz's Hylomorphic Monad." *History of Philosophy Quarterly* 19 no. 1 (2002): 21–42.

Snyders, Georges. *La pédagogie en France aux XVIIe et XVIIIe siècles.* Paris: Presses universitaires de France, 1965.

Sobel, Dava. *Longitude: The True Story of a Lone Genius Who Solved the Greatest Scientific Problem of his Time.* New York: Walker, 1995.

Sokal, Alan. "What the *Social Text* Affair Does and Does Not Prove." In *A House Built on Sand: Flaws in the Cultural Studies Account of Science.* Edited by Noretta Koertege. Oxford: Oxford University Press, 1998.

Solnon, Jean-François. *La Cour de France.* Paris: Fayard, 1987.

Solomon, Howard M. *Public Welfare, Science, and Propaganda in Seventeenth-Century France: The Innovation of Theophraste Renaudot.* Princeton: Princeton University Press, 1972.

Solomon-Bayet, Claire. "Un préambule théorique à une académie des arts." *Revue d'histoire des sciences* 13 (1970): 229–250.

Sortais, Gaston. *Le cartésianisme chez les Jésuites français au XVIIe et au XVIIIe siècle.* Paris: Beauchesne, 1929.

Spencer, Samia, ed. *French Women and the Age of Enlightenment.* Bloomington: Indiana University Press, 1984.

Spink, John Stephenson. *French Free-Thought from Gassendi to Voltaire.* London: University of London, Athlone Press, 1960.

Stafford, Barbara Maria. *Artful Science: Enlightenment Entertainment and the Eclipse of Visual Education.* Cambridge: MIT Press, 1994.

———. *Body Criticism: Imaging the Unseen in Enlightenment Art and Medicine.* Cambridge: MIT Press, 1991.

Stewart, Larry. *The Rise of Public Science: Rhetoric, Technology, and Natural Philosophy in Newtonian Britain, 1660–1750.* Cambridge: Cambridge University Press, 1992.

Strosetzki, Christoph. *Rhetorique de la conversation: sa dimension littéraire et linguistique dans la société française du XVIIe siècle.* Translated by Sabine Seubert. In *Biblio 17–20, Papers on French Seventeenth-Century Literature.* Paris: Papers on French Seventeenth Century Literaure, 1984.

Stroup, Alice. *A Company of Scientists: Botany, Patronage, and Community at the Seventeenth-Century Parisian Royal academy of Sciences.* Berkeley: University of California Press, 1990.

———. *Royal Funding of the Parisian Académie royale des sciences during the 1690s.* Transactions of the American Philosophical Society, 77. Philadelphia: American Philosophical Society, 1987.

Sturdy, David J. *Science and Social Status: The Members of the Académie des Sciences, 1666–1750.* Woodbridge: Boydell Press, 1995.

Sutton, Geoffrey V. *Science for a Polite Society: Gender, Culture, and the Demonstration of Enlightenment.* Boulder: Westview Press, 1995.

Symcox, Geoffrey. *The Crisis of French Sea Power, 1688–1697: From the guerre d'escadre to the guerre de course.* The Hague: M. Nijhoff, 1974.

Taillemite, Étienne. *Colbert, secrétaire d'état de la Marine et les reformes de 1669.* Paris: Académie de marine, 1970.

———. *L'Histoire ignorée de la marine française.* Paris: Perrin, 1988.

Taton, Rene, ed. *Enseignement et diffusion des sciences en France au dix-huitième siècle.* Paris: Hermann, 1986.

Terrall, Mary. "Emilie du Châtelet and the Gendering of Science." *History of Science* 33 (1995): 283–310.

———. "Gendered Spaces, Gendered Audiences: Inside and outside the Paris Academy of Sciences." *Configurations* 2 (1995): 207–232.

———. *The Man Who Flattened the Earth: Maupertuis and the Sciences in the Enlightenment.* Chicago: University of Chicago Press, 2002.

———. "Representing the Earth's Shape: The Polemics Surrounding Maupertuis' Expedition to Lapland." *Isis* 83 (1992): 218–237.

Thackray, Arnold. *Atoms and Powers: An Essay on Newtonian Matter-Theory and the Development of Chemistry.* Cambridge: Harvard University Press, 1970.

Todhunter, I. *A History of the Mathematical Theories of Attraction and the Figure of the Earth from the Time of Newton to that of Laplace.* London: Macmillan and Co., 1873.

———. *History of the Mathematical Theory of Probability from the Time of Pascal to that of Laplace.* Cambridge: Macmillan and Co., 1865.

Toland, John. *Letters to Serena*. London, 1704. Facsimile edition edited by René
Wallek. New York Garland Pub., 1976.

———. *Pantheisticon, or The Form of Celebrating the Socratic Society*. London,
1751. Facsimile edition edited by René Wallek. New York: Sam Paterson,
1976.

———. *Pantheisticon sive formula celebrandae sodalitatis socraticae*. London:
Cosmopoli, 1720.

Tonelli, Giorgio. "La nécessité des lois de la nature au XVIIIe siècle et chez Kant en
1762." *Revue d'histoire des sciences* 12 (1959): 225–241.

———. *La pensee philosophique de Maupertuis son milieu et ses sources*. New York:
G. Olms, 1987.

Torlais, Dr. Jean-Henri-Louis-Marie. "L'Abbé Nollet, 1700–1770, et la physique
expérimentale au XVIIIe siècle." *Les Conférences du Palais de la découverte*
Série D, no. 60 (1959).

———. "Les medecins et l'électricité au XVIIIe siècle." *Société Française de l'Histoire
de la Médicine* (1953): 63–70.

———. *Un esprit encyclopédique en dehors de "l'Encyclopédie": Réaumur, d'après des
documents inédits*. Paris: Desclée, de Brouwer et cie, 1936.

———. *Un physicien au siécle des lumières, l'Abbé Nollet, 1700–1770*. Paris: Sipuco,
1954.

———. *Un rochelais grand-maître de la Franc-maçonnerie et physicien au XVIIIe
siècle: Le Révérend J. T. Desaguliers*. La Rochelle: F. Pijollet, 1937.

Treat, Ida Frances. *Un Cosmopolite italien du XVIIIe siècle: Francesco Algarotti*.
Trévoux: Jeannin, 1913.

Truesdell, C. A. "A Program toward Rediscovering the Rational Mechanics of the
Age of Reason." *Archive for the History of the Exact Sciences* 1 (1960): 3–36.

Trystram, Florence. *Le procès des étoiles: Récit de la prestigieuse expédition de trois
savants français en Amérique du Sud et des aventures qui s'ensuivirent (1735–1771)*.
Paris: Seghers, 1989.

Udank, Jack. "Portrait of the Philosopher as Tramp." In *A New History of French
Literature*. Edited by Dennis Hollier. Cambridge: Harvard University Press,
1989.

Ultee, Martin. "The Republic of Letters: Learned Correspondence, 1680–1720."
Seventeenth Century 2 (1987): 95–112.

Vailati, Ezio. *Leibniz and Clarke: A Study of Their Correspondence*. Oxford: Oxford
University Press, 1997.

Vaillot, René. *Avec Madame du Châtelet, 1734–1749*. Oxford: Voltaire Foundation,
1988.

———. *Madame du Châtelet*. Paris: Albin Michel, 1978.

Van Damme, Stéphane. *Descartes: Essai d'histoire culturelle d'une grandeur
philosophique (xviie–xxe siècle)*. Paris: Presses de Sciences Po, 2002.

Van Kley, Dale. *The Jansenists and the Expulsion of the Jesuits in France, 1757–1767*.
New Haven: Yale University Press, 1975.

———. *The Religious Origins of the French Revolution.* New Haven: Yale University Press, 1997.

Varignon, Pierre. *Nouvelles conjectures sur la pesanteur.* Paris: J. Boudot, 1690.

———. *Traité de mécanique.* Paris: Jombert, 1725.

Vartanian, Aram. *Diderot and Descartes: A Study of Scientific Naturalism in the Enlightenment.* Princeton: Princeton University Press, 1953.

Venturi, Franco. *La jeunesse de Diderot (de 1713 à 1753).* Paris: Albert Skira, 1939.

Vercruysse, Jeroom. *Voltaire et la Holland.* Geneva: Institut et Musée Voltaire 1966.

Vernière, Paul. *Spinoza et la pensée française avant la Révolution.* Paris: Presses universitaires de France, 1954.

Vidal, Mary. *Watteau's Painted Conversations: Art, Literature, and Talk in Seventeenth- and Eighteenth-Century France.* New Haven: Yale University Press, 1992.

Villemot. *Nouveau système ou nouvelle explication du mouvement des planètes.* Lyon: L. Declaustre, 1707.

Vincent, Monique. *Donneau de Visé et le Mercure galant.* Lille: Atelier national Reproduction des thèses, Université Lille III, 1987.

Voltaire, François Marie d'Arouet. *Correspondence and Related Documents.* Vols. 85–135 of *Les oeuvres complètes de Voltaire.* Edited by Theodore Besterman. Geneva: Institut et Musée Voltaire, 1968–1976.

———. *Eléments de la philosophie de Newton, mise à la portée de tout le monde.* N.p., 1738.

———. *Épîtres sur le bonheur.* Paris: Chez Prault fils, 1738.

———. "Essai sur la nature du feu et sur sa propagation." *Recueil des Pièces qui ont remporté les Prix de l'Académie des Sciences,* vol. 4. Paris: Chez Claude Jombert, 1739.

———. *Lettres philosophiques. Édition critique avec un introduction et un commentaire.* Edited by Gustave Lanson. 2 vols. Paris: Edouard Cornély et Cie., 1909.

———. *Oeuvres complètes de Voltaire.* Edited by A. Beuchot. 72 vols. Paris: 1829–1840.

———. *Oeuvres complètes.* Edited by L. E. D. Moland and G. Bengesco. 52 vols. Paris: Garnier frères, 1877–1885.

———. *Oeuvres complètes de Voltaire.* Edited by Theodore Besterman. 135 vols. Geneva, Banbury, and Oxford: Voltaire Foundation at the Taylor Institution, 1968–1977.

———. *Philosophical Letters.* Translated and edited by Ernest Dilworth. Indianapolis: Bobbs-Merrill, 1961.

———. *Le préservatif, ou Critique des observations sur les écrits modernes.* La Haye: J. Neaulme, 1738.

———. *Réponse à toutes les objections principales qu'on a faites en France contre la philosophie de Newton.* Paris, 1739.

Wade, Ira Owen. *The Clandestine Organization and Diffusion of Philosophic Ideas in France from 1700 to 1750.* Princeton: Princeton University Press, 1938.

———. *The Intellectual Devevelopment of Voltaire.* Princeton: Princeton University Press, 1969.

———. *Studies on Voltaire with Some Unpublished Papers of Mme. du Châtelet.* Princeton: Princeton University Press, 1947.

———. *Voltaire and Madame du Châtelet: An Essay on the Intellectual Activity at Cirey.* Princeton: Princeton University Press, 1941.

Waddicor, Mark H., ed. *Voltairomanie.* Exeter: University of Exeter, 1983.

Waff, Craig B. "Alexis Clairaut and His Proposed Modification of Newton's Inverse-Square Law of Gravitation." In *Avant avec après Copernic. La réprésentation de l'univers et ses conséquences épistemologiques.* Paris: A. Blanchard, 1975.

———. "Universal Gravitation and the Motion of the Moon's Apogee: The Establishment and Reception of Newton's Inverse-Square Law, 1687–1749." Ph.D. dissertation, Johns Hopkins University, 1976.

Wagner, Jacques. *Marmontel journaliste et le Mercure de France (1725–1761).* Grenoble: Presses universitaires de Grenoble, 1975.

Wallis, Peter and Ruth. *Newton and Newtoniana, 1672–1975: A Bibliography.* Surrey: Dawson, 1977.

Walters, Robert L., and W. H. Barber. Introduction to *Eléments de la philosophie de Newton.* In vol. 15 of *Oeuvres complètes de Voltaire.* Edited by Theodore Besterman. Oxford: Voltaire Foundation, 1992.

Waquet, Françoise. "Qu'est-ce que la République des Lettres? Essai de sémantique historique." *Bibliothèque de l'ecole des chartes* 147 (1989): 480–481.

Watson, Richard A. *The Downfall of Cartesianism, 1673–1712: A Study of Epistemological Issues in Late Seventeenth-Century Cartesianism.* The Hague: Martinus Nijhoff, 1966.

Watson, Richard A., and Marjorie Greene. *Malebranche's First and Last Critics: Simon Foucher and Dortous de Mairan.* Introductions and translations by Richard A. Watson and Marjorie Greene. Carbondale and Edwardsville: Southern Illinois University Press, 1995.

Wellman, Kathleen Anne. *Making Social Science: The Conferences of Theophraste Renaudot, 1633–1642.* Norman: University of Oklahoma Press, 2003.

———. *La Mettrie: Medicine, Philosophy and Enlightenment.* Durham: Duke University Press, 1992.

Westfall, Richard. *Force in Newton's Physics: The Science of Dynamics in the Seventeenth Century.* New York: American Elsevier, 1971.

———. *Never at Rest: A Biography of Isaac Newton.* Cambridge: Cambridge University Press, 1980.

Westman, Robert S. "The Astronomer's Role in the Sixteenth Century: A Preliminary Study." *History of Science* 18 (1980): 105–147.

Whitesides, D. T. *The Mathematical Principles Underlying Newton's "Principia Mathematica."* Glasgow: University of Glasgow, 1970.

Williams, Elizabeth. *A Cultural History of Medical Vitalism in Enlightenment Montpellier.* Aldershot: Ashgate, 2003.

Wilson, Arthur. *Diderot*. Oxford: Oxford University Press, 1972.

Wilson, Catherine. "The Reception of Leibniz in the Eighteenth Century." In *The Cambridge Companion to Leibniz*. Edited by Nicolas Jolley. Cambridge: Cambridge University Press, 1995.

Wilson, Curtis. "From Kepler's Laws So-called, to Universal Gravitation: Empirical Fractors." *Archive for the History of Exact Sciences* 6 (1969–1970): 65–103.

Wolf, C. *Histoire de l'Observatoire de Paris de sa fondation à 1793*. Paris: Gauthier-Villars, 1902.

Yoder, Joella G. *Unrolling Time: Christian Huygens and the Mathematization of Nature*. Cambridge: Cambridge University Press, 1988.

Zinsser, Judith. *La Dame d'Esprit. A Biography of the Marquise du Châtelet*. New York: Penguin, 2006.

———. "Entrepreneur of the Republic of Letters: Emilie de Breteuil, Marquise du Châtelet and Bernard Mandeville's *Fable of the Bees*." *French Historical Studies* 25 (Fall 2002): 595–624.

Index